T0325677

Fixed Point Theory and Graph Theory

Fixed Point Theory and Graph Theory

Foundations and Integrative Approaches

Edited by

Monther Rashed Alfuraidan
King Fahd University of Petroleum & Minerals, Department of
Mathematics and Statistics, Dhahran, Saudi Arabia

Qamrul Hasan Ansari
Aligarh Muslim University, Department of Mathematics, Aligarh,
India, and
King Fahd University of Petroleum & Minerals, Department
of Mathematics and Statistics, Dhahran, Saudi Arabia

AMSTERDAM • BOSTON • HEIDELBERG • LONDON
NEW YORK • OXFORD • PARIS • SAN DIEGO
SAN FRANCISCO • SINGAPORE • SYDNEY • TOKYO
Academic Press is an imprint of Elsevier

Academic Press is an imprint of Elsevier
125 London Wall, London EC2Y 5AS, UK
525 B Street, Suite 1800, San Diego, CA 92101-4495, USA
50 Hampshire Street, 5th Floor, Cambridge, MA 02139, USA
The Boulevard, Langford Lane, Kidlington, Oxford OX5 1GB, UK

Notices
Knowledge and best practice in this field are constantly changing. As new research and experience broaden our understanding, changes in research methods, professional practices, or medical treatment may become necessary.

Practitioners and researchers must always rely on their own experience and knowledge in evaluating and using any information, methods, compounds, or experiments described herein. In using such information or methods they should be mindful of their own safety and the safety of others, including parties for whom they have a professional responsibility.

To the fullest extent of the law, neither the Publisher nor the authors, contributors, or editors, assume any liability for any injury and/or damage to persons or property as a matter of products liability, negligence or otherwise, or from any use or operation of any methods, products, instructions, or ideas contained in the material herein.

Library of Congress Cataloging-in-Publication Data
A catalog record for this book is available from the Library of Congress

British Library Cataloguing in Publication Data
A catalogue record for this book is available from the British Library

ISBN: 978-0-12-804295-3

For information on all Academic Press publications
visit our website at http://www.elsevier.com

Working together
to grow libraries in
developing countries

www.elsevier.com • www.bookaid.org

Publisher: Nikki Levy
Acquisition Editor: Graham Nisbet
Editorial Project Manager: Susan Ikeda
Production Project Manager: Poulouse Joseph
Designer: Matthew Limbert

CONTENTS

FOREWORD

The present book differs in important ways from previous texts which deal largely with metric fixed point theory. Foremost, the topics collected here are not found in any single previous treatments and most are entirely new. These include detailed discussions of Caristi's theorem in a number of settings, as well as numerous approximation techniques in Banach spaces, hyperbolic spaces, and more general geodesic spaces. In recent years metric fixed point theory has increasingly moved away from traditional functional analytic settings, thus opening up new avenues of application. The present book reflects this trend by discussing in detail aspects of the theory which involve a blend of metric fixed point theory and graph theory, as well as how retractions are used to prove fixed point results in ordered sets. Many examples are included as well as potential applications. Because of the comprehensive background material this book is self-contained and easily accessible. A notable feature is the thoroughness of the bibliographic citations, many of which are not found in any previous sources.

This book should be a worthwhile resource for students and researchers interested in almost any aspect of metric fixed point theory, and especially for those interested in many of the more recent trends.

William A. Kirk
Iowa City, IA, USA

ACKNOWLEDGMENTS

We would like to thank our illustrious friends and colleagues in the Fixed Point Theory Research Group at the Department of Mathematics and Statistics at King Fahd University of Petroleum & Minerals who, through their encouragement and help, were instrumental in the development of this book. In particular, we are grateful to Prof. Khalid Al-Sultan, the Rector of King Fahd University of Petroleum & Minerals, and Prof. Mohammad Al-Homoud, the Vice Rector for Academic Affairs of King Fahd University of Petroleum & Minerals, for their unflinching support in the organization of the International Workshop on Fixed Point Theory and Applications during December 22–24, 2014. This Workshop provided the platform to contact and invite the delegates to contribute to this book as authors. It is heartening to note that most of them accepted our offer and generously contributed their research for inclusion in the book. The book in its present form is, in large measure, a fruit of their wholehearted cooperation. We are profoundly grateful to them.

The combination of Fixed Point Theory and Graph Theory is the brainchild of our friend Prof. M. A. Khamsi, University of Texas at El Paso. He is the prime mover behind the international workshop on this topic and the subsequent plan to publish a book which combines these two subjects. We are deeply thankful and grateful to him for his encouragement, support and help. Thanks are also due to Prof. Suliman Al-Homidan, Dean, College of Science, King Fahd University of Petroleum & Minerals, for his motivating advice and help to publish this book. We would like to convey our special thanks to Mr. Graham Nisbet, Senior Acquisitions Editor, Mathematical Sciences, Elsevier, for taking a keen interest in publishing this book.

Monther Rashed Alfuraidan
Qamrul Hasan Ansari
December 2015

PREFACE

Fixed point theory is without doubt one of the most important tools of modern mathematics as attested by Browder, who is considered as one of the pioneers in the development of the nonlinear functional analysis. The flourishing field of fixed point theory started in the early days of topology through the work of Poincaré, Lefschetz-Hopf and Leray-Schauder for example. Fixed point theory is widely used in different areas such as ordinary and partial differential equations, economics, logic programming, convex optimization, control theory, etc. In metric fixed point theory, successive approximations are rooted in the work of Cauchy, Fredholm, Liouville, Lipschitz, Peano and Picard. It is well accepted among experts of this subarea that Banach is responsible for laying the ground for an abstract framework well beyond the scope of elementary differential and integral equations. The fixed point theory of multivalued mappings is as important as the theory of single-valued mappings. After Nadler gave the multivalued version of the Banach fixed point theorem for contraction mappings, many authors generalized his fixed point theorem in different ways. A constructive proof of a fixed point theorem makes the conclusion more valuable in view of the fact that it yields an algorithm for computing a fixed point.

The last decade has seen some excitement about monotone mappings after the publication of Ran and Reurings of the analogue to the Banach fixed point theorem in metric spaces endowed with a partial order. Their result was motivated by the problem of finding the solutions to some matrix equations, which often arise in the analysis of ladder networks, dynamic programming, control theory, stochastic filtering, statistics and many other applications. Jachymski is recognized for making the connection between classical metric fixed point theory and graph theory. This book can be treated as a foundation for both theories and presents the most up-to-date results bridging the two theories.

Each chapter is written by different author(s) who attempt to render the major results understandable to a wide audience, including nonspecialists, and at the same time to provide a source for examples, references, open questions, and sometimes new approaches for those currently working in these areas of mathematics. This book should be of interest to graduate students seeking a field of interest, to mathematicians interested in learning about the subject, and to specialists. All the chapters are self-contained.

Chapter 1 describes the classical Caristi fixed point theorem and its various versions, and studies Caristi-Browder operators in the setting of metric spaces, $\mathbb{R}_+^m$-metric spaces, $s(\mathbb{R}_+)$-metric spaces, and Kasahara spaces. It also provides some new research directions in the Caristi-Browder operator theory.

Chapter 2 is devoted to the fixed point theorems for self and nonself almost contraction mappings, and common fixed point theorems for almost contraction mappings. Iterative approximation of fixed points is also studied for implicit almost contraction mappings.

The approximate fixed points are useful when a computational approach is involved. This subject is addressed in Chapter 3. Some known results about the concept of approximate fixed points of a mapping are presented. Most of the approximate fixed points discussed in this chapter are generated by an algorithm that allows its computational implementation. An application of these results to the case of a nonlinear semigroup of mappings is given.

Chapter 4 presents the mathematical idea of viscosity methods for solving some applied nonlinear analysis problems, namely, fixed point problems, split common fixed point problems, split equilibrium problems, hierarchical fixed point problems, etc. together with examples in linear programming, semi-coercive elliptic problems and the area of finance. Some numerical experiments for an inverse heat equation are also given.

Chapter 5 deals with several kinds of extragradient iterative algorithms for some nonlinear problems, namely, fixed point problems, variational inequality problems, hierarchical variational inequality problems and split feasibility problems. Several extragradient iterative algorithms for finding a common solution of a fixed point problem and a variational inequality problem are also presented and investigated.

Chapter 6 concerns with the fixed point (common fixed point) problems for nonexpansive (asymptotically quasi-nonexpansive) mappings in Banach spaces and certain important classes of metric spaces. Some results about weak convergence, $\triangle$-convergence and strong convergence of explicit and multistep schemes of nonexpansive-type mappings to a common fixed point in the context of uniformly convex Banach spaces, CAT(0) spaces, hyperbolic spaces and convex metric spaces are presented.

Chapter 7 discusses a new area that overlaps between metric fixed point theory and graph theory. This new area yields interesting generalizations of the Banach contraction principle in metric and modular spaces endowed with a graph. The bridge between both theories is mainly motivated by the applications of matrix and differential equations.

The last chapter gives an overview how retractions are used to prove fixed point theorems for relation-preserving maps in continuous as well as discrete settings. After first discussing comparative retractions in the context of infinite ordered sets and analysis, the remainder of the chapter focuses on retractions that remove a single point in finite ordered sets, graphs and simplicial complexes.

Monther Rashed Alfuraidan
Qamrul Hasan Ansari
Dhahran, Saudi Arabia; and Aligarh, India
December 2015

ABOUT THE AUTHORS

Monther Rashed Alfuraidan is Associate Professor of Mathematics in the Department of Mathematics & Statistics at King Fahd University of Petroleum & Minerals at Dhahran, Saudi Arabia. He obtained his Ph.D. (Mathematics) from Michigan State University. He has written more than twenty articles on graph theory, algebraic graph theory and metric fixed point theory. He has peer-reviewed many articles (among others) for: *Algebraic Journal of Combinatorics*, *Arabian Journal of Mathematics*, *Fixed Point Theory and Applications* and *Journal of Inequality and Applications*.

Qamrul Hasan Ansari is Professor of Mathematics at Aligarh Muslim University, Aligarh, India. He is also joint professor at King Fahd University of Petroleum & Minerals, Dhahran, Saudi Arabia. He obtained his Ph.D. (Mathematics) from Aligarh Muslim University, India. He is an associate editor of *Journal of Optimization Theory and Applications*, *Fixed Point Theory and Applications*, *Fixed Point Theory* and *Carpathian Journal of Mathematics*. He has also edited several special issues of several journals, namely, *Journal of Global Optimization*, *Fixed Point Theory and Applications*, *Abstract and Applied Analysis*, *Journal of Inequalities and Applications*, *Applicable Analysis*, *Positivity*, *Filomat*, etc. He has written more than 180 articles on variational inequalities, fixed point theory and applications, vector optimization, etc. in various international peer-reviewed journals. He has edited six books for Springer, Taylor & Francis and Narosa, India. He is an author of a book on metric spaces published by Narosa, India and has co-authored one book on variational inequalities and nonsmooth optimization for Taylor & Francis.

Mostafa Bachar received his Ph.D. from the University of Pau (France) in Applied Mathematics; he is currently working as associate professor at King Saud University (KSA). He has previously held the position of Research Associate at the Institute of Mathematics and Scientific Computing at the University of Graz (Austria). Recently he has edited two books in *Lecture Notes in Mathematics: Mathematical Biosciences Subseries*, Springer. His current research is in mathematical modeling in mathematical biology and mathematical analysis.

Vasile Berinde is Professor of Mathematics and Director of the Department of Mathematics and Computer Science in the Faculty of Sciences, North University Center at Baia Mare, Technical University of Cluj-Napoca, Romania. He obtained his B.Sc., M.Sc. (in Computer Science) and Ph.D. (in Mathematics) from the Babes-Bolyai University of Cluj-Napoca, under the supervision of Professor Ioan A. Rus. His main scientific interests are in nonlinear analysis and fixed point theory. He is the founder and current Editor-in-Chief of the journals *Carpathian Journal of Mathematics* and *Creative Mathematics and Informatics* and editorial board member for the journals *Fixed Point Theory, Fixed Point Theory and Applications, Journal of Applied Mathematics, General Mathematics, Linear and Nonlinear Analysis, Carpathian Mathematical Publications, Didactica Mathematica, Gazeta Matematică, Revista de Matematică din Timişoara*, etc. He has published more than 150 scientific articles on the iterative approximation of fixed points, fixed point theory and its applications to differential and integral equations. He has published several books, textbooks and monographs with Romanian publishers. Two of them have been undertaken by international publishers: *Iterative Approximation of Fixed Points* (Springer, 2007) and *Exploring, Investigating and Discovering in Mathematics* (Birkhäuser, 2004).

Buthinah A. Bin Dehaish is Associate Professor of Mathematics at King Abdullaziz University. She obtained her Ph.D. (Mathematics) from King Abdullaziz University. She has written more than 15 articles on harmonic analysis and fixed point theory and its applications in various international peer-reviewed journals.

Hafiz Fukhar-ud-din is an associate professor at the Department of Mathematics and Statistics, King Fahd University of Petroleum and Minerals, Dhahran, Saudi Arabia. He obtained his Ph.D. (Mathematics) from Tokyo Institute of Technology, Japan. He has supervised three Ph.D. and five M.Phil. students. He has written more than 40 articles on BCH-algebras and fixed point iterative methods in Banach spaces, hyperbolic spaces and convex metric spaces. He has contributed a chapter to the book *Fixed Point Theory, Variational Analysis, and Optimization*, Chapman & Hall/CRC, 2014.

Mohamed Amine Khamsi is Professor of Mathematics at the University of Texas at El Paso. He obtained his Ph.D. (Mathematics) from the University of Paris. He has written more than 70 articles on fixed point theory and its applications in various international peer-reviewed journals. He has co-authored five books.

Abdul Rahim Khan is Professor of Mathematics in the Department of Mathematics and Statistics, King Fahd University of Petroleum and Minerals, Dhahran, Saudi Arabia. He obtained a Ph.D. (Mathematics) from the University of Wales, UK. He has supervised five Ph.D. and eleven M.Phil. students. He is an editor of *Carpathian Journal of Mathematics* and *Arabian Journal of Mathematics*. He has written more than 115 articles on fixed point theory and iterative methods of nonlinear mappings on a Banach space, CAT(0) space and convex metric space, in various peer-reviewed journals. He has contributed a book chapter in *Advances in Non-Archimedean Analysis, Contemporary Mathematics, American Mathematical Society, 2011* and *Fixed Point Theory, Variational Analysis, and Optimization*, Chapman & Hall/CRC, 2014.

Paul-Emile Maingé is an assistant professor at the Université des Antilles, Campus de Schoelcher, Martinique, France. He holds a Ph.D. degree from Bordeaux University, France (1996) in the field of Applied Mathematics and an accreditation to supervise research in Mathematics (Habilitation à Diriger des Recherches) from French West Indies University, France (2008). He has vast experience of teaching and research at university level in various institutions including Bordeaux, Paris, Guadeloupe and Martinique. His field covers several areas of Applied Mathematics such as optimization, variational analysis, convex analysis, dynamical systems, variational inequalities, fixed-point problems and partial differential equations (slow and fast diffusion problems). He has obtained the Grant of Scientific Excellence for the periods 2006/2010 and 2011/2015. He already has more than 30 research papers to his credit which were published in leading world class journals and acts as a referee for many international journals.

Abdellatif Moudafi is a Full Professor, first class. He obtained a Ph.D. degree from Montpellier University, France (1991) in the field of Applied Mathematics (Nonlinear Analysis and Optimization), a doctorate (Doctorat d'état) from Casablanca University, Morocco (1995) and an accreditation to supervise research in Mathematics (Habilitation à Diriger des Recherches) from Montpellier University, France (1997). He has vast experience of teaching and research at university level in various institutions including Montpellier, Clermont-Ferrand, Limoges, Marrakech, Guadeloupe, Montreal, Martinique, New York and Aix-Marseille. His field covers many areas of Applied Mathematics such as optimization, applied nonlinear analysis, variational analysis, convex analysis, asymptotic analysis, discretization of gradient dynamical systems, proximal algorithms, variational inequalities, approximation and regularization of fixed-point problems and equilibria. He has supervised successfully several Ph.D. and M.S./M.Phil. students. He has obtained the Grant of Scientific Excellence for the periods 2001/2005, 2006/2010, and 2011/2015. He is currently member of the editorial board of several reputed international journals of mathematics. He has more than 100 research papers to his credit which were published in leading world class journals and acts as a referee for over 50 international journals.

Mădălina Păcurar is Associate Professor of Mathematics, Department of Statistics, Forecast and Mathematics, Faculty of Economics and Business Administration, Babes-Bolyai University of Cluj-Napoca, Romania. She obtained her Ph.D. in Mathematics from the Babes-Bolyai University of Cluj-Napoca, under the supervision of Professor Ioan A. Rus. Her scientific interest is in fixed point theory. She has published more than 40 scientific articles in the field of fixed point theory on product spaces. She has co-authored some textbooks (*Mathematics for Economics*) and had a monograph published (Iterative methods for fixed point approximation, Editura Risoprint, Cluj-Napoca, 2009) on her main research interest area.

Ioan A. Rus is Emeritus Professor, Faculty of Mathematics and Computer Science, Babes-Bolyai University of Cluj-Napoca, Romania. He is the founder (1969) of the fixed point research group in Cluj-Napoca, one of the oldest and most important research groups still active in this area. He supervised 26 Ph.D. theses at Babes-Bolyai University of Cluj-Napoca with topics in fixed point theory and its applications to differential and integral functional equations. He is the author of about 200 scientific papers. He has also had several books and monographs published, amongst which we mention the impressive monograph fixed point theory (Cluj University Press, Cluj-Napoca, 2008, xx+509 pp.), co-authors A. Petruşel and G. Petruşel. He is the founder of the first journal devoted entirely to fixed point theory: *Seminar on Fixed Point Theory* (1983–2002), now known as *Fixed Point Theory: An International Journal on Fixed Point Theory, Computation and Applications* since vol. 3 (2003). He is currently its Editor-in-Chief and also an editorial board member of *Carpathian Journal of Mathematics*, and *Studia Universitatis Babeş-Bolyai. Mathematica*.

D. R. Sahu is professor at the Department of Mathematics at Banaras Hindu University, Varanasi, India. He obtained his Ph.D. from Pt. Ravi Shankar Shukla University, Raipur, India. He is the co-author of a book on fixed point theory published by Springer. He has written more than 100 articles on existence, approximation and applications of fixed point problems of nonlinear operators in various peer-reviewed publications. He has contributed chapters for various edited books.

Bernd S. W. Schröder has served as Professor and Chair of the Department of Mathematics at the University of Southern Mississippi since 2014. He obtained his Ph.D. in Mathematics (Probability Theory and Harmonic Analysis) from Kansas State University in 1992. He held a position at Hampton University from 1992 to 1997, first as a research associate from 1992 to 1993 and then as an assistant professor from 1993 to 1997. Subsequently, he moved to Louisiana Tech University, where he was an associate professor from 1997 to 2003 and a professor from 2003 to 2014. At Louisiana Tech, he served as Program Chair for Mathematics and Statistics from 2003 to 2014 and as Academic Director for Mathematics and Statistics from 2008 to 2014. He has written 42 refereed papers on ordered sets, graphs, analysis and engineering education and he has written six textbooks, ranging from first year calculus to a monograph on ordered sets.

CHAPTER 1

Caristi-Browder Operator Theory in Distance Spaces

Vasile Berinde[*][†] and Ioan A. Rus[‡]

† North University Center at Baia Mare, Department of Mathematics and Computer Science, Victoriei nr. 76, 430122 Baia Mare, Romania
† King Fahd University of Petroleum and Minerals, Department of Mathematics and Statistics, Dhahran, Saudi Arabia
‡ Babeş-Bolyai University of Cluj-Napoca, Department of Mathematics, Kogălniceanu nr. 1, Cluj-Napoca, Romania
* Corresponding: `vasile.berinde@gmail.com`

Contents

Abstract

Starting from the classical Caristi fixed point theorem and its various versions, the aim of this chapter is to study Caristi-Browder operators in the following settings: (1) metric spaces; (2) $\mathbb{R}_+^m$-metric spaces; (3) $s(\mathbb{R}_+)$-metric spaces; (4) Kasahara spaces. Some new research directions in the Caristi-Browder operator theory are also indicated.

1.1. From the Caristi Fixed Point Theorems to Caristi, Caristi-Kirk and Caristi-Browder Operators

The following results are well known (see Refs. [35, 37, 98, 156]).

Theorem 1.1.1 (Caristi-Kirk Theorem). *Let (X,d) be a complete metric space and let $f : X \to X$ be such that there exists a lower semicontinuous (l.s.c.) functional*

http://dx.doi.org/10.1016/B978-0-12-804295-3.50001-2
1

$\varphi : X \to \mathbb{R}_+$ *with*

$$d(x, f(x)) \leq \varphi(x) - \varphi(f(x)), \quad \text{for all } x \in X.$$

Then

$$F_f := \{x \in X : f(x) = x\} \neq \emptyset.$$

Theorem 1.1.2 (Caristi-Browder Theorem). *Let* $f : X \to X$ *be with closed graphic and there exists* $\varphi : X \to \mathbb{R}_+$ *such that*

$$d(x, f(x)) \leq \varphi(x) - \varphi(f(x)), \quad \text{for all } x \in X.$$

Then

(a) $F_f \neq \emptyset$;

(b) $f^n(x) \to x^*(x) \in F_f$ *as* $n \to \infty$ *for all* $x \in X$.

Theorem 1.1.3 (Caristi Theorem). *Let* $Y \subset X$ *be a nonempty closed subset and* $f : Y \to X$ *a metrically inward contraction (i.e., for all* $x \in Y$, *there exists* $u \in Y$ *such that* $d(x, u) + d(u, f(x)) = d(x, f(x))$, *where* $u = x \Leftrightarrow x = f(x)$*). Then* f *has a fixed point.*

Remark 1.1.1. The first theorem (Caristi-Kirk Theorem in this chapter) is also known as: Caristi Theorem, Caristi-Kirk Theorem, Kirk-Caristi Theorem, Caristi-Ekeland Theorem, Caristi-Brøndsted Theorem, Caristi-Browder Theorem, Caristi-Kirk-Browder Theorem, etc. For some of the most relevant generalizations, further developments and related results (see Refs. [19, 33–35, 57, 63, 79, 92, 95, 98, 101, 134, 137, 161, 166, 183]).

We summarize here some of the early contributions on the Caristi-Kirk Theorem, following the presentation by Caristi himself [38]: Wong [183] has simplified the original transfinite induction argument, Browder [35] has given a proof which avoids the use of transfinite induction, Siegel [161] has presented a simple constructive proof, Brézis and Browder [33] have given a general theorem on ordered sets which includes Caristi-Kirk Theorem, Kasahara [92] has proved an analog of the Caristi-Kirk Theorem in the setting of partially ordered sets.

On the other hand, the Caristi-Kirk Theorem is actually implicit in Brøndsted's work [34] and is essentially equivalent to Ekeland's variational principle [63] (which is not formulated as a fixed point theorem). A generalization of Caristi-Kirk Theorem was given by Downing and Kirk [57]. Moreover, Kirk [98] has shown that the Caristi-Kirk Theorem characterizes completeness.

Remark 1.1.2. An operator $f : X \to X$ which satisfies (a) and (b) in Theorem 1.1.2 is, by definition, *a weakly Picard operator* and we denote (see Ref. [151]) $f^\infty(x) := x^*(x)$,

$x \in X$. It is clear that $f^\infty(X) = F_f$ and $f^\infty : X \to F_f$ is a set retraction.

The above results give rise to the following notions:

Definition 1.1.1. An operator $f : (X,d) \to (X,d)$ is a *Caristi operator* if there exists a functional $\varphi : X \to \mathbb{R}_+$ such that

$$d(x, f(x)) \leq \varphi(x) - \varphi(f(x)), \quad \text{for all } x \in X. \tag{1.1.1}$$

Definition 1.1.2. An operator $f : (X,d) \to (X,d)$ is a *Caristi-Kirk operator* if f is a Caristi operator with φ a lower semicontinuous functional.

Definition 1.1.3. By definition, an operator $f : (X,d) \to (X,d)$ is a *Caristi-Browder operator* if it is a Caristi operator with closed graphic.

Remark 1.1.3. If we consider on (X,d) the following partial order corresponding to $\varphi : X \to \mathbb{R}_+$ (see Refs. [34, 92, 95, 101]):

$$x \leq_{d,\varphi} y \quad \Leftrightarrow \quad d(x,y) \leq \varphi(x) - \varphi(y),$$

then, $f : (X, \leq_{d,\varphi}) \to (X, \leq_{d,\varphi})$ is progressive (i.e., $x \leq_{d,\varphi} f(x)$ for all $x \in X$). If f is a progressive operator, then for the maximal element set, $\text{Max}(X, \leq_{d,\varphi})$, we have the inclusion $\text{Max}(X, \leq_{d,\varphi}) \subset F_f$, so the problem is to give conditions on d and φ such that $\text{Max}(X, \leq_{d,\varphi}) \neq \emptyset$.

The Caristi-Kirk Theorem and its extensions are developments on these topics. This aspect has been largely studied (see Refs. [1, 95, 101, 80]; see also Refs. [4, 19, 23, 33, 34, 57, 63, 71, 73, 100, 115, 137, 141, 161, 166, 177, 183, 187, 122, 134].

For some other extensions of Caristi Theorem, see Ref. [155] and the references therein (see also Ref. [36]).

The aim of this chapter is to study the Caristi-Browder operators. The plan of the rest of the chapter is the following:
- List of notations / symbols
- Weakly Picard operators on L-spaces
- Caristi-Browder operators on metric spaces
- Caristi-Browder operators on $\mathbb{R}_+^m$-metric spaces
- Caristi-Browder operators on $s(\mathbb{R}_+)$-metric spaces
- Caristi-Browder operators on Kasahara spaces
- Research directions in the Caristi-Browder operator theory
  - Caristi-Browder operator on partial metric spaces
  - Caristi-Browder operator on b-metric spaces

- Nonself Caristi-Browder operators
- Multivalued Caristi-Browder operators

Several other important aspects related to Caristi's fixed point theorem, like its equivalence to Ekeland's variational principle, Takahashi's minimization theorem, lower semicontinuity, drop theorem, etc., are not covered in this survey, but an almost comprehensive bibliographic coverage is given in the list of references, see in particular Refs. [2, 3, 5–18, 20, 22, 24, 39–56, 58–60, 64–66, 69, 70, 72, 74, 76, 78, 81–83, 85, 86, 88–91, 93, 94, 96, 97, 99, 102–109, 111, 116, 120, 121, 123–125, 127, 130–139, 141–144, 146–149, 159–182, 184, 185, 187–193].

1.2. List of Notations

- $\mathbb{N} = \{0,1,2,\ldots\}$; $\mathbb{N}^* = \mathbb{N} \setminus \{0\}$;
- $\mathbb{R}$ denotes the set of real numbers: $\mathbb{R}^* = \mathbb{R} \setminus \{0\}$; $\mathbb{R}_+ = [0,\infty)$;
- Let X be a nonempty set and let $f : X \to X$ be an operator. Then
  - 1_X denotes the identity operator;
  - $F_f := \{x \in X : f(x) = x\}$ – the set of fixed points of f;
  - $f^0 := 1_X$, $f^1 := f,\ldots,f^n := f \circ f^{n-1}$ ($n \in \mathbb{N}^*$) – the iterates of f;
- Let $(X,+,\mathbb{R},\|\cdot\|)$ be a real normed space. Then
  - $\|\cdot\|$ or $\to$ denotes the strong convergence;
  - $\rightharpoonup$ denotes the weak convergence;
- Let (X,d) be a metric space. Then
  - $\mathscr{P}(X) := \{A : A \subset X\}$;
  - $\mathscr{P}_{cb}(X) := \{A : A \in \mathscr{P}(X), \text{i.e.}, A \text{ is nonempty, closed and bounded}\}$;
  - For $A,B \in \mathscr{P}(X)$,

  $$D(A,B) = \inf\{d(a,b) : a \in A, b \in B\} \text{ denotes the distance between } A \text{ and } B,$$

  - For $A,B \in \mathscr{P}_{cb}(X)$,

  $$H_d(A,B) = \max\{\sup\{D(a,B) : a \in A\}, \sup\{D(b,A) : b \in B\}\}$$

  denotes the *Pompeiu-Hausdorff metric* on $\mathscr{P}_{cb}(X)$ induced by the metric d.

1.3. Weakly Picard Operators on L-Spaces

Let X be a nonempty set and

$$s(X) := \{\{x_n\}_{n\in\mathbb{N}} : x_n \in X, n \in \mathbb{N}\}.$$

Let $c(X) \subset s(X)$ be a nonempty subset of $s(X)$ and $\text{Lim} : c(X) \to X$ be an operator. We consider on the triple $(X,c(X),\text{Lim})$ the following Fréchet axioms:

(c_1) If $x_n = x$ for all $n \in \mathbb{N}$, then $\{x_n\}_{n\in\mathbb{N}} \in c(X)$ and $\text{Lim}\{x_n\}_{n\in\mathbb{N}} = x$.

(c$_2$) If $\{x_n\}_{n\in\mathbb{N}} \in c(X)$ and $\mathrm{Lim}\{x_n\}_{n\in\mathbb{N}} = x$, then all its subsequences are in $c(X)$ and have the same limit, x.

By definition, a triple $(X, c(X), \mathrm{Lim})$ which satisfies the above axioms is called an *L-space* (see Refs. [151, 154] and the references therein).

In what follows we shall use the notation $(X, \rightarrow)$, instead of $(X, c(X), \mathrm{Lim})$ and

$$x_n \rightarrow x \text{ as } n \rightarrow \infty,$$

instead of $\mathrm{Lim}\{x_n\}_{n\in\mathbb{N}} = x$.

Example 1.3.1. Let (X, τ) be a topological Hausdorff space. Then $(X, \xrightarrow{\tau})$ is an *L*-space.

Example 1.3.2. Let (X, d) be a metric space. Then $(X, \xrightarrow{d})$ is an *L*-space.

Example 1.3.3. Let $(X, +, \mathbb{R}, \|\cdot\|)$ be a normed space. Then, $(X, \xrightarrow{\|\cdot\|})$ and $(X, \rightharpoonup)$ are *L*-spaces.

Example 1.3.4. Let (X, d) be an $\mathbb{R}_+^m$-metric space, i.e., $d : X \times X \rightarrow \mathbb{R}_+^m$, and d satisfies the standard axioms of a metric. Then, $(X, \xrightarrow{d})$ is an *L*-space.

Example 1.3.5. Let (X, d) be a *b*-metric space (or quasi-metric space), i.e., $d : X \times X \rightarrow \mathbb{R}_+$ satisfies the first two standard axioms of a metric, while the triangle inequality is replaced by

$$d(x,y) \leq s[d(x,z) + d(z,y)], \quad \text{for all } x,y,z \in X, \text{ where } s > 1 \text{ is given,}$$

(see Refs. [21, 26, 27]). Then $(X, \xrightarrow{d})$ is an *L*-space.

Example 1.3.6. Let X be a nonempty set. The functional $d : X \times X \rightarrow \mathbb{R}_+$ is called a *semimetric* if it satisfies the first two standard axioms of a metric, while the pair (X, d) is called a *semimetric space*. If $d : X \times X \rightarrow \mathbb{R}_+$ is continuous, then $(X, \xrightarrow{d})$ is an *L*-space.

Example 1.3.7. Let X be a nonempty set, $(G, +, \leq, \rightarrow)$ an *L*-space ordered group and $G_+ := \{g \in G \,|\, g \geq 0\}$. Let $d : X \times X \rightarrow G_+$ be a G_+-metric, i.e., d satisfies the standard axioms of a metric. Then $(X, \xrightarrow{d})$ is an *L*-space.

Let $(X, \rightarrow)$ be an *L*-space and $f : X \rightarrow X$ be an operator.

By definition, f is a *weakly Picard operator* (WPO) if the sequence $\{f^n(x)\}_{n\in\mathbb{N}}$ of its iterates at x converges for all $x \in X$ and its limit (which may depend on x) is a fixed point of f.

If f is a WPO, then we consider the operator $f^\infty : X \to X$ defined by

$$f^\infty(x) := \mathrm{Lim} f^n(x), \quad \text{for all } x \in X.$$

If f is a WPO with $F_f = \{x^*\}$, then, by definition, f is called a *Picard operator* (PO).

If (X,d) is a metric space, $f : X \to X$ is an operator and $\psi : \mathbb{R}_+ \to \mathbb{R}_+$ is a function then, by definition, f is a *ψ-WPO* if

(i) ψ is increasing, continuous at 0 and $\psi(0) = 0$;

(ii) $d(x, f^\infty(x)) \leq \psi(d(x, f(x)))$ for all $x \in X$.

For more considerations on weakly Picard operator theory (see Refs. [28, 30, 151, 156, 158]).

1.4. Caristi-Browder Operators on Metric Spaces

We start with some remarks an Caristi operators on a metric space (X,d) (see Refs. [35, 61, 62, 77, 154]).

Remark 1.4.1. An operator $f : X \to X$ is a Caristi operator if and only if

(c) $\displaystyle\sum_{k=0}^{\infty} d\left(f^k(x), f^{k+1}(x)\right)$ is convergent for all $x \in X$,

(see Ref. [62]). Indeed, if f is a φ-Caristi operator then

$$\sum_{k=0}^{n} d\left(f^k(x), f^{k+1}(x)\right) \leq \varphi(x) - \varphi(f^{n+1}(x)) \leq \varphi(x).$$

If we have (c) satisfied, then

$$d(x, f(x)) = \theta(x) - \theta(f(x)), \quad \text{for all } x \in X,$$

where

$$\theta(x) = \sum_{k=0}^{\infty} d\left(f^k(x), f^{k+1}(x)\right),$$

and hence (1.1.1) is satisfied.

Remark 1.4.1 gives rise to the following problem.

Problem 1.4.1. Which generalized contractions are Caristi operators?

Example 1.4.1. Let (X,d) be a complete metric space and $f : X \to X$ a *Banach contraction*, i.e., a map satisfying

$$d(f(x), f(y)) \le a\, d(x,y), \quad \text{for all } x, y \in X, \tag{1.4.2}$$

where $0 \le a < 1$ is a constant. Then f is a Caristi operator (see Refs. [30, 150, 156] for details).

Example 1.4.2. An operator $f : X \to X$ is called a *graphic contraction* if there exists a constant $0 \le a < 1$ such that

$$d(f^2(x), f(x)) \le a\, d(x, f(x)), \quad \text{for all } x \in X.$$

Then f is a Caristi operator (see Refs. [30, 150, 156] for details).

Example 1.4.3. Let (X,d) be a complete metric space and $f : X \to X$ a *Kannan contraction*, i.e., a map for which there exists $b \in \left(0, \frac{1}{2}\right)$ such that

$$d(f(x), f(y)) \le b\,[d(x, f(x)) + d(y, f(y))], \quad \text{for all } x, y \in X. \tag{1.4.3}$$

Then f is a Caristi operator (see Refs. [30, 90, 150, 156] for details).

Example 1.4.4. Let (X,d) be a complete metric space and $f : X \to X$ a *Chatterjea contraction*, i.e., a map for which there exists $c \in \left(0, \frac{1}{2}\right)$ such that

$$d(f(x), f(y)) \le c\,[d(x, f(y)) + d(y, f(x))], \quad \text{for all } x, y \in X. \tag{1.4.4}$$

Then f is a Caristi operator (see Refs. [30, 42, 150, 156] for details).

Example 1.4.5. Let (X,d) be a complete metric space and $f : X \to X$ be a *Zamfirescu contraction*, i.e., a map having the property that, for each pair $x, y \in X$, at least one of the conditions (1.4.2), (1.4.3), and (1.4.4) is satisfied. Then f is a Caristi operator (see Refs. [30, 150, 156, 186] for details).

Example 1.4.6. Let (X,d) be a complete metric space and let $f : X \to X$ be a *Ćirić quasi-contraction*, i.e., a map for which there exists $0 < h < 1$ such that

$$d(f(x), f(y)) \le h \cdot \max\{d(x,y), d(x, f(x)), d(y, f(y)), d(x, f(y)), d(y, f(x))\}, \tag{1.4.5}$$

for all $x, y \in X$. Then f is a Caristi operator (see Refs. [30, 46, 150, 156] for details).

Example 1.4.7. Let (X,d) be a complete metric space and $f : X \to X$ a *φ-contraction*, i.e., a map for which there exists a *comparison function*, i.e., a function $\varphi : \mathbb{R}_+ \to \mathbb{R}_+$

satisfying

(i_φ) φ is nondecreasing;

(ii_φ) the sequence $\{\varphi^n(t)\}$ converges to zero, for each $t \in \mathbb{R}_+$, such that

$$d(f(x), f(y)) \le \varphi(d(x,y)), \quad \text{for all } x, y \in X. \tag{1.4.6}$$

If φ is a strong comparison function, then f is a Caristi operator (see Refs. [30, 150, 156] for more details).

Example 1.4.8. Let (X, d) be a complete metric space and let $f : X \to X$ be an *almost contraction*, i.e., a map for which there exist a constant $\delta \in (0, 1)$ and some $L \ge 0$ such that

$$d(f(x), f(y)) \le \delta \cdot d(x,y) + Ld(y, f(x)), \quad \text{for all } x, y \in X. \tag{1.4.7}$$

Then f is a Caristi operator (see Refs. [30, 29, 150, 156] for details).

Remark 1.4.2. We note that the Caristi operators in Examples 1.4.1–1.4.7 are POs, while the Caristi operator in Example 1.4.8 is a WPO.

Remark 1.4.3. For $f : X \to X$, let us denote

$$\mathscr{F}_f(X, \mathbb{R}_+) := \{\varphi : X \to \mathbb{R}_+ : d(x, f(x)) \le \varphi(x) - \varphi(f(x)) \text{ for all } x \in X\}.$$

We note that

(a) f is Caristi operator if and only if $\mathscr{F}_f(X, \mathbb{R}_+) \ne \emptyset$;

(b) $\varphi_f(x) := \sum_{k=0}^{\infty} d\left(f^k(x), f^{k+1}(x)\right)$ is the first element of the poset $(\mathscr{F}_f(X, \mathbb{R}_+), \le)$ if and only if $\mathscr{F}_f(X, \mathbb{R}_+) \ne \emptyset$.

Remark 1.4.4. For $\varphi : X \to \mathbb{R}_+$ let us denote

$$\mathscr{F}_\varphi(X, X) := \{f : X \to X \,|\, d(x, f(x)) \le \varphi(x) - \varphi(f(x)) \text{ for all } x \in X\}.$$

Let $f, g \in \mathscr{F}_\varphi(X, X)$ and $X = X_1 \cup X_2$ be a partition of X. Let $h : X \to X$ be defined by

$$h(x) := \begin{cases} f(x), & \text{for } x \in X_1, \\ g(x), & \text{for } x \in X_2. \end{cases}$$

Then $h \in \mathscr{F}_\varphi(X, X)$.

Remark 1.4.5. As $1_X \in \mathscr{F}_\varphi(X, X)$, it is impossible to control the set $F_f \setminus \text{Max}(X, \le_{d,\varphi}$

) in $\mathscr{F}_\varphi(X,X)$. So, the following problem arises.

Problem 1.4.2. Let $(X,\leq)$ be a poset with $\mathrm{Max}(X,\leq) \neq \emptyset$. Let $f : X \to X$ be a progressive operator, i.e., $x \leq f(x)$ for all $x \in X$. The problem is to study the set $F_f \setminus \mathrm{Max}(X,\leq)$.

Remark 1.4.6. Let $CO(X,X) := \{f : X \to X \mid f \text{ is a Caristi operator}\}$. We remark that if $f \in CO(X,X)$, then $f^k \in CO(X,X)$ for all $k \in \mathbb{N}$.

Theorem 1.4.1. *Let (X,d) be a complete metric space and let $f : X \to X$ be a Caristi-Browder (CB) operator. Then*

(a) *f is a WPO;*

(b) *if f is a φ-CB operator with $\varphi(x) \leq c \cdot d(x,f(x))$ for some $c > 0$, then f is ψ-WPO with $\psi(t) = c \cdot t$, $t \in \mathbb{R}_+$.*

Proof. (a) This is just the Caristi-Browder Theorem.

(b) We have

$$d\left(x,f^\infty(x)\right) \leq \sum_{k=0}^{n} d\left(f^k(a),f^{k+1}(x)\right) + d\left(f^{n+1}(x),f^\infty(x)\right), \quad \text{for all } x \in X.$$

From this relation we have that

$$d\left(x,f^\infty(x)\right) \leq \sum_{k=0}^{\infty} d\left(f^k(x),f^{k+1}(x)\right) \leq \varphi(x) \leq c \cdot d(x,f(x)).$$

$\square$

Theorem 1.4.2. *Let (X,d) be a complete metric space and let $f_i : X \to X$, $i = 1,2$, be φ_i-CB operators with $\varphi_i(x) \leq c_i d(x,f_i(x))$, $i = 1,2$, for some $c_i > 0$. In addition, we suppose that there exists $\eta > 0$ such that*

$$d(f_1(x),f_2(x)) \leq \eta, \quad \text{for all } x \in X.$$

Then

$$H_d(F_{f_1},F_{f_2}) \leq \eta \max\{c_1,c_2\}.$$

Proof. By Theorem 1.4.1 we have that f_i is a c_i-WPO, $i = 1,2$. This implies that

$$d\left(x,f_i^\infty(x)\right) \leq c_i d(x,f_i(x)), \quad \text{for all } x \in X \text{ and all } i = 1,2.$$

Let $x_2 \in F_{f_2}$. Then

$$d\left(x_2, f_1^\infty(x_2)\right) \leq c_1 \, d(x_2, f_1(x_2)) = c_1 d(f_2(x_2), f_1(x_2)) \leq c_1 \, \eta.$$

If $x_1 \in F_{f_1}$, then in a similar way we have that

$$d(x_1, f_2^\infty(x_1)) \leq c_2 \, \eta.$$

Now the proof follows by Lemma 8.1.3(e) in Ref. [150].   □

In order to prove Theorem 1.4.3 we shall need the following lemma [151].

Lemma 1.4.1 [151, Abstract Gronwall Lemma]. *Let $(X, d, \leq)$ be an ordered metric space (i.e., a set X endowed with a metric d and a partial order relation $\leq$ which is closed with respect to d) and let $f : X \to X$ be an operator satisfying the following two conditions:*

(i) *f is a PO ($F_f = \{x_f^*\}$);*

(ii) *f is increasing.*

Then

(a) *$x \leq f(x) \Rightarrow x \leq x_f^*$;*

(b) *$x \geq f(x) \Rightarrow x \geq x_f^*$.*

Theorem 1.4.3. *Let $(X, d, \leq)$ be a complete ordered metric space and let $f : X \to X$ be an operator. We suppose that*

(i) *$f : (X, d) \to (X, d)$ is a Caristi-Browder operator;*

(ii) *$f : (X, \leq) \to (X, \leq)$ is an increasing operator.*

Then

(a) *$x \in X, \ x \leq f(x) \Rightarrow x \leq f^\infty(x)$;*

(b) *$x \in X, \ x \geq f(x) \Rightarrow x \geq f^\infty(x)$.*

Proof. By Theorem 1.4.1, f is a WPO with respect to $\xrightarrow{d}$. On the other hand, the operator $f : (X, \leq) \to (X, \leq)$ is increasing. So, we are in the conditions of the Abstract Gronwall Lemma and the conclusion follows.   □

Using an abstract result for WPOs [157] (see also Ref. [238, Theorem 6.3.3]), from the above result we obtain the following Gronwall type results.

Corollary 1.4.1. *Let (X,d) be a complete ordered metric space and let $f : X \to X$ be an increasing almost contraction. Then*

(a) $x \in X$, $x \le f(x) \Rightarrow x \le f^\infty(x)$;

(b) $x \in X$, $x \ge f(x) \Rightarrow x \ge f^\infty(x)$.

Proof. From Ref. [156, Theorem 6.3.3], we know that the following two statements are equivalent:

(WP$_1$) There exists an *L*-space structure on the set X, "$\to$", such that the mapping $f : (X, \to) \to (X, \to)$ is a WPO;

(WP$_4$) There exists a complete metric d on X and a number $\alpha \in (0,1)$ such that

    (i) $f : (X,d) \to (X,d)$ has closed graphic;
    (ii) $d\left(f^2(x), f(x)\right) \le \alpha \cdot d(x, f(x))$, for all $x, y \in X$.

On the other hand, from Ref. [30, Theorem 2.11], we know that (WP$_1$) holds, with the *L*-space structure $\overset{d}{\to}$. Hence, (WP$_4$) also holds. Thus, by (ii) and Example 1.4.2, f is a Caristi operator and, by (i), f is a Caristi-Browder operator. The conclusion now follows by Theorem 1.4.3.   □

Example 1.4.9. Let $X = \mathbb{R}$ with the usual metric $d(x,y) = |x - y|$ and let $f : \mathbb{R} \to \mathbb{R}$ be defined by $f(x) = 0$, if $x \in (-\infty, 3]$ and $f(x) = -\frac{1}{2}$, if $x \in (3, +\infty)$. Then

(a) f is a Kannan operator with $b = \frac{1}{7}$ (see Ref. [129, Example 1.3.2]); hence f is an almost contraction too.

(b) f is a Caristi operator (by Example 1.4.3);

(c) f is not a Caristi-Browder operator (with respect to d).

Indeed, f has no closed graphic, since for $x_n = 3 + \frac{1}{n} \in (3, +\infty)$ we have $x_n \to 3$ as $n \to \infty$, $f(x_n) = -\frac{1}{2} \to -\frac{1}{2}$ as $n \to \infty$, but $f(3) = 0 \ne -\frac{1}{2}$.

Remark 1.4.7. Corollary 1.4.1 provides only sufficient conditions for an operator to have properties (a) and (b). Indeed, as shown by Example 1.4.9, f is nonincreasing and is not a Caristi-Browder operator, hence Corollary 1.4.1 cannot be applied.

However, both conclusions of Corollary 1.4.1 hold since $f^\infty(x) = 0$ for all $x \in \mathbb{R}$ and

$$\{x \in \mathbb{R} : x \le f(x)\} = (-\infty, 0), \quad \{x \in \mathbb{R} : x \ge f(x)\} = [0, \infty).$$

For more considerations on weakly Picard operator theory on metric spaces, see Refs. [19, 30, 151, 156, 158].

1.5. Caristi-Browder Operators on $\mathbb{R}_+^m$-Metric Spaces

Let (X,d) be an $\mathbb{R}_+^m$-metric space. By definition an operator $f : X \to X$ is a *Caristi operator* if there exists $\varphi : \mathbb{R}_+^m \to \mathbb{R}_+^m$ such that

$$d(x, f(x)) \leq \varphi(x) - \varphi(f(x)), \quad \text{for all } x \in X.$$

If, in addition, f has closed graphic, then f is called a *Caristi-Browder operator*.

As in the case of ordinary metric spaces (i.e., $d(x,y) \in \mathbb{R}_+$) we have the following.

Remark 1.5.1. An operator $f : X \to X$ is a Caristi operator if and only if

$$\sum_{k=0}^{\infty} d\left(f^k(x), f^{k+1}(x) \right) \text{ is convergent for all } x \in X.$$

This remark is useful to give examples of Caristi operators in an $\mathbb{R}_+^m$-metric space.

Example 1.5.1. A matrix $S \in \mathbb{R}_+^{m \times m}$ is said to be convergent to zero if and only if $S^n \to 0$ as $n \to \infty$. Also, by definition, an operator $f : X \to X$ is an *S-contraction* if there exists a matrix $S \in \mathbb{R}_+^{m \times m}$ convergent to zero such that

$$d(f(x), f(y)) \leq S d(x,y), \quad \text{for all } x,y \in X.$$

By the properties of matrices convergent to 0 (see Ref. [156]), and by Remark 1.5.1, it follows that any S-contraction is a Caristi operator.

Example 1.5.2. An operator $f : X \to X$ is called a *graphic contraction* if there exists a matrix $S \in \mathbb{R}_+^{m \times m}$ convergent to zero such that

$$d(f^2(x), f(x)) \leq S d(x, f(x)), \quad \text{for all } x \in X.$$

By Remark 1.5.1, it follows that each graphic contraction is a Caristi operator.

Example 1.5.3. Let X be a partially ordered set such that every pair $x,y \in X$ has a lower and an upper bound. Furthermore, let d be a metric on X such that (X,d) is a complete $\mathbb{R}_+^m$-metric space and suppose also that $f : X \to X$ is continuous, monotone (i.e., increasing or decreasing) and satisfies the following assumptions:

(i) there exists a matrix $A \in \mathbb{R}_+^{m \times m}$ convergent to zero such that

$$d(f(x), f(y)) \leq A d(x,y), \quad \text{for all } x \geq y;$$

(ii) there exists $x_0 \in X$ such that $x_0 \leq f(x_0)$ or $x_0 \geq f(x_0)$.

Then f is a Caristi-Browder operator (see Theorem 2.1 in Ref. [87] for details).

Example 1.5.4. Let X be a partially ordered set such that every pair $x, y \in X$ has a lower and an upper bound. Let d and ρ be two $\mathbb{R}_+^m$-metrics on X. Let $f : X \to X$ and suppose the following assumptions are satisfied:

  (i) $d(x,y) \leq \rho(x,y)$ for all $x \geq y$;

  (ii) (X,d) is a generalized ordered complete $\mathbb{R}_+^m$-metric space;

  (iii) $f : (X,d) \to (X,d)$ is a continuous mapping;

  (iv) f is a monotone mapping;

  (v) there exists a matrix $A \in \mathbb{R}_+^{m \times m}$ convergent to zero such that
$$d(f(x), f(y)) \leq A d(x,y), \quad \text{for all } x \geq y;$$

  (vi) there exists $x_0 \in X$ such that $x_0 \leq f(x_0)$ or $x_0 \geq f(x_0)$.

Then f is a Caristi-Browder operator (see Ref. [87, Theorem 3.1] for details).

Remark 1.5.2. For more details on the matrices convergent to zero and for the fixed point theory of operators on $\mathbb{R}_+^m$-metric spaces, see Ref. [156, pp. 82–86] and the references therein (see also Ref. [25]).

Theorem 1.5.1. *Let (X,d) be a complete $\mathbb{R}_+^m$-metric space and $f : X \to X$ a Caristi-Browder operator. Then the following assertions hold.*

  (a) *f is a WPO;*

  (b) *If f is a φ-CB operator with $\varphi(x) \leq C d(x, f(x))$ for some $C \in \mathbb{R}_+^{m \times m}$, then*
$$d(x, f^\infty(x)) \leq C d(x, f(x)), \quad \text{for all } x \in X.$$

Proof. (a) Let f be a φ-CB operator. Then we have
$$\sum_{k=0}^{n} d\left(f^k(x), f^{k+1}(x)\right) \leq \varphi(x) - \varphi\left(f^{n+1}(x)\right) \leq \varphi(x), \quad \text{for all } x \in X.$$

Since $\mathbb{R}_+^m$ is regular (i.e., $\{x_n\}_{n \in \mathbb{N}} \subset \mathbb{R}_+^m$ increasing and bounded with respect to $\leq$ $\Rightarrow \{x_n\}$ converges w. r. t. $\overline{\mathbb{R}^m} \to$), it follows that
$$\sum_{k=0}^{\infty} d\left(f^k(x), f^{k+1}(x)\right) \text{ converges for all } x \in X.$$

This implies that
$$f^n(x) \to x^*(x) \text{ as } n \to \infty, \quad \text{for all } x \in X.$$

But f has a closed graphic, so $x^*(x) \in F_f$.

(b) We have

$$d(x, f^\infty(x)) \leq \sum_{k=0}^{n} d\left(f^k(x), f^{k+1}(x)\right) + d\left(f^{n+1}(x), f^\infty(x)\right), \quad \text{for all } x \in X.$$

This implies that

$$d(x, f^\infty(x)) \leq \sum_{k=0}^{\infty} d\left(f^k(x), f^{k+1}(x)\right) \leq \varphi(x) \leq Cd(x, f(x)).$$

$\square$

In a similar way to the considerations in Section 1.4 we have the following.

Theorem 1.5.2. *Let (X,d) be a complete $\mathbb{R}_+^m$-metric space and f_i, $i = 1,2$, φ_i-CB operator with $\varphi_i(x) \leq C_i d(x, f_i(x))$ for some $c_i \in \mathbb{R}_+^{m \times m}$. In addition we suppose that there exists $\eta_i \in \mathbb{R}_+^m$ such that*

$$d(f_1(x), f_2(x)) \leq \eta, \quad \text{for all } x \in X.$$

Then

$$H_d(F_{f_1}, F_{f_2}) \leq Cd(x, f(x)),$$

where $C \in \mathbb{R}_+^{m \times m}$ such that $C_i \leq C$, $i = 1,2$.

Theorem 1.5.3. *Let $(X, d, \leq)$ be a complete ordered $\mathbb{R}_+^m$-metric space and $f : X \to X$ an operator. We suppose that*

(i) $f : (X,d) \to (X,d)$ *is a Caristi-Browder operator;*

(ii) $f : (X, \leq) \to (X, \leq)$ *is an increasing operator.*

Then

(a) $x \in X$, $x \leq f(x) \Rightarrow x \leq f^\infty(x)$;

(b) $x \in X$, $x \geq f(x) \Rightarrow x \geq f^\infty(x)$.

By Theorem 1.5.3 and Examples 1.5.1–1.5.4 we have the following results.

Corollary 1.5.1. *Let (X,d) be a complete ordered $\mathbb{R}_+^m$-metric space and let $f : X \to X$ be an increasing S-contraction. Then*

(a) $x \in X$, $x \leq f(x) \Rightarrow x \leq f^\infty(x)$;

(b) $x \in X$, $x \geq f(x) \Rightarrow x \geq f^{\infty}(x)$.

Corollary 1.5.2. *Let* (X,d) *be a complete ordered* $\mathbb{R}_+^m$-*metric space and let* $f : X \to X$ *be an increasing graphic contraction. If* f *has closed graph, then*

(a) $x \in X$, $x \leq f(x) \Rightarrow x \leq f^{\infty}(x)$;

(b) $x \in X$, $x \geq f(x) \Rightarrow x \geq f^{\infty}(x)$.

Corollary 1.5.3. *Let* X *be a partially ordered set such that every pair* $x,y \in X$ *has a lower and an upper bound. Furthermore, let d be a metric on X such that* (X,d) *is a complete* $\mathbb{R}_+^m$-*metric space and suppose also that* $f : X \to X$ *is continuous, increasing and satisfies the following assumptions:*

(i) *there exists a matrix* $A \in \mathbb{R}_+^{m \times m}$ *convergent to zero such that*
$$d(f(x), f(y)) \leq A d(x,y), \quad \text{for all } x \geq y;$$

(ii) *there exists* $x_0 \in X$ *such that* $x_0 \leq f(x_0)$ *or* $x_0 \geq f(x_0)$.

Then

(a) $x \in X$, $x \leq f(x) \Rightarrow x \leq f^{\infty}(x)$;

(b) $x \in X$, $x \geq f(x) \Rightarrow x \geq f^{\infty}(x)$.

Proof. By Example 1.5.3, f is a Caristi-Browder operator. The conclusion now follows by Theorem 1.5.3. $\qquad\qquad\square$

Corollary 1.5.4. *Let* X *be a partially ordered set such that every pair* $x,y \in X$ *has a lower and an upper bound. Let d and* ρ *be two* $\mathbb{R}_+^m$-*metrics on X. Let* $f : X \to X$ *and suppose the following assumptions are satisfied:*

(i) $d(x,y) \leq \rho(x,y)$, *for all* $x \geq y$;

(ii) (X,d) *is a generalized ordered complete* $\mathbb{R}_+^m$-*metric space;*

(iii) $f : (X,d) \to (X,d)$ *is a continuous mapping;*

(iv) f *is increasing;*

(v) *there exists a matrix* $A \in \mathbb{R}_+^{m \times m}$ *convergent to zero such that*
$$d(f(x), f(y)) \leq A d(x,y), \quad \text{for all } x \geq y;$$

(vi) *there exists* $x_0 \in X$ *such that* $x_0 \leq f(x_0)$ *or* $x_0 \geq f(x_0)$.

Then

(a) $x \in X$, $x \le f(x) \Rightarrow x \le f^\infty(x)$;

(b) $x \in X$, $x \ge f(x) \Rightarrow x \ge f^\infty(x)$.

For more considerations on weakly Picard operator theory on $\mathbb{R}_+^m$-metric spaces, see Refs. [28, 30, 151, 156, 158].

1.6. Caristi-Browder Operators on $s(\mathbb{R}_+)$-Metric Spaces

Let $s(\mathbb{R}) := \{\{x_n\}_{n \in \mathbb{N}} : x_n \in \mathbb{R}, n \in \mathbb{N}\}$. Then $(s(\mathbb{R}), +, \mathbb{R}, \to)$ is a linear ordered L-space, where $\to$ is the termwise convergence.

Let X be a nonempty set. By definition, a functional $d : X \times X \to s(\mathbb{R}_+)$ is an $s(\mathbb{R}_+)$-metric if d satisfies the corresponding standard Fréchet's axioms of a metric.

Remark 1.6.1. A functional $d : X \times X \to s(\mathbb{R}_+)$, $(x,y) \mapsto (d_k(x,y))_{k \in \mathbb{N}}$ is an $s(\mathbb{R}_+)$-metric on X if and only if

(i) d_k is a pseudometric, for all $k \in \mathbb{N}$, i.e., d_k satisfies:
   (a) $d_k(x,x) = 0$ for all $x,y \in X$;
   (b) $d_k(x,y) = d_k(y,x)$ for all $x,y \in X$;
   (c) $d_k(x,z) \le d_k(x,y) + d_k(y,z)$ for all $x,y,z \in X$;

(ii) for all $x,y \in X$, $x \ne y$, $\exists k \in \mathbb{N}$ such that $d_k(x,y) \ne 0$.

Example 1.6.1. $X := s(\mathbb{R})$, $d(x,y) := (|x_0, -y_0|, \ldots, |x_n - y_n|, \ldots)$.

Example 1.6.2. $X := C([a,b], s(\mathbb{R}))$, $d(x,y) := (\|x_0 - y_0\|_\infty, \ldots, \|x_n - y_n\|_\infty, \ldots)$.

Example 1.6.3. $X := C(\mathbb{R}_+, \mathbb{R})$,
$$d(x,y) := (|x(0) - y(0)|, \|x|_{[0,1]} - y|_{[0,1]}\|_\infty, \ldots, \|x|_{[0,n]} - y|_{[0,n]}\|_\infty, \ldots).$$

Definition 1.6.1 [156, Definition 6.2.2]. An $s(\mathbb{R}_+)$-metric space (X,d) is complete (in the Weierstrass sense) if, for any $\{x_n\} \subset X$,
$$\sum_{n \in \mathbb{N}^*} d(x_n, x_{n+1}) \text{ converges} \Rightarrow \{x_n\} \text{ converges}.$$

Let (X,d) be an $s(\mathbb{R}_+)$-metric space. By definition, an operator $f : X \to X$ is a *Caristi operator* if there exists $\varphi : X \to s(\mathbb{R}_+)$ such that
$$d(x, f(x)) \le \varphi(x) - \varphi(f(x)), \quad \text{for all } x \in X. \tag{c}$$

If, in addition, f has a closed graphic, then f is called a *Caristi-Browder operator*.

Remark 1.6.2. Let $\varphi : X \to s(\mathbb{R}_+)$, $x \mapsto (\varphi_0(x), \ldots, \varphi_n(x), \ldots)$. Then condition (c) take the following form:

$$d_k(x, f(x)) \le \varphi_k(x) - \varphi_k(f(x)), \quad \text{for all } x \in X, \ k \in \mathbb{N}.$$

For a more restrictive condition (c), see for example Ref. [110].

Theorem 1.6.1. *Let (X, d) be a complete $s(\mathbb{R}_+)$-metric space and let $f : X \to X$ be a Caristi-Browder operator. Then the following assertions hold.*

(a) *f is WPO;*

(b) *If f is a φ-CB operator with*

$$\varphi_k(x) \le c_k d_k(x, f(x)), \quad \text{for all } x \in X, k \in \mathbb{N},$$

for some $c_k \ge 0$, $k \in \mathbb{N}$, then

$$d_k(x, f^\infty(x)) \le c_k d_k(x, f(x)), \quad \text{for all } x \in X, k \in \mathbb{N}.$$

Proof. (a) Let f be a φ-CB operator. Then we have

$$\sum_{i=0}^{n} d_k\left(f^i(x), f^{i+1}(x)\right) \le \varphi_k(x) - \varphi_k\left(f^{n+1}(x)\right) \le \varphi_k(x), \quad \text{for all } x \in X.$$

This means that

$$\sum_{i=0}^{\infty} d_k\left(f^i(x), f^{i+1}(x)\right) \text{ converges for all } x \in X, \text{ for all } k \in \mathbb{N},$$

and this implies that

$$f^n(x) \xrightarrow{d} x^*(x) \text{ as } n \to \infty, \text{ for all } x \in X.$$

But f has a closed graphic (as a Caristi-Browder operator), so $x^*(x) \in F_f$.

(b) We have

$$d_k\left(x, f^\infty(x)\right) \le \sum_{i=0}^{n} d_k\left(f^i(x), f^{i+1}(x)\right) + d_k\left(f^{n+1}(x), f^\infty(x)\right), \text{ for all } x \in X \text{ and all } k \in \mathbb{N}$$

and this implies that

$$d_k\left(x, f^\infty(x)\right) \le \sum_{i=0}^{\infty} d_k\left(f^i(x), f^{i+1}(x)\right) \le \varphi(x) \le c_k d_k(x, f(x)), \quad \text{for all } k \in \mathbb{N}.$$

$\square$

Theorem 1.6.2. *Let (X,d) be a complete $s(\mathbb{R}_+)$-metric space and $f,g : X \to X$ be two operators. Suppose that the following conditions hold.*

(i) *There exists $\varphi : X \to s(\mathbb{R}_+)$ such that*

   (1) $d(x,f(x)) \leq \varphi(x) - \varphi(f(x))$ *for all $x \in X$;*
   (2)   *There exists $c_{1,k} > 0 : \varphi_k(x) \leq c_{1k} d_k(x, f(x))$ for all $x \in X$.*

(ii) *There exists $\theta : X \to s(\mathbb{R}_+)$ such that:*

   (3) $d(x,g(x)) \leq \theta(x) - \theta(g(x))$ *for all $x \in X$;*
   (4)   *There exists $c_{2k} > 0 : \theta_k(x) \leq c_{2k} d_k(x, g(x))$ for all $x \in X$.*

(iii) *f and g are with closed graphic.*

(iv) *There exists $\eta \in s(\mathbb{R}_+) : d(f(x), g(x)) \leq \eta$ for all $x \in X$.*

Then the following assertions hold.

(a) *f and g are WPOs;*

(b) *$H_{dk}(F_f, F_g) \leq \max(c_{1k}, c_{2k}) \cdot \eta$ for all $k \in \mathbb{N}$, where H_{d_k} is the Pompeiu-Hausdorff functional corresponding to d_k.*

Proof. (a) We remark that f and g are as in Theorem 1.6.1. So, f and g are WPOs.

(b) From Theorem 1.6.1 we have that
- $d_k(x, f^\infty(x)) \leq c_{1k} d_k(x, f(x))$ for all $x \in X, k \in \mathbb{N}$,
- $d_k(x, g^\infty(x)) \leq c_{2k} d_k(x, g(x))$ for all $x \in X, k \in \mathbb{N}$.

Now the proof follows from a similar result to Lemma 8.1.3(e) in Ref. [150], but established for the case of a pseudometric (instead of a metric).   □

Theorem 1.6.3. *Let $(X,d,\leq)$ be a complete ordered $s(\mathbb{R}_+)$-metric space and $f : X \to X$ an operator. We suppose that*

(i) *$f : (X,d) \to (X,d)$ is a Caristi-Browder operator;*

(ii) *$f : (X,\leq) \to (X,\leq)$ is an increasing operator.*

Then

(a) *$x \in X, x \leq f(x) \Rightarrow x \leq f^\infty(x)$;*

(b) *$x \in X, x \geq f(x) \Rightarrow x \geq f^\infty(x)$.*

Proof. By Theorem 1.6.2, f is a WPO with respect to $\xrightarrow{d}$. On the other hand, the operator $f : (X,\leq) \to (X,\leq)$ is increasing. So, we have the conditions of the Abstract Gronwall Lemma (Lemma 1.4.1) and the conclusion follows.   □

1.7. Caristi-Browder Operators on Kasahara Spaces

Let X be a nonempty set, $\to$ an L-space structure on X and $d : X \times X \to \mathbb{R}_+$ a functional. By definition (see Ref. [153]) the triple $(X, \to, d)$ is a *Kasahara space* if we have the following compatibility condition between $\to$ and d satisfied:

$$x_n \in X, \sum_{n \in \mathbb{N}} d(x_n, x_{n+1}) < +\infty \quad \Rightarrow \quad \{x_n\}_{n \in \mathbb{N}} \text{ converges in } (X, \to).$$

We give some examples of Kasahara spaces, taken from Ref. [153] (see also Refs. [67, 68]).

Example 1.7.1 [153]. Let (X, d) be a complete metric space and let d be the convergence structure induced by the metric $\xrightarrow{d}$ on X. Then $(X, \xrightarrow{d}, d)$ is a Kasahara space.

Example 1.7.2 [153]. Let (X, ρ) be a complete semimetric space (i.e., $\rho : X \times X \to \mathbb{R}_+$ satisfies the following two axioms:

(i) $d(x, y) = 0$ if and only if $x = y$;

(ii) $d(x, z) \leq d(x, y) + d(y, z)$ for all $x, y, z \in X$).

Let $d : X \times X \to \mathbb{R}_+$ be a functional such that there exists $c > 0$ with $\rho(x, y) \leq c \cdot d(x, y)$, for all $x, y \in X$. If d is continuous, then $(X, \xrightarrow{\rho}, d)$ is a Kasahara space.

Example 1.7.3 [153]. Let (X, ρ) be a complete quasi-metric space (i.e., $\rho : X \times X \to \mathbb{R}_+$ satisfies the following two axioms:

(i) $d(x, y) = d(y, x) = 0$ if and only if $x = y$;

(ii) $d(x, z) \leq d(x, y) + d(y, z)$ for all $x, y, z \in X$).

Let $d : X \times X \to \mathbb{R}_+$ be a functional such that there exists $c > 0$ with $\rho(x, y) \leq c \cdot d(x, y)$, for all $x, y \in X$. Then $(X, \xrightarrow{\rho}, d)$ is a Kasahara space.

Definition 1.7.1. Let $(X, \to, d)$ be a Kasahara space. An operator $f : X \to X$ is a *Caristi operator* if there exists $\varphi : X \to \mathbb{R}_+$ such that

$$d(x, f(x)) \leq \varphi(x) - \varphi(f(x)), \quad \text{for all } x \in X.$$

If in addition $f : (X, \to) \to (X, \to)$ has a closed graphic then, by definition, f is a *Caristi-Browder operator*.

Theorem 1.7.1. *Let $(X, \to, d)$ be a Kasahara space and let $f : X \to X$ be a Caristi-Browder operator. Then*

(a) $f : (X, \rightarrow) \rightarrow (X, \rightarrow)$ is a WPO;

(b) if

(i) f is a φ-CB operator with $\varphi(x) \le c\, d(x, f(x))$ for all $x \in X$, for some $c > 0$;

(ii) d satisfies triangle inequality.

Then

$$d(x, f^{\infty}(x)) \le c\, d(x, f(x)), \quad \text{for all } x \in X.$$

Proof. (a) From the condition

$$d(x, f(x)) \le \varphi(x) - \varphi(f(x)), \quad \text{for all } x \in X,$$

we have that

$$\sum_{k=0}^{n} d\left(f^k(x), f^{k+1}(x)\right) \le \varphi(x) - \varphi\left(f^{n+1}(x)\right) \le \varphi(x), \quad \text{for all } x \in X.$$

Since $(X, \rightarrow, d)$ is a Kasahara space it follows that

$$f^n(x) \to x^*(x) \text{ as } n \to \infty.$$

But $f : (X, \rightarrow) \rightarrow (X, \rightarrow)$ has a closed graphic. This implies that $x^*(x) \in F_f$.

(b) We remark that

$$d\left(x, f^{\infty}(x)\right) \le \sum_{k=0}^{n} d\left(f^k(x), f^{k+1}(x)\right) + d\left(f^{n+1}(x), f^{\infty}(x)\right).$$

This implies

$$d\left(x, f^{\infty}(x)\right) \le \sum_{k=0}^{\infty} d\left(f^k(x), f^{k+1}(x)\right) \le c\, d(x, f(x)).$$

This completes the proof.   □

Theorem 1.7.2. *Let $(X, d, \rightarrow)$ be a Kasahara space and $\le$ a partial order on X such that $(X, \rightarrow, \le)$ is an ordered L-space. Let $f : X \to X$ be an operator and we suppose that*

(i) $f : (X, d, \rightarrow) \rightarrow (X, d, \rightarrow)$ *is a Caristi-Browder operator;*

(ii) $f : (X, \le) \rightarrow (X, \le)$ *is increasing.*

Then

(a) $x \in X, x \le f(x) \Rightarrow x \le f^{\infty}(x)$;

(b) $x \in X, x \ge f(x) \Rightarrow x \ge f^{\infty}(x)$.

Proof. By Theorem 1.7.1, f is a WPO with respect to $\overset{d}{\to}$. On the other hand, the operator $f : (X, \leq) \to (X, \leq)$ is increasing. So, we have the conditions of the Abstract Gronwall Lemma and the conclusion follows. □

From the above consideration the following question arises.

Problem 1.7.1. The following notion is given in Ref. [153].

Let X be a nonempty set, $\to$ an L-space structure on X, $(G, +, \leq, \overline{G}\to)$ an L-space ordered semigroup with the unity 0, the least element in $(G, \leq)$ and $d_G : X \times X \to G$ an operator. By definition the triple $(X, d_G, \to)$ is a generalized Kasahara space if and only if we have the following compatibility condition between $\to$ and d_G:

$$x_n \in X, \sum_{n \in \mathbb{N}} d(x_n, x_{n+1}) < \infty \quad \Rightarrow \quad \{x_n\}_{n \in \mathbb{N}} \text{ converges in } (X, \to).$$

The problem is to study the Caristi-Browder operators on a generalized Kasahara space

$$(X, d_G, \to)(d_G := \mathbb{R}_+^m\text{-metric}, d_G := s(\mathbb{R}_+)\text{-metric, etc.}).$$

For more considerations, see Refs. [67, 68].

1.8. Research Directions in the Caristi-Browder Operator Theory

In this chapter our main aim was to study single-valued self Caristi-Browder operators, from the point of view of Picard operators, weakly Picard operators and Gronwall lemma.

A large area of research is left open for the case of single-valued self and non-self Caristi-Browder operators in various settings, as well as for multivalued self and nonself Caristi-Browder operators.

Here we mention just a few directions of research directly related to the material in this chapter:

1. Caristi-Browder operators on partial metric spaces

2. Caristi-Browder operators on b-metric spaces

3. Nonself Caristi-Browder operators

4. Multivalued Caristi-Browder operators.

For other directions of research, essentially related to Caristi type fixed point theorems, like Ekeland's variational principle, Takahashi's minimization theorem, lower semicontinuity, drop theorem, etc., see Refs. [2–18, 20, 22, 24, 39–56, 58–60,

64–66, 69, 70, 72, 74, 76, 78, 81–83, 85, 86, 88, 89, 91, 93, 94, 96, 97, 99, 102–109, 111, 116, 120, 121, 123–125, 127, 130–139, 141–144, 146–149, 159–182, 184, 185, 187–193].

REFERENCES

1. Aamri M, El Moutawakil D, Quelques nouveaux théorèmes du point fixe dans espaces topologiques de type F. Extracta Math. 2002;17:211–219.
2. Abdeljawad T, Karapinar E, Quasicone metric spaces and generalizations of Caristi Kirk's theorem. Fixed Point Theory Appl. 2009;2009:Article ID 574387.
3. Acar Ö, Altun I, Some generalizations of Caristi type fixed point theorem on partial metric spaces. Filomat 2012;26:833–837.
4. Acar Ö, Altun I, Romaguera S, Caristi's type mappings on complete partial metric spaces, Fixed Point Theory 2013;14:3–10.
5. Agarwal RP, Khamsi MA, Extension of Caristi's fixed point theorem to vector valued metric spaces. Nonlinear Anal. 2011;74:141–145.
6. Alamri B, Suzuki T, Khan L, Caristi's fixed point theorem and Subrahmanyam's fixed point theorem in v-generalized metric spaces. J. Funct. Spaces 2015;2015:Article ID 709391.
7. Alfuraidan MR, Remarks on Caristi's fixed point theorem in metric spaces with a graph. Fixed Point Theory Appl. 2014:2014:Article ID 240.
8. Alfuraidan MR, Khamsi MA, Caristi fixed point theorem in metric spaces with a graph. Abstr. Appl. Anal. 2014;2014:Article ID 303484.
9. Alsiary T, Latif A, Generalized Caristi fixed point results in partial metric spaces. J. Nonlinear Convex Anal. 2015;16:119–125.
10. Amini-Harandi A, Farajzadeh AP, O'Regan D, On generalizations of Ekeland's variational principle and Takahashi's minimization theorem and applications. Nonlinear Funct. Anal. Appl. 2009;14:653–664.
11. Amini-Harandi A, Some generalizations of Caristi's fixed point theorem with applications to the fixed point theory of weakly contractive set-valued maps and the minimization problem. Nonlinear Anal. 2010;72:4661–4665.
12. Angelov VG, An extension of Kirk-Caristi theorem to uniform spaces. Antarct. J. Math. 2004;1: 47–51.
13. Antony J, Subrahmanyam PV, Quasi-gauge and Caristi's theorem. J. Math. Phys. Sci. 1991;25: 35–36.
14. Araya Y, Ekeland's variational principle and its equivalent theorems in vector optimization. J. Math. Anal. Appl. 2008;346:9–16.
15. Araya Y, Tanaka T, On generalizing Caristi's fixed point theorem. In: Nonlinear Anal. Convex Anal. 41–46, Yokohama:Yokohama Publ.; 2007.
16. Aruffo A, Bottaro G, Generalizations of sequential lower semicontinuity. Boll. Unione Mat. Ital. 2008;9(1):293–318.
17. Aydi H, Karapinar E, Kumam P, A note on 'Modified proof of Caristi's fixed point theorem on partial metric spaces' J. Inequal. Appl. 2013; 2013:Article ID 210, and J. Inequal. Appl. 2013;2013:Article ID 355.
18. Bae JS, Fixed point theorems for weakly contractive multivalued maps. J. Math. Anal. Appl. 2003;284:690–697.
19. Bae JS, Park S, Remarks on the Caristi-Kirk fixed point theorem. Bull. Korean Math. Soc. 1983;19:57–60.
20. Bae JS, Cho EW, Yeom, SH, A generalization of the Caristi-Kirk fixed point theorem and its applications to mapping theorems. J. Korean Math. Soc. 1994;31:29–48.
21. Bakhtin IA, The contraction mapping principle in almost metric space [in Russian]. Functional Anal. Ul'yanovsk. Gos. Ped. Inst., Ul'yanovsk. 1989;30:26–37.
22. Basu T, Some fixed point theorems for multivalued mappings using Caristi and Kirk type condition. J. Indian Acad. Math. 1986;8:58–63.

23. Bata M, Molnár A, Varga C, On Ekeland's variational principle in *b*-metric spaces. Fixed Point Theory 2011;12:21–28.
24. Beg I, Abbas M, Random fixed point theorems for Caristi type random operators. J. Appl. Math. Comput. 2007;25:425–434.
25. Belitskii GR, Lyubich YI, Matrix norms and their applications. Translated from the Russian by A. Iacob. In: Operator Theory. Advances and Applications, 36. Basel: Birkhäuser Verlag; 1988.
26. Berinde V, Generalized contractions in quasimetric spaces. Seminar on Fixed Point Theory. "Babeş-Bolyai" Univ. Cluj-Napoca. 1993;93.3:3–9.
27. Berinde V, Sequences of operators and fixed points in quasimetric spaces. Studia Univ. Babeş-Bolyai Math. 1996;41:23–27.
28. Berinde V, Contracţii generalizate şi aplicaţii. Baia Mare:Cub Press; 1997.
29. Berinde V, Approximating fixed points of weak contractions using the Picard iteration. Nonlinear Anal. Forum. 2004;9:43–53.
30. Berinde V, Iterative Approximation of Fixed Points. Berlin:Springer; 2007.
31. Bîrsan T, New generalizations of Caristi's fixed point theorem via Brézis-Browder principle. Math. Morav. 2004;8:1–5.
32. Bollenbacher A, Hicks TL, A fixed point theorem revisited. Proc. Am. Math. Soc. 1988;102:898–900.
33. Brézis H, Browder FE, A general principle on ordered sets in nonlinear functional analysis. Adv. Math. 1976;21:355–364.
34. Brøndsted A, Fixed points and partial orders. Proc. Am. Math. Soc. 1976;60:365–366.
35. Browder FE, On a theorem of Caristi and Kirk. In: Fixed Point Theory and Its Applications (Proc. Sem., Dalhousie Univ., Halifax, N.S., 1975). New York:Academic Press; 1976:23–27.
36. Cardinali T, Rubbioni P, A generalization of the Caristi fixed point theorem in metric spaces. Fixed Point Theory 2010;11:3–10.
37. Caristi J, Fixed point theorems for mapping satisfying inwardness conditions. Trans. Am. Math. Soc. 1976;215:241–251.
38. Caristi J, Fixed point theory and inwardness conditions. Applied nonlinear analysis (Proc. Third Internat. Conf., Univ. Texas, Arlington, TX, 1978). New York:Academic Press; 1979: 479–483.
39. Chandel RS, Fixed point theorem for multivalued mappings under the Caristi-Kirk type condition. Jñānābha. 1988;18:117–124.
40. Chang SS, Luo Q, Caristi's fixed point theorem for fuzzy mappings and Ekeland's variational principle. Fuzzy Sets and Systems 1994;64:119–125.
41. Chang SS, Cho YJ, Kim, JK, Ekeland's variational principle and Caristi's coincidence theorem for set-valued mappings in probabilistic metric spaces. Period. Math. Hungar. 1996;33:83–92.
42. Chatterjea SK, Fixed-point theorems. C.R. Acad. Bulgare Sci. 1972;25:727–730.
43. Checa E, Romaguera S, A revision of Caristi's fixed-point theorem for nonsymmetric distance functions. (Spanish) Proceedings of the XIVth Spanish-Portuguese Conference on Mathematics, Vol. IIII (Spanish) (Puerto de la Cruz, 1989), 309–312, Univ. La Laguna, La Laguna, 1990.
44. Chen SB, Tian SP, Mao ZY, On Caristi's fixed point theorem in quasi-metric spaces. Dyn. Contin. Discrete Impuls. Syst. Ser. A Math. Anal. 2006;13A:1150–1157.
45. Chikkala R, Baisnab AP, An analogue of Caristi-fixed-point theorem in a quasi-metric space. Proc. Nat. Acad. Sci. India Sect. A 1991;61:237–243.
46. Ćirić LB, A generalization of Banach's contraction principle. Proc. Am. Math. Soc. 1974;45:267–273.
47. Ćirić LB, A generalization of Caristi's fixed point theorem. Math. Pannon. 1992;3:51–57.
48. Cho YJ, Park KS, Chang SS, Fixed point theorems in metric spaces and probabilistic metric spaces. Int. J. Math. Math. Sci. 1996;19:243–252.
49. Cho SH, Some generalizations of Caristi's fixed point theorem with applications. Int. J. Math. Anal. (Ruse) 2013;7:557–564.
50. Cho SH, Bae JS, Na KS, Fixed point theorems for multivalued contractive mappings and multivalued Caristi type mappings in cone metric spaces. Fixed Point Theory Appl. 2012;2012:Article ID 133.

51. Chung KJ, On weakly nonlinear contractions. Kodai Math. J. 1983;6:301–307.
52. Dan ND, Remark on a question of Kirk about Caristi's fixed point theorem. Math. Operationsforsch. Statist. Ser. Optim. 1981;12:177–179.
53. Dancs S, Hegedüs M, Medvegyev P, A general ordering and fixed-point principle in complete metric space. Acta Sci. Math. (Szeged) 1983;46:381–388.
54. Daneš J, Equivalence of some geometric and related results of nonlinear functional analysis. Comment. Math. Univ. Carolin. 1985;26:443–454.
55. de Blasi FS, Myjak J, Reich S, Zaslavski AJ, Generic existence and approximation of fixed points for nonexpansive set-valued maps. Set-Valued Var. Anal. 2009;17:97–112.
56. Diallo MO, Oudadess M, Extensions of Caristi-Kirk's theorem. Turkish J. Math. 1996;20:153–158.
57. Downing D, Kirk WA, A generalization of Caristi's theorem with applications to nonlinear mapping theory. Pacific J. Math. 1977;69:339–346.
58. Du WS, On Caristi type maps and generalized distances with applications. Abstr. Appl. Anal. 2013; 2013:Article ID 407219.
59. Du WS, Karapinar E, A note on Caristi-type cyclic maps: related results and applications. Fixed Point Theory Appl. 2013;2013:Article ID 344.
60. Du WS, On Latif's fixed point theorems. Taiwanese J. Math. 2011;15:1477–1485.
61. Eisenfeld J, Lakshmikantham V, Fixed point theorems through abstract cones. J. Math. Anal. 1975;52:25–35.
62. Eisenfeld J, Lakshmikantham V, Fixed point theorems on closed sets through abstract cones. Appl. Math. Comput. 1977;3:155–167.
63. Ekeland I, On the variational principle. J. Math. Anal. Appl. 1974;47:324–353.
64. Eshghinezhad S, Fakhar M, Some generalizations of Ekeland's variational principle with applications to fixed point theory. Math. Comput. Modelling 2013;57:1250–1258.
65. Fang JX, The variational principle and fixed point theorems in certain topological spaces. J. Math. Anal. Appl. 1996;202:398–412.
66. Feng YJ, Liu SY, Fixed point theorems for multi-valued contractive mappings and multi-valued Caristi type mappings. J. Math. Anal. Appl. 2006;317:103–112.
67. Filip AD, Fixed point theorems in Kasahara spaces with respect to an operator and applications. Fixed Point Theory 2011;12:329–340.
68. Filip AD, Fixed Point Theory in Kasahara Spaces. Ph.D. Thesis. "Babeş-Bolyai" University of Cluj-Napoca, 2011, Cluj-Napoca, Romania.
69. Frigon M, On some generalizations of Ekeland's principle and inward contractions in gauge spaces. J. Fixed Point Theory Appl. 2011;10:279–298.
70. Glab S, On the converse of Caristi's fixed point theorem. Bull. Pol. Acad. Sci. Math. 2004;52:411–416.
71. Goebel K, Kirk WA, Topics in Metric Fixed Point Theory. Cambridge:Cambridge Univ. Press; 1990.
72. Guo TX, Yang YJ, Ekeland's variational principle for an L^0-valued function on a complete random metric space. J. Math. Anal. Appl. 2012;389:1–14.
73. Granas A, Horvath C, On the order theoretic Cantor theorem. Taiwanese J. Math. 2004;4:203–213.
74. Hadžić O, Žikić T, On Caristi's fixed point theorem in F-type topological spaces. Novi Sad J. Math. 1998;28:91–98.
75. Haghi RH, Rezapour S, Shahzad N, Be careful on partial metric fixed point results. Topol. Appl. 2013;160:450–454.
76. He F, Qiu JH, A vectorial Ekeland's variational principle on bornological vector spaces. Adv. Math. (China) 2013;42:889–895.
77. Hicks TL, Fixed point theorems for quasi-metric spaces. Math. Japonicae 1988;33:231–236.
78. Hoáng T, A fixed point theorem involving a hybrid inwardness-contraction condition. Math. Nachr. 1981;102:271–275.
79. Husain SA, Sehgal VM, A remark on a fixed point theorem of Caristi. Math. Japon. 1980;25: 27–30.
80. Hyers DH, Isac G, Rassias TM, Topics in Nonlinear Analysis and Applications. River Edge: Word Scientific; 1997.

81. Isac G, A fixed-point theorem of Caristi type in locally convex spaces. Applic. Univ. u Novom Sadu Zb. Rad. Prirod.-Mat. Fak. Ser. Mat. 1985;15:31–42.
82. Jachymski J, Caristi's fixed point theorem and selections of set-valued contractions. J. Math. Anal. Appl. 1998;227:55–67.
83. Jachymski J, Converses to fixed point theorems of Zermelo and Caristi. Nonlinear Anal. 2003;52:1455–1463.
84. Jachymski J, A stationary point theorem characterizing metric completeness. Appl. Math. Lett. 2011;24:169–171.
85. Jleli M, Samet B, Vetro C, Vetro F, From Caristi's theorem to Ekeland's variational principle in 0_σ-complete metric-like spaces. Abstr. Appl. Anal. 2014;2014:Article ID 319619.
86. Jung JS, Cho YJ, Kim JK, Minimization theorems for fixed point theorems in fuzzy metric spaces and applications. Fuzzy Sets and Systems 1994;61:199–207.
87. Jurja N, A Perov-type fixed point theorem in generalized ordered metric spaces. Creat. Math. Inform. 2008;17:427–430.
88. Kada O, Suzuki T, Takahashi W, Nonconvex minimization theorems and fixed point theorems in complete metric spaces. Math. Japon. 1996;44:381–391.
89. Kadelburg Z, Radenović S, Simić S, Abstract metric spaces and Caristi-Nguyen-type theorems. Filomat 2011;25:111–124.
90. Kannan R, Some results on fixed points. Bull. Calcutta Math. Soc. 1968;10:71–76.
91. Karapinar E, Generalizations of Caristi Kirk's theorem on partial metric spaces. Fixed Point Theory Appl. 2011;2011:Article ID 4.
92. Kasahara S, On fixed point in partially ordered sets and Kirk-Caristi theorem. Math. Seminar Notes. 1975;35:Article ID 3.
93. Khamsi MA, Remarks on Caristi's fixed point theorem. Nonlinear Anal. 2009;71:227–231.
94. Khamsi MA, Misane D, Compactness of convexity structures in metric spaces. Math. Japon. 1995;41:321–326.
95. Khamsi MA, Kirk WA, An Introduction to Metric Spaces and Fixed Point Theory. New York:Wiley; 2001.
96. Khamsi MA, Wojciechowski PJ, On the additivity of the Minkowski functionals. Numer. Funct. Anal. Optim. 2013;34:635–647.
97. Kim, TH, Kim ES, Shin SS, Minimization theorems relating to fixed point theorems on complete metric spaces. Math. Japon. 1997;45:97–102.
98. Kirk WA, Caristi's fixed point theorem and metric convexity. Colloq. Math. 1976;36:81–86.
99. Kirk WA, Caristi's fixed point theorem and the theory of normal solvability. In: Fixed Point Theory and Its Applications (Proc. Sem., Dalhousie Univ., Halifax, N.S., 1975), pp. 109–120. New York: Academic Press; 1976.
100. Kirk WA, Saliga LM, The Brézis-Browder order principle and extensions of Caristi's theorem. Nonlinear Anal. 2001;47:2765–2778.
101. Kirk WA, Shahzad N, Fixed Point Theory in Distance Space. Berlin:Springer; 2014.
102. Kirk WA, Shahzad N, Generalized metrics and Caristi's theorem. Fixed Point Theory Appl. 2013; 2013:Article ID 129.
103. Kirk WA, Shahzad N, Erratum to: Generalized metrics and Caristi's theorem. Fixed Point Theory Appl. 2014;2014:177.
104. Klin-eam C, Modified proof of Caristi's fixed point theorem on partial metric spaces. J. Inequal. Appl. 2013;2013:Article ID 210.
105. Kumar L, Som T, Metiya N, Extension of Caristi's theorem to cone metric space. Int. J. Funct. Anal. Oper. Theory Appl. 2012;4:97–107.
106. Lael F, Nourouzi K, The role of regularity to reach the vector valued version of Caristi's fixed point theorem. J. Nonlinear Convex Anal. 2015;16:937–942.
107. Lahrech S, Benbrik A, Kirk-Caristi stationary point theorem in uniform spaces. Int. J. Appl. Math. 2004;16:365–373.
108. Latif A, Generalized Caristi's fixed point theorems. Fixed Point Theory Appl. 2009;2009:Article ID 170140.

109. Latif A, Hussain N, Kutbi MA, Fixed points for w-contractive multimaps. Int. J. Math. Math. Sci. 2009;2009:Article ID 769467.
110. Latif A, Hussain N, Kutbi MA, Applications of Caristi's fixed point results. J. Ineq. Appl. 2012;20:12–40.
111. Latif A, Kutbi MA, On multivalued Caristi type fixed point theorems. Acta Univ. Apulensis Math. Inform. 2011;28:179–188.
112. Lazăr TA, Teoreme de punct fix pentru operatori non-self şi aplicaţii. Cluj-Napoca:Casa Cărţii de Ştiinţă;2014.
113. Lazăr TA, Fixed point for non-self nonlinear contractions and non-self Caristi type operators. Creat. Math. Inform. 2008;17:446–451.
114. Lazăr TA, Petruşel A, Shahzad N, Fixed point for non-self operators and domain invariance theorems. Nonlinear Anal. 2009;70:117–125.
115. Li Z, Remarks on Caristi's fixed point theorem and Kirk problem. Nonlinear Anal. 2010;73:3751–3755.
116. Li Z, Jiang SJ, Maximal and minimal point theorems and Caristi's fixed point theorem. Fixed Point Theory Appl. 2011;2011:Article ID 103.
117. Lin LJ, Chuang CS, Wang SY, From quasivariational inclusion problems to Stampacchia vector quasiequilibrium problems. Stampacchia set-valued vector Ekeland's variational principle and Caristi's fixed point theorem. Nonlinear Anal. 2009;71:179–185.
118. Lin LJ, Wang SY, Ekeland's variational principle for vectorial multivalued mappings in a uniform space. Nonlinear Anal. 2011;74:5057–5068.
119. Liu Z, Chen ZS, Kang SM, Kim HG, Common stationary point theorems for a pair of multi-valued mappings involving Caristi and contractive type conditions. Int. J. Appl. Math. Sci. 2006;3:69–81.
120. Madhusudana Rao J, An extension of Caristi's theorem to multifunctions. Bull. Math. Soc. Sci. Math. R. S. Roumanie (N.S.) 1985;29(77):79–80.
121. Maciejewski M, Inward contractions on metric spaces. J. Math. Anal. Appl. 2007;330:1207–1219.
122. Marika R, Some forms of the axiom of choice. J. Buch. Kurt-Gödel-Ges. Wien 1988;1:24–34.
123. Mizoguchi N, A generalization of Brøndsted's results and its applications. Proc. Am. Math. Soc. 1990;108:707–714.
124. Mizoguchi N, Takahashi W, Fixed point theorems for multivalued mappings on complete metric spaces. J. Math. Anal. Appl. 1989;141:177–188.
125. Mukherjee RN, Som T, Yadav SR, An application of Caristi's fixed point theorem. Bull. Calcutta Math. Soc. 1988;80:44–46.
126. Mureşan, AS, Some fixed point theorems of Maia type. Seminar on Fixed Point Theory. Preprint, Univ. "Babeş-Bolyai" Cluj-Napoca. 1988;88-3:35–42.
127. Nimana N, Petrot N, Saksirikun W, Fixed points of set-valued Caristi-type mappings on semi-metric spaces via partial order relations. Fixed Point Theory Appl. 2014;2014:Article ID 27.
128. Obama T, Kuroiwa D, Common fixed point theorems of Caristi type mappings with w-distance. Sci. Math. Jpn. 2010;72:41–48.
129. Păcurar M, Iterative Methods for Fixed Point Approximation. Cluj-Napoca:Risoprint; 2010.
130. Park S, On Kasahara's extension of the Caristi-Kirk fixed point theorem. Math. Japon. 1982;27:509–512.
131. Park S, On extensions of the Caristi-Kirk fixed point theorem. J. Korean Math. Soc. 1983;19:143–151.
132. Park S, Bae JS, On the Ray-Walker extension of the Caristi-Kirk fixed point theorem. Nonlinear Anal. 1985;9:1135–1136.
133. Park S, Rhoades BE, An equivalent form of the Caristi-Kirk-Browder theorem. Bull. Math. Soc. Sci. Math. R. S. Roumanie (N.S.) 1988;32(80):159–160.
134. Pasicki L, A short proof of the Caristi theorem. Comment. Math. Prace Mat. 1977/1978;20:427–428.
135. Pathak HK, Khan MS, Fixed points of fuzzy mappings under Caristi-Kirk type condition. J. Fuzzy Math. 1998;6:119–126.
136. Pathak, HK, Mishra SN, Some results related to Caristi's fixed point theorem and Ekeland's variational principle. Demonstr. Math. 2001;34:859–872.
137. Penat JP, A short constructive proof of Caristi's fixed point theorem. Publ. Math. Univ. Paris 1976;10:1–3.

138. Petruşel A, Caristi type operators and applications. Bul. Ştiinţ. Univ. Baia Mare Ser. B Fasc. Mat.-Inform. 2002;18:297–302.
139. Petruşel A, Caristi type operators and applications. Studia Univ. Babeş-Bolyai. Math. 2003;48:115–123.
140. Petruşel A, Rus IA, Şerban MA, Basic problems of the metric fixed point theorem for multivalued operator. J. Nonlinear Convex Anal. 2014;15:493–513.
141. Petruşel A, Sîntămărian A, Single-valued and multi-valued Caristi type operators. Publ. Math. Debrecen. 2002;60:167–177.
142. Phan QK, On Caristi-Kirk's theorem and Ekeland's variational principle for Pareto extrema. Bull. Polish Acad. Sci. Math. 1989(1990);37:33–39.
143. Popa V, A Caristi selection theorem for a multifunction satisfying an implicit contractive condition of Latif-Beg type. Sci. Stud. Res. Ser. Math. Inform. 2010;20:55–59.
144. Precup R, On the continuation method and the nonlinear alternative for Caristi-type non-self-mappings. J. Fixed Point Theory Appl. 2014;16:3–10.
145. Proinov PD, A unified theory of cone metric spaces and its applications to the fixed point theory. Fixed Point Theory Appl. 2013;2013:103.
146. Qiu JH, Ekeland's variational principle in locally convex spaces and the density of extremal points. J. Math. Anal. Appl. 2009;360:317–327.
147. Qiu JH, He F, A general vectorial Ekeland's variational principle with a p-distance. Acta Math. Sin. (Engl. Ser.) 2013;29:1655–1678.
148. Qiu JH, Li B, He F, Vectorial Ekeland's variational principle with a w-distance and its equivalent theorems. Acta Math. Sci. Ser. B Engl. Ed. 2012;32:2221–2236.
149. Qiu JH, Rolewicz S, Ekeland's variational principle in locally p-convex spaces and related results. Studia Math. 2008;186:219–235.
150. Rus IA, Generalized Contractions and Applications. Cluj-Napoca:Cluj Univ. Press; 2001.
151. Rus IA, Picard operators and applications. Sci. Math. Jpn. 2003;58:191–219.
152. Rus IA, Fixed point theory in partial metric spaces. Analele Univ. de Vest, Timişoara. Mat.-Info. 2008;46:149–160.
153. Rus IA, Kasahara spaces. Sci. Math. Jpn. 2010;72:101–110.
154. Rus IA, Heuristic introduction to weakly Picard operator theory. Creat. Math. Inform. 2014;23:243–252.
155. Rus IA, The generalized retraction methods in fixed point theory for nonself operators. Fixed Point Theory 2014;15:559–578.
156. Rus IA, Petruşel, A, Petruşel, G., Fixed Point Theory. Cluj-Napoca:Cluj Univ. Press; 2008.
157. Rus IA, Petruşel A, Şerban MA, Weakly Picard operators: equivalent definitions, applications and open problems. Fixed Point Theory 2006;7:3–22.
158. Rus IA, Şerban MA, Basic problems of the metric fixed point theory and the relevance of a metric fixed point theorem. Carpathian J. Math. 2013;29:239–258.
159. Shi C, Caristi type hybrid fixed point theorems in Menger probabilistic metric space. Appl. Math. Mech. 1997;18:185–192.
160. Shu Q, Li XS, Some τ-distance version of Caristi type coincidence point theorems for set-valued mappings. Nonlinear Anal. Forum. 2012;17:73–79.
161. Siegel J, A new proof of Caristi's fixed point theorem. Proc. Am. Math. Soc. 1977;66:54–56.
162. Singh SL, Extensions of the Caristi-Kirk fixed point theorem. J. Nat. Phys. Sci. 1990;4:1–11.
163. Sitthikul K, Saejung S, Common fixed points of Caristi's type mappings via w-distance. Fixed Point Theory Appl. 2015;2015:Article ID 6.
164. Squassina M, Symmetry in variational principles and applications. J. Lond. Math. Soc. 2012;85:323–348.
165. Sun JL, Sun JX, A generalization of Caristi's fixed point theorem and its applications. J. Math. Res. Exposition 2006;26:199–206.
166. Suzuki T, Generalized Caristi's fixed point theorems by Bae and others. J. Math. Anal. Appl. 2005;302:502–508.
167. Suzuki T, Counterexamples on τ-distance versions of generalized Caristi's fixed point theorems. Bull. Kyushu Inst. Technol. Pure Appl. Math. 2005;52:15–20.

168. Száz A, Some easy to remember abstract forms of Ekeland's variational principle and Caristi's fixed point theorem. Appl. Anal. Discrete Math. 2007;1:335–339.
169. Szilágyi T, A characterization of complete metric spaces and other remarks to a theorem of Ekeland. Ann. Univ. Sci. Budapest Eötvös Sect. Math. 1984(1985);27:103–106.
170. Tasković MR, Extension of theorems by Krasnoselskij, Stečenko, Dugundji, Granas, Kiventidis, Romaguera, Caristi and Kirk. Math. Morav. 2002;6:109–118.
171. Turinici M, Remarks about some Caristi-Kirk fixed point theorems. An. Ştiinţ. Univ. Ovidius Constanţa. Ser. Mat. 1999;7:81–92.
172. Turinici M, Pseudometric versions of the Caristi-Kirk fixed point theorem. Fixed Point Theory 2004;5:147–161.
173. Turinici M, Functional type Caristi-Kirk theorems. Libertas Math. 2005;25:1–12.
174. Turinici M, Function pseudometric variants of the Caristi-Kirk theorem. Fixed Point Theory 2006;7:341–363.
175. Turinici M, Functional versions of the Caristi-Kirk theorem. Rev. Un. Mat. Argentina 2009;50: 87–97.
176. Turinici M, Common fixed points for Banach-Caristi contractive pairs. ROMAI J. 2013;9:197–203.
177. Turinici M, Selected Topics in Metrical Fixed Point Theory. Iaşi:Editura Olm; 2014.
178. Ume JS, Some existence theorems generalizing fixed point theorems on complete metric spaces. Math. Japon. 1994;40:109–114.
179. Vetro C, Vetro F, Caristi type selections of multivalued mappings. J. Funct. Spaces 2015;2015:Article ID 941856.
180. Yie S, A generalizaton of fixed point theorems. J. Korean Math. Soc. 1988;25:77–82.
181. Wlodarczyk K, Plebaniak R, Maximality principle and general results of Ekeland and Caristi types without lower semicontinuity assumptions in cone uniform spaces with generalized pseudodistances. Fixed Point Theory Appl. 2010;2010:Article ID 175453.
182. Wlodarczyk K, Plebaniak R, Dynamic processes, fixed points, endpoints, asymmetric structures, and investigations related to Caristi, Nadler, and Banach in uniform spaces. Abstr. Appl. Anal. 2015;2015:Article ID 942814.
183. Wong CS, On a fixed point of contractive type. Proc. Am. Math. Soc. 1976;57:283–284.
184. Wong CS, A drop theorem without vector topology. J. Math. Anal. Appl. 2007;329:452–471.
185. Wu Z, Equivalent extensions to Caristi-Kirk's fixed point theorem, Ekeland's variational principle, and Takahashi's minimization theorem. Fixed Point Theory Appl. 2010;2010:Article ID 970579.
186. Zamfirescu T, Fix point theorems in metric spaces. Arch. Math. (Basel) 1972;23:292–298.
187. Zhang G, Jiang D, On the fixed point theorems of Caristi type. Fixed Point Theory 2013;14: 523–529.
188. Zhang SS, Huang NJ, On the principle of randomization of fixed points for set-valued mappings with applications. Northeast. Math. J. 1991;7:486–491.
189. Zhang SS, Chen YQ, Guo JL, Ekeland's variational principle and Caristi's fixed point theorem in probabilistic metric space. Acta Math. Appl. Sinica (English Ser.) 1991;7:217–228.
190. Zhang SS, Kang SK, Chen RD, Wang XD, Caristi-type fixed point theorems and applications. J. Chengdu Univ. Sci. Tech. 1995;5:70–75.
191. Zhang SS, Luo Q, Set-valued Caristi's fixed point theorem and Ekeland's variational principle. Appl. Math. Mech. (English Ed.) 1989;10:119–121.
192. Zhu J, Li Q, Generalized Ekeland's variational principle on generalized metric space. Southeast Asian Bull. Math. 2008;32:601–615.
193. Zhu J, Wei L, Zhu CC, Caristi type coincidence point theorem in topological spaces. J. Appl. Math. 2013;2013:Article ID 902692.

CHAPTER 2

Iterative Approximation of Fixed Points of Single-valued Almost Contractions

Vasile Berinde[a][*][†] and Mădălina Păcurar[‡]

[*] North University Center at Baia Mare, Department of Mathematics and Computer Science, Victoriei nr. 76, 430122 Baia Mare, Romania

[†] King Fahd University of Petroleum and Minerals, Department of Mathematics and Statistics, Dhahran, Saudi Arabia

[‡] Babeş-Bolyai University of Cluj-Napoca, Department of Statistics, Analysis, Forecast and Mathematics Faculty of Economics and Bussiness Administration, 56–60 T. Mihali St., 400591 Cluj-Napoca, Romania

[*] Corresponding: vasile.berinde@gmail.com

Contents

Abstract

The aim of this chapter is to survey the most relevant developments done in the last decade around the concept of almost contraction, introduced in Berinde V, Approximating fixed points of weak contractions using the Picard iteration. Nonlinear Anal. Forum 2004;9(1):43–53.

2.1. Introduction

Metrical fixed point theory developed around Banach's contraction principle, which, in the case of a metric space setting, can be briefly stated as follows.

Theorem 2.1.1. *Let (X,d) be a complete metric space and $T : X \to X$ a strict contraction, i.e., a map satisfying*

$$d(Tx,Ty) \le a\,d(x,y), \quad \text{for all } x,y \in X, \tag{2.1.1}$$

http://dx.doi.org/10.1016/B978-0-12-804295-3.50002-4

where $0 \leq a < 1$ is constant. Then

(p1) *T has a unique fixed point p in X (i.e., $Tp = p$);*

(p2) *The Picard iteration $\{x_n\}_{n=0}^{\infty}$ defined by*

$$x_{n+1} = Tx_n, \quad n = 0, 1, 2, \ldots, \tag{2.1.2}$$

converges to p, for any $x_0 \in X$.

Remark 2.1.1. A map satisfying (p1) and (p2) in Theorem 2.1.1 is said to be a *Picard operator* (see Refs. [235, 236, 239, 240] for more details).

Theorem 2.1.1, which was established in a complete linear normed space in 1922 by Stefan Banach [49] (see also Ref. [50]), is in fact a formalization of the method of successive approximation that has previously been systematically used by Picard in 1890 [210] to study differential and integral equations.

Being a simple and versatile tool in establishing existence and uniqueness theorems for operator equations, Theorem 2.1.1 plays a very important role in nonlinear analysis. This fact motivated researchers to try to extend and generalize Theorem 2.1.1 in such a way that its area of applications should be enlarged as much as possible.

Most of these generalizations considered only continuous mappings, like in the original case of the contraction mapping in Theorem 2.1.1. It was natural to ask whether there exist or not alternative contractive conditions that ensure the conclusions of Theorem 2.1.1 but which do not force implicitly or explicitly that T is continuous.

This question was answered in the affirmative by Kannan in 1968 [149], who proved a fixed point theorem which extends Theorem 2.1.1 to mappings that need not be continuous. Kannan considered instead of (2.1.1) the following condition: there exists $b \in \left(0, \frac{1}{2}\right)$ such that

$$d(Tx, Ty) \leq b[d(x, Tx) + d(y, Ty)], \quad \text{for all } x, y \in X. \tag{2.1.3}$$

Following Kannan's theorem, many papers were devoted to obtaining fixed point theorems for various classes of contractive type conditions that do not require the continuity of T (see for example Refs. [60, 235, 239] and the references therein).

One of these results is actually a sort of dual of Kannan's fixed point theorem, and is due to Chatterjea [85]. It makes use of a condition similar to (2.1.3): there exists $c \in \left(0, \frac{1}{2}\right)$ such that

$$d(Tx, Ty) \leq c[d(x, Ty) + d(y, Tx)], \quad \text{for all } x, y \in X. \tag{2.1.4}$$

Based on the fact (established later by Rhoades [226]) that the contractive conditions (2.1.1), (2.1.3), and (2.1.4) are independent, Zamfirescu [280] obtained a very interesting fixed point theorem, by combining (2.1.1), (2.1.3) and (2.1.4).

Theorem 2.1.2. *Let (X,d) be a complete metric space and $T : X \to X$ a map for which there exist the real numbers a, b and c satisfying $0 \le a < 1$, $0 < b$, $c < 1/2$ such that for each pair x, y in X, at least one of the following is true:*

(z_1) $d(Tx, Ty) \le a\, d(x,y)$;

(z_2) $d(Tx, Ty) \le b[d(x, Tx) + d(y, Ty)]$;

(z_3) $d(Tx, Ty) \le c[d(x, Ty) + d(y, Tx)]$.

Then T is a Picard operator.

The class of *almost contractions*, the central concept surveyed in this chapter, is closely related to Zamfirescu's contractions. Indeed, by conditions (z_2) and (z_3) one obtains that T satisfies the conditions:

$$d(Tx, Ty) \le \frac{b}{1-b} d(x,y) + \frac{2b}{1-b} d(y, Tx), \quad \text{for all } x, y \in X, \qquad (2.1.5)$$

and

$$d(Tx, Ty) \le \frac{c}{1-c} d(x,y) + \frac{2c}{1-c} d(y, Tx), \quad \text{for all } x, y \in X, \qquad (2.1.6)$$

respectively.

Thus, the birth of *almost contractions* could be dated to the very moment when we realized that (2.1.5) and (2.1.6) share the same property, i.e., $0 < \frac{b}{1-b} < 1, 0 < \frac{c}{1-c} < 1$, and that ($z_1$), ($z_2$) and ($z_3$) could be unified within a single condition of the form:

$$d(Tx, Ty) \le \delta\, d(x,y) + L d(y, Tx), \qquad (2.1.7)$$

with the constants δ and L satisfying $0 < \delta < 1$ and $L \ge 0$.

Soon after the publication of the first papers devoted to *almost contractions* [53–55], various researchers were attracted by the novelty that this class of mappings has brought to fixed point theory, see the rich list of references [2–16, 18–30, 40–48, 51, 60–76, 78–80, 84, 88–93, 102–104, 106, 107, 109–123, 125–143, 147, 148, 152–163, 166, 168–189, 194, 195, 197–199, 202–209, 216, 217, 219–225, 231, 232, 241–248, 250–258, 260–263, 266, 268–271, 276, 277].

It is therefore the main aim of this chapter to survey some of the most relevant developments in the last ten years or so on the concept of *almost contraction*.

The main aspects considered in the chapter are as follows:

- Fixed point theorems for single-valued self almost contractions
- Iterative approximation of the fixed point of implicit almost contractions
- Common fixed point theorems for almost contractions
- Almost contractive type mappings on product spaces
- Fixed point theorems for single-valued nonself almost contractions.

2.2. Fixed Point Theorems for Single-valued Self Almost Contractions

Definition 2.2.1 [55]. Let (X,d) be a metric space. A map $T : X \to X$ is called an *almost contraction* if there exist the constants $\delta \in (0,1)$ and $L \geq 0$ such that

$$d(Tx,Ty) \leq \delta \cdot d(x,y) + L d(y,Tx), \quad \text{for all } x, y \in X. \qquad (2.2.8)$$

In order to be more precise, we shall also call T a (δ, L)-*almost contraction*.

Remark 2.2.1. Note that the almost contraction condition (2.2.8) is not symmetric. But, due to the symmetry of the distance, (2.2.8) implicitly includes the following dual one:

$$d(Tx,Ty) \leq \delta \cdot d(x,y) + L \cdot d(x,Ty), \quad \text{for all } x, y \in X, \qquad (2.2.9)$$

and so, by (2.2.8) and (2.2.9), we obtain the following *symmetric* condition:

$$d(Tx,Ty) \leq \delta \cdot d(x,y) + L_1 \cdot [d(x,Ty) + d(x,Ty)], \quad \text{for all } x, y \in X, \qquad (2.2.10)$$

where $L_1 = L/2$. As shown in Ref. [73], (2.2.10) does not imply either (2.2.8) or (2.2.9).

Remark 2.2.2. Note that at the beginning (see Refs. [53–55]) and in some other subsequent papers, the author adopted the name *weak contraction* to designate an *almost contraction*. Soon after these papers had been published, we discovered that the term *weak contraction* had been used previously by other authors in different contexts.

Indeed, in 1967 Sz.-Nagy and Foiaş [267] used the concept of *weak contraction* in the context of the spectral theory of operators.

Later, Dugundji and Granas [124] also considered the concept of *weak contraction*, this time in the field of metrical fixed point theory. Dugundji and Granas called *weak contraction* a mapping $T : X \to X$ that satisfies the following condition:

$$d(Tx,Ty) \leq d(x,y) - \phi(d(x,y)), \quad \text{for all } x, y \in X, \qquad (2.2.11)$$

where $\phi : \mathbb{R}_+ \to \mathbb{R}_+$ is a compactly positive function. They also obtained some applications of weak contractions, including a domain invariance theorem (see Ref. [124] for details).

Apparently not aware of the paper by Dugundji and Granas (the paper [124] is not cited in Ref. [17]), in 1997 Alber and Guerre-Delabriere [17] used exactly the same condition for a map T defined from a closed convex subset C of a Banach space X into C and called T *weakly contractive* if

$$\|Tx - Ty\| \leq \|x - y\| - \psi(\|x - y\|), \quad \text{for all } x, y \in C,$$

where $\psi : \mathbb{R}_+ \to \mathbb{R}_+$ is continuous and nondecreasing, ψ is positive on $\mathbb{R}_+ \setminus 0$, $\psi(0) = 0$ and $\lim\limits_{t \to \infty} \psi(t) = +\infty$.

The concept of weak contraction by Alber and Guerre-Delabriere has been extremely successful, as shown by Google Scholar, where we found more than 340 papers citing this reference.

So, the existence of different concepts of weak contraction in fixed point theory led us to change the name adopted in 2003 (see Refs. [53–55] and the subsequent papers) to that of *almost contraction*. This name has been adopted for the first time in our paper from 2008 [62], five years after the original use of almost contractions.

It is therefore not surprising that the authors that studied this class of mappings used various names to designate an almost contraction, depending on the source of documentation: weak contraction [3, 5, 25, 27, 44, 110, 137, 140, 147, 169, 177, 178, 186, 188, 205, 209, 224, 242, 248, 254, 255, 256], almost contraction [2, 6, 7, 11, 14, 16, 19, 20, 23, 24, 28, 40, 42, 43, 44, 91, 134, 183, 198, 199, 203, 204, 232, 246, 253, 257, 258, 260, 262, 271], (δ, L)-almost contraction [10, 128, 138, 277], Berinde mapping [9, 11, 22, 41, 47, 89, 90, 152, 202, 245, 251, 255, 266], etc.

Remark 2.2.3. Obviously, any classical contraction (2.1.1) will satisfy (2.2.8) with $\delta = a$ and $L = 0$, and hence the class of almost contractions (properly) includes the Banach contractions (see Examples 2.2.1 and 2.2.2).

Other examples of almost contractions are given in the following propositions.

Proposition 2.2.1. *Let (X, d) be a metric space. Any Kannan contraction, i.e., any mapping $T : X \to X$ satisfying the contractive condition (2.1.3), is an almost contraction.*

Proof. By condition (2.1.3) and the triangle inequality, we get

$$
\begin{aligned}
d(Tx, Ty) &\leq b\big[d(x, Tx) + d(y, Ty)\big] \\
&\leq b\Big\{\big[d(x, y) + d(y, Tx)\big] + \big[d(y, Tx) + d(Tx, Ty)\big]\Big\},
\end{aligned}
$$

which yields

$$(1 - b)d(Tx, Ty) \leq b\,d(x, y) + 2b \cdot d(y, Tx)$$

and this implies

$$d(Tx, Ty) \leq \frac{b}{1 - b} d(x, y) + \frac{2b}{1 - b} d(y, Tx), \quad \text{for all } x, y \in X.$$

Hence, in view of the condition $0 < b < \frac{1}{2}$, (2.2.8) holds with $\delta = \frac{b}{1-b}$ and $L = \frac{2b}{1-b}$. $\quad\square$

Proposition 2.2.2. *Let* (X,d) *be a metric space. Any Kannan contraction, i.e., any mapping* $T : X \to X$ *satisfying the contractive condition (2.1.4), is an almost contraction.*

Proof. Using $d(x,Ty) \le d(x,y) + d(y,Tx) + d(Tx,Ty)$ by (2.1.4) we get, after simple computations,

$$d(Tx,Ty) \le \frac{c}{1-c} d(x,y) + \frac{2c}{1-c} d(y,Tx),$$

which is (2.2.8), with $\delta = \frac{c}{1-c} < 1$ (since $c < 1/2$) and $L = \frac{2c}{1-c} \ge 0$. $\square$

From Remark 2.2.3 and Propositions 2.2.1 and 2.2.2, we have the following proposition.

Proposition 2.2.3. *Let* (X,d) *be a metric space. Any Zamfirescu contraction, i.e., any mapping* $T : X \to X$ *satisfying the assumptions in Theorem 2.1.2, is an almost contraction.*

Proposition 2.2.4 [197]. *Let* (X,d) *be a metric space. Any Ćirić-Reich-Rus contraction, i.e., any mapping* $T : X \to X$ *satisfying the condition*

$$d(Tx,Ty) \le \alpha d(x,y) + \beta[d(x,Tx) + d(y,Ty)], \quad \text{for all } x,y \in X,$$

where $\alpha, \beta \in \mathbb{R}_+$ *and* $\alpha + 2\beta < 1$, *is an almost contraction.*

Proposition 2.2.5. *Let* (X,d) *be a metric space. Let* $T : X \to X$ *be a quasi-contraction [96], i.e., an operator for which there exists* $0 < h < 1$ *such that*

$$d(Tx,Ty) \le h \cdot M(x,y), \quad \text{for all } x,y \in X, \tag{2.2.12}$$

where

$$M(x,y) = \max \left\{ d(x,y), d(x,Tx), d(y,Ty), d(x,Ty), d(y,Tx) \right\}. \tag{2.2.13}$$

If $0 < h < 1/2$, *then* T *is an almost contraction.*

Proof. Let T satisfy (2.2.13) and let $x,y \in X$ be arbitrarily taken. We have to discuss five possible cases.

CASE 1. $M(x,y) = d(x,y)$. In this case, by virtue of (2.2.12), conditions (2.2.8) and (2.2.9) are obviously satisfied (with $\delta = h$ and $L = 0$).

CASE 2. $M(x,y) = d(x,Tx)$. In this case, by (2.2.12) and triangle inequality one obtains

$$d(Tx,Ty) \leq h\,d(x,Tx) \leq h[d(x,y)+d(y,Tx)],$$

and so (2.2.8) holds with $\delta = h$ and $L = h$.

Since $d(x,Tx) \leq d(x,Ty)+d(Ty,Tx)$, we get

$$d(Tx,Ty) \leq \frac{h}{1-h}d(x,Ty) \leq \delta d(x,y)+\frac{h}{1-h}d(x,Ty), \quad \text{for all } \delta \in (0,1).$$

So (2.2.9) also holds.

CASE 3. $M(x,y) = d(y,Ty)$, when (2.2.8) and (2.2.9) follow by Case 2, by virtue of the symmetry of $M(x,y)$.

CASE 4. $M(x,y) = d(x,Ty)$, when (2.2.9) is obviously true and (2.2.8) is obtained only if $h < \frac{1}{2}$. Indeed, since by (2.2.12), $d(Tx,Ty) \leq h \cdot d(x,Ty)$ and

$$d(x,Ty) \leq d(x,y)+d(y,Tx)+d(Tx,Ty),$$

one obtains

$$d(Tx,Ty) \leq \frac{h}{1-h}d(x,y)+\frac{h}{1-h}d(y,Tx),$$

which is (2.2.8) with $\delta = \frac{h}{1-h} < 1$ (since $h < \frac{1}{2}$) and $L = \frac{h}{1-h} > 0$.

CASE 5. $M(x,y) = d(y,Tx)$, which reduces to Case 4.   □

Remark 2.2.4. Proposition 2.2.5 shows that the quasi-contractions with $0 < h < 1/2$ are almost contractions. It appears then that $h < \frac{1}{2}$ is not a necessary condition for a quasi-contraction to be an almost contraction, as there exist quasi-contractions with $h \geq \frac{1}{2}$, which are still almost contractions, as shown by Example 2.2.2.

There are many other examples of contractive conditions which imply the almost contractiveness condition (see the list in Rhoades' classification [226]).

The first main result of this section is the following theorem.

Theorem 2.2.1 [55, Theorem 2]. *Let (X,d) be a complete metric space and let $T : X \to X$ be a (δ,L)-almost contraction. Then*

(a) *$Fix(T) = \{x \in X : Tx = x\} \neq \emptyset$;*

(b) *For any $x_0 \in X$, the Picard iteration $\{x_n\}_{n=0}^{\infty}$, $x_n = T^n x_0$, converges to some $x^* \in Fix(T)$;*

(c) *The following estimate holds:*

$$d(x_{n+i-1},x^*) \leq \frac{\delta^i}{1-\delta}d(x_n,x_{n-1}), \quad n = 0,1,2,\ldots; i = 1,2,\ldots. \quad (2.2.14)$$

Proof. We shall prove that T satisfying (2.2.8) has at least a fixed point in X. To this end, let $x_0 \in X$ be arbitrary but fixed and let $\{x_n\}_{n=0}^{\infty}$ be the Picard iteration defined by (2.1.2).

Take $x := x_{n-1}$, $y := x_n$ in (2.2.8) to obtain

$$d(Tx_{n-1}, Tx_n) \leq \delta \cdot d(x_{n-1}, x_n),$$

which shows that

$$d(x_n, x_{n+1}) \leq \delta \cdot d(x_{n-1}, x_n). \tag{2.2.15}$$

Using (2.2.15), we obtain by induction

$$d(x_n, x_{n+1}) \leq \delta^n d(x_0, x_1), \quad n = 0, 1, 2, \dots,$$

and then

$$
\begin{aligned}
d(x_n, x_{n+p}) &\leq \delta^n \left(1 + \delta + \cdots + \delta^{p-1}\right) d(x_0, x_1) \\
&= \frac{\delta^n}{1 - \delta} (1 - \delta^p) \cdot d(x_0, x_1), \quad n, p \in \mathbb{N}, \ p \neq 0.
\end{aligned}
\tag{2.2.16}
$$

Since $0 < \delta < 1$, (2.2.16) shows that $\{x_n\}_{n=0}^{\infty}$ is a Cauchy sequence and hence it is convergent. Let us denote

$$x^* = \lim_{n \to \infty} x_n. \tag{2.2.17}$$

Then

$$d(x^*, Tx^*) \leq d(x^*, x_{n+1}) + d(x_{n+1}, Tx^*) = d(x_{n+1}, x^*) + d(Tx_n, Tx^*).$$

By (2.2.8), we have

$$d(Tx_n, Tx^*) \leq \delta d(x_n, x^*) + L d(x^*, Tx_n)$$

and hence

$$d(x^*, Tx^*) \leq (1 + L) d(x^*, x_{n+1}) + \delta \cdot d(x_n, x^*), \tag{2.2.18}$$

which is valid for all $n \geq 0$. Letting $n \to \infty$ in (2.2.18), we obtain

$$d(x^*, Tx^*) = 0,$$

i.e., x^* is a fixed point of T.

By letting $p \to \infty$ in (2.2.16), one obtains the *a priori* estimate

$$d(x_n, x^*) \leq \frac{\delta^n}{1 - \delta} d(x_0, x_1), \quad n = 0, 1, 2, \dots, \tag{2.2.19}$$

where δ is the constant appearing in (2.2.8).

On the other hand, let us observe that by (2.2.15) we inductively obtain

$$d(x_{n+k}, x_{n+k+1}) \leq \delta^{k+1} \cdot d(x_{n-1}, x_n), \quad k, n \in \mathbb{N},$$

and hence, similarly to deriving (2.2.16), we get

$$d(x_n, x_{n+p}) \leq \frac{\delta(1 - \delta^p)}{1 - \delta} d(x_{n-1}, x_n), \quad n \geq 1, \ p \in \mathbb{N}^*. \tag{2.2.20}$$

Now, by letting $p \to \infty$ in (2.2.20), the *a posteriori* estimate follows:

$$d(x_n, x^*) \leq \frac{\delta}{1 - \delta} d(x_{n-1}, x_n), \quad n = 1, 2, \dots. \tag{2.2.21}$$

Now, (2.2.19) and (2.2.21) can be merged to obtain the estimate (2.2.14).   □

Remark 2.2.5.   (a) Theorem 2.2.1 is a significant extension of Theorem 2.1.1, Theorem 2.1.2 and many other related results in metrical fixed point theory.

(b) Note that for all the fixed point theorems mentioned in (a) as particular cases of Theorem 2.2.1, the fixed point is unique. However, in general, almost contractions need not have a unique fixed point, as shown by Examples 2.2.1 and 2.2.2.

(c) Recall (see Refs. [235, 236, 239]), that an operator $T : X \to X$ is said to be a *weakly Picard operator* if the sequence $\{T^n x_0\}_{n=0}^{\infty}$ converges for all $x_0 \in X$ and its limit is a fixed point of T. So, almost contractions are weakly Picard operators, while Banach contraction, Kannan contraction and Ćirić quasi-contractions are Picard operators.

(d) Note also that condition (2.2.8) implies the so-called Banach orbital condition or graphic contraction condition

$$d(Tx, T^2 x) \leq a d(x, Tx), \quad \text{for all } x \in X, \tag{2.2.22}$$

studied by various authors (see Ref. [239, p. 39], for some historical remarks). As shown by the Graphic Contraction Principle (see Ref. [239, p. 35]), a graphic contraction that has closed graph is a weakly Picard operator.

So, in this context, the merit of condition (2.2.8) is that no other additional hypothesis is needed in order for an almost contraction to be a weakly Picard operator.

Remark 2.2.6. From a numerical point of view, a fixed point theorem is valuable if, apart from the conclusion regarding the existence (and, possibly, uniqueness) of the fixed point:

(a) it provides a method (generally, *iterative*) for constructing the fixed point(s);

(b) it is able to provide information on the error estimate or / and rate of convergence of the iterative process used to approximate the fixed point; and

(c) it can give concrete information on the stability of this procedure, that is, on the data dependence of the fixed point(s).

As shown above, Theorem 2.2.1 does possess all these features (see also Theorem 2.2.2), for the rate of convergence.

The next examples illustrate the diversity of almost contractions.

Example 2.2.1. Let $X = [0,1]$ be the unit interval with the usual norm and let $T : [0,1] \rightarrow [0,1]$ be given by $Tx = \frac{1}{2}$, for $x \in [0,2/3)$ and $Tx = 1$, for $x \in [2/3,1]$.

As T has two fixed points, that is, $Fix(T) = \{\frac{1}{2},1\}$, it does not satisfy either Banach contraction condition (2.1.1), Kannan contraction condition (2.1.3), Chatterjea contraction condition (2.1.4), Zamfirescu contractive conditions (z_1) and (z_2), or Ćirić's quasi-contraction condition (2.2.12), but T satisfies the almost contraction condition (2.2.8).

Indeed, for $x,y \in [0,2/3)$ or $x,y \in [2/3,1]$, (2.2.8) is obvious. For $x \in [0,2/3)$ and $y \in [2/3,1]$ or $y \in [0,2/3)$ and $x \in [2/3,1]$ we have $d(Tx,Ty) = 1/2$ and $d(y,Tx) = |y-1/2| \in [1/6,1/2]$, in the first case, and $d(y,Tx) = |y-1| \in [1/3,1]$, in the second case, which show that it suffices to take $L = 3$ in order to ensure that (2.2.8) holds for $0 < \delta < 1$ and $L \geq 0$ arbitrary and all $x,y \in X$.

Example 2.2.2. Let $X = [0,1]$ with the usual metric and let $T : [0,1] \rightarrow [0,1]$ be defined by

$$Tx = \frac{2}{3}x, \text{ if } 0 \leq x < \frac{1}{2} \text{ and } Tx = \frac{2}{3}x + \frac{1}{3}, \text{ if } \frac{1}{2} \leq x \leq 1.$$

Then:

(a) T is an almost contraction with constants $\delta = \frac{2}{3}$, $L = 6$ and $Fix(T) = \{0;1\}$;

(b) T does not satisfy any of the contraction conditions of Banach, Kannan, Chatterjea and Zamfirescu, and is not a Ćirić quasi-contraction, as T has two fixed points.

Proof. We have to discuss the following possible cases:

I. If $x,y \in [0,\frac{1}{2})$, then $T(x) = \frac{2}{3}x$ and $T(y) = \frac{2}{3}y$. Then condition (2.2.8) becomes

$$\left|\frac{2}{3}x - \frac{2}{3}y\right| \leq \delta|x-y| + L\left|y - \frac{2}{3}x\right|,$$

which obviously holds for $\delta \geq \frac{2}{3}$ and any $L \geq 0$.

II. If $x, y \in [\frac{1}{2}, 1]$, then $T(x) = \frac{2}{3}x + \frac{1}{3}$ and $T(y) = \frac{2}{3}y + \frac{1}{3}$. Similarly to case I, this holds for $\delta \geq \frac{2}{3}$ and any $L \geq 0$.

III. If $x \in [0, \frac{1}{2})$ and $y \in [\frac{1}{2}, 1]$, then $T(x) = \frac{2}{3}x$ and $T(y) = \frac{2}{3}y + \frac{1}{3}$. Condition (2.2.8) becomes

$$\left| \frac{2}{3}x - \frac{2}{3}y - \frac{1}{3} \right| \leq \delta |x - y| + L \left| y - \frac{2}{3}x \right|. \tag{2.2.23}$$

On the left-hand side we have that:

$$-1 \leq \frac{2}{3}x - \frac{2}{3}y - \frac{1}{3} < -\frac{1}{3} \Rightarrow \left| \frac{2}{3}x - \frac{2}{3}y - \frac{1}{3} \right| \in \left(\frac{1}{3}, 1 \right],$$

while on the right-hand side we have that:

$$\frac{1}{6} < y - \frac{2}{3}x \leq 1 \Rightarrow \left| y - \frac{2}{3}x \right| \in \left(\frac{1}{6}, 1 \right].$$

Then (2.2.23) holds for $\delta \in [0, 1)$ and $L \geq 6$.

IV. If $x \in [\frac{1}{2}, 1]$ and $y \in [0, \frac{1}{2})$, then $f(x) = \frac{2}{3}x + \frac{1}{3}$ and $f(y) = \frac{2}{3}y$.

Condition (2.2.8) becomes

$$\left| \frac{2}{3}x - \frac{2}{3}y + \frac{1}{3} \right| \leq \delta |x - y| + L \left| y - \frac{2}{3}x - \frac{1}{3} \right|. \tag{2.2.24}$$

On the left-hand side we have that:

$$\left| \frac{2}{3}x - \frac{2}{3}y - \frac{1}{3} \right| \in \left(\frac{1}{3}, 1 \right],$$

while on the right-hand side we have that:

$$-1 < y - \frac{2}{3}x - \frac{1}{3} \leq -\frac{1}{6} \Rightarrow \left| y - \frac{2}{3}x - \frac{1}{3} \right| \in \left[\frac{1}{6}, 1 \right).$$

Then (2.2.24) holds for $\delta \in [0, 1)$ and $L \geq 6$.

The conclusion is that T satisfies (2.2.8) for any $x, y \in X$ if $\delta \in [\frac{2}{3}, 1)$ and $L \geq 6$. Taking $\delta = \frac{2}{3}$ and $L = 6$, notice that $\delta + L > 1$.   □

Example 2.2.3. Let $[0, 1]$ be the unit interval with the usual norm. Let $T : [0, 1] \rightarrow [0, 1]$ be the identity map, i.e., $Tx = x$, for all $x \in [0, 1]$. Then

(a) T does not satisfy Ciric's contractive condition (2.2.12);

(b) T satisfies the almost contraction condition (2.2.8) with $\delta \in (0, 1)$ arbitrary and

$L \geq 1 - \delta$. Indeed, conditions (2.2.8) and (2.2.9) lead to

$$|x-y| \leq \delta|x-y| + L \cdot |y-x|,$$

which is true for all $x, y \in [0,1]$ if we take $\delta \in (0,1)$ arbitrary and $L \geq 1 - \delta$.

(c) The set of fixed points of T is the interval $[0,1]$, i.e., $F(T) = [0,1]$.

It is possible to force an almost contraction to be a Picard operator by imposing an additional contractive condition, quite similar to (2.2.8), as shown by the next theorem.

Theorem 2.2.2 [55, Theorem 2.2]. *Let (X,d) be a complete metric space and $T : X \to X$ be a (δ, L)- almost contraction for which there exist $\delta_u \in (0,1)$ and $L_u \geq 0$ such that*

$$d(Tx, Ty) \leq \delta_u \cdot d(x,y) + L_u \cdot d(x, Tx), \quad \text{for all } x, y \in X. \tag{2.2.25}$$

Then

(a) *T has a unique fixed point, i.e., $F(T) = \{x^*\}$;*

(b) *The Picard iteration $\{x_n\}_{n=0}^{\infty}$ given by (2.1.2) converges to x^*, for any $x_0 \in X$;*

(c) *The following estimate holds:*

$$d(x_{n+i-1}, x^*) \leq \frac{\delta^i}{1-\delta} d(x_n, x_{n-1}), \quad n = 0,1,2,\ldots; i = 1,2,\ldots; \tag{2.2.26}$$

(d) *The rate of convergence of the Picard iteration is given by*

$$d(x_n, x^*) \leq \delta_u d(x_{n-1}, x^*), \quad n = 1,2,\ldots. \tag{2.2.27}$$

Proof. Assume T has two distinct fixed points $x^*, y^* \in X$. Then, by (2.2.25), with $x := x^*$, $y := y^*$ we get

$$d(x^*, y^*) \leq \delta_u \cdot d(x^*, y^*) \quad \Leftrightarrow \quad (1 - \delta_u) d(x^*, y^*) \leq 0,$$

so contradicting $d(x^*, y^*) > 0$.

Letting $y := x_n$, $x := x^*$ in (2.2.25), we obtain the estimate (2.2.27).

The rest of the proof follows by Theorem 2.2.1. □

Remark 2.2.7.   (a) Note that the uniqueness condition (2.2.25) has been used by Osilike [190 – 192] to prove stability results for certain fixed point iteration procedures. It is shown there (see Refs. [190, 191]), that condition (2.2.25) alone does not imply T has a fixed point. But if a mapping T satisfying (2.2.25) has a fixed point, then this fixed point is certainly unique.

(b) It is easy to check that any operator T satisfying one of the conditions (2.1.1), (2.1.3), (2.1.4), or the conditions in Theorem 2.1.2, also satisfies the uniqueness conditions (2.2.25). Therefore, in view of Examples 2.2.1–2.2.3, Theorem 2.2.1 and Theorem 2.2.2 properly generalize Theorem 2.1.1, Kannan's fixed point theorem [149], Theorem 2.1.2, and many other related results.

(c) Rus [236] has shown that, if T is a weakly Picard operator, then there exists a partition of X,

$$X = \bigcup_{\lambda \in \Lambda} X_\lambda$$

such that $T|X_\lambda$ is a Banach contraction. In the case of the almost contractions in Examples 2.2.1–2.2.3, we have the following partitions.

For T in Example 2.2.1, we have

$$[0,1] = [0,2/3) \cup [2/3,1];$$

for T in Example 2.2.2, we have

$$[0,1] = [0,1/2) \cup [1/2,1];$$

while for T in Example 2.2.3, we have

$$[0,1] = \bigcup_{\lambda \in [0,1]} \{\lambda\}.$$

(d) As can easily be seen, Theorem 2.2.2 (as well as Theorem 2.2.1, except for the uniqueness of the fixed point) preserves all conclusions in the Banach contraction principle in its complete form [60, Theorem 2.1] under significantly weaker contractive conditions.

Indeed, the metrical contractive conditions known in the literature (see Ref. [226]) that involve on the right-hand side the displacements

$$d(x,y), d(x,Tx), d(y,Ty), d(x,Ty), d(y,Tx)$$

with nonnegative coefficients, say

$$a(x,y), b(x,y), c(x,y), d(x,y), e(x,y),$$

respectively, are commonly based on the very restrictive assumption

$$0 < a(x,y) + b(x,y) + c(x,y) + d(x,y) + e(x,y) \le 1,$$

while, in condition (2.2.8), which involves only the displacements

$$d(x,y), d(y,Tx)$$

the constant coefficients δ and L are not required to satisfy

$$\delta + L \leq 1.$$

This is obvious for T in Example 2.2.2, where we have

$$\delta + L = 2/3 + 6.$$

It is possible to extend significantly Theorems 2.2.1 and 2.2.2, by replacing $d(x,y)$ in (2.2.8) by a certain expression of the displacements $d(x,y)$, $d(x,Tx)$, $d(y,Ty)$, $d(x,Ty)$, $d(y,Tx)$, first used by Ćirić [95]:

$$M_1(x,y) = \max\left\{d(x,y), d(x,Tx), d(y,Ty), \frac{1}{2}[d(x,Ty) + d(y,Tx)]\right\}. \quad (2.2.28)$$

We thus have the following result, taken from Ref. [64].

Theorem 2.2.3. *Let (X,d) be a complete metric space and let $T : X \to X$ be a strong Ćirić almost contraction, that is, a mapping for which there exist two constants $\alpha \in [0,1)$ and $L \geq 0$ such that*

$$d(Tx,Ty) \leq \alpha \cdot M_1(x,y) + Ld(y,Tx), \quad \text{for all } x,y \in X, \quad (2.2.29)$$

where $M_1(x,y)$ is given by (2.2.28). Then

(a) *$Fix(T) = \{x \in X : Tx = x\} \neq \emptyset$;*

(b) *For any $x_0 = x \in X$, the Picard iteration $\{x_n\}_{n=0}^{\infty}$ given by (2.1.2) converges to some $x^* \in Fix(T)$;*

(c) *The following estimate holds:*

$$d(x_{n+i-1}, x^*) \leq \frac{\alpha^i}{1-\alpha} d(x_n, x_{n-1}), \quad n = 0,1,2,\ldots; i = 1,2,\ldots. \quad (2.2.30)$$

Proof. Let $x \in X$ be arbitrary and let $\{x_n\}_{n=0}^{\infty}$ be the Picard iteration defined by (2.1.2) with $x_0 = x$. By taking $x := x_{n-1}$, $y := x_n$ in (2.2.29), we obtain

$$d(x_n, x_{n+1}) = d(Tx_{n-1}, Tx_n) \leq \alpha \cdot M_1(x_{n-1}, x_n),$$

that is,

$$d(x_n, x_{n+1}) \leq \alpha \max\left\{d(x_{n-1}, x_n), d(x_n, x_{n+1}), \frac{1}{2}[d(x_{n-1}, x_{n+1}) + 0]\right\},$$

since $d(x_n, Tx_{n-1}) = 0$. Now, by the triangle inequality

$$d(x_{n-1}, x_{n+1}) \leq d(x_{n-1}, x_n) + d(x_n, x_{n+1})$$

and using the inequality $\frac{a+b}{2} \leq \max\{a,b\}$, we deduce that either

$$\max\left\{d(x_{n-1},x_n),d(x_n,x_{n+1}),\frac{1}{2}d(x_{n-1},x_{n+1})\right\}=d(x_{n-1},x_n) \qquad (2.2.31)$$

or

$$\max\left\{d(x_{n-1},x_n),d(x_n,x_{n+1}),\frac{1}{2}d(x_{n-1},x_{n+1})\right\}=d(x_n,x_{n+1}). \qquad (2.2.32)$$

The case (2.2.32) cannot hold because it would lead to the contradiction

$$d(x_n,x_{n+1}) \leq h\,d(x_n,x_{n+1}).$$

Hence, (2.2.31) must always hold, and this leads to

$$d(x_n,x_{n+1}) \leq h\,d(x_{n-1},x_n).$$

The rest of the proof is similar to that of Theorem 2.2.1. $\qquad\qquad\qquad\square$

Like in the case of Theorem 2.2.1, it is possible to force the uniqueness of the fixed point of a Ćirić strong almost contraction by imposing an additional contractive condition, quite similar to (2.2.29), as shown by the next theorem.

Theorem 2.2.4. *Let (X,d) be a complete metric space and let $T : X \to X$ be a Ćirić strong almost contraction for which there exist $\theta \in [0,1)$ and some $L_1 \geq 0$ such that*

$$d(Tx,Ty) \leq \theta \cdot d(x,y)+L_1 \cdot d(x,Tx), \quad \text{for all } x,y \in X. \qquad (2.2.33)$$

Then

(a) *T has a unique fixed point, i.e., $Fix(T) = \{x^*\}$;*

(b) *The Picard iteration $\{x_n\}_{n=0}^{\infty}$ given by (2.1.2) converges to x^*, for any $x_0 \in X$;*

(c) *The error estimate (2.2.30) holds.*

(d) *The rate of convergence of the Picard iteration is given by*

$$d(x_n,x^*) \leq \theta\,d(x_{n-1},x^*), \quad n=1,2,\ldots. \qquad (2.2.34)$$

Proof. Assume T has two distinct fixed points, say $x^*,y^* \in X$. Then by (2.2.25), with $x := x^*$, $y := y^*$ we get

$$d(x^*,y^*) \leq \theta \cdot d(x^*,y^*) \quad \Leftrightarrow \quad (1-\theta)\,d(x^*,y^*) \leq 0,$$

so contradicting $d(x^*,y^*) > 0$.

Now letting $y := x_n$, $x := x^*$ in (2.2.33), we obtain the estimate (2.2.34).

The rest of the proof follows by Theorem 2.2.3. $\qquad\qquad\qquad\square$

Remark 2.2.8. Note that $M_1(x,y)$ given by (2.2.28) cannot be replaced by $M(x,y)$ appearing on the right-hand side of quasi-contraction condition (2.2.12), i.e.,

$$M(x,y) = \max\{d(x,y), d(x,Tx), d(y,Ty), d(x,Ty), d(y,Tx)\}.$$

This would lead to the contraction condition

$$d(Tx,Ty) \le \alpha \cdot M(x,y) + Ld(y,Tx), \quad \text{for all } x,y \in X, \tag{2.2.35}$$

with $\alpha \in [0,1)$ and $L \ge 0$, which is too weak to ensure the existence of a fixed point (see the next example, taken from Ref. [63]; see also Ref. [62]).

Example 2.2.4. Let $X = \mathbb{N} = \{0,1,2,\dots\}$ with the usual norm and let T be defined by $T(n) = n+1$. Then T does satisfy (2.2.35) with $\alpha = \frac{1}{2}$ and $L = 2$ but T is fixed point free. Indeed, if we take $x = n$, $y = m$, $m > n$, then $d(Tx,Ty) = m - n$, $M(x,y) = m - n + 1$, $d(y,Tx) = m - n - 1$. Thus condition (2.2.35) reduces to

$$m - n \le \alpha(m - n + 1) + 2(m - n - 1) = \frac{5}{2}(m - n) - \frac{3}{2},$$

which is true, since $m - n \ge 1$.

An equivalent (see Ref. [199]) contractive condition that ensures the uniqueness of the fixed point has been obtained by Babu et al. [48] for almost contractions.

We state the fixed point theorem corresponding to this uniqueness condition in the case of Ćirić strong almost contractions.

Theorem 2.2.5. *Let (X,d) be a complete metric space and let $T : X \to X$ be a mapping for which there exist $\alpha \in [0,1)$ and some $L \ge 0$ such that for all $x,y \in X$*

$$d(Tx,Ty) \le \alpha \cdot M_1(x,y) + L\min\{d(x,Tx), d(y,Ty), d(x,Ty), d(y,Tx)\}. \tag{2.2.36}$$

Then

(a) *T has a unique fixed point, i.e., $\text{Fix}(T) = \{x^*\}$;*

(b) *The Picard iteration $\{x_n\}_{n=0}^{\infty}$ given by (2.1.2) converges to x^*, for any $x_0 \in X$;*

(c) *The error estimate (2.2.30) holds.*

For more details and results on Ćirić strong almost contractions, see Refs. [62, 63].

Starting from the fact that φ-contractions are natural generalizations of Banach contractions, we can extend the previous results from almost contractions to the more general class of almost φ-contractions. The same extension can be done for Ćirić strong almost contractions.

To do this, the central concept is that of the comparison function (see Ref. [52] and references therein for more details, results and proofs).

Definition 2.2.2. A map $\varphi : \mathbb{R}_+ \to \mathbb{R}_+$ is called a *comparison function* if it satisfies:

(i_φ) φ is monotone increasing, i.e., $t_1 < t_2 \Rightarrow \varphi(t_1) \leq \varphi(t_2)$;

(ii_φ) the sequence $\{\varphi^n(t)\}_{n=0}^\infty$ converges to zero, for all $t \in \mathbb{R}_+$, where φ^n stands for the nth iterate of φ.

If φ satisfies (i_φ) and

(iii_φ) $\displaystyle\sum_{k=0}^\infty \varphi^k(t)$ converges for all $t \in \mathbb{R}_+$, then φ is said to be a *(c)-comparison function*.

It has been shown (see for example Ref. [60]), that φ satisfies (iii_φ) if and only if there exist $0 < c < 1$ and a convergent series of positive terms, $\displaystyle\sum_{n=0}^\infty u_n$, such that

$$\varphi^{k+1}(t) \leq c\varphi^k(t) + u_k, \quad \text{for all } t \in \mathbb{R}_+ \quad \text{and } k \geq k_0 \text{ (fixed)}.$$

It is also known that if φ is a (c)-comparison function, then the sum of the comparison series, i.e.,

$$s(t) = \sum_{k=0}^\infty \varphi^k(t), \quad t \in \mathbb{R}_+, \tag{2.2.37}$$

is monotone increasing and continuous at zero, and that any (c)-comparison function is a comparison function.

A prototype for comparison functions is

$$\varphi(t) = at, \quad t \in \mathbb{R}_+ \quad (0 \leq a < 1)$$

but, as shown by Example 2.2.5, the comparison functions need not be either linear or continuous.

Note however that any comparison function is continuous at zero.

Example 2.2.5. Let $\varphi_1(t) = \frac{t}{t+1}, t \in \mathbb{R}_+$ and $\varphi_2(t) = \frac{1}{2}t$, if $0 \leq t < 1$ and $\varphi_2(t) = t - \frac{1}{3}$, if $t \geq 1$. Then φ_1 is a nonlinear comparison function, which is not a (c)-comparison function, while φ_2 is a discontinuous (c)-comparison function.

By replacing the well-known strict contractive condition (2.1.1) appearing in Banach's fixed point theorem, i.e.,

$$d(Tx, Ty) \leq ad(x, y), \quad \text{for all } x, y \in X,$$

by a more general one

$$d(Tx,Ty) \leq \varphi\big(d(x,y)\big), \quad \text{for all } x,y \in X, \tag{2.2.38}$$

where φ is a certain comparison function, several fixed point theorems have been obtained (see Ref. [235] and references therein).

Recall that an operator T which satisfies a condition of the form (2.2.38) is commonly named a φ-*contraction*.

Following the way in which the Banach contractions were extended to φ-contractions, in the following we would like to extend Theorems 2.2.1 and 2.2.2 from almost contractions to almost φ-contractions.

Definition 2.2.3. Let (X,d) be a metric space. A mapping $T : X \to X$ is said to be an *almost φ-contraction* or a (φ,L)-*almost contraction* provided that there exist a comparison function φ and some $L \geq 0$, such that

$$d(Tx,Ty) \leq \varphi\big(d(x,y)\big) + Ld(y,Tx), \quad \text{for all } x,y \in X. \tag{2.2.39}$$

Remark 2.2.9. Clearly, any almost contraction is an almost φ-contraction, with $\varphi(t) = \delta t$, $t \in \mathbb{R}_+$ and $0 < \delta < 1$. Also, all φ-contractions are almost φ-contractions with $L \equiv 0$ in (2.2.39).

The next theorem, taken from Ref. [54], extends Theorem 2.2.1 from almost contractions to almost φ-contractions.

Theorem 2.2.6. *Let (X,d) be a complete metric space and $T : X \to X$ an almost φ-contraction with φ a (c)-comparison function. Then*

(a) $F(T) = \{x \in X : Tx = x\} \neq \phi$;

(b) *For any $x_0 \in X$, the Picard iteration $\{x_n\}_{n=0}^{\infty}$ defined by $x_0 \in X$ and*

$$x_{n+1} = Tx_n, \quad n = 0,1,2,\ldots, \tag{2.2.40}$$

converges to a fixed point x^* of T;

(c) *The following estimate:*

$$d(x_n,x^*) \leq s\big(d(x_n,x_{n+1})\big), \quad n = 0,1,2,\ldots, \tag{2.2.41}$$

holds, where $s(t)$ is given by (2.2.37).

Proof. We shall prove that T has at least one fixed point in X. To this end, let $x_0 \in X$ be arbitrary and $\{x_n\}_{n=0}^{\infty}$ be the Picard iteration defined by (2.2.40).

Since T is an almost φ-contraction, there exist a (c)-comparison function φ and some $L \geq 0$, such that

$$d(Tx, Ty) \leq \varphi\big(d(x,y)\big) + L \cdot d(y, Tx) \qquad (2.2.42)$$

holds, for all $x, y \in X$.

Take $x := x_{n-1}$, $y := x_n$ in (2.2.42). We get

$$d(x_n, x_{n+1}) \leq \varphi\big(d(x_{n-1}, x_n)\big), \quad \text{for all } n = 1, 2, \ldots. \qquad (2.2.43)$$

Since φ is not decreasing, by (2.2.43) we have

$$d(x_{n+1}, x_{n+2}) \leq \varphi\big(d(x_n, x_{n+1})\big),$$

which inductively yields

$$d(x_{n+k}, x_{n+k+1}) \leq \varphi^k\big(d(x_n, x_{n+1})\big), \quad k = 0, 1, 2, \ldots.$$

By the triangle rule, we have

$$\begin{aligned}
d(x_n, x_{n+p}) &\leq d(x_n, x_{n+1}) + d(x_{n+1}, x_{n+2}) + \cdots + d(x_{n+p-1}, x_{n+p}) \\
&\leq r + \varphi(r) + \cdots + \varphi^{n+p-1}(r),
\end{aligned} \qquad (2.2.44)$$

where $r := d(x_n, x_{n+1})$.

Again by (2.2.43), we find

$$d(x_n, x_{n+1}) \leq \varphi^n\big(d(x_0, x_1)\big), \quad n = 0, 1, 2, \ldots, \qquad (2.2.45)$$

which, by property (ii_φ) of a comparison function, implies

$$\lim_{n \to \infty} d(x_n, x_{n+1}) = 0. \qquad (2.2.46)$$

As φ is positive, it is obvious that

$$r + \varphi(r) + \cdots + \varphi^{n+p-1}(r) < s(r), \qquad (2.2.47)$$

where $s(r)$ is the sum of the series $\sum\limits_{k=0}^{\infty} \varphi^k(r)$.

Then by (2.2.44) and (2.2.47), we get

$$d(x_n, x_{n+p}) \leq s\big(d(x_n, x_{n+1})\big), \quad n \in \mathbb{N}, \; p \in \mathbb{N}. \qquad (2.2.48)$$

Since s is continuous at zero, (2.2.46) and (2.2.47) imply that $\{x_n\}_{n=0}^{\infty}$ is a Cauchy sequence. As X is complete, $\{x_n\}_{n=0}^{\infty}$ is convergent.

Let $x^* = \lim\limits_{n \to \infty} x_n$. We shall prove that x^* is a fixed point of T. Indeed,

$$d(x^*, Tx^*) \leq d(x^*, x_{n+1}) + d(x_{n+1}, Tx^*) = d(x_{n+1}, x^*) + d(Tx_n, Tx^*).$$

By (2.2.42), we have

$$d(Tx_n, Tx^*) \leq \varphi(d(x_n, x^*)) + Ld(x^*, Tx_n),$$

and hence

$$d(x^*, Tx^*) \leq (1+L)d(x_{n+1}, x^*) + \varphi(d(x_n, x^*)), \quad \text{for all } n \geq 0. \qquad (2.2.49)$$

Now letting $n \to \infty$ in (2.2.49) and using the continuity of φ at zero, we obtain

$$d(x^*, Tx^*) = 0,$$

i.e., x^* is a fixed point of T. The estimate (2.2.41) follows by (2.2.44) by letting $p \to \infty$. $\qquad \square$

Remark 2.2.10.   (a) Using the *a posteriori* error estimate (2.2.41) and (2.2.45) we easily obtain

$$d(x_n, x^*) \leq s\left(\varphi^n(d(x_0, x_1))\right), \quad n = 0, 1, 2, \ldots,$$

which is the *a priori* estimate for the Picard iteration $\{x_n\}_{n=0}^{\infty}$.

(b) If we take $\varphi(t) = \delta \cdot t$, $t \in \mathbb{R}_+$, $0 < \delta < 1$, by Theorem 2.2.6 we obtain the corresponding result for almost contractions, i.e., Theorem 2.2.1.

Like in the case of almost contractions, in order to guarantee the uniqueness of the fixed point of T, we have to consider an additional contractive type condition, as in the next theorem.

Theorem 2.2.7. *Let X and T be as in Theorem 2.2.6. Suppose T also satisfies the following condition: there exist a comparison function ψ and some $L_1 \geq 0$ such that*

$$d(Tx, Ty) \leq \psi(d(x,y)) + L_1 d(x, Tx), \quad \text{for all } x, y \in X. \qquad (2.2.50)$$

Then

(a) *T has a unique fixed point, i.e., $F(T) = \{x^*\}$;*

(b) *The estimate (2.2.41) holds;*

(c) *The rate of convergence of the Picard iteration is expressed by*

$$d(x_n, x^*) \leq \varphi(d(x_{n-1}, x^*)), \quad n = 1, 2, \ldots. \qquad (2.2.51)$$

Proof. Assume there exist two distinct fixed points $x^*, y^* \in X$. Then by (2.2.50) with $x := x^*$ and $y := y^*$, we get

$$d(x^*, y^*) \leq \psi(d(x^*, y^*)),$$

which by induction yields

$$d(x^*,y^*) \leq \psi^n(d(x^*,y^*)), \quad n = 1,2,\dots. \qquad (2.2.52)$$

Now, letting $n \to \infty$ in (2.2.52) we get

$$d(x^*,y^*) = 0,$$

i.e., $x^* = y^*$, a contradiction. Therefore, T has a unique fixed point.

To obtain (2.2.51), we let $x := x^*$, $y := x_n$ in (2.2.50).   □

The next Maia type extension of Theorem 2.2.7 is very natural (see Ref. [54]).

Theorem 2.2.8. *Let X be a nonempty set and d,ρ be two metrics on X such that (X,d) is complete. Let $T : X \to X$ be a self operator satisfying the following.*

(i) *There exists a (c)-comparison function φ and $L \geq 0$ such that*

$$d(Tx,Ty) \leq \varphi(d(x,y)) + Ld(y,Tx), \quad \text{for all } x,y \in X.$$

(ii) *There exists a comparison function ψ and $L_1 \geq 0$ such that*

$$\rho(Tx,Ty) \leq \psi(\rho(x,y)) + L_1\rho(x,Tx), \quad \text{for all } x,y \in X.$$

Then

(a) *T has a unique fixed point x^*;*

(b) *The Picard iteration $\{x_n\}_{n=0}^{\infty}$, $x_{n+1} = Tx_n$, $n \geq 0$, converges to x^*, for all $x_0 \in X$;*

(c) *The a posteriori error estimate*

$$d(x_n,x^*) \leq s(d(x_n,x_{n+1})), \quad n = 0,1,2,\dots,$$

holds, where $s(t) = \sum_{k=0}^{\infty} \varphi^k(t)$;

(d) *The rate of convergence of the Picard iteration is given by*

$$\rho(x_n,x^*) \leq \psi(\rho(x_{n-1},x^*)), \quad n \geq 1.$$

For other extensions of single-valued almost contractions, see Refs. [6–9, 11, 12, 14, 18, 20, 22, 28, 30, 41, 42, 44, 45, 47, 51, 89, 91, 125, 128, 138, 140, 161, 169, 181, 198, 199, 202, 203, 208, 216, 217, 243–248, 251, 252, 254, 257, 260, 262–266, 271, 277].

Final note. In Ref. [261] the authors tried to prove that Theorems 2.2.1 and 2.2.2 are false. Their claims were shown to be false and based on some wrong calculations (see the reply paper [73]).

Another attempt to diminish the merits of almost contractions is contained in a series of two recent papers [274, 275], where the author claims that "the almost contraction condition is *almost covered*" by a contractive condition studied in Ref. [155].

To be more specific, we present the statement of the main result in the paper [157], and then show that the above claim is not valid either. The authors of Ref. [157] established the following result (Theorem 2).

Theorem 2.2.9. *Let (X,d) be a complete metric space, T a self map of X, and $\varphi : \mathbb{R}_+ \to \mathbb{R}_+$ an increasing and continuous function with the property $\varphi(t) = 0$ if and only if $t = 0$. Furthermore, let a,b,c be three decreasing functions from $\mathbb{R}_+ \setminus \{0\}$ into $[0,1)$ such that $a(t) + 2b(t) + c(t) < 1$ for every $t > 0$. Suppose that T satisfies the following condition:*

$$
\begin{aligned}
\varphi(d(Tx,Ty)) \leq{} & a(x,y)\varphi(d(x,y)) \\
& + b(x,y)[\varphi(d(x,Tx)) + \varphi(d(y,Ty))] \\
& + c(d(x,y))\min\{\varphi(d(x,Ty)),\varphi(d(y,Tx))\},
\end{aligned}
\tag{2.2.53}
$$

where $x,y \in X$ and $x \neq y$. Then T has a unique fixed point.

Because the coefficients $a(t)$, $b(t)$, $c(t)$ are subject to the strong condition $a(t) + 2b(t) + c(t) < 1$ (see Remark 2.2.7), the class of mappings satisfying (2.2.53) cannot cover the class of almost contractions, for which the corresponding nonnegative coefficient $c(t)$ should be free of any restriction.

2.3. Implicit Almost Contractions

A simple and natural way to unify and prove in a simple manner several metrical fixed point theorems is to consider an implicit contraction type condition instead of the usual explicit contractive conditions.

It appears that Turinici [272] was the first to consider fixed point theorems for contractions defined by implicit relations. If (X,d) is a metric space, in Ref. [272] there are considered mappings $T : X \to X$, satisfying the implicit contraction condition

$$
d(Tx,Ty) \leq f(d(x,y),d(x,Tx),d(y,Ty)), \quad \text{for all } x,y \in X,
$$

where $f : \mathbb{R}_+^3 \to \mathbb{R}_+$.

Later, Popa [211, 212], initiated a systematic study of the contractions defined by implicit relations of the form:

$$
F(d(Tx,Ty),d(x,y),d(x,Tx),d(y,Ty),d(x,Ty),d(y,Tx)) \leq 0,
$$

where $F : \mathbb{R}_+^6 \to \mathbb{R}_+$.

This direction of research led to a comprehensive literature (that cannot be completely cited here) on fixed point, common fixed point and coincidence point theorems, both for single-valued and multivalued mappings, in various ambient spaces.

So, the aim of this section is to obtain general fixed point theorems for implicit almost contractions, largely by following Ref. [69].

Let $\mathscr{F}$ be the set of all continuous real functions $F : \mathbb{R}_+^6 \to \mathbb{R}_+$, for which we consider the following conditions:

(F_{1a}) F is nonincreasing in the fifth variable and $F(u,v,v,u,u+v,0) \le 0$ for $u,v \ge 0$ $\Rightarrow \exists h \in [0,1)$ such that $u \le hv$;

(F_{1b}) F is nonincreasing in the fourth variable and $F(u,v,0,u+v,u,v) \le 0$ for $u,v \ge 0 \Rightarrow \exists h \in [0,1)$ such that $u \le hv$;

(F_{1c}) F is nonincreasing in the third variable and $F(u,v,u+v,0,v,u) \le 0$ for $u,v \ge 0$ $\Rightarrow \exists h \in [0,1)$ such that $u \le hv$;

(F_2) $F(u,u,0,0,u,u) > 0$, for all $u > 0$.

The following functions are related to well-known fixed point theorems and satisfy most of the conditions (F_{1a})–(F_2) above.

Example 2.3.1. The function $F \in \mathscr{F}$, given by

$$F(t_1,t_2,t_3,t_4,t_5,t_6) = t_1 - at_2,$$

where $a \in [0,1)$, satisfies (F_2) and (F_{1a})–(F_{1c}), with $h = a$.

Example 2.3.2. Let $b \in \left[0,\frac{1}{2}\right)$. Then the function $F \in \mathscr{F}$, given by

$$F(t_1,t_2,t_3,t_4,t_5,t_6) = t_1 - b(t_3 + t_4),$$

satisfies (F_2) and (F_{1a})–(F_{1c}), with $h = \frac{b}{1-b} < 1$.

Example 2.3.3. Let $c \in \left[0,\frac{1}{2}\right)$. Then the function $F \in \mathscr{F}$, given by

$$F(t_1,t_2,t_3,t_4,t_5,t_6) = t_1 - c(t_5 + t_6),$$

satisfies (F_2) and (F_{1a})–(F_{1c}), with $h = \frac{c}{1-c} < 1$.

Example 2.3.4. The function $F \in \mathscr{F}$, given by

$$F(t_1,t_2,t_3,t_4,t_5,t_6) = t_1 - a\max\left\{t_2, \frac{t_3+t_4}{2}, \frac{t_5+t_6}{2}\right\},$$

where $a \in [0,1)$, satisfies (F_2) and (F_{1a})–(F_{1c}), with $h = a$.

Example 2.3.5. The function $F \in \mathscr{F}$, given by

$$F(t_1,t_2,t_3,t_4,t_5,t_6) = t_1 - at_2 - b(t_3 + t_4) - c(t_5 + t_6),$$

where $a, b, c \in [0, 1)$ and $a + 2b + 2c < 1$, satisfies (F_2) and (F_{1a})–(F_{1c}), with $h = \frac{a+b+c}{1-b-c} < 1$.

Example 2.3.6. The function $F \in \mathscr{F}$, given by

$$F(t_1, t_2, t_3, t_4, t_5, t_6) = t_1 - a \max \left\{ t_2, \frac{t_3 + t_4}{2}, t_5, t_6 \right\},$$

where $a \in [0, 1)$, satisfies (F_2) and (F_{1b}), (F_{1c}), with $h = a$ and (F_{1a}), with $h = \frac{a}{1-a} < 1$, if $a < 1/2$.

Example 2.3.7. The function $F \in \mathscr{F}$, given by

$$F(t_1, t_2, t_3, t_4, t_5, t_6) = t_1 - at_2 - Lt_3,$$

where $a \in [0, 1)$ and $L \geq 0$, satisfies (F_2) and (F_{1b}), with $h = a$, but, in general, does not satisfy (F_{1a}) and (F_{1c}).

Example 2.3.8. The function $F \in \mathscr{F}$, given by

$$F(t_1, t_2, t_3, t_4, t_5, t_6) = t_1 - at_2 - Lt_6,$$

where $a \in [0, 1)$ and $L \geq 0$, satisfies (F_{1a}), with $h = a$, but, in general, does not satisfy (F_{1b}), (F_{1c}) and (F_2).

The following theorem, which is an enriched version of Theorem 3 of Popa [211] and unifies the most important metrical fixed point theorems for contractive mappings in Rhoades' classification [226], was given in Ref. [69].

Theorem 2.3.1. *Let (X, d) be a complete metric space, $T : X \to X$ a self mapping for which there exists $F \in \mathscr{F}$ such that for all $x, y \in X$,*

$$F\left(d(Tx, Ty), d(x, y), d(x, Tx), d(y, Ty), d(x, Ty), d(y, Tx)\right) \leq 0. \qquad (2.3.54)$$

If F satisfies (F_{1a}) and (F_2) then:

(p1) T has a unique fixed point $\bar{x}$ in X;

(p2) The Picard iteration $\{x_n\}_{n=0}^{\infty}$ defined by

$$x_{n+1} = Tx_n, \quad n = 0, 1, 2, \ldots, \qquad (2.3.55)$$

converges to $\bar{x}$, for any $x_0 \in X$.

(p3) *The following estimate holds:*

$$d(x_{n+i-1},\bar{x}) \leq \frac{h^i}{1-h}d(x_n,x_{n-1}), \quad n=0,1,2,\ldots; i=1,2,\ldots, \qquad (2.3.56)$$

where h is the constant appearing in (F_{1a}).

(p4) *If, additionally, F satisfies (F_{1c}), then the rate of convergence of Picard iteration is given by:*

$$d(x_{n+1},\bar{x}) \leq hd(x_n,\bar{x}), \quad n=0,1,2,\ldots. \qquad (2.3.57)$$

Remark 2.3.1. (a) If F is the function in Example 2.3.1, then by Theorem 2.3.1 we obtain the Banach contraction mapping principle, in its complete form (see Theorem 2.1.1).

(b) If F is the function in Example 2.3.2, then by Theorem 2.3.1 we obtain Theorem 1 in Ref. [58], that extends the well-known Kannan fixed point theorem [149].

(c) If F is the function in Example 2.3.3, then by Theorem 2.3.1 we obtain a fixed point theorem that extends the Chatterjea fixed point theorem [85].

(d) If F is the function in Example 2.3.4, then by Theorem 2.3.1 we obtain Theorem 2 in Ref. [58], that extends the well-known Zamfirescu fixed point theorem [280].

(e) If F is the function in Example 2.3.5, then by Theorem 2.3.1 we obtain a fixed point theorem that extends the Reich-Rus fixed point theorem [239].

For other important particular cases of Theorem 2.3.1, see Refs. [53, 69] and the references therein.

The first main result of this section extends Theorem 2.3.1 in such a way to also include some known fixed point theorems for explicit almost contractions.

Theorem 2.3.2. *Let (X,d) be a complete metric space, $T: X \to X$ a self mapping for which there exists $F \in \mathscr{F}$, satisfying (F_{1a}), such that for all $x,y \in X$,*

$$F(d(Tx,Ty),d(x,y),d(x,Tx),d(y,Ty),d(x,Ty),d(y,Tx)) \leq 0. \qquad (2.3.58)$$

Then

(p1) *$Fix(T) \neq \emptyset$;*

(p2) *For any $x_0 \in X$, the Picard iteration $\{x_n\}_{n=0}^{\infty}$ converges to a fixed point $\bar{x}$ of T.*

(p3) *The following estimate holds:*

$$d(x_{n+i-1}, \bar{x}) \leq \frac{h^i}{1-h} d(x_n, x_{n-1}), \quad n = 0, 1, 2, \ldots; i = 1, 2, \ldots, \qquad (2.3.59)$$

where h is the constant appearing in (F_{1a}).

(p4) *If, additionally, F satisfies (F_{1c}), then the rate of convergence of Picard iteration is given by:*

$$d(x_{n+1}, \bar{x}) \leq h d(x_n, \bar{x}), \quad n = 0, 1, 2, \ldots. \qquad (2.3.60)$$

Proof. (p1) Let x_0 be an arbitrary point in X and $x_{n+1} = T x_n, n = 0, 1, \ldots$, be the Picard iteration. If we take $x := x_{n-1}$ and $y := x_n$ in (2.3.58) and denote $u := d(x_n, x_{n+1})$, $v := d(x_{n-1}, x_n)$ we get

$$F(u, v, v, u, d(x_{n-1}, x_{n+1}), 0) \leq 0.$$

By the triangle inequality, $d(x_{n-1}, x_{n+1}) \leq d(x_{n-1}, x_n) + d(x_n, x_{n+1}) = u + v$ and, since F is nonincreasing in the fifth variable, we have

$$F(u, v, v, u, u+v, 0) \leq F(u, v, v, u, d(x_{n-1}, x_{n+1}), 0) \leq 0$$

and hence, in view of assumption (F_{1a}), there exists $h \in [0, 1)$ such that $u \leq hv$, that is,

$$d(x_n, x_{n+1}) \leq h d(x_{n-1}, x_n), \qquad (2.3.61)$$

which, in a straightforward way, leads to the conclusion that $\{x_n\}_{n=0}^{\infty}$ is a Cauchy sequence.

Since (X, d) is complete, there exists a $\bar{x}$ in X such that

$$\lim_{n \to \infty} x_n = \bar{x}. \qquad (2.3.62)$$

By taking $x := x_n$ and $y := \bar{x}$ in (2.3.58) we get

$$F(d(T x_n, T\bar{x}), d(x_n, \bar{x}), d(x_n, T x_n), d(\bar{x}, T\bar{x}), d(x_n, T\bar{x}), d(\bar{x}, T x_n)) \leq 0. \qquad (2.3.63)$$

As F is continuous, by letting $n \to \infty$ in (2.3.63), we obtain

$$F(d(\bar{x}, T\bar{x}), 0, 0, d(\bar{x}, T\bar{x}), d(\bar{x}, T\bar{x}), 0)) \leq 0,$$

which, by assumption (F_{1a}), yields $d(\bar{x}, T\bar{x}) \leq 0$, that is, $\bar{x} = T\bar{x}$.

(p2) This follows by the proof of (p1).

(p3) This follows by a double inductive process by means of (2.3.61).

(p4) By taking $x := x_n$ and $y := \bar{x}$ in (2.3.58) we get

$$F(d(T x_n, \bar{x}), d(x_n, \bar{x}), d(x_n, T x_n), d(\bar{x}, \bar{x}), d(x_n, \bar{x}), d(\bar{x}, T x_n)) \leq 0,$$

that is,

$$F\left(d(x_{n+1},\bar{x}),d(x_n,\bar{x}),d(x_n,x_{n+1}),0,d(x_n,\bar{x}),d(\bar{x},x_{n+1})\right) \leq 0. \tag{2.3.64}$$

Denote $u := d(x_{n+1},\bar{x})$, $v := d(x_n,\bar{x})$. Then, by the triangle inequality we have $d(x_n,x_{n+1}) \leq d(x_n,\bar{x}) + d(x_{n+1},\bar{x}) = u+v$ and hence, in view of assumption (F_{1c}), by (2.3.64) we obtain

$$F\left(u,v,u+v,0,v,u\right) \leq F\left(u,v,d(x_n,x_{n+1}),0,v,u\right) \leq 0,$$

which again by (F_{1c}) implies the existence of $h \in [0,1)$ such that $u \leq hv$, which is exactly the desired estimate (2.3.60). □

Remark 2.3.2. (a) If F in Theorem 2.3.2 also satisfies (F_2), then by Theorem 2.3.2 we obtain Theorem 2.3.1.

(b) If F is the function in Example 2.3.7, then by Theorem 2.3.2 (but not by Theorem 2.3.1) we obtain Theorem 2.2.1, i.e., the existence theorem [55, Theorem 2.1].

(c) If F is the function in Example 2.3.8, then by Theorem 2.3.2 (or by Theorem 2.3.1) we obtain Theorem 2.2.2, i.e., the existence and uniqueness theorem [55, Theorem 2.2].

Remark 2.3.3. From the unifying error estimates (2.3.59), inspired by Ref. [278], we get both the *a priori* estimate

$$d(x_n,\bar{x}) \leq \frac{h^n}{1-h}d(x_0,x_1), \quad n=0,1,2,\ldots,$$

and the *a posteriori* estimate

$$d(x_n,\bar{x}) \leq \frac{h}{1-h}d(x_n,x_{n-1}), \quad n=1,2,\ldots,$$

which are extremely important in applications, especially when approximating the solutions of nonlinear equations.

One can also obtain an existence and uniqueness fixed point theorem, corresponding to Theorem 2.3.2.

Theorem 2.3.3. *Let (X,d) be a complete metric space, $T : X \to X$ a self mapping for which there exists $F \in \mathscr{F}$, satisfying (F_{1a}), such that for all $x,y \in X$, (2.3.58) holds, and there exists $G \in \mathscr{F}$, satisfying (F_2), such that for all $x,y \in X$,*

$$G\left(d(Tx,Ty),d(x,y),d(x,Tx),d(y,Ty),d(x,Ty),d(y,Tx)\right) \leq 0. \tag{2.3.65}$$

Then

(p1) *T has a unique fixed point $\bar{x}$ in X;*

(p2) *For any $x_0 \in X$, the Picard iteration $\{x_n\}_{n=0}^{\infty}$ converges to $\bar{x}$.*

(p3) *The error estimate (2.3.59) holds;*

(p4) *If, additionally, F or G satisfies (F_{1c}), then the rate of convergence of the Picard iteration is given by:*

$$d(x_{n+1}, \bar{x}) \leq h d(x_n, \bar{x}), \quad n = 0, 1, 2, \ldots. \qquad (2.3.66)$$

Proof. The existence of the fixed point as well as the estimates (2.3.59) and (2.3.66) follow as in the proof of Theorem 2.3.2.

In order to prove the uniqueness of $\bar{x}$, assume the contrary, i.e., there exists $\bar{y} \in Fix(T), \bar{x} \neq \bar{y}$. Then by taking $x := \bar{x}$ and $y := \bar{y}$ in (2.3.65) and by denoting $\delta := d(\bar{x}, \bar{y}) > 0$ we get

$$G(\delta, \delta, 0, 0, \delta, \delta) \leq 0,$$

which contradicts (F_2).

This proves that T has a unique fixed point. □

Remark 2.3.4. (a) If F is the function in Example 2.3.8 and G is the function in Example 2.3.7, then by Theorem 2.3.3 we obtain Theorem 2.2.2, i.e., Theorem 2.2 in Ref. [55].

(b) If $F \equiv G$ is the function in Example 2.3.9, then by Theorem 2.3.3 one obtains the main result (Theorem 2.3) in Ref. [47].

(c) If F is the function in Example 2.3.11, then by Theorem 2.3.3 one obtains the second uniqueness result [63, Theorem 2.4].

Theorem 2.3.3 can now be significantly extended by considering two metrics on the set X, similarly to Theorem 5 in Ref. [54].

Theorem 2.3.4. *Let X be a nonempty set and d, ρ two metrics on X such that (X, d) is complete. Let $T : X \rightarrow X$ be a self operator for which*

(i) *there exists $F \in \mathscr{F}$ satisfying (F_{1a}) such that for all $x, y \in X$,*

$$F(d(Tx, Ty), d(x, y), d(x, Tx), d(y, Ty), d(x, Ty), d(y, Tx)) \leq 0;$$

(ii) *there exists $G \in \mathscr{F}$ satisfying (F_{1c}) and (F_2) such that for all $x, y \in X$,*

$$G(\rho(Tx, Ty), \rho(x, y), \rho(x, Tx), \rho(y, Ty), \rho(x, Ty), \rho(y, Tx)) \leq 0.$$

Then

(a) *T has a unique fixed point $\bar{x}$;*

(b) *The Picard iteration $\{x_n\}_{n=0}^{\infty}$, $x_{n+1} = Tx_n$, $n \geq 0$, converges to $\bar{x}$, for all $x_0 \in X$;*

(c) *The error estimate (2.3.59) holds;*

(d) *The rate of convergence of the Picard iteration is given by*

$$\rho(x_n, x^*) \leq h\rho(x_{n-1}, x^*), \quad n \geq 1. \tag{2.3.67}$$

Proof. The existence of the fixed point as well as the estimates (2.3.59) and (2.3.67) follow as in the proof of Theorem 2.3.2.

In order to prove the uniqueness of $\bar{x}$, assume the contrary, i.e., there exists $\bar{y} \in Fix(T), \bar{x} \neq \bar{y}$. Then by taking $x := \bar{x}$ and $y := \bar{y}$ in (2.3.65) and by denoting $\delta := \rho(\bar{x}, \bar{y}) > 0$ we get

$$G(\delta, \delta, 0, 0, \delta, \delta) \leq 0,$$

which contradicts (F_2). This proves that T has a unique fixed point. $\square$

In order to illustrate the generality of Theorems 2.3.3 and 2.3.4, we consider three more examples of functions $F \in \mathscr{F}$.

Example 2.3.9. The function $F \in \mathscr{F}$, given by

$$F(t_1, t_2, t_3, t_4, t_5, t_6) = t_1 - at_2 - L\min\{t_3, t_4, t_5, t_6\},$$

where $a \in [0, 1)$ and $L \geq 0$, satisfies (F_2) and (F_{1a})–(F_{1c}), with $h = a$.

Example 2.3.10. The function $F \in \mathscr{F}$, given by

$$F(t_1, t_2, t_3, t_4, t_5, t_6) = t_1 - a\max\left\{t_2, t_3, t_4, \frac{t_5 + t_6}{2}\right\} - Lt_6,$$

where $a \in [0, 1)$ and $L \geq 0$, satisfies (F_{1a}), with $h = a$, but, in general, does not satisfy (F_{1b}), (F_{1c}) and (F_2). To prove (F_{1a}) let us observe that with $F(u, v, v, u, u + v, 0) \leq 0$ one obtains

$$u - a\max\left\{v, v, u, \frac{u+v}{2}\right\} \leq 0.$$

If one admits that $u > v$, then by the previous inequality one obtains $u - au \leq 0 \Leftrightarrow (1 - a)u \leq 0$, a contradiction. Hence $u \leq v$ and thus (F_{1a}) is satisfied with $h = a$.

Example 2.3.11. The function $F \in \mathscr{F}$, given by

$$F(t_1, t_2, t_3, t_4, t_5, t_6) = t_1 - a \max \left\{ t_2, t_3, t_4, \frac{t_5 + t_6}{2} \right\} - L \min \{ t_3, t_4, t_5, t_6 \},$$

where $a \in [0, 1)$ and $L \geq 0$, satisfies (F_2) and (F_{1a})–(F_{1c}), with $h = a$.

Remark 2.3.5. (a) If we set $d \equiv \rho$, by Theorem 2.3.4 we obtain Theorem 2.3.3.

(b) If F is the function in Example 2.3.8 and G is the function in Example 2.3.7, then by Theorem 2.3.3 we obtain Theorem 2.2.2, i.e., Theorem 2.2 in Ref. [55].

(c) If F is the function given by

$$F(t_1, t_2, t_3, t_4, t_5, t_6) = t_1 - a \max \left\{ t_2, t_3, t_4, \frac{t_5 + t_6}{2} \right\} - L t_6,$$

where $a \in [0, 1)$ and $L \geq 0$, and G is as in Example 2.3.7, then F satisfies (F_2) and (F_{1a})–(F_{1c}), with $h = a$, and hence by Theorem 2.3.3 one obtains the first uniqueness result [63, Theorem 2.3].

All contractive conditions considered in this chapter are defined by *linear* functions $F \in \mathscr{F}$ (see Examples 2.3.1–2.3.11 and Remark 2.3.5), but generally, in Theorems 2.3.2, 2.3.3, and 2.3.4, neither F nor G is assumed to be linear.

This ensures a great generality to the results obtained in the present section. Several *nonlinear* contractive conditions associated with similar fixed point theorems can be found, for example, in Refs. [211, 212].

It is very important to note that, to our best knowledge, the contraction conditions defined by the functions in Examples 2.3.7–2.3.11 in this chapter have not been considered in any other papers devoted to fixed point theorems for mappings defined by implicit relations.

2.4. Common Fixed Point Theorems for Almost Contractions

The Banach contraction mapping principle (Theorem 2.1.1) has been extended in another direction than the ones illustrated in the previous sections by Jungck [144] to obtain the following common fixed point theorem.

Theorem 2.4.1 [144]. *Let (X, d) be a complete metric space. Let S be a continuous self map on X and T be any self map on X that commutes with S. Further let S and T satisfy $T(X) \subset S(X)$ and there exists a constant $\lambda \in (0, 1)$ such that for every $x, y \in X$,*

$$d(Tx, Ty) \leq \lambda d(Sx, Sy), \quad \text{for all } x, y \in X. \tag{2.4.68}$$

Then S and T have a unique common fixed point.

Note that Theorem 2.4.1 reduces to Theorem 2.1.1 in the case $S = I$ (the identity map on X).

Other common fixed point results for the Kannan, Chatterjea and Zamfirescu contractive conditions have been recently obtained in Refs. [1] and [63] respectively, while the corresponding common fixed point result for Ciric's fixed point theorem has been derived by Das and Naik [105].

Theorem 2.4.2 [105]. *Let* (X,d) *be a complete metric space. Let S be a continuous self map on X and T be any self map on X that commutes with S. Further let S and T satisfy* $T(X) \subset S(X)$. *If there exists a constant* $h \in (0,1)$ *such that for every* $x, y \in X$,

$$d(Tx, Ty) \leq hM(x,y), \tag{2.4.69}$$

where

$$M(x,y) = \max \big\{ d(Sx, Sy), d(Sx, Tx), d(Sy, Ty), d(Sx, Ty), d(Sy, Tx) \big\},$$

then S and T have a unique common fixed point.

Due to the fact that Theorem 2.4.2 requires S and T to be commuting mappings, an extension of this result to weakly commuting generalized quasi-contractions has been given in Ref. [56].

As all Banach, Kannan, Chatterjea and Zamfirescu contractive conditions imply the almost contractive condition (2.2.8), it is the main purpose of the present section to extend Theorem 1 in Ref. [56], and thus all its subsequent results, to the case of a pair of mappings (S, T) satisfying an almost contraction condition.

To this end we need some notions and results from Refs. [1] and [145].

Definition 2.4.1 [1]. Let S and T be self maps of a nonempty set X. If there exists $x \in X$ such that $Sx = Tx$ then x is called a *coincidence point* of S and T, while $y = Sx = Tx$ is called a *point of coincidence* (or *coincidence value*) of S and T. If $Sx = Tx = x$, then x is called a *common fixed point* of S and T.

Definition 2.4.2 [145]. Let S and T be self maps of a nonempty set X. The pair of mappings S and T is said to be *weakly compatible* if they commute at their coincidence points.

The next proposition will be needed to prove the last part in our main results.

Proposition 2.4.1 [1, Proposition 1.4]. *Let S and T be weakly compatible self maps of a nonempty set X. If S and T have a unique coincidence point x, then x is the unique* *common fixed point of S and T.*

For some other recent related results, see Refs. [72, 146].

We start this section by presenting a coincidence point theorem for almost contraction type mappings.

Theorem 2.4.3 [63]. *Let (X,d) be a metric space and let $T, S : X \to X$ be two mappings for which there exist a constant $\delta \in (0,1)$ and some $L \geq 0$ such that*

$$d(Tx, Ty) \leq \delta \cdot d(Sx, Sy) + Ld(Sy, Tx), \quad \text{for all } x, y \in X. \tag{2.4.70}$$

If the range of S contains the range of T and $S(X)$ is a complete subspace of X, then T and S have a coincidence point in X.

Moreover, for any $x_0 \in X$, the iteration $\{Sx_n\}$ defined by (2.4.72) converges to some coincidence point x^* of T and S, with the following error estimate:

$$d(Sx_{n+i-1}, x^*) \leq \frac{\delta^i}{1-\delta} d(Sx_n, Sx_{n-1}), \quad n = 0, 1, 2, \ldots; i = 1, 2, \ldots. \tag{2.4.71}$$

Proof. Let x_0 be an arbitrary point in X. Since $T(X) \subset S(X)$, we can choose a point x_1 in X such that $Tx_0 = Sx_1$. Continuing in this way, for a value x_n in X, we can find $x_{n+1} \in X$ such that

$$Sx_{n+1} = Tx_n, \quad n = 0, 1, \ldots. \tag{2.4.72}$$

If $x := x_n, y := x_{n-1}$ are two successive terms of the sequence defined by (2.4.72), then by (2.4.70) we have

$$d(Sx_n, Sx_{n+1}) = d(Tx_{n-1}, Tx_n) \leq L \cdot d(Sx_n, Tx_{n-1}) + \delta \cdot d(Sx_{n-1}, Sx_n),$$

which in view of (2.4.72) yields

$$d(Sx_{n+1}, Sx_n) \leq \delta \cdot d(Sx_n, Sx_{n-1}), \quad n = 0, 1, 2, \ldots. \tag{2.4.73}$$

Now by induction, from (2.4.73) we obtain

$$d(Sx_{n+k}, Sx_{n+k-1}) \leq \delta^k \cdot d(Sx_n, Sx_{n-1}), n, k = 0, 1, \ldots (k \neq 0),$$

and then, for $p > i$, we get after straightforward calculations

$$d(Sx_{n+p}, Sx_{n+i-1}) \leq \frac{\delta^i(1 - \delta^{p-i+1})}{1-\delta} \cdot d(Sx_n, Sx_{n-1}), \quad n \geq 0; i \geq 1. \tag{2.4.74}$$

Take $i = 1$ in (2.4.74) and then, by an inductive process, we get

$$d(Sx_{n+p}, Sx_n) \leq \frac{\delta}{1-\delta} \cdot d(Sx_n, Sx_{n-1}) \leq \frac{\delta^n}{1-\delta} \cdot d(Sx_1, Sx_0), \quad n = 0, 1, 2 \ldots,$$

which shows that $\{Sx_n\}$ is a Cauchy sequence.

Since $S(X)$ is complete, there exists a value of x^* in $S(X)$ such that

$$\lim_{n\to\infty} Sx_{n+1} = x^*. \tag{2.4.75}$$

We can find $p \in X$ such that $Sp = x^*$. By (2.4.72) and (2.4.73) we further have

$$d(Sx_n, Tp) \leq \delta d(Sx_{n-1}, Sp) \leq \delta^{n-1} d(Sx_1, Sp),$$

which shows that we also have

$$\lim_{n\to\infty} Sx_n = Tp. \tag{2.4.76}$$

Now by (2.4.75) and (2.4.76) we find that $Tp = Sp$, that is, p is a coincidence point of T and S (or x^* is a point of coincidence of T and S). The estimate (2.4.71) is obtained from (2.4.74) by letting $p \to \infty$. $\qquad\square$

Remark 2.4.1. Let us note that the coincidence point ensured by Theorem 2.4.3 is not generally unique (see Ref. [55, Example 1]).

In order to obtain a common fixed point theorem from the coincidence Theorem 2.4.3, we need the uniqueness of the coincidence point, which can be obtained by imposing an additional contractive condition, similar to (2.4.70).

Theorem 2.4.4. *Let (X, d) be a metric space and let $T, S : X \to X$ be two mappings satisfying (2.4.70) for which there exist a constant $\theta \in (0,1)$ and some $L_1 \geq 0$ such that*

$$d(Tx, Ty) \leq \theta \cdot d(Sx, Sy) + L_1 d(Sx, Tx), \quad \text{for all } x, y \in X. \tag{2.4.77}$$

If the range of S contains the range of T and $S(X)$ is a complete subspace of X, then T and S have a unique coincidence point in X. Moreover, if T and S are weakly compatible, then T and S have a unique common fixed point in X.

In both cases, for any $x_0 \in X$, the iteration $\{Sx_n\}$ defined by (2.4.72) converges to the unique common fixed point (coincidence point) x^* of S and T, with the error estimate (2.4.71).

The convergence rate of the iteration $\{Sx_n\}$ is given by

$$d(Sx_n, x^*) \leq \theta \cdot d(Sx_{n-1}, x^*), \quad n = 1, 2, \ldots. \tag{2.4.78}$$

Proof. By the proof of Theorem 2.4.3, we have that T and S have at least one point of coincidence. Now let us show that T and S actually have a unique point of coincidence.

Assume there exists $q \in X$ such that $Tq = Sq$. Then, by (2.4.77) we get

$$d(Sq,Sp) = d(Tq,Tp) \leq 2\delta d(Sq,Tq) + \delta d(Sq,Tp) = \delta d(Sq,Sp),$$

which shows that $Sq = Sp = x^*$, that is, T and S have a unique point of coincidence, x^*.

Now if T and S are weakly compatible, by Proposition 2.4.1 it follows that x^* is their unique common fixed point. The estimate (2.4.78) is obtained from (2.4.77) by taking $x = x_n$ and $y = x^*$. $\qquad\square$

An equivalent (see Ref. [199]) but simpler contractive condition that ensures the uniqueness of the coincidence point and which actually unifies (2.4.70) and (2.4.77) has been very recently obtained by Babu et al. [48]. We state in the following the common fixed point theorem corresponding to this fixed point result.

Theorem 2.4.5. *Let (X,d) be a metric space and let $T, S : X \rightarrow X$ be two mappings for which there exist a constant $\delta \in (0,1)$ and some $L \geq 0$ such that*

$$d(Tx,Ty) \leq \delta \cdot d(Sx,Sy)$$
$$+ L\min\{d(Sx,Tx) + d(Sy,Ty) + d(Sx,Ty) + d(Sy,Tx)\}, \quad (2.4.79)$$

for all $x,y \in X$. If the range of S contains the range of T and $S(X)$ is a complete subspace of X, then T and S have a unique coincidence point in X. Moreover, if T and S are weakly compatible, then T and S have a unique common fixed point in X.

In both cases, for any $x_0 \in X$, the iteration $\{Sx_n\}$ defined by (2.4.72) converges to the unique common fixed point (coincidence point) x^* of S and T, with the error estimate (2.4.71) and convergence rate given by (2.4.78).

Proof. If $x := x_n$, $y := x_{n-1}$ are two successive terms of the sequence defined by (2.4.72), then by (2.4.79) we have

$$d(Sx_n,Sx_{n+1}) = d(Tx_{n-1},Tx_n) \leq \delta \cdot d(Sx_{n-1},Sx_n) + L \cdot M,$$

where

$$M = \min\{d(Sx_n,Tx_n) + d(Sx_{n-1},Tx_{n-1}) + d(Sx_n,Tx_{n-1}) + d(Sx_{n-1},Tx_n)\} = 0,$$

since $d(Sx_n,Tx_{n-1}) = 0$. The rest of the proof follows from that of Theorem 2.4.4. $\qquad\square$

Remark 2.4.2. (a) If $S = I$ (the identity map on X), then by Theorem 2.4.3 we obtain the existence fixed point theorem given in Ref. [55] for almost contractions (Theorem 2.2.1).

If $S = I$, then by Theorem 2.4.4 we obtain the existence and uniqueness fixed point theorem given in Ref. [55] for almost contractions (Theorem 2.2.2).

If $S = I$, then by Theorem 2.4.5 we obtain the existence and uniqueness fixed point theorem given in Ref. [48] for strict almost contractions.

(b) If $S = I$ and $L = 0$ in condition (2.4.70), then by Theorem 2.4.3 we obtain a result that extends Jungck's common fixed point theorem [144] from commuting mappings to weakly compatible mappings.

Three other particular cases that are obtained from our main results are given in the following as corollaries.

Corollary 2.4.1. *Let (X,d) be a metric space and let $T, S : X \to X$ be two mappings for which there exist $b \in [0, \frac{1}{2})$ such that, for all $x, y \in X$,*

$$(z_2) \qquad\qquad d(Tx, Ty) \leq b[d(Sx, Tx) + d(Sy, Ty)].$$

If the range of S contains the range of T and $S(X)$ is a complete subspace of X, then T and S have a unique coincidence point in X. Moreover, if T and S are weakly compatible, then T and S have a unique common fixed point in X.

In both cases, the iteration $\{Sx_n\}$ defined by (2.4.72) converges to the unique (coincidence) common fixed point x^* of S and T, for any $x_0 \in X$, with the following error estimate:

$$d(Sx_{n+i-1}, x^*) \leq \frac{\delta^i}{1 - \delta} d(Sx_n, Sx_{n-1}), \quad n, i = 0, 1, 2, \ldots (i \neq 0), \qquad (2.4.80)$$

where $\delta = \frac{b}{1-b}$.

The convergence rate of the iteration $\{Sx_n\}$ is given by

$$d(Sx_n, x^*) \leq \delta \cdot d(Sx_{n-1}, x^*), \quad n = 1, 2, \ldots. \qquad (2.4.81)$$

Proof. By condition (z_2) and the triangle rule, we get

$$\begin{aligned}
d(Tx, Ty) &\leq b[d(x, Tx) + d(y, Ty)] \leq \\
&\leq b\Big\{ [d(x, y) + d(y, Tx)] + [d(y, Tx) + d(Tx, Ty)] \Big\},
\end{aligned}$$

which yields

$$(1 - b)d(Tx, Ty) \leq bd(x, y) + 2b \cdot d(y, Tx)$$

and which implies

$$d(Tx, Ty) \leq \frac{b}{1-b} d(x, y) + \frac{2b}{1-b} d(y, Tx), \quad \text{for all } x, y \in X.$$

Now, in view of $0 < b < \frac{1}{2}$, (2.4.70) holds with $\delta = \frac{b}{1-b}$ and $L = \frac{2b}{1-b}$. The uniqueness condition (2.4.81) follows similarly. To obtain the conclusion, apply Theorem 2.4.4.

$\square$

Corollary 2.4.2. *Let (X,d) be a metric space and let $T, S : X \to X$ be two mappings for which there exist $c \in [0, \frac{1}{2})$ such that, for all $x, y \in X$,*

(z_3) $$d(Tx, Ty) \leq c\big[d(Sx, Ty) + d(Sy, Tx)\big].$$

If the range of S contains the range of T and $S(X)$ is a complete subspace of X, then T and S have a unique coincidence point in X. Moreover, if T and S are weakly compatible, then T and S have a unique common fixed point in X.

In both cases, the iteration $\{Sx_n\}$ defined by (2.4.72) converges to the unique (coincidence) common fixed point x^* of S and T, for any $x_0 \in X$, with the following error estimate:

$$d(Sx_{n+i-1}, x^*) \leq \frac{\delta^i}{1 - \delta} d(Sx_n, Sx_{n-1}), \quad n, i = 0, 1, 2, \ldots \ (i \neq 0), \qquad (2.4.82)$$

where $\delta = \frac{c}{1-c}$.

The convergence rate of the iteration $\{Sx_n\}$ is given by

$$d(Sx_n, x^*) \leq \delta \cdot d(Sx_{n-1}, x^*), \quad n = 1, 2, \ldots. \qquad (2.4.83)$$

Proof. By condition (z_3) and the triangle rule, we get

$$d(Tx, Ty) \leq \frac{c}{1 - c} d(x, y) + \frac{2c}{1 - c} d(y, Tx),$$

which is (2.4.70), with $\delta = \frac{c}{1-c} < 1$ and $L = \frac{2c}{1-c} \geq 0$.

The uniqueness condition (2.4.81) follows similarly. Now apply Theorem 2.4.4 to obtain the conclusion.

$\square$

Since any Banach contraction condition implies (2.4.70) (with $L=0$), by Corollaries 2.4.1 and 2.4.2 we obtain in particular the main result in Ref. [66].

Corollary 2.4.3. *Let (X,d) be a metric space and let $T, S : X \to X$ be two mappings for which there exist $a \in [0, 1), b, c \in [0, \frac{1}{2})$ such that, for all $x, y \in X$, at least one of the following conditions is true:*

(z_1) $d(Tx, Ty) \leq a\, d(Sx, Sy)$;

(z_2) $d(Tx, Ty) \leq b\big[d(Sx, Tx) + d(Sy, Ty)\big]$;

(z_3) $d(Tx, Ty) \leq c\big[d(Sx, Ty) + d(Sy, Tx)\big]$.

If the range of S contains the range of T and S(X) is a complete subspace of X, then T and S have a unique coincidence point in X. Moreover, if T and S are weakly compatible, then T and S have a unique common fixed point in X.

In both cases, the iteration $\{Sx_n\}$ defined by (2.4.72) converges to the unique (co-incidence) common fixed point x^* of S and T, for any $x_0 \in X$, with the following error estimate:

$$d(Sx_{n+i-1}, x^*) \le \frac{\delta^i}{1-\delta} d(Sx_n, Sx_{n-1}), \quad n = 0, 1, 2, \ldots; i = 1, 2, \ldots,$$

where $\delta = \max\left\{a, \frac{b}{1-b}, \frac{c}{1-c}\right\}$.

The convergence rate of the iteration $\{Sx_n\}$ is given by

$$d(Sx_n, x^*) \le \delta \cdot d(Sx_{n-1}, x^*), \quad n = 1, 2, \ldots.$$

Remark 2.4.3. It is important to note that all our results established here are important from a computational point of view, due to the fact that they offer a method for computing the common fixed points (the coincidence points, respectively).

Moreover, for the iterative method thus obtained, we have *a priori* and *a posteriori* error estimates, both contained in the unified estimates of the form (2.4.71). Note that in (2.2.19) and (2.4.70) we can have $\delta = 0$, provided that in this case we also have $L = 0$, which ensures that Theorems 2.2.1 and 2.4.3 also include the Banach contraction mapping principle.

Several other fixed point results can be obtained as particular cases of our main results in this section (see Refs. [85, 149]).

For common fixed point theorems of almost contractive mappings in cone metric spaces, see Ref. [67].

For other developments concerning common fixed point theorems or coincidence theorems for almost contractive type mappings, see Refs. [13, 46, 123, 137, 146, 166, 185, 196, 204, 222–224, 232, 253, 255, 256, 258, 268, 276], etc.

2.5. Almost Contractive type Mappings on Product Spaces

Banach's contraction principle (Theorem 2.1.1) for a self mapping $T : X \to X$ has been generalized by Prešić [218] to the case of self mappings defined on a product space, $f : X^k \to X$. This generalization has a direct connection with a dynamic field of research devoted today to the study of nonlinear difference equations, with applications in economics, biology, ecology, genetics, psychology, sociology, probability theory and others (see for example Refs. [81, 86, 108, 127, 164, 165, 167, 193, 234, 264, 265] and others). Some examples of such equations (see Refs. [234, 264] and the papers referred to therein) are:

- the generalized Beddington-Holt stock recruitment model:

$$x_{n+1} = ax_n + \frac{bx_{n-1}}{1+cx_{n-1}+dx_n}, \quad x_0, x_1 > 0, n \in \mathbb{N},$$

  where $a \in (0,1)$, $b \in \mathbb{R}_+^*$ and $c,d \in \mathbb{R}_+$, with $c+d > 0$;
- the delay model of a perennial grass:

$$x_{n+1} = ax_n + (b+cx_{n-1})e^{x_n}, \quad n \in \mathbb{N},$$

  where $a,c \in (0,1)$ and $b \in \mathbb{R}_+$;
- the flour beetle population model:

$$x_{n+3} = ax_{n+2} + bx_n e^{-(cx_{n+2}+dx_n)}, \quad n \in \mathbb{N},$$

  where $a,b,c,d \geq 0$ and $c+d > 0$.

These suggest considering the following kth order nonlinear difference equation, corresponding to a k-step iteration method:

$$x_{n+k} = f(x_n, \ldots, x_{n+k-1}), \quad n \in \mathbb{N}, \tag{2.5.84}$$

with the initial values $x_0, \ldots, x_k \in X$, where (X,d) is a metric space, $k \in \mathbb{N}$, $k \geq 1$ and $f : X^k \to X$.

Equation (2.5.84) can be studied by means of fixed point theory in view of the fact that $x^* \in X$ is a solution of (2.5.84) if and only if x^* is a *fixed point of f*, that is,

$$x^* = f(x^*, \ldots, x^*).$$

Remark 2.5.1. For any operator $f : X^k \to X$, k a positive integer, we can define its *associate operator $F : X \to X$* (see Refs. [234, 249]) by

$$F(x) = f(x, \ldots, x), \quad x \in X.$$

Obviously, $x \in X$ is a fixed point of $f : X^k \to X$ if and only if it is a fixed point of its associate operator F. This enables the study of f by means of the operator F.

One of the pioneering results in this direction is due to Prešić [218] (see Refs. [279]). We begin by defining the following concept.

Definition 2.5.1. Let (X,d) be a metric space, k a positive integer, $\alpha_1, \alpha_2, \ldots, \alpha_k \in \mathbb{R}_+$, $\sum_{i=1}^{k} \alpha_i = \alpha < 1$ and $f : X^k \to X$ a mapping satisfying

$$d(f(x_0, \ldots, x_{k-1}), f(x_1, \ldots, x_k)) \leq \sum_{i=1}^{k} \alpha_i d(x_{i-1}, x_i), \tag{2.5.85}$$

for all $x_0, \ldots, x_k \in X$. Then f is called a *Prešić operator*.

It is obvious that for $k = 1$ the above definition reduces to the definition of classical Banach contractions.

The result of Prešić [218], enriched with some quantitative information regarding the rate of convergence of the k-step iterative method, is given below.

Theorem 2.5.1. *Let (X, d) be a complete metric space, k a positive integer and $f : X^k \to X$ a Prešić operator. Then*

(a) *f has a unique fixed point x^*;*

(b) *the sequence $\{y_n\}_{n \geq 0}$,*

$$y_{n+1} = f(y_n, y_n, \ldots, y_n), \quad n \geq 0, \qquad (2.5.86)$$

converges to x^*;

(c) *the sequence $\{x_n\}_{n \geq 0}$ with $x_0, \ldots, x_{k-1} \in X$ and*

$$x_n = f(x_{n-k}, x_{n-k+1}, \ldots, x_{n-1}), \quad n \geq k, \qquad (2.5.87)$$

also converges to x^*, with a rate estimated by

$$d(x_{n+1}, x^*) \leq \alpha d(x_n, x^*) + M \cdot \theta^n, \quad n \geq 0, \qquad (2.5.88)$$

where $M > 0$ and $\theta \in (0, 1)$ are constants.

As already said, Prešić's result is a generalization of the contraction Banach principle (Theorem 2.1.1), by considering contractions defined on product spaces. It was then natural to search for similar Prešić type extensions also for other classes of generalized contractions. This has been done for Prešić-Rus operators in Ref. [233] (see also Ref. [234]), for Ćirić-Prešić operators in Ref. [101], for Prešić-Kannan operators in Ref. [201], for almost Prešić operators in Ref. [203] and so on. The study of Prešić type results has been very dynamic lately, as the great number of very recent papers on the topic shows (see references in Ref. [76] for a few of them).

In the following we shall briefly present only the results strictly related to almost contractions, i.e., those concerning almost Prešić operators. The name of this class suggests a Prešić type extension starting from the class of almost contractions. Actually only a subclass is referred to, namely that of *strict almost contractions*, by this meaning those almost contractions, as defined in the first section of this chapter, which satisfy the additional condition which ensures the uniqueness of the fixed point.

Definition 2.5.2. Let (X,d) be a metric space. An operator $f : X \to X$ is called a *strict almost contraction* if it satisfies both conditions

$$d(f(x), f(y)) \leq \delta d(x,y) + L d(y, f(x)), \quad \text{for any } x, y \in X, \qquad (2.5.89)$$

and

$$d(f(x), f(y)) \leq \delta_u d(x,y) + L_u d(x, f(x)), \quad \text{for any } x, y \in X, \qquad (2.5.90)$$

with some real constants $\delta \in [0,1)$, $L \geq 0$ and $\delta_u \in [0,1)$, $L_u \geq 0$, respectively.

Having in view Definition 2.5.2, we can restate (part of) Theorem 2.2.2 as follows (see also Ref. [197]):

Theorem 2.5.2. *Let (X,d) be a complete metric space and $f : X \to X$ a strict almost contraction with constants $\delta \in [0,1)$, $L \geq 0$ and $\delta_u \in [0,1)$, $L_u \geq 0$, respectively. Then f has a unique fixed point, say x^*, that can be approximated by means of the Picard iteration $\{x_n\}_{n \geq 0}$ of f, starting from any $x_0 \in X$.*

Remark 2.5.2. An equivalent definition of strict almost contractions, first introduced in Ref. [47], is studied in Ref. [199], where it is shown that an operator $f : X \to X$ is a strict almost contraction if and only if there exist two constants $\delta_B \in [0,1)$ and $L_B \geq 0$ such that

$$d(f(x), f(y)) \leq \delta_B d(x,y)$$
$$+ L_B \min\{d(x, f(x)), d(y, f(y)), d(x, f(y)), d(y, f(x))\}, \quad (2.5.91)$$

for any $x, y \in X$.

Having in view this equivalent definition of strict almost contractions, we introduce:

Definition 2.5.3. Let (X,d) be a metric space and k a positive integer. An operator $f : X^k \to X$ for which there exist some real constants $\delta_1, \ldots, \delta_k \in \mathbb{R}_+$ with $\sum_{i=1}^{k} \delta_i = \delta < 1$ and $L \geq 0$ such that

$$d(f(x_0, \ldots, x_{k-1}), f(x_1, \ldots, x_k)) \leq \sum_{i=1}^{k} \delta_i d(x_{i-1}, x_i) + \overline{M}(x_0, x_k), \qquad (2.5.92)$$

where

$$\overline{M}(x_0, x_k) = L \min\{d(x_0, f(x_0, \ldots, x_0)), d(x_k, f(x_k, \ldots, x_k)),$$
$$d(x_0, f(x_k, \ldots, x_k)), d(x_k, f(x_0, \ldots, x_0)), d(x_k, f(x_0, x_1, \ldots, x_{k-1}))\},$$

for any $x_0, \ldots, x_k \in X$, is called an *almost Prešić operator*.

It is easy to check that for $k = 1$ the terms $d(x_k, f(x_0, \ldots, x_0))$ and $d(x_k, f(x_0, x_1, \ldots, x_{k-1}))$ actually coincide, and Definition 2.5.3 reduces to the equivalent definition of strict almost contractions mentioned above. Also for $L = 0$, from the above condition (2.5.92), we obtain condition (2.5.85) which defines Prešić operators.

Remark 2.5.3. Considering f as in Definition 2.5.3 and its associate operator F, for any $x_0, \ldots, x_k \in X$, we have

$$
\begin{aligned}
\overline{M}(x_0, x_k) &= L\min\{d(x_0, f(x_0, \ldots, x_0)), d(x_k, f(x_k, \ldots, x_k)), \\
&\quad d(x_0, f(x_k, \ldots, x_k)), d(x_k, f(x_0, \ldots, x_0)), d(x_k, f(x_0, x_1, \ldots, x_{k-1}))\} \\
&\leq L\min\{d(x_0, f(x_0, \ldots, x_0)), d(x_k, f(x_k, \ldots, x_k)), d(x_0, f(x_k, \ldots, x_k)), \\
&\quad d(x_k, f(x_0, \ldots, x_0))\} \\
&= L\min\{d(x_0, F(x_0)), d(x_k, F(x_k)), d(x_0, F(x_k)), d(x_k, F(x_0))\}.
\end{aligned}
$$

In order to prove our main result we shall also need the following lemmas:

Lemma 2.5.1 [218]. *Let $k \in \mathbb{N}, k \neq 0$ and $\alpha_1, \alpha_2, \ldots, \alpha_k \in \mathbb{R}_+$ such that $\sum\limits_{i=1}^{k} \alpha_i = \alpha < 1$. If $\{\Delta_n\}_{n \geq 1}$ is a sequence of positive numbers satisfying*

$$\Delta_{n+k} \leq \alpha_1 \Delta_n + \alpha_2 \Delta_{n+1} + \ldots + \alpha_k \Delta_{n+k-1}, \quad n \geq 1, \tag{2.5.93}$$

then there exist $L > 0$ and $\theta \in (0,1)$ such that $\Delta_n \leq L \cdot \theta^n$ for all $n \geq 1$.

Lemma 2.5.2 [52]. *Let $\{a_n\}_{n \geq 0}$, $\{b_n\}_{n \geq 0}$ be two sequences of positive real numbers and $q \in (0,1)$ such that $a_{n+1} \leq q a_n + b_n, n \geq 0$ and $b_n \to 0$ as $n \to \infty$. Then $\lim\limits_{n \to \infty} a_n = 0$.*

In the following we shall prove the convergence of the Prešić type method constructed by means of almost Prešić operators, also providing the rate of convergence for this iterative procedure.

Theorem 2.5.3. *Let (X,d) be a complete metric space, k a positive integer and $f : X^k \to X$ an almost Prešić operator with constants $\delta_1, \ldots, \delta_k \in \mathbb{R}_+$, $\sum\limits_{i=1}^{k} \delta_i = \delta < 1$ and $L \geq 0$. Then*

(a) *f has a unique fixed point, say $x^* \in X$;*

(b) *the sequence $\{y_n\}_{n\geq 0}$ defined by*

$$y_n = f(y_{n-1}, \ldots, y_{n-1}), \quad n \geq 1,$$

converges to x^* for any starting point $y_0 \in X$;

(c) *the sequence $\{x_n\}_{n\geq 0}$ defined by $x_0, \ldots, x_{k-1} \in X$ and (2.5.84) also converges to x^*, with a rate estimated by*

$$d(x_n, x^*) \leq E_{n-k} + \delta d(x_{n-1}, x^*), \quad n \geq 0, \tag{2.5.94}$$

where

$$E_{n-k} := \delta_1 d(x_{n-k}, x_{n-k+1}) + (\delta_1 + \delta_2) d(x_{n-k+1}, x_{n-k+2})$$
$$+ \cdots + (\delta_1 + \cdots + \delta_{k-1}) d(x_{n-2}, x_{n-1}). \tag{2.5.95}$$

Proof. (a) and (b) We consider the associate operator $F : X \to X$ defined by $F(x) = f(x, \ldots, x)$, $x \in X$. For any $x, y \in X$ we have that:

$$
\begin{aligned}
d(F(x), F(y)) &= d(f(x, \ldots, x), f(y, \ldots, y)) \\
&\leq d(f(x, \ldots, x), f(x, \ldots, x, y)) + \cdots + d(f(x, y, \ldots, y), f(y, \ldots, y)).
\end{aligned}
$$

By (2.5.92) and Remark 2.5.3, this implies:

$$
\begin{aligned}
d(F(x), F(y)) &\leq \delta_k d(x, y) + L \min\{d(x, F(x)), d(y, F(y)), d(x, F(y)), d(y, F(x))\} \\
&+ \delta_{k-1} d(x, y) + L \min\{d(x, F(x)), d(y, F(y)), d(x, F(y)), d(y, F(x))\} \\
&+ \cdots + \\
&+ \delta_1 d(x, y) + L \min\{d(x, F(x)), d(y, F(y)), d(x, F(y)), d(y, F(x))\},
\end{aligned}
$$

which is equivalent to

$$d(F(x), F(y)) \leq \delta d(x, y) + + kL \min\{d(x, F(x)), d(y, F(y)), d(x, F(y)), d(y, F(x))\},$$

that is, F satisfies condition (2.5.91) above, with constants $\delta \in [0, 1)$ and $kL \geq 0$, so by Theorem 2.5.2 and Remark 2.5.2 above it has a unique fixed point, say $x^* \in X$, that can be obtained as the limit of the successive approximations of F starting from any $x \in X$.

Having in view the definition of F and considering the sequence of successive approximations of F, $\{y_n\}_{n\geq 0}$ defined by

$$y_n = F(y_{n-1}) = f(y_{n-1}, y_{n-1}, \ldots, y_{n-1}), \quad n \geq 1,$$

this leads exactly to conclusions (a) and (b).

(c) Now let us prove that the k-step iterative method $\{x_n\}_{n\geq 0}$ given by (2.5.84) converges to x^* as well.

Let $x_0, \ldots, x_{k-1} \in X$ and $x_n = f(x_{n-k}, \ldots, x_{n-1})$, $n \geq k$. Then

$$
\begin{aligned}
d(x_k, x^*) &= d(f(x_0, \ldots, x_{k-1}), f(x^*, \ldots, x^*)) \qquad (2.5.96) \\
&\leq d(f(x_0, \ldots, x_{k-1}), f(x_1, \ldots, x_{k-1}, x^*)) \\
&\quad + \cdots + d(f(x_{k-1}, x^*, \ldots, x^*), f(x^*, \ldots, x^*)).
\end{aligned}
$$

When applying (2.5.92) and Remark 2.5.3 for each term of the sum on the right-hand side of (2.5.96), we get

$$
\begin{aligned}
&d(f(x_0, \ldots, x_{k-1}), f(x_1, \ldots, x_{k-1}, x^*)) \\
&\leq \delta_1 d(x_0, x_1) + \cdots + \delta_{k-1} d(x_{k-2}, x_{k-1}) + \delta_k d(x_{k-1}, x^*) \\
&\quad + L\min\{d(x_0, F(x_0)), d(x^*, F(x^*)), d(x_0, F(x^*)), d(x^*, F(x_0))\}
\end{aligned}
$$

and so on,

$$
\begin{aligned}
&d(f(x_{k-2}, x_{k-1}, x^*, \ldots, x^*), f(x_{k-1}, x^*, \ldots, x^*)) \\
&\leq \delta_1 d(x_{k-2}, x_{k-1}) + \delta_2 d(x_{k-1}, x^*) + \\
&\quad + L\min\{d(x_{k-2}, F(x_{k-2})), d(x^*, F(x^*)), d(x_{k-2}, F(x^*)), d(x^*, F(x_{k-2}))\},
\end{aligned}
$$

respectively

$$
\begin{aligned}
&d(f(x_{k-1}, x^*, \ldots, x^*), f(x^*, \ldots, x^*)) \leq \delta_1 d(x_{k-1}, x^*) \\
&\quad + L\min\{d(x_{k-1}, F(x_{k-1})), d(x^*, F(x^*)), d(x_{k-1}, F(x^*)), d(x^*, F(x_{k-1}))\}. \quad (2.5.97)
\end{aligned}
$$

As $d(x^*, F(x^*)) = 0$, (2.5.96) finally leads to

$$
\begin{aligned}
d(x_k, x^*) &\leq \delta_1 d(x_0, x_1) + (\delta_1 + \delta_2) d(x_1, x_2) + \cdots + \\
&\quad + (\delta_1 + \cdots + \delta_{k-1}) d(x_{k-2}, x_{k-1}) + \delta d(x_{k-1}, x^*). \quad (2.5.98)
\end{aligned}
$$

Since k is a fixed positive integer, so are the coefficients δ_1, $\delta_1 + \delta_2$, $\ldots$, $\delta_1 + \cdots + \delta_{k-1}$, δ. Therefore we may denote

$$
E_0 := \delta_1 d(x_0, x_1) + (\delta_1 + \delta_2) d(x_1, x_2) + \cdots + (\delta_1 + \cdots + \delta_{k-1}) d(x_{k-2}, x_{k-1}),
$$

so (2.5.98) can be written as

$$
d(x_k, x^*) \leq E_0 + \delta d(x_{k-1}, x^*). \qquad (2.5.99)
$$

Similarly, we get

$$
\begin{aligned}
d(x_{k+1}, x^*) &\leq \delta_1 d(x_1, x_2) + (\delta_1 + \delta_2) d(x_2, x_3) + \cdots + \\
&\quad + (\delta_1 + \cdots + \delta_{k-1}) d(x_{k-1}, x_k) + \delta d(x_k, x^*). \quad (2.5.100)
\end{aligned}
$$

Denoting

$$
E_1 := \delta_1 d(x_1, x_2) + (\delta_1 + \delta_2) d(x_2, x_3) + \cdots + (\delta_1 + \cdots + \delta_{k-1}) d(x_{k-1}, x_k),
$$

inequality (2.5.100) can be written as

$$d(x_{k+1}, x^*) \leq E_1 + \delta d(x_k, x^*). \qquad (2.5.101)$$

In this manner we obtain, for $n \geq k$, that

$$d(x_n, x^*) \leq \delta_1 d(x_{n-k}, x_{n-k+1}) + (\delta_1 + \delta_2) d(x_{n-k+1}, x_{n-k+2})$$
$$+ \cdots + (\delta_1 + \cdots + \delta_{k-1}) d(x_{n-2}, x_{n-1}) + \delta d(x_{n-1}, x^*).$$

Denoting

$$E_{n-k} := \delta_1 d(x_{n-k}, x_{n-k+1}) + (\delta_1 + \delta_2) d(x_{n-k+1}, x_{n-k+2}) \qquad (2.5.102)$$
$$+ \cdots + (\delta_1 + \cdots + \delta_{k-1}) d(x_{n-2}, x_{n-1}),$$

inequality (2.5.98) becomes

$$d(x_n, x^*) \leq E_{n-k} + \delta d(x_{n-1}, x^*), \text{ for } n \geq k. \qquad (2.5.103)$$

In order to apply Lemma 2.5.2, we still have to prove that the sequence $\{E_n\}_{n \geq 0}$ given by

$$E_n = \delta_1 d(x_n, x_{n+1}) + (\delta_1 + \delta_2) d(x_{n+1}, x_{n+2}) + \cdots +$$
$$+ (\delta_1 + \cdots + \delta_{k-1}) d(x_{n+k-2}, x_{n+k-1}), \quad n \geq 0,$$

converges to 0 as $n \to \infty$.

For $n \geq k$, we have

$$d(x_n, x_{n+1}) = d(f(x_{n-k}, \ldots, x_{n-1}), f(x_{n-k+1}, \ldots, x_n)). \qquad (2.5.104)$$

By (2.5.92) this yields

$$d(x_n, x_{n+1}) \leq \delta_1 d(x_{n-k}, x_{n-k+1}) + \cdots + \delta_k d(x_{n-1}, x_n) +$$
$$+ L \min\{d(x_{n-k}, F(x_{n-k})), d(x_n, F(x_n)), d(x_{n-k}, F(x_n)),$$
$$d(x_n, F(x_{n-k})), d(x_n, f(x_{n-k}, \ldots, x_{n-1}))\}. \qquad (2.5.105)$$

As $d(x_n, f(x_{n-k}, \ldots, x_{n-1})) = 0$, (2.5.104) finally leads to

$$d(x_n, x_{n+1}) \leq \delta_1 d(x_{n-k}, x_{n-k+1}) + \cdots + \delta_k d(x_{n-1}, x_n), \text{ for } n \geq k. \qquad (2.5.106)$$

According to Lemma 2.5.1, this implies the existence of $\theta \in (0, 1)$ and $K \geq 0$ such that

$$d(x_n, x_{n+1}) \leq K\theta^{n+k}, n \geq 0.$$

Since k is fixed, it is evident that the sequence $\{E_n\}_{n \geq 0}$ converges to 0 as $n \to \infty$.

Denoting $\overline{E}_n := E_{n-k}$, (2.5.103) is written as:

$$d(x_n, x^*) \leq \overline{E}_n + \delta d(x_{n-1}, x^*). \tag{2.5.107}$$

Now taking $a_n = d(x_n, x^*)$, $n \geq k$ and $b_n = \overline{E}_n$, $n \geq k$ in Lemma 2.5.2, by (2.5.107) it follows that $d(x_n, x^*) \to 0$ as $n \to \infty$, that is, the multistep iterative method $\{x_n\}_{n \geq 0}$ converges to x^*, the unique fixed point of f.   □

Remark 2.5.4. Note that for $L = 0$, from Theorem 2.5.3, we get the result due to Prešić [218], while for $k = 1$, Theorem 2.2.2 for strict almost contractions in metric spaces is obtained. For $k = 1$ and $L = 0$, Theorem 2.5.3 reduces to the contraction mapping principle (Theorem 2.1.1).

Other results and remarks concerning almost Prešić operators can be found in Ref. [203].

In Section 2.4 of this chapter the existence of coincindence points and common fixed points for single-valued almost contractions was discussed. In the following we shall present some common fixed point results for pairs of mappings where at least one of them is an almost Prešić operator. These were introduced in Ref. [204].

We shall start with the case where only one of the mappings is defined on a product space. In this respect the following extensions of classical notions and results, given in Section 2.4 above (see also Ref. [204]), have to be considered:

Definition 2.5.4. Let X be a nonempty set, k a positive integer and $f : X^k \to X$, $g : X \to X$ two operators.

An element $p \in X$ is called a *coincidence point* of f and g if it is a coincidence point of F and g, where F is defined by $F(x) = f(x, x, \ldots, x)$.

Similarly, $s \in X$ is a *coincidence value* of f and g if it is a coincidence value of F and g.

An element $p \in X$ is a *common fixed point* of f and g if it is a common fixed point of F and g.

Definition 2.5.5. Let X be a nonempty set, k a positive integer and $f : X^k \to X$, $g : X \to X$. The operators f and g are said to be *weakly compatible* if F and g are weakly compatible.

The following result is a generalization of Proposition 1.4 in Ref. [1], included in the previous section as Proposition 2.4.1. For its proof see Ref. [204].

Lemma 2.5.3. *Let X be a nonempty set, k a positive integer and $f : X^k \to X$, $g : X \to X$*

two weakly compatible operators. If f and g have a unique coincidence value $x^* = f(p, \ldots, p) = g(p)$*, then* x^* *is the unique common fixed point of f and g.*

Definition 2.5.2 can be extended as follows:

Definition 2.5.6. Let (X,d) be a metric space, k a positive integer, $\delta_1, \ldots, \delta_k \in \mathbb{R}_+$, with $\sum\limits_{i=1}^{k} \delta_i = \delta < 1$ and $L \geq 0$ constants and $f : X^k \to X$, $g : X \to X$ two operators satisfying:

$$d(f(x_0, \ldots, x_{k-1}), f(x_1, \ldots, x_k)) \leq \sum_{i=1}^{k} \delta_i d(g(x_{i-1}), g(x_i)) + \overline{M_g}(x_0, x_k), \quad (2.5.108)$$

for any $x_0, x_1, \ldots, x_k \in X$, where

$$\begin{aligned}
\overline{M_g}(x_0, x_k) \quad = \quad & L \min\{d(g(x_0), f(x_0, \ldots, x_0)), d(g(x_k)), f(x_k, \ldots, x_k), \\
& d(g(x_k)), f(x_0, \ldots, x_0)), d(g(x_0), f(x_k, \ldots, x_k)), \\
& d(g(x_k), f(x_0, \ldots, x_{k-1}))\}.
\end{aligned}$$

Then f is said to be an *almost Prešić operator w. r. t. g*.

Remark 2.5.5. For any $x_0, x_1, \ldots, x_k \in X$ we have that

$$\begin{aligned}
\overline{M_g}(x_0, x_k) \quad \leq \quad & L \min\{d(g(x_0), f(x_0, \ldots, x_0)), d(g(x_k)), f(x_k, \ldots, x_k), \\
& d(g(x_k)), f(x_0, \ldots, x_0)), d(g(x_0), f(x_k, \ldots, x_k))\},
\end{aligned}$$

that is,

$$\begin{aligned}
\overline{M_g}(x_0, x_k) \quad \leq \quad & L \min\{d(g(x_0), F(x_0)), d(g(x_k)), F(x_k)), \\
& d(g(x_k)), F(x_0)), d(g(x_0), F(x_k))\}.
\end{aligned}$$

Remark 2.5.6. Considering Remarks 2.5.2 and 2.5.5 above, it is easy to see that for $k = 1$ Definition 2.5.6 reduces to the definition of strict almost contractions. Besides, for $g = 1_X$, f is an almost Prešić operator – see Ref. [197] for more results.

A general common fixed point result regarding almost Prešić operators is presented below:

Theorem 2.5.4. *Let* (X,d) *be a metric space,* k *a positive integer and* $f : X^k \to X$, $g : X \to X$ *two operators such that:*

(i) *f is an almost Prešić operator with respect to g;*

(ii) *there exists a complete subspace $Y \subseteq X$ such that $f(X^k) \subseteq Y \subseteq g(X)$.*

Then

(a) *f and g have a unique coincidence value, say $x^* \in X$;*

(b) *the sequence $\{g(z_n)\}_{n \geq 0}$ defined by $z_0 \in X$ and*

$$g(z_n) = f(z_{n-1}, \ldots, z_{n-1}), \quad n \geq 1, \tag{2.5.109}$$

   converges to x^*;

(c) *the sequence $\{g(x_n)\}_{n \geq 0}$ defined by $x_0, \ldots, x_{k-1} \in X$ and*

$$g(x_n) = f(x_{n-k}, \ldots, x_{n-1}), \quad n \geq k, \tag{2.5.110}$$

   converges to x^* as well, with a rate estimated by

$$d(g(x_n), x^*) \leq E_{n-k} + \delta d(g(x_{n-1}), x^*), \tag{2.5.111}$$

   where E_{n-k} is given by (2.5.127);

(d) *if in addition f and g are weakly compatible, then x^* is their unique common fixed point.*

Proof. (a) and (b) Let $z_0 \in X$. Then $f(z_0, \ldots, z_0) \in f(X^k) \subset g(X)$, so there exists $z_1 \in X$ such that

$$f(z_0, \ldots, z_0) = g(z_1).$$

Similarly, $f(z_1, \ldots, z_1) \in f(X^k) \subset g(X)$, so there exists $z_2 \in X$ such that

$$f(z_1, \ldots, z_1) = g(z_2).$$

In this manner we construct the sequence $\{g(z_n)\}_{n \geq 0}$ with $z_0 \in X$ and

$$g(z_n) = f(z_{n-1}, \ldots, z_{n-1}), n \geq 1. \tag{2.5.112}$$

Due to the manner $\{g(z_n)\}_{n \geq 0}$ was constructed, it is easy to recognize that

$$\{g(z_n)\}_{n \geq 0} \subseteq f(X^k) \subseteq Y \subseteq g(X). \tag{2.5.113}$$

For $n \geq 1$, we have

$$\begin{aligned} d(g(z_n), g(z_{n+1})) &= d(f(z_{n-1}, \ldots, z_{n-1}), f(z_n, \ldots, z_n)) \tag{2.5.114} \\ &\leq d(f(z_{n-1}, \ldots, z_{n-1}), f(z_{n-1}, \ldots, z_{n-1}, z_n)) \\ &\quad + \cdots + d(f(z_{n-1}, z_n, \ldots, z_n), f(z_n, \ldots, z_n)). \end{aligned}$$

Applying relation (2.5.108) and then Remark 2.5.5 to each of the distances on the right-hand side of (2.5.114), we obtain

$$d(f(z_{n-1},\ldots,z_{n-1}),f(z_{n-1},\ldots,z_{n-1},z_n)) \tag{2.5.115}$$
$$\leq \delta_k d(g(z_{n-1}),g(z_n)) + L\min\{d(g(z_{n-1}),F(z_{n-1})),d(g(z_n),F(z_n)),$$
$$d(g(z_{n-1}),F(z_n)),d(g(z_n),F(z_{n-1}))\}$$

and so on,

$$d(f(z_{n-1},z_n,\ldots,z_n),f(z_n,\ldots,z_n)) \tag{2.5.116}$$
$$\leq \delta_l d(g(z_{n-1}),g(z_n)) + L\min\{d(g(z_{n-1}),F(z_{n-1})),d(g(z_n),F(z_n)),$$
$$d(g(z_{n-1}),F(z_n)),d(g(z_n),F(z_{n-1}))\}.$$

As $d(g(z_{n-1}),F(z_n)) = d(g(z_{n-1}),f(z_n,\ldots,z_n)) = 0$, (2.5.114) finally becomes

$$d(g(z_n),g(z_{n+1})) \leq \delta d(g(z_{n-1}),g(z_n)). \tag{2.5.117}$$

By induction we get that

$$d(g(z_n),g(z_{n+1})) \leq \delta^n d(g(z_0),g(z_1)), \quad n \geq 0.$$

Using the triangle inequality, for $p \geq 1$ we obtain that:

$$d(g(z_n),g(z_{n+p})) \leq \delta^n \frac{1-\delta^p}{1-\delta} d(g(z_0),g(z_1)), \quad n \geq 0, \tag{2.5.118}$$

where $\delta \in [0,1)$.

Letting $n \to \infty$ in (2.5.118), we find that $\{g(z_n)\}_{n\geq 0}$ is a Cauchy sequence included, by (2.5.113), in the complete subspace Y. Consequently, there exists $x^* \in Y$ such that $x^* = \lim\limits_{n\to\infty} g(z_n)$ and, since $Y \subset g(X)$, there exists $p \in X$ such that

$$g(p) = x^* = \lim\limits_{n\to\infty} g(z_n). \tag{2.5.119}$$

Now we shall prove that $f(p,\ldots,p) = x^*$. We have that

$$d(g(z_{n+1}),f(p,\ldots,p)) = d(f(z_n,\ldots,z_n),f(p,\ldots,p))$$
$$\leq d(f(z_n,\ldots,z_n),f(z_n,\ldots,z_n,p)) + \cdots + d(f(z_n,p,\ldots,p),f(p,\ldots,p)). \tag{2.5.120}$$

It is obvious that the minimum among five quantities is less or equal to any of these quantities, which we may conveniently choose.

Thus, when applying (2.5.108) to the distances on the right-hand side of (2.5.120), each time we choose this quantity to be $d(g(p),f(z_n,\ldots,z_n))$. In this manner (2.5.120) becomes:

$$d(g(z_{n+1}),f(p,\ldots,p)) \leq \delta d(g(z_n),g(p)) + kLd(g(p),f(z_n,\ldots,z_n)),$$

i.e.,

$$d(g(z_{n+1}), f(p, \ldots, p)) \leq \delta d(g(z_n), x^*) + kLd(x^*, g(z_{n+1})), \quad n \geq 0.$$

Letting $n \longrightarrow \infty$, by (2.5.119) it immediately follows that

$$f(p, \ldots, p) = x^* = g(p),$$

that is, x^* is a coincidence value for f and g.

In order to prove its uniqueness, we suppose there would be some $q \in X$ such that

$$f(q, \ldots, q) = g(q) \neq x^*.$$

Then

$$d(g(p), g(q)) = d(f(p, \ldots, p), f(q, \ldots, q))$$
$$\leq d(f(p, \ldots, p), f(p, \ldots, p, q)) + \cdots + d(f(p, q, \ldots, q), f(q, \ldots, q)). \quad (2.5.121)$$

We now use a similar reasoning as before, applying (2.5.108) to each of the distances on the right-hand side of inequality (2.5.121). This time we choose the minimum less or equal to $d(g(p), f(p, \ldots, p)) = d(g(q), f(q, \ldots, q)) = 0$. Thus (2.5.121) becomes:

$$d(g(p), g(q)) \leq \delta d(g(p), g(q)),$$

or

$$(1 - \delta)d(g(p), g(q)) \leq 0.$$

As $\delta \in [0, 1)$, this implies that $d(g(p), g(q)) = 0$, so x^* is the unique coincidence value of f and g.

(c) Let $x_0, \ldots, x_{k-1} \in X$. Then $f(x_0, \ldots, x_{k-1}) \in f(X^k) \subset g(X)$, so there exists $x_k \in X$ such that

$$f(x_0, \ldots, x_{k-1}) = g(x_k).$$

Similarly, $f(x_1, \ldots, x_k) \in f(X^k) \subset g(X)$, so there exists $x_{k+1} \in X$ such that

$$f(x_1, \ldots, x_k) = g(x_{k+1}).$$

In this manner we construct the sequence $\{g(x_n)\}_{n \geq 0}$ defined by $x_0, \ldots, x_{k-1} \in X$ and

$$g(x_n) = f(x_{n-k}, \ldots, x_{n-1}), \quad n \geq k.$$

Again we notice that, by construction,

$$\{g(x_n)\}_{n \geq 0} \subset f(x^k) \subset g(X). \quad (2.5.122)$$

We shall prove that $\{g(x_n)\}_{n\geq 0}$ converges to x^* as well. We have that:

$$\begin{aligned} d(g(x_k),x^*) &= d(f(x_0,\ldots,x_{k-1}),f(p,\ldots,p)) \quad &(2.5.123)\\ &\leq d(f(x_0,\ldots,x_{k-1}),f(x_1,\ldots,x_{k-1},p))+\cdots+\\ &\quad +d(f(x_{k-1},p,\ldots,p),f(p,\ldots,p)). \end{aligned}$$

As $d(g(p),f(p,\ldots,p))=0$, by applying (2.5.108) to each of the distances on the right-hand side of the above inequality (2.5.123), it follows that

$$d(g(x_k),x^*) \leq \delta_1 d(g(x_0),g(x_1))+(\delta_1+\delta_2)d(g(x_1),g(x_2))+\ldots+$$
$$+(\delta_1+\ldots+\delta_{k-1})d(g(x_{k-2}),g(x_{k-1}))+\delta d(g(x_{k-1}),g(p)). \quad (2.5.124)$$

Since k is a fixed positive integer, the coefficients δ_1, $\delta_1+\delta_2$, $\ldots$, $\delta_1+\cdots+\delta_{k-1}$, δ are also fixed, so we may denote

$$E_0 := \delta_1 d(g(x_0),g(x_1))+(\delta_1+\delta_2)d(g(x_1),g(x_2))+\ldots+$$
$$+(\delta_1+\ldots+\delta_{k-1})d(g(x_{k-2}),g(x_{k-1})),$$

and (2.5.124) can be written as

$$d(g(x_k),x^*) \leq E_0+\delta d(g(x_{k-1}),x^*). \quad (2.5.125)$$

In the same manner, we obtain

$$d(g(x_{k+1}),x^*) \leq E_1+\delta d(g(x_k),x^*),$$

where

$$E_1 := \delta_1 d(g(x_1),g(x_2))+(\delta_1+\delta_2)d(g(x_2),g(x_3))+\ldots+$$
$$+(\delta_1+\ldots+\delta_{k-1})d(g(x_{k-1}),g(x_k)).$$

By induction, for $n \geq k$ we obtain:

$$d(g(x_n),x^*) \leq E_{n-k}+\delta d(g(x_{n-1}),x^*), \quad (2.5.126)$$

where

$$E_{n-k} := \delta_1 d(g(x_{n-k}),g(x_{n-k+1}))+(\delta_1+\delta_2)d(g(x_{n-k+1}),g(x_{n-k+2}))+$$
$$+\ldots+(\delta_1+\ldots+\delta_{k-1})d(g(x_{n-2}),g(x_{n-1})), \quad n\geq k. \quad (2.5.127)$$

Inequality (2.5.126) leads us to the estimation of the rate of convergence (2.5.111).

In order to apply Lemma 2.5.2, we have to prove that $\{E_n\}_{n\geq 0}$, defined by

$$E_n := \delta_1 d(g(x_n),g(x_{n+1}))+(\delta_1+\delta_2)d(g(x_{n+1}),g(x_{n+2}))$$
$$+\ldots+(\delta_1+\ldots+\delta_{k-1})d(g(x_{n+k-2}),g(x_{n+k-1})), \quad n\geq 0, \quad (2.5.128)$$

converges to 0. For $n \geq k$ we have that

$$d(g(x_n), g(x_{n+1})) = d(f(x_{n-k}, \ldots, x_{n-1}), f(x_{n-k+1}, \ldots, x_n)),$$

which by (2.5.108) yields

$$
\begin{aligned}
d(g(x_n), g(x_{n+1})) \leq{}& \delta_1 d(g(x_{n-k}), g(x_{n-k+1})) + \cdots + \delta_k d(g(x_{n-1}), g(x_n)) \\
&+ L \min \{ d(g(x_{n-k}), f(x_{n-k}, \ldots, x_{n-k})), d(g(x_n), f(x_n, \ldots, x_n)), \\
&\quad d(g(x_{n-k}), f(x_n, \ldots, x_n)), d(g(x_n), f(x_{n-k}, \ldots, x_{n-k})), \\
&\quad d(g(x_n), f(x_{n-k}, x_{n-k+1} \ldots, x_{n-1})) \} .
\end{aligned}
$$

Since $d(g(x_n), f(x_{n-k}, \ldots, x_{n-1})) = 0$, it follows that

$$d(g(x_n), g(x_{n+1})) \leq \delta_1 d(g(x_{n-k}), g(x_{n-k+1})) + \cdots + \delta_k d(g(x_{n-1}), g(x_n)).$$

By Lemma 2.5.1 this implies the existence of $\theta \in (0,1)$ and $K \geq 0$ such that

$$d(g(x_n), g(x_{n+1})) \leq K \theta^n, \quad n \geq k.$$

It is then immediate that the sequence $\{E_n\}_{n \geq 0}$ converges to 0 as $n \to \infty$. Denoting $\overline{E}_n = E_{n-k}$, from (2.5.126) we get:

$$d(g(x_n), x^*) \leq \overline{E}_n + \delta d(g(x_{n-1}), x^*). \tag{2.5.129}$$

Now taking $a_n = d(g(x_n), x^*)$, $n \geq k$, and $b_n = \overline{E}_n$, $n \geq k$, by (2.5.129) and Lemma 2.5.2 it follows that

$$d(g(x_n), x^*) \to 0, \text{ as } n \to \infty,$$

so $\{g(x_n)\}_{n \geq 0}$ converges to the unique coincidence value x^* as well.

(d) If f and g are weakly compatible, then by Lemma 2.5.3 their unique coincidence value is actually their unique common fixed point. □

Going further, we can establish common fixed point results for the more general case $f : X^k \to X$ and $g : X^l \to X$, with k and l positive integers. In this respect we have to extend the notions mentioned above (see also Refs. [196, 200, 204]).

Definition 2.5.7. Let X be a metric space, k, l positive integers and $f : X^k \to X$, $g : X^l \to X$ two operators.

An element $p \in X$ is called a *coincidence point* of f and g if it is a coincidence point of F and G, where $F, G : X \to X$ are the associate operators of f and g respectively (see Remark 2.5.1).

An element $s \in X$ is called a *coincidence value* of f and g if it is a coincidence value of F and G.

An element $p \in X$ is called a *common fixed point* of f and g if it is a common fixed point of F and G.

Definition 2.5.8. Let (X,d) be a metric space, k,l positive integers and $f : X^k \to X$, $g : X^l \to X$. The operators f and g are said to be *weakly compatible* if F and G are weakly compatible.

The following extends Definition 2.5.6:

Definition 2.5.9. Let (X,d) be a metric space, k,l positive integers and $f : X^k \to X$, $g : X^l \to X$ two operators such that f is an almost Prešić operator w.r.t. $G : X \to X$, the associated operator of g. Then f is said to be an *almost Prešić operator w.r.t. g.*

Using this definition one can prove the following extension of Theorem 2.5.4:

Theorem 2.5.5. *Let (X,d) be a metric space, k and l positive integers and $f : X^k \to X$, $g : X^l \to X$ two operators such that:*

(i) *f is an almost Prešić operator with respect to g;*

(ii) *there exists a complete subspace $Y \subseteq X$ such that $f(X^k) \subseteq Y \subseteq G(X)$, where $G : X \to X$ is the associated operator of g.*

Then

(a) *f and g have a unique coincidence value, say x^*, in X;*

(b) *the sequence $\{G(z_n)\}_{n\geq 0}$ defined by $z_0 \in X$ and*
$$G(z_n) = f(z_{n-1}, \ldots, z_{n-1}), \quad n \geq 1,$$
converges to x^*;

(c) *the sequence $\{G(x_n)\}_{n\geq 0}$ defined by $x_0, \ldots, x_{k-1} \in X$ and*
$$G(x_n) = f(x_{n-k}, \ldots, x_{n-1}), \quad n \geq k,$$
converges to x^* as well, with a rate estimated by
$$d(G(x_n), x^*) \leq \overline{E}_n + \delta d(G(x_{n-1}), x^*),$$
where
$$\overline{E}_n := \delta_1 d(G(x_{n-k}), G(x_{n-k+1})) + (\delta_1 + \delta_2)d(G(x_{n-k+1}), G(x_{n-k+2}))$$
$$+ \ldots + (\delta_1 + \ldots + \delta_{k-1})d(G(x_{n-2}), G(x_{n-1})), \quad n \geq k;$$

(d) *if in addition f and g are weakly compatible, then x^* is their unique common fixed point.*

For the study of stability of k-step fixed point iterative schemes associated with contractive type mappings defined on product spaces, see Ref. [75]. For some other developments on Prešić operators, see a partial list in Ref. [76].

2.6. Fixed Point Theorems for Single-valued Nonself Almost Contractions

In a natural continuation and completion of the abundant fixed point theory for self mappings, produced in the last five decades, it was also an important and challenging research topic to obtain fixed point theorems for *nonself mappings*.

In 1972, Assad and Kirk [38] extended the Banach contraction mapping principle to nonself multivalued contraction mappings $T : K \to \mathscr{P}(X)$ in the case where (X,d) is a convex metric space in the sense of Menger and K is a nonempty closed subset of X such that T maps ∂K (the boundary of K) into K. In 1976, by using an alternative and weaker condition, i.e., T is metrically inward, Caristi [82] has shown that any nonself single-valued contraction has a fixed point.

Next, in 1978, Rhoades [228] proved a fixed point result in Banach spaces for single-valued nonself mapping satisfying the following contraction condition:

$$d(Tx,Ty) \le \lambda \max \left\{ \frac{d(x,y)}{2}, d(x,Tx), d(y,Ty), \frac{d(x,Ty)+d(y,Tx)}{1+2\lambda} \right\}, \quad (2.6.130)$$

for all $x,y \in K$, where $0 < \lambda < 1$.

Rhoades' result [228] has been slightly extended by Ćirić [98]. Note that although the class of mappings satisfying (2.6.130) is large enough to include some discontinuous mappings, it does not include contraction mappings satisfying (2.1.1) for $\frac{1}{2} \le \lambda < 1$.

A more general result, which also solved a very difficult problem that was open for more than 20 years, has been obtained by Ćirić [99], who considered instead of (2.6.130) the quasi-contraction condition that he previously introduced and studied in Ref. [96]:

$$d(Tx,Ty) \le \lambda \max \left\{ d(x,y), d(x,Tx), d(y,Ty), d(x,Ty), d(y,Tx) \right\}, \quad (2.6.131)$$

for all $x,y \in K$, where $0 < \lambda < 1$. More recently, Ćirić, Ume, Khan and Pathak [100] have considered a contraction condition which is more general than (2.6.130) and (2.6.131), i.e.,

$$d(Tx,Ty) \le \max \left\{ \varphi(d(x,y)), \varphi(d(x,Tx)), \varphi(d(y,Ty)), \varphi(d(x,Ty)), \varphi(d(y,Tx)) \right\}, \quad (2.6.132)$$

for all $x,y \in K$, where $\varphi : \mathbb{R}_+ \to \mathbb{R}_+$ is a certain comparison function.

For some other fixed point results for nonself mappings, see also Refs. [33–37, 71] and Problem 5 in Ref. [238].

As shown in Section 2.2, quasi-contractions and almost contractions are independent classes of mappings as the latter have a unique fixed point, while the former do not.

Starting from these facts, the aim of the present section is to obtain fixed point theorems for nonself almost contractions. Thus, we shall give a solution to Problem 5 in Ref. [238] in the case of almost contractions. The material is adapted from Ref. [74].

In order to do so, we first present some aspects and results related to self almost contractions and then we extend them to nonself almost contractions.

Let X be a Banach space, K a nonempty closed subset of X and $T : K \to X$ a nonself mapping. If $x \in K$ is such that $Tx \notin K$, then we can always choose an $y \in \partial K$ (the boundary of K) such that $y = (1 - \lambda)x + \lambda Tx (0 < \lambda < 1)$, which actually expresses the fact that

$$d(x, Tx) = d(x, y) + d(y, Tx), \quad \text{for } y \in \partial K, \qquad (2.6.133)$$

where we denoted $d(x, y) = \|x - y\|$.

In general, the set Y of points y satisfying condition (2.6.133) above may contain more than one element.

In this context we shall need the following concept.

Definition 2.6.1. Let X be a Banach space, K a nonempty closed subset of X and $T : K \to X$ a nonself mapping. Let $x \in K$ with $Tx \notin K$ and let $y \in \partial K$ be the corresponding elements given by (2.6.133). If, for any such elements x, we have

$$d(y, Ty) \le d(x, Tx), \qquad (2.6.134)$$

for all corresponding $y \in Y$, then we say that T has property (M).

Very recently we found that a condition quite similar to (2.6.134) had been used in Ref. [126] (see also Ref. [83]).

Note also that the nonself mapping T in the next example has property (M).

Example 2.6.1. Let X be the set of real numbers with the usual metric, $K = [0, 1]$, and let $T : K \to X$ be defined (see Ref. [100, Remark 1.3]) by $Tx = -0.1$ if $x = 0.9$ and $Tx = \frac{x}{x+1}$ if $x \ne 0.9$.

Then T satisfies condition (2.6.132), T is discontinuous, 0 is the unique fixed point of T and T is continuous at 0, T has property (M) but T does not satisfy the almost contraction condition (2.6.135) below. Indeed, the only $x \in K$ with $Tx \notin K$ is $x = 0.9$ and the corresponding $y \in \partial K$ is $y = 0$. It is now easy to check that (2.6.134) holds.

To prove the last claim take $x \neq 0.9$ and $y = \frac{x}{1+x}$ in (2.6.135) to get, for any $x > 0$,

$$\frac{1+x}{1+2x} \leq \delta < 1, \quad x > 0.$$

If we now let $x \to 0$ in the previous double inequality, we get the contradiction

$$1 \leq \delta < 1.$$

We now state and prove our main result in this section, which is taken from Ref. [74].

Theorem 2.6.1. *Let X be a Banach space, K a nonempty closed subset of X and $T : K \to X$ a nonself almost contraction, that is, a mapping for which there exist two constants $\delta \in [0,1)$ and $L \geq 0$ such that*

$$d(Tx, Ty) \leq \delta \cdot d(x,y) + Ld(y, Tx), \quad \text{for all } x, y \in K. \tag{2.6.135}$$

If T has property (M) and satisfies Rothe's boundary condition

$$T(\partial K) \subset K, \tag{2.6.136}$$

then T has a fixed point in K.

Proof. If $T(K) \subset K$, then T is actually a self mapping on the closed set K and the conclusion follows by Theorem 2.2.1 for $X = K$. Therefore, we consider the case $T(K) \not\subset K$. Let $x_0 \in \partial K$. By (2.6.136) we know that $Tx_0 \in K$. Denote $x_1 = Tx_0$. Now, if $Tx_1 \in K$, set $x_2 = Tx_1$. If $Tx_1 \notin K$, we can choose an element x_2 on the segment $[x_1, Tx_1]$ which also belongs to ∂K, that is,

$$x_2 = (1 - \lambda)x_1 + \lambda Tx_1 \quad (0 < \lambda < 1).$$

Continuing in this way we obtain a sequence $\{x_n\}$ whose terms satisfy one of the following properties:

(i) $x_n = Tx_{n-1}$, if $Tx_{n-1} \in K$;

(ii) $x_n = (1 - \lambda)x_{n-1} + \lambda Tx_{n-1} \in \partial K \, (0 < \lambda < 1)$, if $Tx_{n-1} \notin K$.

To simplify the argumentation in the proof, let us denote

$$P = \{x_k \in \{x_n\} : x_k = Tx_{k-1}\}$$

and

$$Q = \{x_k \in \{x_n\} : x_k \neq Tx_{k-1}\}.$$

Note that $\{x_n\} \subset K$ and that, if $x_k \in Q$, then both x_{k-1} and x_{k+1} belong to the set P. Moreover, by virtue of (2.6.136), we cannot have two consecutive terms of $\{x_n\}$ in the set Q (but we can have two consecutive terms of $\{x_n\}$ in the set P) .

We claim that $\{x_n\}$ is a Cauchy sequence. To prove this, we must discuss three different cases:

CASE I. $x_n, x_{n+1} \in P$;

CASE II. $x_n \in P$, $x_{n+1} \in Q$;

CASE III. $x_n \in Q$, $x_{n+1} \in P$.

CASE I. $x_n, x_{n+1} \in P$.

In this case we have $x_n = Tx_{n-1}$, $x_{n+1} = Tx_n$ and by (2.6.135), we get

$$d(x_{n+1}, x_n) = d(Tx_n, Tx_{n-1}) \leq \delta d(x_n, x_{n-1}) + Ld(x_n, Tx_{n-1}),$$

that is,

$$d(x_{n+1}, x_n) \leq \delta d(x_n, x_{n-1}), \tag{2.6.137}$$

since $x_n = Tx_{n-1}$.

CASE II. $x_n \in P$, $x_{n+1} \in Q$.

In this case we have $x_n = Tx_{n-1}$ but $x_{n+1} \neq Tx_n$ and

$$d(x_n, x_{n+1}) + d(x_{n+1}, Tx_n) = d(x_n, Tx_n).$$

Hence

$$d(x_n, x_{n+1}) \leq d(x_n, Tx_n) = d(Tx_{n-1}, Tx_n)$$

and so by using (2.6.135), we get

$$d(x_n, x_{n+1}) \leq \delta d(x_n, x_{n-1}) + Ld(x_n, Tx_{n-1}) = \delta d(x_n, x_{n-1}),$$

which yields again inequality (2.6.137).

CASE III. $x_n \in Q$, $x_{n+1} \in P$.

In this situation, we have $x_{n-1} \in P$. Having in view that T has property (M), it follows that

$$d(x_n, x_{n+1}) = d(x_n, Tx_n) \leq d(x_{n-1}, Tx_{n-1}).$$

Since $x_{n-1} \in P$ we have $x_{n-1} = Tx_{n-2}$ and by (2.6.135) we get

$$d(Tx_{n-2}, Tx_{n-1}) \leq \delta d(x_{n-2}, x_{n-1}) + Ld(x_{n-1}, Tx_{n-2}) = \delta d(x_{n-2}, x_{n-1}),$$

which shows that

$$d(x_n, x_{n+1}) \leq \delta d(x_{n-2}, x_{n-1}). \tag{2.6.138}$$

Therefore, by summarizing all three cases and using (2.6.137) and (2.6.138), it follows that the sequence $\{x_n\}$ satisfies the inequality

$$d(x_n, x_{n+1}) \leq \delta \max\{d(x_{n-2}, x_{n-1}), d(x_{n-1}, x_n)\}, \quad \text{for all } n \geq 2. \tag{2.6.139}$$

Now, by induction for $n \geq 2$, from (2.6.139) one obtains

$$d(x_n, x_{n+1}) \leq \delta^{[n/2]} \max\{d(x_0, x_1), d(x_1, x_2)\},$$

where $[n/2]$ denotes the greatest integer not exceeding $n/2$.

Further, for $m > n > N$,

$$d(x_n, x_m) \leq \sum_{i=N}^{\infty} d(x_i, x_{i-1}) \leq 2\frac{\delta^{[N/2]}}{1-\delta} \max\{d(x_0, x_1), d(x_1, x_2)\},$$

which shows that $\{x_n\}$ is a Cauchy sequence.

Since $\{x_n\} \subset K$ and K is closed, $\{x_n\}$ converges to some point in K.

Denote

$$x^* = \lim_{n\to\infty} x_n, \tag{2.6.140}$$

and let $\{x_{n_k}\} \subset P$ be an infinite subsequence of $\{x_n\}$ (such a subsequence always exists) that we denote in the following for simplicity by $\{x_n\}$ too.

Then

$$d(x^*, Tx^*) \leq d(x^*, x_{n+1}) + d(x_{n+1}, Tx^*) = d(x_{n+1}, x^*) + d(Tx_n, Tx^*).$$

By (2.6.135), we have

$$d(Tx_n, Tx^*) \leq \delta d(x_n, x^*) + Ld(x^*, Tx_n),$$

and hence

$$d(x^*, Tx^*) \leq (1+L)d(x^*, x_{n+1}) + \delta \cdot d(x_n, x^*), \quad \text{for all } n \geq 0. \tag{2.6.141}$$

Letting $n \to \infty$ in (2.6.141) we obtain

$$d(x^*, Tx^*) = 0,$$

which shows that x^* is a fixed point of T. $\qquad\square$

Remark 2.6.1. Note that although T satisfying (2.6.135) may be discontinuous (see Example 2.6.2), T is also continuous at the fixed point. Indeed, if $\{y_n\}$ is a sequence in K convergent to $x^* = Tx^*$, then by (2.6.135) we have

$$d(Ty_n, x^*) = d(Tx^*, Ty_n) \leq \delta d(x^*, y_n) + Ld(y_n, Tx^*),$$

and letting $n \to \infty$ in the previous inequality, we get exactly the continuity of T at the fixed point x^*:

$$d(Ty_n, x^*) \to 0 \text{ as } n \to \infty, \text{ that is, } Ty_n \to x^*.$$

Example 2.6.2. Let X be the set of real numbers with the usual norm, $K = [0, 1]$ be the unit interval and let $T : [0,1] \to \mathbb{R}$ be given by $Tx = \frac{2}{3}x$ for $x \in [0,1/2)$, $T\left(\frac{1}{2}\right) = -1$, and $Tx = \frac{2}{3}x + \frac{1}{3}$, for $x \in (1/2,1]$.

As T has two fixed points, that is, $Fix(T) = \{0,1\}$, it does not satisfy either of Ćirić's conditions (2.6.131) and (2.6.132), or the Banach, Kannan, Chatterjea, Zamfirescu or Ćirić [94] contractive conditions in the corresponding nonself form, but T satisfies the contraction condition (2.6.135).

Indeed, for the cases

(1) $x \in [0,1/2)$, $y \in (1/2,1]$;
(2) $y \in [0,1/2)$, $x \in (1/2,1]$;
(3) $x,y \in [0,1/2)$; and
(4) $x,y \in (1/2,1]$,

we have by Example 1.3.10 in Ref. [197, pp. 28–29] that (2.6.135) is satisfied with $\delta = 2/3$ and $L \geq 6$.

We have to cover the remaining four cases:

(5) $x = 1/2$, $y \in [0,1/2)$;
(6) $x \in [0,1/2)$, $y = 1/2$;
(7) $x = 1/2$, $y \in (1/2,1]$; and
(8) $x \in (1/2,1]$, $y = 1/2$.

Case (5), $x = 1/2$, $y \in [0,1/2)$. In this case, (2.6.135) reduces to

$$\left|-1 - \frac{2}{3}y\right| \leq \delta \left|\frac{1}{2} - y\right| + L|y + 1|, \quad y \in [0,1/2).$$

Since $\left|-1 - \frac{2}{3}y\right| \leq \frac{4}{3}$ and $1 \leq |y+1|$, in order to have the previous inequality satisfied, we simply need to take $L \geq \frac{4}{3}$.

Case (6), $x \in [0,1/2)$, $y = 1/2$. In this case, (2.6.135) reduces to

$$\left|\frac{2}{3}x + 1\right| \leq \delta \left|x - \frac{1}{2}\right| + L\left|\frac{1}{2} - \frac{2}{3}x\right|, \quad x \in [0,1/2).$$

Since $\left|\frac{2}{3}x + 1\right| \leq \frac{4}{3}$ and $\left|\frac{1}{2} - \frac{2}{3}x\right| \geq \frac{1}{6}$, to have the previous inequality satisfied, it is enough to take $L \geq 8$.

Case (7), $x = 1/2$, $y \in (1/2,1]$. In this case, (2.6.135) reduces to

$$\left|-1 - \frac{2}{3}y - \frac{1}{3}\right| \leq \delta \left|\frac{1}{2} - y\right| + L|y + 1|, \quad y \in (1/2,1].$$

Since $\left|1 + \frac{2}{3}y + \frac{1}{3}\right| \leq 2$ and $|y + 1| > \frac{3}{2}$, to have the previous inequality satisfied, it is enough to take $L \geq \frac{4}{3}$.

Case (8), $x \in (1/2,1]$, $y = 1/2$. Similarly, we find that (2.6.135) holds with $L \geq 8$ and $0 < \delta < 1$ arbitrary.

By summarizing all possible cases, we conclude that T satisfies (2.6.135) with $\delta = 2/3$ and $L = 8$.

As we have shown in Section 2.2, it is possible to force the uniqueness of the fixed point of an almost contraction by imposing an additional contractive condition, quite similar to (2.6.135), as shown by the next theorem.

Theorem 2.6.2. *Let X be a Banach space, K a nonempty closed subset of X and $T : K \to X$ a nonself almost contraction for which there exist $\theta \in (0,1)$ and some $L_1 \geq 0$ such that*

$$d(Tx,Ty) \leq \theta \cdot d(x,y) + L_1 \cdot d(x,Tx), \quad \text{for all } x,y \in K. \tag{2.6.142}$$

If T has property (M) and satisfies Rothe's boundary condition

$$T(\partial K) \subset K,$$

then T has a unique fixed point in K.

Remark 2.6.2. By the considerations in the first part of this section we could immediately obtain various fixed point results as corollaries of Theorems 2.6.1 and 2.6.2, for T satisfying one of the conditions of the type (2.6.130).

REFERENCES

1. Abbas M, Jungck G, Common fixed point results for noncommuting mappings without continuity in cone metric spaces. J. Math. Anal. Appl. 2008;341:416–420.
2. Abbas M, Coincidence points of multivalued f-almost nonexpansive mappings. Fixed Point Theory 2012;13:3–10.
3. Abbas M, Babu GVR, Alemayehu GN, On common fixed points of weakly compatible mappings satisfying 'generalized condition (B)'. Filomat 2011;25:9–19.
4. Abbas M, Damjanović B, Lazović R, Fuzzy common fixed point theorems for generalized contractive mappings. Appl. Math. Lett. 2010;23:1326–1330.
5. Abbas M, Hussain N, Rhoades BE, Coincidence point theorems for multivalued f-weak contraction mappings and applications. Rev. R. Acad. Cienc. Exactas Fs. Nat. Ser. A Math. RACSAM 2011;105:261–272.
6. Abbas M, Ilić D, Common fixed points of generalized almost nonexpansive mappings. Filomat 2010;24:11–18.
7. Abbas M, Kim JK, Nazir T, Common fixed point of mappings satisfying almost generalized contractive condition in partially ordered G-metric spaces. J. Comput. Anal. Appl. 2015;19:928–938.
8. Abbas M, Turkoglu D, Fixed point theorem for a generalized contractive fuzzy mapping. J. Intell. Fuzzy Systems 2014;26:33–36.
9. Abbas M, Vetro P, Khan SH, On fixed points of Berinde's contractive mappings in cone metric spaces. Carpathian J. Math. 2010;26:121–133.
10. Abbas M, Ali B, Romaguera S, Coincidence points of generalized multivalued (f,L)-almost F-contraction with applications. J. Nonlinear Sci. Appl. 2015;8:919–934.
11. Acar Ö, Berinde V, Altun I, Fixed point theorems for Ćirić-type strong almost contractions on partial metric spaces. J. Fixed Point Theory Appl. 2012;12:247–259.
12. Acar Ö, Altun I, Durmaz G, A fixed point theorem for new type contractions on weak partial metric spaces. Vietnam J. Math. DOI 10.1007/s10013-014-0112-0.
13. Aghajani A, Radenović S, Roshan JR, Common fixed point results for four mappings satisfying almost generalized (S,T)-contractive condition in partially ordered metric spaces. Appl. Math. Comput. 2012;218:5665–5670.

14. Akinbo G, Mewomo O, Fixed point theorems for a general class of almost contractions in metric spaces. Acta Math. Acad. Paedagog. Nyházi. (N.S.) 2011;27:299–305.
15. Al-Badarneh AA, On bi-shadowing of subclasses of almost contractive type mappings. Canadian J. Pure Appl. Sci. 2015;9:3449–3453.
16. Al-Badarneh AA, Bi-shadowing of some classes of single-valued almost contractions. Appl. Math. Sci. 2015;9:2859–2869.
17. Alber YI, Guerre-Delabriere S, Principle of weakly contractive maps in Hilbert spaces. In: Gohberg I, Lyubich Y, editors, New results in operator theory and its applications, Basel: Birkhäuser; 1997;7–22.
18. Allahyari R, Arab R, Shole Haghighi A, A generalization on weak contractions in partially ordered b-metric spaces and its application to quadratic integral equations. J. Inequal. Appl. 2014;2014:Article ID 355.
19. Alghamdi MA, Berinde V, Shahzad N, Fixed points of multivalued nonself almost contractions. J. Appl. Math. 2013;2013:Article ID 621614.
20. Alghamdi MA, Berinde V, Shahzad N, Fixed points of non-self almost contractions. Carpathian J. Math. 2014;30:7–14.
21. Ali MU, Kamran T, Hybrid generalized contractions. Math. Sci. (Springer) 2013;7:5 pp.
22. Altun I, Acar Ö, Fixed point theorems for weak contractions in the sense of Berinde on partial metric spaces. Topology Appl. 2012;159:2642–2648.
23. Altun I, Acar Ö, Multivalued almost contractions in metric space endowed with a graph. Creat. Math. Inform. 2015;24:1–8.
24. Altun I, Durmaz G, Minak G, Romaguera S, Multivalued almost F-contractions on complete metric spaces. Filomat (in press).
25. Altun I, Hancer HA, Minak G, On a general class of weakly Picard operators. Miskolc Math. Notes (in press).
26. Altun I, Minak G, Dağ H, Multivalued F-contractions on complete metric spaces. J. Nonlinear Convex Anal. 2015;16:659–666.
27. Altun I, Olgun M, Minak G, On a new class of multivalued weakly Picard operators on complete metric spaces. Taiwanese J. Math. 2015;19:659–672.
28. Altun I, Sadarangani K, Fixed point theorems for generalized almost contractions in partial metric spaces. Math. Sci. (Springer) 2014;8:6 pp.
29. Amini-Harandi A, Fakhar M, Hajisharifi HR, Hussain N, Some new results on fixed and best proximity points in preordered metric spaces. Fixed Point Theory Appl. 2013;2013:Article ID 263.
30. Ariza-Ruiz D, Convergence and stability of some iterative processes for a class of quasinonexpansive type mappings. J. Nonlinear Sci. Appl. 2012;5:93–103.
31. Assad NA, On a fixed point theorem of Iséki. Tamkang J. Math. 1976;7:19–22.
32. Assad NA, On a fixed point theorem of Kannan in Banach spaces. Tamkang J. Math. 1976;7:91–94.
33. Assad NA, On some nonself nonlinear contractions. Math. Japon. 1988;33:17–26.
34. Assad NA, On some nonself mappings in Banach spaces. Math. Japon. 1988;33:501–515.
35. Assad NA, Approximation for fixed points of multivalued contractive mappings. Math. Nachr. 1988;139:207–213.
36. Assad NA, A fixed point theorem in Banach space. Publ. Inst. Math. (Beograd) (N.S.) 1990;47(61):137–140.
37. Assad NA, A fixed point theorem for some non-self-mappings. Tamkang J. Math. 1990;21:387–393.
38. Assad NA, Kirk WA, Fixed point theorems for set-valued mappings of contractive type. Pacific J. Math. 1972;43:553–562.
39. Assad NA, Sessa S, Common fixed points for nonself compatible maps on compacta. Southeast Asian Bull. Math. 1992;16:91–95.
40. Aydi H, Felhi A, Sahmim S, Fixed points of multivalued nonself almost contractions in metric-like spaces. Math. Sci. 2015;9:103–108.
41. Aydi H, Hadj Amor S, Karapinar E, Berinde-type generalized contractions on partial metric spaces. Abstr. Appl. Anal. 2013;2013:Article ID 312479.
42. Aydi H, Hadj Amor S, Karapinar E, Some almost generalized (ψ, ϕ)-contractions in G-metric spaces. Abstr. Appl. Anal. 2013;2013:Article ID 165420.

43. Aydi H, Karapinar E, Mustafa Z, Some tripled coincidence point theorems for almost generalized contractions in ordered metric spaces. Tamkang J. Math. 2013;44:233–251.
44. Babu GVR, Babu DR, Rao KN, Kumar BVS, Fixed points of (ψ, ϕ)-almost weakly contractive maps in G-metric spaces. Appl. Math. E-Notes 2014;14:69–85.
45. Babu GVR, Kidane KT, Fixed points of almost generalized $\alpha - \Psi$-contractive maps. Int. J. Math. Sci. Comput. 2013;3:30–38.
46. Babu GVR, Sailaja PD, Existence of common fixed points of generalized almost weakly contractive maps. Proc. Jangjeon Math. Soc. 2013;16:71–86.
47. Babu GVR, Sandhya ML, Kameswari MVR, A note on a fixed point theorem of Berinde on weak contractions. Carpathian J. Math. 2008;24:8–12.
48. Babu GVR, Subhashini P, Coupled coincidence points of ϕ almost generalized contractive mappings in partially ordered metric spaces. J. Adv. Res. Pure Math. 2013;5:1–16.
49. Banach S, Sur les opérations dans les ensembles abstraits et leur applications aux equations intégrales. Fund. Math. 1922;3:133–181.
50. Banach S, Théorie des Operations Linéaires. Monografie Matematyczne. Warszawa-Lwow;1932.
51. Bekeshie T, Naidu GA, Recent fixed point theorems for T-contractive mappings and T-weak (almost) contractions in metric and cone metric spaces are not real generalizations. J. Nonlinear Anal. Optim. 2013;4:219–225.
52. Berinde V, Contracţii generalizate şi aplicaţii. Baia Mare:Editura Cub Press 22;1997.
53. Berinde V, On the approximation of fixed points of weak contractive mappings. Carpathian J. Math. 2003;19:7–22.
54. Berinde V, Approximating fixed points of weak ϕ-contractions using the Picard iteration. Fixed Point Theory 2003;4:131–142.
55. Berinde V, Approximating fixed points of weak contractions using the Picard iteration. Nonlinear Anal. Forum 2004;9:43–53.
56. Berinde V, A common fixed point theorem for nonself mappings. Miskolc Math. Notes. 2004;5:137–144.
57. Berinde V, Approximation of fixed points of some nonself generalized ϕ-contractions. Math. Balkanica (N.S.) 2004;18:85–93.
58. Berinde V, Berinde M, On Zamfirescu's fixed point theorem. Rev. Roumaine Math. Pures Appl. 2005;50:443–453.
59. Berinde V, A convergence theorem for some mean value fixed point iteration procedures. Demonstratio Math. 2005;38:177–184.
60. Berinde V, Iterative Approximation of Fixed Points. Berlin:Springer;2007.
61. Berinde V, A convergence theorem for Mann iteration in the class of Zamfirescu operators. An. Univ. Vest Timiş. Ser. Mat.-Inform. 2007;45:33–41.
62. Berinde V, General constructive fixed point theorems for Ćirić-type almost contractions in metric spaces. Carpathian J. Math. 2008;24:10–19.
63. Berinde V, Approximating common fixed points of noncommuting discontinuous weakly contractive mappings in metric spaces. Carpathian J. Math. 2009;25:13–22.
64. Berinde V, Some remarks on a fixed point theorem for Ćirić-type almost contractions. Carpathian J. Math. 2009;25:157–162.
65. Berinde V, Approximating common fixed points of noncommuting almost contractions in metric spaces. Fixed Point Theory. 2010;11:179–188.
66. Berinde V, Common fixed points of noncommuting discontinuous weakly contractive mappings in cone metric spaces. Taiwanese J. Math. 2010;14:1763–1776.
67. Berinde V, Common fixed points of noncommuting almost contractions in cone metric spaces. Math. Commun. 2010;15:229–241.
68. Berinde V, Stability of Picard iteration for contractive mappings satisfying an implicit relation. Carpathian J. Math. 2011;27:13–23.
69. Berinde V, Approximating fixed points of implicit almost contractions. Hacet. J. Math. Stat. 2012;41:93–102.
70. Berinde M, Berinde V, On a general class of multi-valued weakly Picard mappings. J. Math. Anal. Appl. 2007;326:772–782.

71. Berinde V, Măruşter Ş, Rus IA, An abstract point of view on iterative approximation of fixed points of nonself operators. J. Nonlinear Convex Anal. 2014;15:851–865.
72. Berinde V, Păcurar M, Fixed points and continuity of almost contractions. Fixed Point Theory 2008;9:23–34.
73. Berinde V, Păcurar M, A note on the paper "Remarks on fixed point theorems of Berinde". Nonlinear Anal. Forum. 2009;14:119–124.
74. Berinde V, Păcurar M, Fixed point theorems for nonself single-valued almost contractions. Fixed Point Theory 2013;14:301–311.
75. Berinde V, Păcurar M, Stability of k-step fixed point iterative methods for some Prešić type contractive mappings. J. Inequal. Appl. 2014;2014:Article ID 149.
76. Berinde V, Păcurar M, A constructive approach to coupled fixed point theorems in metric spaces. Carpathian J. Math. 2015;31:277–287.
77. Berinde V, Păcurar M, Rus IA, From a Dieudonné theorem concerning the Cauchy problem to an open problem in the theory of weakly Picard operators. Carpathian J. Math. 2014;30:283–292.
78. Berinde V, Vetro F, Common fixed points of mappings satisfying implicit contractive conditions. Fixed Point Theory Appl. 2012;2012:Article ID 105.
79. Bose RK, Some Suzuki type fixed point theorems for generalized contractive multifunctions. Int. J. Pure Appl. Math. 2013;84:13–27.
80. Bota MF, Karapinar E, A note on "Some results on multi-valued weakly Jungck mappings in b-metric space". Cent. Eur. J. Math. 2013;11:1711–1712.
81. Camouzis E, Chatterjee E, Ladas G, On the dynamics of $x_{n+1} = (\delta x_{n-2} + x_{n-3})/(A + x_{n-3})$. J. Math. Anal. Appl. 2007;331:230–239.
82. Caristi J, Fixed point theorems for mappings satisfying inwardness conditions. Trans. Am. Math. Soc. 1976;215:241–251.
83. Caristi J, Fixed point theory and inwardness conditions. In: Applied nonlinear analysis (Proc. Third Int. Conf., Univ. Texas, Arlington, Tex., 1978), New York:Academic Press;1979; 479–483.
84. Chandok S, Choudhury BS, Metiya N, Fixed point results in ordered metric spaces for rational type expressions with auxiliary functions. J. Egyptian Math. Soc. 2015;23:95–101.
85. Chatterjea SK, Fixed-point theorems. C.R. Acad. Bulgare Sci. 1972;25:727–730.
86. Chen YZ, A Presić type contractive condition and its applications. Nonlinear Anal. 2009;71:2012–2017.
87. Chifu C, Petruşel G, Generalized contractions in metric spaces endowed with a graph. Fixed Point Theory Appl. 2012;2012:Article ID 161.
88. Cho SH, Fixed point theorems for weak $\alpha_* - (\Phi, L)$-contractive set-valued maps in cone metric spaces. Int. J. Math. Anal. 2013; 7:2967–2979.
89. Cho SH, A fixed point theorem for a Ćirić-Berinde type mapping in orbitally complete metric spaces. Carpathian J. Math. 2014;30:63–70.
90. Cho SH, Fixed point theorems for Ćirić-Berinde type contractive multivalued mappings. Abstr. Appl. Anal. 2015;2015:Article ID 768238.
91. Choudhury BS, Metiya N, Fixed point theorems for almost contractions in partially ordered metric spaces. Ann. Univ. Ferrara Sez. VII Sci. Mat. 2012;58:21–36.
92. Choudhury BS, Metiya N, Coincidence point theorems for a family of multivalued mappings in partially ordered metric spaces. Acta Univ. M. Belii Ser. Math. 2013;10–23.
93. Choudhury BS, Metiya N, Som T, Bandyopadhyay C, Multivalued fixed point results and stability of fixed point sets in metric spaces. Facta Univ. Ser. Math. Inform. 2015;30:501–512.
94. Ćirić LB, Generalized contractions and fixed-point theorems. Publ. l'Inst. Math. (Beograd) 1971;12:19–26.
95. Ćirić LB, On contraction type mappings. Math. Balkanica. 1971;1:52–57.
96. Ćirić LB, A generalization of Banach's contraction principle. Proc. Am. Math. Soc. 1974;45:267–273.
97. Ćirić LB, Convergence theorems for a sequence of Ishikawa iteration for nonlinear quasi-contractive mappings. Indian J. Pure Appl. Appl. Math. 1999;30:425–433.
98. Ćirić LB, A remark on Rhoades' fixed point theorem for non-self mappings. Int. J. Math. Math. Sci. 1993;16:397–400.

99. Ćirić LB, Quasi contraction non-self mappings on Banach spaces. Bull. Cl. Sci. Math. Nat. Sci. Math. 1998;23:25–31.
100. Ćirić LB, Ume JS, Khan MS, Pathak HK, On some nonself mappings. Math. Nachr. 2003;251:28–33.
101. Ćirić LB, Prešić SB, On Prešić type generalization of the Banach contraction mapping principle. Acta Math. Univ. Comenianae 2007;76:143–147.
102. Ćirić LB, Abbas M, Damjanović B, Saadati R, Common fuzzy fixed point theorems in ordered metric spaces. Math. Comput. Modelling 2011;53:1737–1741.
103. Ćirić LB, Abbas M, Saadati R, Hussain N, Common fixed points of almost generalized contractive mappings in ordered metric spaces. Appl. Math. Comput. 2011;217:5784–5789.
104. Cvetkovic M, Rakocevic V, Extensions of Perov theorem. Carpathian J. Math. 2015;31:181–188.
105. Das KM, Naik KV, Common fixed point theorems for commuting maps on metric spaces. Proc. Am. Math. Soc. 1979;77:369–373.
106. De la Sen M, Agarwal RP, Ibeas A, Results on proximal and generalized weak proximal contractions including the case of iteration-dependent range sets. Fixed Point Theory Appl. 2014;2014:Article ID 169.
107. De la Sen M, Some further results on weak proximal contractions including the case of iteration-dependent image sets. In: Mathematical Methods in Engineering and Economics, Proceedings of the 2014 International Conference on Applied Mathematics and Computational Methods in Engineering II (AMCME '14). Proceedings of the 2014 International Conference on Economics and Business Administration II (EBA '14), Prague, Czech Republic April 2–4, 2014, pp. 15–20.
108. Devault R, Dial G, Kocic VL, Ladas G, Global behavior of solutions of $x_{n+1} = ax_n + f(x_n, x_{n-1})$. J. Difference Eq. Appl. 1998;3:311–330.
109. Di Bari C, Vetro P, Common fixed points for three or four mappings via common fixed point for two mappings. arXiv:1302.3816.
110. Du WS, Fixed point theorems for generalized multivalued weak contractions. Int. J. Math. Anal. (Ruse) 2008;2:181–186.
111. Du WS, Some new results and generalizations in metric fixed point theory. Nonlinear Anal. 2010;73:1439–1446.
112. Du WS, New cone fixed point theorems for nonlinear multivalued maps with their applications. Appl. Math. Lett. 2011;24:172–178.
113. Du WS, On coincidence point and fixed point theorems for nonlinear multivalued maps. Topology Appl. 2012;159:49–56.
114. Du WS, On approximate coincidence point properties and their applications to fixed point theory. J. Appl. Math. 2012; 2012:Article ID 302830.
115. Du WS, On generalized weakly directional contractions and approximate fixed point property with applications. Fixed Point Theory Appl. 2012;2012:Article ID 6.
116. Du WS, New existence results and generalizations for coincidence points and fixed points without global completeness. Abstr. Appl. Anal. 2013;2012:Article ID 214230.
117. Du WS, He ZH, Chen YL, New existence theorems for approximate coincidence point property and approximate fixed point property with applications to metric fixed point theory. J. Nonlinear Convex Anal. 2012;13:459–474.
118. Du WS, Karapinar E, Shahzad N, The study of fixed point theory for various multivalued non-self-maps. Abstr. Appl. Anal. 2013;2013:Article ID 938724.
119. Du WS, Khojasteh F, New results and generalizations for approximate fixed point property and their applications. Abstr. Appl. Anal. 2014;2014:Article ID 581267.
120. Du WS, Khojasteh F, Chiu YN, Some generalizations of Mizoguchi-Takahashi's fixed point theorem with new local constraints. Fixed Point Theory Appl. 2014;2014:Article ID 31.
121. Du WS, Zheng SX, Nonlinear conditions for coincidence point and fixed point theorems. Taiwanese J. Math. 2012;16:857–868.
122. Du WS, Zheng SX, New nonlinear conditions and inequalities for the existence of coincidence points and fixed points. J. Appl. Math. 2012;2012:Article ID 196759.
123. Dubey AK, Tiwari SK, Dubey RP, Common fixed point theorems for T-weak contraction mapping in a cone metric space. Mathematica Aeterna 2013;3:121–131.

124. Dugundji J, Granas A, Weakly contractive maps and elementary domain invariance theorem. Bull. Greek Math. Soc. 1978;19:141–151.
125. Durmaz G, Minak G, Altun I, Fixed point results for $\alpha - \Psi$-contractive mappings including almost contractions and applications. Abstr. Appl. Anal. 2014;2014:Article ID 869123.
126. Eisenfeld J, Lakshmikantham V, Fixed point theorems on closed sets through abstract cones. Appl. Math. Comput. 1977;3:155–167.
127. El-Metwally H, Grove EA, Ladas G, Levins R, Radin M, On the difference equation $x_{n+1} = \alpha + \beta x_{n-1} e^{-x_n}$. Nonlinear Anal. 2001;47:4623–4634.
128. Erduran A, Kadelburg Z, Nashine HK, Vetro C, A fixed point theorem for (ϕ, L)-weak contraction mappings on a partial metric space. J. Nonlinear Sci. Appl. 2014;7:196–204.
129. Filip AD, Petru PT, Fixed point theorems for multivalued weak contractions. Stud. Univ. Babeş-Bolyai Math. 2009;54:33–40.
130. Filip AD, Petruşel A, Fixed point theorems on spaces endowed with vector-valued metrics. Fixed Point Theory Appl. 2010;2010:Article ID 28138.
131. Gabeleh M, Best proximity point theorems for single- and set-valued non-self mappings. Acta Math. Sci. Ser. B Engl. Ed. 2014;34:1661–1669.
132. Gabeleh M, Best proximity point theorems via proximal non-self mappings. J. Optim. Theory Appl. 2015;164:565–576.
133. George R, Reshma KP, Padmavati A, Fixed point theorems for cyclic contractions in b-metric spaces. J. Nonlinear Funct. Anal. 2015;2015:Article ID 5.
134. Gül, Karapinar E, On almost contractions in partially ordered metric spaces via implicit relations. J. Inequal. Appl. 2012;2012:Article ID 217.
135. He ZH; Du WS, Lin IJ, The existence of fixed points for new nonlinear multivalued maps and their applications. Fixed Point Theory Appl. 2011;2011:Article ID 84.
136. Hussain N, Amini-Harandi A, Cho YJ, Approximate endpoints for set-valued contractions in metric spaces. Fixed Point Theory Appl. 2010;2010:Article ID 614867.
137. Hussain N, Cho YJ, Weak contractions, common fixed points, and invariant approximations. J. Inequal. Appl. 2009;2009:Article ID 390634.
138. Hussain N, Parvaneh V, Roshan JR, Kadelburg Z, Fixed points of cyclic weakly (ψ, φ, L, A, B)-contractive mappings in ordered b-metric spaces with applications. Fixed Point Theory Appl. 2013;2013:Article ID 256.
139. Javahernia M, Razani A, Khojasteh F, Common fixed point of the generalized Mizoguchi-Takahashi's type contractions. Fixed Point Theory Appl. 2014;2014:Article ID 195.
140. Jleli M, Karapinar E, Samet B, Fixed point results for almost generalized cyclic (ψ, ϕ)-weak contractive type mappings with applications. Abstr. Appl. Anal. 2012;2012:Article ID 917831.
141. Jleli M, Karapinar E, Samet B, Further generalizations of the Banach contraction principle. J. Inequal. Appl. 2014;2014:Article ID 439.
142. Jleli M, Samet B, A new generalization of the Banach contraction principle. J. Ineq. Appl. 2014;2014:Article ID 38.
143. Jleli M, Samet B, Vetro C, Fixed point theory in partial metric spaces via ϕ-fixed point's concept in metric spaces. J. Inequal. Appl. 2014;2014:Article ID 426.
144. Jungck G, Commuting maps and fixed points. Am. Math. Monthly 1976;83:261–263.
145. Jungck G, Common fixed points for noncontinuous nonself maps on non-metric spaces. Far East J. Math. Sci. 1996;4:199–215.
146. Kalinde AK, Mishra S. N., Pathak HK, Some results on common fixed points with applications. Fixed Point Theory 2005;6:285–301.
147. Kamran T, Multivalued f-weakly Picard mappings. Nonlinear Anal. 2007;67:2289–2296.
148. Kamran T, Cakić N, Hybrid tangential property and coincidence point theorems. Fixed Point Theory 2008; 9:487–496.
149. Kannan R, Some results on fixed points. Bull. Calcutta Math. Soc. 1968;10:71–76.
150. Kannan R, Some results on fixed points. III. Fund. Math. 1971;70:169–177.
151. Kannan R, Construction of fixed points of a class of nonlinear mappings. J. Math. Anal. Appl. 1973;41:430–438.

152. Karapinar E, Sadarangani K, Berinde mappings in ordered metric spaces. Rev. R. Acad. Cienc. Exactas Fs. Nat. Ser. A Math. RACSAM 2015;109:353–366.
153. Karapinar E, Sintunavarat W, The existence of optimal approximate solution theorems for generalized α-proximal contraction non-self mappings and applications. Fixed Point Theory Appl. 2013;2013:Article ID 323.
154. Karuppiah U, Dharsini AMP, Some fixed point theorems satisfying (s,t)-contractive condition in partially ordered partial metric space. Far East J. Math. Sci. 2014;95:19–50.
155. Khandani H, Vaezpour SM, Sims B, Fixed point and common fixed point theorems of contractive multivalued mappings on complete metric spaces. J. Comput. Anal. Appl. 2011;13:1025–1039.
156. Khan AR, Abbas M, Nazir T, Ionescu C, Fixed points of multivalued contractive mappings in partial metric spaces. Abstr. Appl. Anal. 2014;2014:Article ID 230708.
157. Khan MS, Swaleh M, Sessa S, Fixed point theorems by altering distances between the points. Bull. Aust. Math. Soc. 1984;30:1–9.
158. Khan MS, Berzig M, Samet B, Some convergence results for iterative sequences of Prešić type and applications. Adv. Difference Equ. 2012;2012:Article ID 38.
159. Khan MS, Jhade PK, On a fixed point theorem with PPF dependence in the Razumikhin class. Gazi Univ. J. Sci . 2015;28:211–219.
160. Khan SH, Common fixed points of two quasi-contractive operators in normed spaces by iteration. Int. J. Math. Anal. (Ruse) 2009;3:145–151.
161. Kikina L, Kikina K, Vardhami I, Fixed point theorems for almost contractions in generalized metric spaces. Creat. Math. Inform. 2014;23:65–72.
162. Kiran Q, Kamran T, Nadler's type principle with high order of convergence. Nonlinear Anal. 2008;69:4106–4120.
163. Klanarong C, Suantai S, Coincidence point theorems for some multi-valued mappings in complete metric spaces endowed with a graph. Fixed Point Theory Appl. 2015;2015:Article ID 129.
164. Kocic VL, A note on the non-autonomous Beverton-Holt model. J. Difference Equ. Appl. 2005;11:415–422.
165. Kocic VL, Ladas G, Global Asymptotic Behavior of Nonlinear Difference Equations of Higher Order with Applications. Dordrecht:Kluwer Academic Publishers; 1993.
166. Kumar A, Rathee S, Fixed point and common fixed point results in cone metric space and application to invariant approximation. Fixed Point Theory Appl. 2015;2015:Article ID 1.
167. Kuruklis SA, The asymptotic stability of $x_{n+1} - ax_n + bx_{n-k} = 0$. J. Math. Anal. Appl. 1994;188:719–731.
168. Kutbi MA, Sintunavarat W, On sufficient conditions for the existence of past-present-future dependent fixed point in the Razumikhin class and application. Abstr. Appl. Anal. 2014;2014:Article ID 342687.
169. Latif A, Mongkolkeha C, Sintunavarat W, Fixed point theorems for generalized $\alpha - \beta$-weakly contraction mappings in metric spaces and applications. Scientific World J. 2014;2014:Article ID 784207.
170. Lin IJ, Chen TH, New existence theorems of coincidence points approach to generalizations of Mizoguchi-Takahashi's fixed point theorem. Fixed Point Theory Appl. 2012;2012:Article ID 156.
171. Lin IJ, Jian KR, New fixed point theorems for nonlinear multivalued maps and mt-functions in complete metric spaces. Nonlinear Anal. Diff. Eq. 2013;1:29–41.
172. Lin IJ, Wang TY, New fixed point theorems for generalized distances. Int. J. Math. Anal. 2013;7:1843–1855.
173. Mǎruşter L, Mǎruşter S, On the error estimation and T-stability of the Mann iteration. J. Comput. Appl. Math. 2015;276:110–116.
174. Mehmood N, Azam A, Aleksić S, Topological vector-space valued cone Banach spaces. Int. J. Anal. Appl. 2014;6:205–219.
175. Minak G, Acar Ö, Altun I, Multivalued pseudo-Picard operators and fixed point results. J. Funct. Spaces Appl. 2013;2013:Article ID 827458.
176. Minak G, Altun I, Some new generalizations of Mizoguchi-Takahashi type fixed point theorem. J. Inequal. Appl. 2013;2013:Article ID 493.

177. Minak G, Altun I, Multivalued weakly Picard operators on partial metric spaces. Nonlinear Funct. Anal. Appl. 2014;19:45–59.
178. Minak G, Altun I, Romaguera S, Recent developments about multivalued weakly Picard operators. Bull. Belg. Math. Soc. Simon Stevin 2015;22:411–422.
179. Minak G, Helvaci A, Altun I, Ćirić type generalized F-contractions on complete metric spaces and fixed point results. Filomat 2014;28:1143–1151.
180. Mohsenalhosseini SAM, Approximate best proximity pairs in metric space for contraction maps. Adv. Fixed Point Theory 2014;4:310–324.
181. Mongkolkeha C, Kongban C, Kumam P, Existence and uniqueness of best proximity points for generalized almost contractions. Abstr. Appl. Anal. 2014;2014:Article ID 813614.
182. Morales JR, Rojas EM, Coincidence points for multivalued mappings. An. Ştiinţ. Univ. "Ovidius" Constanţa Ser. Mat. 2011;19:37–150.
183. Mustafa Z, Karapinar E, Aydi H, A discussion on generalized almost contractions via rational expressions in partially ordered metric spaces. J. Inequal. Appl. 2014;2014:Article ID 219.
184. Mustafa Z, Parvaneh V, Roshan JR, Kadelburg Z, b_2-metric spaces and some fixed point theorems. Fixed Point Theory Appl. 2014;2014:Article ID 144.
185. Nashine HK, Kadelburg Z, Common fixed point theorems for a pair of multivalued mappings under weak contractive conditions in ordered metric spaces. Bull. Belg. Math. Soc. Simon Stevin 2012;19:577–596.
186. Olatinwo MO, Some results on multi-valued weakly Jungck mappings in b-metric space. Cent. Eur. J. Math. 2008;6:610–621.
187. Olatinwo MO, An extension of some fixed point theorems for single-valued and multi-valued Picard operators in b-metric spaces. Nonlinear Anal. Forum. 2009;14:103–111.
188. Olatinwo MO, Imoru CO, A generalization of some results on multi-valued weakly Picard mappings in b-metric space. Fasc. Math. 2008;40:45–56.
189. Olatinwo MO, Postolache M, Stability results for Jungck-type iterative processes in convex metric spaces. Appl. Math. Comput. 2012;218:6727–6732.
190. Osilike MO, Stability results for fixed point iteration procedures. J. Nigerian Math. Soc. 1995/96;14/15:17–29.
191. Osilike MO, Stability of the Ishikawa iteration method for quasi-contractive maps. Indian J. Pure Appl. Math. 1997;28(9):1251–1265.
192. Osilike MO, Short proofs of stability results for fixed point iteration procedures for a class of contractive-type mappings. Indian J. Pure Appl. Math. 1999;30(12):1229–1234.
193. Ortega JM, Rheinboldt WC, Iterative Solution of Nonlinear Equations in Several Variables. New York:Academic Press;1970.
194. Pathak HK, Agarwal RP, Cho YJ, Coincidence and fixed points for multi-valued mappings and its application to nonconvex integral inclusions. J. Comput. Appl. Math. 2015;283:201–217.
195. Pathak HK, George R, Nabwey HA, El-Paoumy MS, Reshma KP, Some generalized fixed point results in a b-metric space and application to matrix equations. Fixed Point Theory Appl. 2015;2015:Article ID 101.
196. Păcurar M, Approximating common fixed points of Prešić-Kannan type operators by a multi-step iterative method. An. Ştiinţ. Univ. "Ovidius" Constanţa Ser. Mat. 2009;17:153–168.
197. Păcurar M, Iterative Methods for Fixed Point Approximation. Cluj-Napoca:Editura Risoprint;2009.
198. Păcurar M, Sequences of almost contractions and fixed points in b-metric spaces. An. Univ. Vest Timiş. Ser. Mat.-Inform. 2010;48:125–137.
199. Păcurar M, Remark regarding two classes of almost contractions with unique fixed point. Creat. Math. Inform. 2010;19:178–183.
200. Păcurar M, A multi-step iterative method for approximating common fixed points of Prešić-Rus type operators on metric spaces. Studia Univ. "Babeş-Bolyai". Math. 2010;55:149–162.
201. Păcurar M, A multi-step iterative method for approximating fixed points of Prešić-Kannan operators. Acta Math. Univ. Comen. New Ser. 2010;79:77–88.
202. Păcurar M, Fixed point theory for cyclic Berinde operators. Fixed Point Theory 2011;12:419–428.
203. Păcurar M, Fixed points of almost Prešić operators by a k-step iterative method. An. Ştiinţ. Univ. Al. I. Cuza Iaşi. Mat. (N.S.) 2011;57:199–210.

204. Păcurar M, Common fixed points for almost Prešić type operators. Carpathian J. Math. 2012;28:117–126.
205. Păcurar M, Păcurar RV, Approximate fixed point theorems for weak contractions on metric spaces. Carpathian J. Math. 2007;23:149–155.
206. Petruşel A, Petruşel G, Multivalued contractions of Feng-Liu type in complete gauge spaces. Carpathian J. Math. 2008;24:392–396.
207. Petruşel A, Petruşel G, Urs C, Vector-valued metrics, fixed points and coupled fixed points for nonlinear operators. Fixed Point Theory Appl. 2013;2013:Article ID 218.
208. Phuengrattana W, Suantai S, Comparison of the rate of convergence of various iterative methods for the class of weak contractions in Banach spaces. Thai J. Math. 2013;11:217–226.
209. Phon-on A, Sama-Ae A, Makaje N, Riyapan P, Busaman S, Coincidence point theorems for weak graph preserving multi-valued mapping. Fixed Point Theory Appl. 2014;2014:248.
210. Picard E, Memoire sur la théorie des equations aux derivées partielles et la methode des approximations successives. J. Math. Pures et Appl. 1890;6:145–210.
211. Popa V, Fixed point theorems for implicit contractive mappings. Stud. Cerc. St. Ser. Mat. Univ. Bacău. 1997;7:127–133.
212. Popa V, Some fixed point theorems for compatible mappings satisfying an implicit relation. Demonstratio Math. 1999;32:157–163.
213. Popa V, On some fixed point theorems for implicit almost contractive mappings. Carpathian J. Math. 2013;29:223–229.
214. Popa V, Common (E.A)-property and altering distance in metric spaces. Sci. Stud. Res. Ser. Math. Inform. 2014;24:115–133.
215. Popa V, Patriciu AM, A general fixed point theorem for a pair of self mappings with common limit range property in G-metric spaces. Facta Univ. Ser. Math. Inform. 2014;29:351–370.
216. Popescu O, Picard iteration converges faster than Mann iteration for a class of quasi-contractive operators. Math. Commun. 2007;12:195–202.
217. Popescu O, Comparison of fastness of the convergence among Krasnoselskij iterations. Bull. Transilv. Univ. Braşov Ser. B (N.S.) 2007;14(49):27–32.
218. Prešić SB, Sur une classe d' inéquations aux différences finite et sur la convergence de certaines suites. Publ. Inst. Math. (Beograd) (N.S.) 1965;5(19):75–78.
219. Rao KPR, Bindu HS, Ali MM, Coupled fixed point theorems in d-complete topological spaces. J. Nonlinear Sci. Appl. 2012;5:186–194.
220. Rao KPR, Rao KRK, Karapinar E, Common coupled fixed point theorems in d-complete topological spaces. Ann. Funct. Anal. 2012;3:107–114.
221. Raphael P, Pulickakunnel S, Approximate fixed point theorems for generalized T-contractions in metric spaces. Stud. Univ. Babeş-Bolyai Math. 2012;57:551–559.
222. Rashwan RA, Hammad HA, Common fixed point theorems for weak contraction mapping of integral type in modular spaces. Universal J. Comput. Math. 2014;2:69–86.
223. Rathee S, Kumar A, Some common fixed-point and invariant approximation results with generalized almost contractions. Fixed Point Theory Appl. 2014;2014:Article ID 23.
224. Rathee S, Kumar A, Tas K, Invariant approximation results via common fixed point theorems for generalized weak contraction maps. Abstr. Appl. Anal. 2014;2014:Article ID 752107.
225. Redjel N, On some extensions of Banach's contraction principle and applications to the convergence and stability of some iterative processes. Adv. Fixed Point Theory 2014;4:555–570.
226. Rhoades BE, A comparison of various definitions of contractive mappings. Trans. Am. Math. Soc. 1977;226:257–290.
227. Rhoades BE, Extensions of some fixed theorems of Ciric, Maiti and Pal. Math. Seminar Notes 1978;6:41–46.
228. Rhoades BE, A fixed point theorem for some non-self-mappings. Math. Japon. 1978/79;23:457–459.
229. Rhoades BE, Contractive definitions revisited. Contemporary Math. 1983;21:189–205.
230. Rhoades BE, Contractive definitions and continuity. Contemporary Math. 1988;72:233–245.
231. Romaguera S, On Nadler's fixed point theorem for partial metric spaces. Math. Sci. Appl. E-Notes 2013;1:1–8.

232. Roshan JR, Parvaneh V, Sedghi S, Shobkolaei N, Shatanawi W, Common fixed points of almost generalized (Ψ, Φ) s-contractive mappings in ordered b-metric spaces. Fixed Point Theory Appl. 2013;2013:Article ID 159.

233. Rus IA, An iterative method for the solution of the equation $x = f(x, \ldots, x)$. Anal. Numér. Théor. Approx. 1981;10:95–100.

234. Rus IA, An abstract point of view in the nonlinear difference equations. In: Conf. on An., Functional Equations, App. and Convexity, Cluj-Napoca, October 15–16, 1999, 272–276.

235. Rus IA, Generalized Contractions and Applications. Cluj-Napoca: Cluj Univ. Press;2001.

236. Rus IA, Picard operators and applications. Sci. Math. Jpn. 2003;58:191–219.

237. Rus IA, Heuristic introduction to weakly Picard operator theory. Creat. Math. Inform. 2014;23:243–252.

238. Rus IA, Five open problems in fixed point theory in terms of fixed point structures (I): Singlevalued operators. In: Espinola R, Petruşel A, Prus, S, editors, Fixed Point Theory and Its Applications. Cluj-Napoca:Casa Cărţii de Ştiinţă;2013;39–60.

239. Rus IA, Petruşel A, Petruşel G, Fixed Point Theory. Cluj-Napoca: Cluj Univ. Press;2008.

240. Rus IA, Şerban MA, Basic problems of the metric fixed point theory and the relevance of a metric fixed point theorem. Carpathian J. Math. 2013;29:239–258.

241. Sadeghi-Hafshejani A, Amini-Harandi A, A fixed point result for a new class of set-valued contractions. Int. J. Nonlinear Anal. Appl. 2014;5:64–70.

242. Saha M, Dey D, Some random fixed point theorems for (θ, L)-weak contractions. Hacet. J. Math. Stat. 2012;41:795–812.

243. Samet B, Fixed points for α-ψ contractive mappings with an application to quadratic integral equations. Electron. J. Differential Equations 2014;2915:Article ID 152.

244. Samet B, The class of (α, ψ)-type contractions in b-metric spaces and fixed point theorems. Fixed Point Theory Appl. 2015;2015:Article ID 92.

245. Samet B, Vetro C, Berinde mappings in orbitally complete metric spaces. Chaos Solitons Fractals 2011;44:1075–1079.

246. Sarma KKM, Kumari VA, Fixed points of almost generalized (α, ψ) contractive maps in G-metric spaces. Thai J. Math. (in press).

247. Sarwar M, Rahman MU, Fixed point theorems for Ciric's and generalized contractions in b-metric spaces. Int. J. Anal. Appl. 2015;7:70–78.

248. Sastry KPR, Babu GVR, Sandhya ML, Weak contractions in Menger spaces. J. Adv. Res. Pure Math. 2010;2:26-35.

249. Şerban MA, Teoria punctului fix pentru operatori definiţi pe produs cartezian. Cluj-Napoca:Presa Universitară Clujeană;2002.

250. Shaddad F, Noorani MSM, Alsulami SM, Fixed point results in cone metric spaces for multivalued maps. Malays. J. Math. Sci. 2014;8:83–102.

251. Shaddad F, Noorani MSM, Alsulami SM, Common fixed-point results for generalized Berinde-type contractions which involve altering distance functions. Fixed Point Theory Appl. 2014;2014:Article ID 24.

252. Shatanawi W, Some fixed point results for generalized ψ-weak contraction mappings in orbitally metric spaces. Chaos Solitons Fractals. 2012;45:520–526.

253. Shatanawi W, Al-Rawashdeh A, Common fixed points of almost generalized (ψ, φ)-contractive mappings in ordered metric spaces. Fixed Point Theory Appl. 2012;2012:Article ID 80.

254. Shatanawi W, Postolache M, Some fixed-point results for a G-weak contraction in G-metric spaces. Abstr. Appl. Anal. 2012;2012:Article ID 815870.

255. Shatanawi W, Postolache M, Coincidence and fixed point results for generalized weak contractions in the sense of Berinde on partial metric spaces. Fixed Point Theory Appl. 2013;2013:Article ID 54.

256. Shatanawi W, Postolache M, Common fixed point theorems for dominating and weak annihilator mappings in ordered metric spaces. Fixed Point Theory Appl. 2013;2013:Article ID 271.

257. Shatanawi W, Saadati R, Park C, Almost contractive coupled mapping in ordered complete metric spaces. J. Inequal. Appl. 2013;2013:Article ID 565.

258. Shobkolaei N, Sedghi S, Roshan JR, Altun I, Common fixed point of mappings satisfying almost generalized (S, T)-contractive condition in partially ordered partial metric spaces. Appl. Math. Comput. 2012;219:443–452.

259. Shukla S, Sen R, Set-valued Prešić-Reich type mappings in metric spaces. Rev. R. Acad. Cienc. Exactas Fís. Nat. Ser. A Math. RACSAM 2014;108:431–440.
260. Singh A, Prajapati DJ, Dimri RC, Some fixed point results of almost generalized contractive mappings in ordered metric spaces. Int. J. Pure Appl. Math. 2013;86:779–789.
261. Singh SL, Pant R, Remarks on fixed point theorems of V. Berinde: "Approximating fixed points of weak contractions using the Picard iteration" [Nonlinear Anal. Forum 9 (2004), no. 1, 43–53]. Nonlinear Anal. Forum. 2007;12:231–234.
262. Sintunavarat W, Kim JK, Kumam P, Fixed point theorems for a generalized almost (Φ, φ)-contraction with respect to S in ordered metric spaces. J. Inequal. Appl. 2012;2012:Article ID 263.
263. Sintunavarat W, Kumam P, Common fixed point theorem for hybrid generalized multi-valued contraction mappings. Appl. Math. Lett. 2012;25:52–57.
264. Stević S, Asymptotic behavior of a class of nonlinear difference equations. Discrete Dynamics in Nature and Society 2006;2006:Article ID 47156.
265. Stević S, On the recursive sequence $x_{n+1} = A + (x_n^p)/(x_{n-1}^p)$. Discrete Dynamics in Nature and Society 2007;2007:Article ID 34517.
266. Suzuki T, Fixed point theorems for Berinde mappings. Bull. Kyushu Inst. Technol. Pure Appl. Math. 2011;58:13–19.
267. Sz.-Nagy B, Foiaş C, Analyse harmonique des opérateurs de l'espace de Hilbert. Paris:Masson et Cie;Budapest:Akadémiai Kiadó;1967.
268. Tahat N, Aydi H, Karapinar E, Shatanawi W, Common fixed points for single-valued and multi-valued maps satisfying a generalized contraction in G-metric spaces. Fixed Point Theory Appl. 2012;2012:Article ID 48.
269. Tiammee J, Suantai S, Coincidence point theorems for graph-preserving multi-valued mappings. Fixed Point Theory Appl. 2014;2014:Article ID 70.
270. Timiş I, Stability of the Picard iterative procedure for mappings which satisfy implicit relations. Comm. Appl. Nonlinear Anal. 2012;19:37–44.
271. Tiwari R, Prajapati PB, Bhardwaj R, Fixed point theorems for generalized almost contractive mappings in ordered metric spaces for integral type. Math. Theory Model. 2015;5:112–120.
272. Turinici M, Fixed points in complete metric spaces. Proceedings of the Institute of Mathematics Iaşi (1974), pp. 179–182. Bucharest:Editura Acad. R. S. R.;1976.
273. Turinici M, Weakly contractive maps in altering metric spaces. ROMAI J. 2013;9:175-183.
274. Turinici M, Linear contractions in product ordered metric spaces. Ann. Univ. Ferrara Sez. VII Sci. Mat. 2013;59:187–198.
275. Turinici M, Contractive operators in relational metric spaces. In: Handbook of Functional Equations, New York:Springer;2014;419–458.
276. Turkoglu AD, Özturk V, Common fixed point results for four mappings on partial metric spaces. Abstr. Appl. Anal. 2012;2012:Article ID 190862.
277. Udo-utun X, On inclusion of F-contractions in (δ, k)-weak contractions. Fixed Point Theory Appl. 2014;2014:Article ID 65.
278. Walter W, Remarks on a paper by F. Browder about contraction. Nonlinear Anal. TMA 1981;5:21–25.
279. Weinitschke H, Über eine Klasse von Iterationsverfahren. Numer. Math. 1964;6:395–404.
280. Zamfirescu T, Fix point theorems in metric spaces. Arch. Math. (Basel) 1972;23:292–298.

CHAPTER 3

Approximate Fixed Points

Mostafa Bachar*, Buthinah A. Bin Dehaish[†] and Mohamed Amine Khamsi[a‡§]

* King Saud University, Department of Mathematics, Riyadh, Saudi Arabia
† King Abdulaziz University, Department of Mathematics, Jeddah, Saudi Arabia
‡ University of Texas at El Paso, Department of Mathematical Sciences, El Paso, USA
§ King Fahd University of Petroleum and Minerals, Department of Mathematics and Statistics, Dhahran, Saudi Arabia
a Corresponding: mohamed@utep.edu

Contents

Abstract

In this chapter, we present some of the known results about the concept of approximate fixed points of a mapping. In particular, we discuss some new results on approximating fixed points of monotone mappings. Then we conclude this chapter with an application of these results to the case of a nonlinear semigroup of mappings. It is worth mentioning that approximate fixed points are useful when a computational approach is involved. In particular, most of the approximate fixed points discussed in this chapter are generated by an algorithm that allows its computational implementation.

3.1. Introduction

"The theory of fixed points is one of the most powerful tools of modern mathematics" quoted by Felix Browder, who gave a new impetus to the modern fixed point theory via the development of nonlinear functional analysis as an active and vital branch of mathematics. The flourishing field of fixed point theory started in the early days of topology (the work of Poincaré, Lefschetz-Hopf, and Leray-Schauder). This theory is applied to many areas of current interest in analysis, with topological considerations playing a crucial role, including the relationship with degree theory. For example,

http://dx.doi.org/10.1016/B978-0-12-804295-3.50003-6

existence problems are usually translated into a fixed point problem like the existence of solutions to elliptic partial differential equations, or the existence of closed periodic orbits in dynamical systems, and more recently the existence of answer sets in logic programming.

In metric fixed point theory, we study results that involve properties of an essentially isometric nature. The division between the metric fixed point theory and the more general topological theory is often a vague one. The use of successive approximations to establish the existence and uniqueness of solutions is the origin of the metric theory. It goes back to Cauchy, Liouville, Lipschitz, Peano, Fredholm and, especially, Picard. However, the Polish mathematician Banach is credited with placing the underlying ideas into an abstract framework suitable for broad applications well beyond the scope of elementary differential and integral equations. The fundamental fixed point theorem of Banach has laid the foundation of metric fixed point theory for contraction mappings on a complete metric space. Fixed point theory of certain important single-valued mappings is very interesting in its own right due to their results having constructive proofs and applications in industrial fields such as image processing engineering, physics, computer science, economics and telecommunication.

The theory of multivalued maps has applications in control theory, convex optimization, differential inclusions, and economics. Following the Banach contraction principle, Nadler [49] introduced the concept of multivalued contractions and established that a multivalued contraction possesses a fixed point in a complete metric space. Subsequently many authors generalized Nadler's fixed point theorem in different ways. A constructive proof of a fixed point theorem makes the theorem more valuable in view of the fact that it yields an algorithm for computing a fixed point.

Nonexpansive mappings are those maps which have Lipschitz constant equal to one. These mappings can obviously be viewed as a natural extension of contraction mappings. However, the fixed point problem for nonexpansive mappings differs greatly from that of the contraction mappings in the sense that additional structure of the domain set is needed to insure the existence of fixed points. It took almost four decades to see the first fixed point results for nonexpansive mappings in Banach and metric spaces following the publication in 1965 of the work of Browder [17], Göhde [33], and Kirk [42].

Recently a new direction has been discovered dealing with the extension of the Banach Contraction Principle [10] to metric spaces endowed with a partial order. Ran and Reurings [52] successfully carried out the first attempt. In particular, they showed how this extension is useful when dealing with some special matrix equations. Another similar approach was carried out by Nieto and Rodríguez-López [50] and used such arguments in solving some differential equations. Jachymski [35] gave a more general unified version of these extensions by considering graphs instead of a partial order (see also the recent papers [1–5]). Before we move on from these historical facts, let us point out that in fact the first attempt to generalize the Banach Contraction Principle to partially ordered metric spaces was carried out by Turinici [58, 59].

In this chapter, the focus is on approximating fixed points of mappings. The main tools used are based on the successive approximations generated by Ishikawa-Krasnoselskii-Mann iterations [34, 44, 45]. Note that these iterations provide algorithms which may be implemented in numerical computations.

In terms of content, this chapter overlaps in places with the following popular books on fixed point theory by Aksoy and Khamsi [6], by Goebel and Kirk [30], by Dugundji and Granas [23], by Khamsi and Kirk [40], and by Zeidler [60]. Material on the general theory of Banach space geometry is drawn from many sources but notably the books by Beauzamy [11], and by Diestel [22].

3.2. Approximate Fixed Points of Mappings in Banach Spaces

The core idea behind the concept of approximate fixed points of a mapping finds its root in the proof of the Banach Contraction Principle. Indeed, the key ingredient in the proof of this famous fixed point theorem is the convergence of an iteration procedure. This idea is very old [14]. The Banach Contraction Principle (BCP in short) uses the concept of Lipschitz mappings.

Definition 3.2.1. Let (M,d) be a metric space. The map $T : M \to M$ is said to be *Lipschitzian* if there exists a constant $k > 0$ (called a *Lipschitz constant*) such that

$$d(T(x),T(y)) \leq k\,d(x,y), \quad \text{for all } x,y \in M.$$

(a) A Lipschitzian mapping with a Lipschitz constant $k < 1$ is called *contraction*.

(b) A Lipschitzian mapping with a Lipschitz constant $k = 1$ is called *nonexpansive*.

A point $x \in M$ is a *fixed point* of T if $T(x) = x$. The set of fixed points of T is denoted by $Fix(T)$.

Theorem 3.2.1 (Banach Contraction Principle). *Let (M,d) be a complete metric space and let $T : M \to M$ be a contraction mapping, with Lipschitz constant $k < 1$. Then T has a unique fixed point ω in M, and for each $x \in M$, we have*

$$\lim_{n \to \infty} T^n(x) = \omega.$$

Moreover, for each $x \in M$ and $n,h \in \mathbb{N}$, we have

$$d\left(T^{n+h}(x),T^n(x)\right) \leq \frac{k^n}{1-k}d\left(T(x),x\right),$$

which implies

$$d\left(T^n(x),\omega\right) \le \frac{k^n}{1-k}d\left(T(x),x\right).$$

In this theorem, we have an interesting behavior of the mapping T. Indeed, under the assumptions of Theorem 3.2.1, we have

$$d\left(T^{n+1}(x),T^n(x)\right) \le \frac{k^n}{1-k}d\left(T(x),x\right),$$

for each $x \in M$ and $n \in \mathbb{N}$, which implies $\lim\limits_{n\to+\infty} d\left(T^{n+1}(x),T^n(x)\right) = 0$. This property suggests the following definition.

Definition 3.2.2. Let (M,d) be a metric space. The map $T : M \to M$ is said to be *asymptotically regular* if, for any $x \in M$, we have $\lim\limits_{n\to+\infty} d\left(T^{n+1}(x),T^n(x)\right) = 0$.

Therefore a contraction mapping is asymptotically regular. This behavior brings us to another observation. If $\lim\limits_{n\to+\infty} d\left(T^{n+1}(x_0),T^n(x_0)\right) = 0$, for some $x_0 \in M$, then

$$\inf\{d(x,T(x)); \, x \in M\} = 0.$$

In particular, for any $\varepsilon > 0$, there exists $x \in M$ such that $d(x,T(x)) \le \varepsilon$. Recall that the quantity $\inf\{d(x,T(x)); \, x \in M\}$ is known as the *minimal displacement* of the mapping T.

Definition 3.2.3. Let (M,d) be a metric space and $T : M \to M$ be a map. Let $\varepsilon > 0$. A point $x \in M$ is called an *ε-approximate fixed point* of T if $d(x,T(x)) \le \varepsilon$. The set of ε-approximate fixed points of T is denoted by $Fix_\varepsilon(T)$. A sequence $\{x_n\}$ is called an *approximate fixed point sequence* if $\lim\limits_{n\to+\infty} d\left(T(x_n),x_n\right) = 0$. The set of approximate fixed point sequences of T will be denoted by $AFPS(T)$.

From a computational point of view, an approximate fixed point is as good as the exact fixed point. The ideal case would be to have an algorithm that will generate an approximate fixed point sequence. Much research has been carried out with this idea in mind. As we have seen, the case of contraction mappings is simple and carries most of the good behavior. But when we move to the case of nonexpansive mappings, most of the simple facts become awfully complicated. Obviously the conclusion of Theorem 3.2.1 fails for nonexpansive mappings even if M is compact. As an example, one may take a geometric rotation on the unit circle in the plane $\mathbb{R}^2$. Some positive ideas started to emerge when M was a normed vector space. Indeed, let $(X, \|.\|)$ be a normed vector space. Let C be a nonempty convex subset of X. Let $T : C \to C$ be a nonexpansive mapping, i.e.,

$$\|T(x) - T(y)\| \le \|x - y\|, \quad \text{for any } x, y \in C.$$

Fix $x_0 \in C$ and $\lambda \in (0,1)$. Define the mapping $T_\lambda : C \to C$ by

$$T_\lambda(x) = (1 - \lambda)x_0 + \lambda T(x), \quad \text{for any } x \in C.$$

It is clear that T_λ is a contraction with λ as a Lipschitz constant. In order to use Theorem 3.2.1, we need to assume that C is complete. In order to achieve that, we assume that $(X, \|.\|)$ is complete, i.e., $(X, \|.\|)$ is a Banach space and C is a closed subset of X. In this case, T_λ has a unique fixed point $x_\lambda \in C$, i.e.,

$$x_\lambda = (1 - \lambda)x_0 + \lambda T(x_\lambda).$$

Moreover, $\{T_\lambda^n(x)\}$ converges to x_λ, for any $x \in C$. Another representation of this fact will be useful later on. Indeed, fix $x_0 \in C$ and consider the sequence $\{x_n\}_{n \ge 1}$ in C defined by

$$x_{n+1} = (1 - \lambda)x_0 + \lambda T(x_n), \ n \ge 0.$$

Then $\{x_n\}$ converges to x_λ. Moreover, we have $x_\lambda - T(x_\lambda) = (1 - \lambda)(x_0 - T(x_\lambda))$. Therefore, if C is bounded, i.e., $\delta(C) = diam(C) = \sup\{\|x - y\|; x, y \in C\} < +\infty$, then we have

$$\|x_\lambda - T(x_\lambda)\| \le (1 - \lambda) \, \delta(C),$$

which implies

$$\inf\{\|x - T(x)\|; x \in C\} \le (1 - \lambda) \, \delta(C), \quad \text{for any } \lambda \in (0,1).$$

Hence $\inf\{\|x - T(x)\|; x \in C\} = 0$. Recall that $\inf\{\|x - T(x)\|; x \in C\}$ is known as the *minimal displacement* of T. In particular, we have $Fix_\varepsilon(T) \ne \emptyset$, for any $\varepsilon > 0$. This result is amazing since easy examples may be found such that $Fix(T)$ is empty.

Example 3.2.1. Let $X = \mathscr{C}([0,1])$ be the space of continuous functions defined on $[0,1]$ endowed with the sup-norm. Let $C = \{f \in X; f(1) = 1 \text{ and } 0 \le f(t) \le 1, \text{ for all } t \in [0,1]\}$. Then C is a nonempty closed convex subset of X which is bounded. Define the map $T : C \to C$ by

$$T(f)(t) = tf(t), \quad t \in [0,1].$$

Then T is nonexpansive with an empty fixed point set. Indeed, if $f \in Fix(T)$, then we must have $f(t) = tf(t)$, for all $t \in [0,1]$. In particular, we have $f(t) = 0$ for $t \in [0,1)$. Since $f(t)$ is continuous, then we must have $f(1) = 0$. This is contradictory to the assumption $f \in C$.

Surprisingly the asymptotic regularity, observed for contraction mappings, occurs in a variant fashion for nonexpansive mappings. Indeed, Ishikawa [34] was the first to observe such behavior. Let us first start with the following technical lemma which has many useful applications.

Lemma 3.2.1 [29]. *Let $(X, \|.\|)$ be a Banach space, K a closed and convex nonempty subset of X, and $T : K \to K$ be nonexpansive. Let $\alpha \in (0,1)$ and set $T_\alpha = (1-\alpha)I + \alpha T$, where I is the identity mapping. Fix $x_0 \in K$ and define the sequence $\{x_n\}$ by*

$$x_{n+1} = T_\alpha(x_n) = (1-\alpha)x_n + \alpha T(x_n), \quad n \in \mathbb{N}.$$

Set $y_n = T(x_n)$, $n \in \mathbb{N}$. Then for each $i, n \in \mathbb{N}$, we have

$$(1+n\alpha)\|y_i - x_i\| \le \|y_{i+n} - x_i\| - (1-\alpha)^{-n}[\|y_{i+n} - x_{i+n}\| - \|y_i - x_i\|]. \tag{GK}$$

If K is bounded, then we have $\lim\limits_{n \to +\infty} \|x_n - T(x_n)\| = 0$, i.e., $\{x_n\}$ is an approximate fixed point sequence of T.

Proof. We proceed by induction to prove the inequality (GK). This inequality is trivial when we take $n = 0$, for any $i \in \mathbb{N}$. Assume it holds for n and let us prove it will hold for $n+1$ and $i \in \mathbb{N}$. Let us start with the inequality (GK) with n and $i+1$ to get

$$(1+n\alpha)\|y_{i+1} - x_{i+1}\| \le \|y_{i+n+1} - x_{i+1}\| - (1-\alpha)^{-n}\Big[\|y_{i+n+1} - x_{i+n+1}\| - \|y_{i+1} - x_{i+1}\|\Big].$$

Moreover, since T is nonexpansive, we have

$$\begin{aligned}
\|y_{i+n+1} - x_{i+1}\| &\le (1-\alpha)\|y_{i+n+1} - x_i\| + \alpha\|y_{i+n+1} - y_i\| \\
&\le (1-\alpha)\|y_{i+n+1} - x_i\| + \alpha \sum_{k=0}^{n} \|y_{i+k+1} - y_{i+k}\| \\
&\le (1-\alpha)\|y_{i+n+1} - x_i\| + \alpha \sum_{k=0}^{n} \|x_{i+k+1} - x_{i+k}\|.
\end{aligned}$$

Combining the two inequalities, we get

$$\begin{aligned}
\|y_{i+n+1} - x_i\| &\ge (1-\alpha)^{-(n+1)}\Big[\|y_{i+n+1} - x_{i+n+1}\| - \|y_{i+1} - x_{i+1}\|\Big] \\
&\quad + (1-\alpha)^{-1}(1+n\alpha)\|y_{i+1} - x_{i+1}\| - \alpha(1-\alpha)^{-1} \sum_{k=0}^{n} \|x_{k+i+1} - x_{k+i}\|.
\end{aligned}$$

Since $\|x_{k+i+1} - x_{k+i}\| = \alpha\|y_{k+i} - x_{k+i}\|$, and since the sequence $\{\|y_n - x_n\|\} = \{\alpha^{-1}\|x_{n+1} - x_n\|\}$ is decreasing and $(1+n\alpha) \le (1-\alpha)^{-n}$, we have

$$\|y_{i+n+1} - x_i\| \geq (1-\alpha)^{-(n+1)}\Big[\|y_{i+n+1} - x_{i+n+1}\| - \|y_{i+1} - x_{i+1}\|\Big]$$
$$+ (1-\alpha)^{-1}(1+n\alpha)\|y_{i+1} - x_{i+1}\| - \alpha^2(1-\alpha)^{-1}(n+1)\|y_i - x_i\|$$
$$= (1-\alpha)^{-(n+1)}\Big[\|y_{i+n+1} - x_{i+n+1}\| - \|y_i - x_i\|\Big]$$
$$+ \Big[(1-\alpha)^{-1}(1+n\alpha) - (1-\alpha)^{-(n+1)}\Big]\|y_{i+1} - x_{i+1}\|$$
$$+ \Big[(1-\alpha)^{-(n+1)} - \alpha^2(1-\alpha)^{-1}(n+1)\Big]\|y_i - x_i\|$$
$$\geq (1-\alpha)^{-(n+1)}\Big[\|y_{i+n+1} - x_{i+n+1}\| - \|y_i - x_i\|\Big]$$
$$\Big[(1-\alpha)^{-1}(1+n\alpha) - (1-\alpha)^{-(n+1)}\Big]\|y_i - x_i\|$$
$$+ \Big[(1-\alpha)^{-(n+1)} - \alpha^2(1-\alpha)^{-1}(n+1)\Big]\|y_i - x_i\|$$
$$= (1-\alpha)^{-(n+1)}\Big[\|y_{i+n+1} - x_{i+n+1}\| - \|y_i - x_i\|\Big]$$
$$+ (1+(n+1)\alpha)\|y_i - x_i\|.$$

Thus the inequality (GK) holds for $n+1$ and any $i \in \mathbb{N}$. By induction, it holds for any $i, n \in \mathbb{N}$. Next we assume that K is bounded, i.e.,

$$\delta(K) = diam(K) = \sup\{\|x - y\|; x, y \in K\} < +\infty.$$

Since $\{\|y_n - x_n\|\}$ is decreasing, $\lim_{n \to +\infty} \|y_n - x_n\| = R$ exists. The inequality (GK) implies

$$(1+n\alpha)\|y_i - x_i\| \leq \delta(K) - (1-\alpha)^{-n}\Big[\|y_{i+n} - x_{i+n}\| - \|y_i - x_i\|\Big], \text{ for any } i, n \in \mathbb{N}.$$

Fix $n \in \mathbb{N}$ and let $i \to +\infty$ to get

$$(1+n\alpha)R \leq \delta(K) - (1-\alpha)^{-n}\Big[R - R\Big] = \delta(K).$$

Hence

$$R \leq \frac{\delta(K)}{(1+n\alpha)}, \quad \text{for any } n \in \mathbb{N}.$$

Obviously this will force $R = 0$, i.e.,

$$\lim_{n \to +\infty} \|y_n - x_n\| = \lim_{n \to +\infty} \|T(x_n) - x_n\| = 0.$$

$\square$

Using this beautiful lemma, we are able to prove the following qualitative result.

Theorem 3.2.2 [29]. *Let $(X, \|.\|)$ be a Banach space, and K a bounded, closed and convex nonempty subset of X. Denote by $\mathscr{F}$ the collection of all nonexpansive self* *mappings defined on K. Fix $\alpha \in (0,1)$. For any $T \in \mathscr{F}$, set $T_\alpha = (1-\alpha)I + \alpha T$, where I is the identity mapping. Then T_α is asymptotically regular on K. Moreover, the sequence $\{\|T_\alpha^n(x) - T_\alpha^{n+1}(x)\|\}$ converges to 0 uniformly for $x \in K$ and $T \in \mathscr{F}$. Precisely, if $\varepsilon > 0$, then there exists $N \in \mathbb{N}$, depending only on ε and K such that if $n \geq N$, if $x \in K$, and if $T \in \mathscr{F}$, then*

$$\|T_\alpha^n(x) - T_\alpha^{n+1}(x)\| \leq \varepsilon.$$

As a direct consequence of Theorem 3.2.2, we have the following result [29].

Corollary 3.2.1. *Let $(X, \|.\|)$ be a Banach space, and K a bounded, closed and convex nonempty subset of X. Let $T : K \to K$ be a nonexpansive mapping for which $\overline{T(K)}$ is compact. Then for $\alpha \in (0,1)$, the iterates of the mapping $T_\alpha = (1-\alpha)I + \alpha T$, where I is the identity mapping, converge to a fixed point of T.*

Proof. Fix $x_0 \in K$ and $\alpha \in (0,1)$. Consider the sequence $\{x_n\}$ defined by

$$x_{n+1} = (1-\alpha)x_n + \alpha T(x_n), \quad n \geq 0.$$

Moreover, we have seen in the proof of Lemma 3.2.1 that $\{\|x_n - T(x_n)\|\}$ is decreasing. Hence

$$\|x_n\| \leq \|x_n - T(x_n)\| + \|T(x_n)\| \leq \|x_0 - T(x_0)\| + \|T(x_n)\|, \quad n \geq 0.$$

Since $\overline{T(K)}$ is compact, then $\{T(x_n)\}$ is bounded. The above inequality implies that $\{x_n\}$ is also bounded. Using the inequality (GK), we conclude that $\lim\limits_{n \to +\infty} \|x_n - T(x_n)\| = 0$. Using the compactness of $\overline{T(K)}$, there exists a subsequence $\{T(x_{\phi(n)})\}$ of $\{T(x_n)\}$ which converges to ω. Hence $\{x_{\phi(n)}\}$ also converges to ω. Since T is continuous, then $T(\omega) = \omega$, i.e., ω is a fixed point of T. Next notice that $\{\|x_n - \omega\|\}$ is decreasing. Indeed, we have

$$\begin{aligned}
\|x_{n+1} - \omega\| &\leq (1-\alpha)\|x_n - \omega\| + \alpha\|T(x_n) - \omega\| \\
&\leq (1-\alpha)\|x_n - \omega\| + \alpha\|T(x_n) - T(\omega)\| \\
&\leq (1-\alpha)\|x_n - \omega\| + \alpha\|x_n - \omega\| = \|x_n - \omega\|,
\end{aligned}$$

for any $n \geq 0$. Therefore, the sequence $\{x_n\} = \{T_\alpha^n(x_0)\}$ converges to a fixed point of T. $\qquad \square$

Remark 3.2.1. Lemma 3.2.1 actually holds when we allow α to change. Indeed, let $\{\alpha_n\}$ be a sequence for which $0 \leq \alpha_n \leq b < 1$ and $\sum\limits_{n=0}^{\infty} \alpha_n = +\infty$. Let $T : K \to K$ be nonexpansive and $x_0 \in K$. Define the sequence $\{x_n\}$ by

$$x_{n+1} = (1 - \alpha_n)x_n + \alpha_n T(x_n), \quad n \geq 0.$$

Later on in this chapter, we will prove an analog to the inequality (GK). The earlier version is due to Ishikawa [34]. Edelstein and O'Brien [25] gave a proof of Theorem 3.2.2 for a single mapping. The conclusion of Corollary 3.2.1 was known to Krasnoselskii [45] when X is uniformly convex and $\alpha = 1/2$. Edelstein [24] extended Krasnoselskii's result to the case when X is strictly convex. As for the weakening of the compactness assumption, Genel and Lindenstrauss [28] constructed an example of a weakly compact convex subset K in the Hilbert space ℓ^2 and a nonexpansive mapping $T : K \rightarrow K$ with the property that some iterates of $T_{1/2}$ fail to converge to a fixed point of T. Finally Baillon, Bruck and Reich [9] extended the conclusion of Corollary 3.2.1 in another direction. Indeed, they showed that if $T : K \rightarrow K$ is nonexpansive for an arbitrary convex set K and if $T_\alpha = (1 - \alpha)I + \alpha T$, $\alpha \in (0,1)$, then T_α satisfies

$$\lim_{n \to +\infty} \left\| T_\alpha^{n+1}(x) - T_\alpha^n(x) \right\| = \frac{1}{k} \lim_{n \to +\infty} \left\| T_\alpha^{n+k}(x) - T_\alpha^n(x) \right\| = \lim_{n \to +\infty} \frac{1}{n} \left\| T_\alpha^n(x) \right\|,$$

for any $x \in K$.

3.3. Approximate Fixed Points of Mappings in Hyperbolic Spaces

Most of the fundamental early results discovered for nonexpansive mappings were done in the context of Banach spaces. It is then natural to try to develop a similar theory in the nonlinear spaces. The closest class of sets considered was the class of hyperbolic spaces that enjoys convexity properties very similar to the linear one. This class of metric spaces includes all normed vector spaces, Hadamard manifolds, as well as the Hilbert ball and the Cartesian product of Hilbert balls. Extensive information on hyperbolic spaces and a detailed treatment of examples like the Hilbert ball can be found in Refs. [29, 31, 43, 53, 54].

Let (X,d) be a metric space and let $\mathbb{R}$ denote the real line. We say that a mapping $c : \mathbb{R} \rightarrow X$ is a *metric embedding* of $\mathbb{R}$ into X if

$$d(c(s),c(t)) = |s - t|, \quad \text{for all real } s, t \in \mathbb{R}.$$

The image of $\mathbb{R}$ under a metric embedding will be called a *metric line*. The image $c([a,b]) \subset X$ of a real interval under a metric embedding will be called a *metric segment*. Let $x, y \in X$. A metric segment $c([a,b])$ is said to join x and y if $c(a) = x$ and $c(b) = y$. We will say that (X,d) is of *hyperbolic type* if X contains a family of metric segments such that for each pair of distinct points x and y in X there is a unique metric line which joins x and y. We will denote by $[x,y]$ or $[y,x]$ the unique metric segment joining the two points x and y from X. Clearly, we have $[x,x] = \{x\}$, for any $x \in X$. Next we give some basic facts about metric spaces of hyperbolic type which shows to some extent their natural similarity to the normed vector spaces.

Lemma 3.3.1. *Let $x, y \in X$, $x \neq y$ and $z, w \in [x, y]$. Then*

(a) $d(x, z) \leq d(x, y)$;

(b) *if $d(x, z) = d(x, w)$, then $z = w$.*

The next result brings us closer to the definition of a convex combination in metric spaces of hyperbolic type.

Lemma 3.3.2. *Let $c : \mathbb{R} \to X$ be a metric embedding, $a \leq b \in \mathbb{R}$ and $t \in [a, b]$. Then*

(a) $d(c(a), c((1-t)a + tb)) = t \, d(c(a), c(b))$;

(b) $d(c(b), c((1-t)a + tb)) = (1-t) \, d(c(a), c(b))$.

Proposition 3.3.1. *Let (X, d) be a metric space of hyperbolic type. Let $x, y \in X$. For each $\alpha \in [0, 1]$, there is a unique point $z \in [x, y]$ such that*

$$d(x, z) = \alpha \, d(x, y) \text{ and } d(y, z) = (1 - \alpha) \, d(x, y).$$

This point will be denoted by $(1 - \alpha)x \oplus \alpha y$ for obvious reasons.

Obviously we have $(1 - \alpha)x \oplus \alpha x = x$, and if $x \neq y$, then for any $z \in [x, y]$, we have $z = (1 - \alpha)x \oplus \alpha y$, with $\alpha = \dfrac{d(x, z)}{d(x, y)}$. This will imply that for any $z \in [x, y]$, we have $d(x, z) + d(z, y) = d(x, y)$. Also if, for some $\alpha, \beta \in [0, 1]$, we have $(1 - \alpha)x \oplus \alpha y = (1 - \beta)x \oplus \beta y$, then $\alpha = \beta$. Moreover, we have $(1 - \alpha)x \oplus \alpha y = \alpha y \oplus (1 - \alpha)x$. The following is a direct consequence of all these properties:

$$[x, y] = \{(1 - \alpha)x \oplus \alpha y; \ \alpha \in [0, 1]\}.$$

Definition 3.3.1 [31]. We say that (M, d) is a *hyperbolic metric space* if

$$d\left(\frac{1}{2}x \oplus \frac{1}{2}y, \frac{1}{2}x \oplus \frac{1}{2}z\right) \leq \frac{1}{2}d(y, z), \quad \text{for all } x, y, z \in X.$$

Note that (M, d) is hyperbolic if and only if

$$d((1 - \alpha)x \oplus \alpha y, (1 - \alpha)z \oplus \alpha w) \leq (1 - \alpha)d(x, z) + \alpha d(y, w),$$

for any $\alpha \in [0, 1]$ and all $x, y, z, w \in X$ [31, 44, 47, 53]. Recall that a subset C of a hyperbolic metric space M is said to be *convex* whenever $[x, y] \subset C$ for any $x, y \in C$.

Definition 3.3.2. Let (M,d) be a hyperbolic metric space. We say that M is *uniformly convex* if for any $a \in M$, for every $r > 0$, and for each $\varepsilon > 0$

$$\delta(r,\varepsilon) = \inf \left\{ 1 - \frac{1}{r} d \left(\frac{1}{2}x \oplus \frac{1}{2}y, a \right) ; d(x,a) \le r, d(y,a) \le r, d(x,y) \ge r\varepsilon \right\} > 0.$$

The definition of uniform convexity finds its origin in Banach spaces [19]. To the best of our knowledge, the first attempt to generalize this concept to metric spaces was made in Ref. [32]. The reader may also consult Refs. [31, 38, 53]. From now onwards we assume that M is a hyperbolic metric space and if (M,d) is uniformly convex, then for every $s \ge 0$ and $\varepsilon > 0$, there exists $\eta(s,\varepsilon) > 0$ depending on s and ε such that

$$\delta(r,\varepsilon) > \eta(s,\varepsilon) > 0, \quad \text{for any } r > s.$$

Recall that the hyperbolic metric space (M,d) is said to be *strictly convex* if whenever

$$d(a,x) = d(a,y) = d(a, \lambda x \oplus (1-\lambda)y)$$

for any $a, x, y \in M$ and $\lambda \in (0,1)$, then we must have $x = y$.

Remark 3.3.1 [38, 41].

(a) Let us observe that $\delta(r,0) = 0$, and $\delta(r,\varepsilon)$ is an increasing function of ε for every fixed r.

(b) For $r_1 \le r_2$ there holds

$$1 - \frac{r_2}{r_1} \left(1 - \delta \left(r_2, \varepsilon \frac{r_1}{r_2} \right) \right) \le \delta(r_1, \varepsilon).$$

(c) If (M,d) is uniformly convex, then (M,d) is strictly convex.

Lemma 3.3.3 [38, 41]. *Assume that (M,d) is a uniformly convex hyperbolic metric space. Let $\{C_n\} \subset M$ be a sequence of nonempty, decreasing, convex, bounded and closed sets. Let $x \in M$ be such that*

$$0 < d = \lim_{n \to \infty} d(x, C_n) < \infty.$$

Let $x_n \in C_n$ be such that $d(x, x_n) \to d$. Then $\{x_n\}$ is a Cauchy sequence.

Proof. Assume to the contrary that this is not the case. Passing to a subsequence if necessary, we can assume that there exists $\varepsilon_0 > 0$ such that

$$\varepsilon_0 \le d(x_k, x_p), \quad k \ne p.$$

Set $d_k = d(x,x_k)$, for any $k \geq 1$. Since (M,d) is uniformly convex, we have

$$d\left(x, \frac{1}{2}x_k \oplus \frac{1}{2}x_p\right) \leq \left(1 - \delta\left(d(k,p), \frac{\varepsilon_0}{d(k,p)}\right)\right) d(k,p),$$

where $d(k,p) = \max\{d_k, d_p\}$. Without loss of generality, we may assume that $d(k,p) \leq 2d$ for each $k,p \geq N$, where N is fixed. Hence

$$d\left(x, \frac{1}{2}x_k \oplus \frac{1}{2}x_p\right) \leq \left(1 - \delta\left(d(k,p), \frac{\varepsilon_0}{2d}\right)\right) d(k,p).$$

Using the uniform convexity of (M,d), there exists $\eta = \eta\left(\frac{d}{3}, \frac{\varepsilon_0}{2d}\right) > 0$ such that

$$\delta\left(d(k,p), \frac{\varepsilon_0}{2d}\right) > \eta\left(\frac{d}{3}, \frac{\varepsilon_0}{2d}\right) > 0,$$

which implies, in view of the fact the sets C_n, $n \geq 1$, are convex and nonincreasing, that

$$\min\{d(x,C_k), d(x,C_p)\} \leq d\left(x, \frac{1}{2}x_k \oplus \frac{1}{2}x_p\right) \leq (1-\eta)d(k,p).$$

Hence

$$\min\{d(x,C_k), d(x,C_p)\} \leq (1-\eta)d(k,p).$$

Letting k and p go to infinity, we get that $0 < d \leq (1-\eta)d$, where $\eta > 0$. This is a contradiction. $\qquad \square$

Recall that a hyperbolic metric space (M,d) is said to have the property (R) if any decreasing sequence of nonempty, convex, bounded and closed sets has a nonempty intersection [36].

Our next result deals with the existence and the uniqueness of the best approximants of convex, closed and bounded sets in a uniformly convex metric space. This result is of interest by itself as uniform convexity implies the property (R), which reduces to reflexivity in the linear case.

Theorem 3.3.1 [38, 41]. *Let (M,d) be a complete and uniformly convex hyperbolic metric space. Let $C \subset M$ be nonempty, convex and closed. Let $x \in M$ be such that $d(x,C) < \infty$. Then there exists a unique best approximant of x in C, i.e., there exists a unique $x_0 \in C$ such that*

$$d(x,x_0) = d(x,C).$$

Proof. Denote $d_0 = d(x,C)$. Let $\{x_n\} \subset C$ be such that

$$d_0 = d(x,C) = \lim_{n \to \infty} d(x_n,x).$$

If $d_0 = 0$, then $x \in C$ since C is closed. Then we must have $x_0 = x$. So we can assume then that $d_0 > 0$. Hence, from Lemma 3.3.3 (applied to $C_n = C$), the sequence $\{x_n\}$ is Cauchy. Since M is complete and C is closed, there exists then $x_0 \in C$ such that

$$\lim_{n \to \infty} d(x_n,x_0) = 0.$$

We claim that x_0 is the best approximant we are seeking. Indeed, we have

$$d(x,C) \leq d(x,x_0) = \lim_{n \to \infty} d(x,x_n) = d(x,C).$$

Hence $d(x,x_0) = d(x,C)$, i.e., x_0 is an approximant. Assume there exists another approximant $y \in C$, i.e., $d(x,y) = d(x,C)$. Since C is convex, we get

$$d(x,C) \leq d\left(x, \frac{1}{2}x_0 \oplus \frac{1}{2}y\right) \leq \frac{d(x,x_0) + d(x,y)}{2} = d(x,C).$$

Hence

$$d\left(x, \frac{1}{2}x_0 \oplus \frac{1}{2}y\right) = d(x,x_0) = d(x,y),$$

which implies that $x_0 = y$ since (M,d) is strictly convex. □

The following result gives the analog of the well-known theorem that states any uniformly convex Banach space is reflexive (see Ref. [31, Theorem 2.1]).

Theorem 3.3.2 [38, 41]. *If (M,d) is a complete and uniformly convex hyperbolic metric space, then (M,d) has the property (R).*

Proof. Let $\{C_n\} \subset M$ be a sequence of nonempty, nonincreasing, convex, bounded and closed sets. We need to prove that this sequence of sets has nonempty intersection. Let $x \in M$. Since sets in $\{C_n\}$ are bounded and nonincreasing, the sequence $\{d(x,C_n)\}$ is increasing and bounded. Hence $\lim_{n \to \infty} d(x,C_n) = d_1$ exists. Let $x_n \in C_n$ be an approximant of x, i.e., $d(x,x_n) = d(x,C_n)$, for any $n \geq 1$. Lemma 3.3.3 implies that $\{x_n\}$ is Cauchy. Hence there exists $y \in M$ such that $\lim_{n \to \infty} x_n = y$. Since sets in $\{C_n\}$ are nonincreasing and closed, thus $y \in C_n$, for any $n \geq 1$. Hence $\bigcap_{n \geq 1} C_n$ is not empty. □

Note that any hyperbolic metric space M which satisfies the property (R) is complete. Indeed, let $\{x_n\}$ be a Cauchy sequence in M. Denote

$$\varepsilon_n = \sup\{d(x_m, x_s); \ m, s \geq n\}, \quad n = 1, \cdots$$

Our assumption implies that $\lim_{n \to \infty} \varepsilon_n = 0$. In hyperbolic metric spaces, closed balls are convex. Therefore the property (R) implies that $\bigcap_{n \geq 1} B(x_n, \varepsilon_n) \neq \emptyset$. It is easy to check that this intersection is reduced to one point which is the limit of $\{x_n\}$.

The following technical lemma will be useful throughout.

Lemma 3.3.4 [38, 41]. *Let (M, d) be a uniformly convex hyperbolic metric space. Assume that there exists $R \in [0, +\infty)$ such that*

$$\limsup_{n \to \infty} d(x_n, a) \leq R, \ \limsup_{n \to \infty} d(y_n, a) \leq R, \ and \ \lim_{n \to \infty} d\left(a, \lambda x_n \oplus (1 - \lambda) y_n\right) = R,$$

for some $\lambda \in (0, 1)$. Then $\lim_{n \to \infty} d(x_n, y_n) = 0$.

Proof. Without loss of generality, we may assume that $R > 0$. Let us first prove the Lemma when $\lambda = \frac{1}{2}$. Assume the conclusion is not true. Let $\gamma > 0$ be arbitrarily chosen. For n sufficiently large, passing to subsequences, if necessary, we may assume that there exists $\varepsilon > 0$ such that $d(x_n, a) \leq R + \gamma$, $d(y_n, a) \leq R + \gamma$ and $d(x_n, y_n) \geq R\varepsilon$, $n \geq 1$. Since (M, d) is uniformly convex, we have

$$0 < \eta(R, \varepsilon) < \delta(R + \gamma, \varepsilon) \leq 1 - \frac{1}{R + \gamma} d\left(a, \frac{1}{2} x_n \oplus \frac{1}{2} y_n\right) \to \frac{\gamma}{R + \gamma}.$$

Letting $\gamma \to 0$, we get a contradiction. Therefore the conclusion of Lemma 3.3.4 is true when $\lambda = \frac{1}{2}$. Assume that $\lambda \in (1/2, 1)$ (the case of $\lambda \in (0, 1/2)$ is done in a similar way), and set

$$\bar{y}_n = (2\lambda - 1) x_n \oplus 2(1 - \lambda) y_n, \quad \text{for } n \geq 1.$$

Then we have

$$\limsup_{n \to \infty} d(\bar{y}_n, a) \leq (2\lambda - 1) \limsup_{n \to \infty} d(x_n, a) + 2(1 - \lambda) \limsup_{n \to \infty} d(y_n, a) \leq R,$$

and $\lim_{n \to \infty} d\left(a, \frac{1}{2} x_n \oplus \frac{1}{2} \bar{y}_n\right) = R$. Therefore, we must have

$$\lim_{n \to \infty} d(x_n, \bar{y}_n) = 2(1 - \lambda) \lim_{n \to \infty} d(x_n, y_n) = 0.$$

This completes the proof of Lemma 3.3.4 since $\lambda \neq 1$. $\qquad\qquad\square$

As we have seen, we were able to extend many linear properties to the case of hyperbolic metric spaces. The weak topology is still hard to capture in the nonlinear case. An approach to the weak convergence was offered by Kuczumow [46] and Lim [48] which they called Δ-*convergence* . Their approach was very successful in the case of $CAT(0)$ spaces. First, we need the definition of the asymptotic radius of a sequence.

Definition 3.3.3. Let (M,d) be a metric space and C a nonempty subset of M. Let $\{x_n\}$ be a bounded sequence in M. Define $r(.,\{x_n\}) : C \to [0,\infty)$ by

$$r(x,\{x_n\}) = \limsup_{n\to\infty} d(x,x_n).$$

The *asymptotic radius* ρ_C of $\{x_n\}$ with respect to C is given by

$$\rho_C = \inf\{r(x,\{x_n\}) : x \in C\}.$$

ρ will denote the asymptotic radius of $\{x_n\}$ with respect to M. A point $\xi \in C$ is said to be an *asymptotic center* of $\{x_n\}$ with respect to C if $r(\xi,\{x_n\}) = r(C,\{x_n\}) = \min\{r(x,\{x_n\}); x \in C\}$.

The set of all asymptotic centers of $\{x_n\}$ with respect to C will be denoted by $A(C,\{x_n\})$. When $C = M$, we we use the notation $A(\{x_n\})$ instead of $A(M,\{x_n\})$. In general, the set $A(C,\{x_n\})$ of asymptotic centers of a bounded sequence $\{x_n\}$ may be empty or contain more than one point. Note that the asymptotic radius is also known in the literature as a type function. For more on this, see Ref. [38].

Definition 3.3.4. Let (M,d) be a metric space. A bounded sequence $\{x_n\}$ in M is said to Δ-*converge* to $x \in M$ if and only if x is the unique asymptotic center of every subsequence $\{u_n\}$ of $\{x_n\}$. We write $x_n \overset{\Delta}{\to} x$ whenever $\{x_n\}$ Δ-converges to x.

In the sequel, the following results will be needed.

Lemma 3.3.5 [27, 38]. *Let (M,d) be a hyperbolic metric space. Assume that M is uniformly convex. Let C be a nonempty, closed and convex subset of M. Then every bounded sequence $\{x_n\} \in M$ has a unique asymptotic center with respect to C.*

Lemma 3.3.6 [27, 38]. *Let (M,d) be a hyperbolic metric space. Assume that M is uniformly convex. Let C be a nonempty, closed and convex subset of M. Let $\{x_n\}$ be a bounded sequence in C such that $A(\{x_n\}) = \{y\}$ and $r(\{x_n\}) = \rho$. If $\{y_m\}$ is a sequence in C such that $\lim_{m\to\infty} r(y_m,\{x_n\}) = \rho$, then $\lim_{m\to\infty} y_m = y$.*

The following result is similar to the demi-closed principle discovered by Göhde in uniformly convex Banach spaces [33].

Lemma 3.3.7 [27]. *Let C be a nonempty, closed and convex subset of a complete uniformly convex hyperbolic space (M,d). Let $T : C \to C$ be a nonexpansive mapping. Let $\{x_n\} \in C$ be an approximate fixed point sequence of T, i.e., $\lim\limits_{n \to \infty} d(x_n, Tx_n) = 0$. If $x \in C$ is the asymptotic center of $\{x_n\}$ with respect to C, then x is a fixed point of T. In particular, if $\{x_n\} \in C$ is an approximate fixed point sequence of T such that $x_n \overset{\Delta}{\to} x$, then x is a fixed point of T.*

Once we lay down the foundations of most interesting known geometric properties in hyperbolic metric spaces, we turn our attention to the extension of the successive iterations studied in Banach spaces that allowed us to generate approximate fixed points. First we need some notations. For any sequence (λ_n) in $[0,1)$, i.e., $0 \le \lambda_n < 1$, denote

$$S_{i,n} = \sum_{s=i}^{i+n-1} \lambda_s \text{ and } P_{i,n} = \prod_{s=i}^{i+n-1} (1 - \lambda_s)^{-1}, \quad \text{for any } i, n \in \mathbb{N}.$$

The following important result is very useful.

Proposition 3.3.2 [29]. *Let (M,d) be a hyperbolic metric space. Let $\{x_n\}$ and $\{y_n\}$ be two sequences in (M,d) such that*

$$x_{n+1} = (1 - \lambda_n)x_n \oplus \lambda_n y_n,$$

with $\lambda_n \in [0,1)$, for any $n \in \mathbb{N}$. Suppose that

$$d(y_n, y_{n+1}) \le d(x_n, x_{n+1}), \quad n \in \mathbb{N}.$$

Then $\{d(x_n, y_n)\}$ is decreasing and for all $i, n \in \mathbb{N}$, we have

$$(1 + S_{i,n})d(x_i, y_i) \le d(x_i, y_{i+n}) + P_{i,n}\Big[d(x_i, y_i) - d(x_{i+n}, y_{i+n})\Big].$$

Note that under the assumptions of Proposition 3.3.2, we have

$$S_{i,n}\, d(x_i, y_i) \le d(x_i, x_{i+n}) + P_{i,n}[d(x_i, y_i) - d(x_{i+n}, y_{i+n})].$$

Indeed, we have

$$d(x_i, y_{i+n}) - d(x_i, y_i) \le d(x_i, x_{i+n}) + d(x_{i+n}, y_{i+n}) - d(x_i, y_i) \le d(x_i, x_{i+n}),$$

since $\{d(x_n, y_n)\}$ is decreasing which implies $d(x_{i+n}, y_{i+n}) - d(x_i, y_i) \le 0$, for any $i, n \in \mathbb{N}$. Using Proposition 3.3.2, we get the following beautiful result.

Proposition 3.3.3 [29]. *In addition to the assumptions of Proposition 3.3.2, assume that*

(i) *the set $\{d(x_i, y_{i+n}); i, n \in \mathbb{N}\}$ is bounded;*

(ii) $\sum\limits_{n=1}^{+\infty} \lambda_n = +\infty;$

(iii) *there exists $b \in [0,1)$ such that $\lambda_n \leq b$, for any $n \in \mathbb{N}$.*

Then we have

$$\lim_{n \to +\infty} d(x_n, y_n) = 0.$$

Proof. Since $d(y_n, y_{n+1}) \leq d(x_n, x_{n+1})$, for any $n \in \mathbb{N}$, then $\{d(x_n, y_n)\}$ is decreasing. Indeed, we have

$$
\begin{aligned}
d(x_{n+1}, y_{n+1}) &\leq (1 - \lambda_n)d(x_n, y_{n+1}) + \lambda_n d(y_n, y_{n+1}) \\
&\leq (1 - \lambda_n)d(x_n, y_n) + (1 - \lambda_n)d(y_n, y_{n+1}) + \lambda_n d(y_n, y_{n+1}) \\
&= (1 - \lambda_n)d(x_n, y_n) + d(y_n, y_{n+1}) \\
&\leq (1 - \lambda_n)d(x_n, y_n) + d(x_n, x_{n+1}) \\
&= (1 - \lambda_n)d(x_n, y_n) + \lambda_n d(x_n, y_n) = d(x_n, y_n),
\end{aligned}
$$

for any $n \in \mathbb{N}$. Assume that $\lim\limits_{n \to +\infty} d(x_n, y_n) = R > 0$. Let $\varepsilon > 0$ such that

$$\varepsilon \exp\left[(1-b)^{-1}\left(\frac{d}{1+R}\right)\right] < R,$$

where $d = \sup\{d(x_i, y_{i+n}); i, n \in \mathbb{N}\} < +\infty$. Choose $i \in \mathbb{N}$ such that for all $n \geq 1$, we have

$$d(x_i, y_i) - d(x_{i+n}, y_{i+n}) \leq \varepsilon.$$

Now choose $N \geq 1$ such that

$$\sum_{s=i}^{i+N-2} \lambda_s \leq \frac{d}{R} \text{ while } \sum_{s=i}^{i+N-1} \lambda_s \geq \frac{d}{R}.$$

It follows that $R \sum\limits_{s=i}^{i+N-1} \lambda_s < d + R$, and therefore

$$
\begin{aligned}
P_{i,N} &= \prod_{s=i}^{i+N-1} (1 - \lambda_s)^{-1} = \prod_{s=i}^{i+N-1}\left(1 + \lambda_s(1 - \lambda_s)^{-1}\right) \\
&= \exp\left[\sum_{s=i}^{i+N-1} \log\left(1 + \lambda_s(1 - \lambda_s)^{-1}\right)\right] \\
&\leq \exp\left[\sum_{s=i}^{i+N-1} \lambda_s(1 - \lambda_s)^{-1}\right] \\
&\leq \exp\left[(1-b)^{-1} \sum_{s=i}^{i+N-1} \lambda_s\right].
\end{aligned}
$$

Proposition 3.3.2 and the above choices for ε, i and N yield

$$
\begin{aligned}
d + R &\leq \left(1 + \sum_{s=i}^{i+N-1} \lambda_s\right) R \leq \left(1 + \sum_{s=i}^{i+N-1} \lambda_s\right) d(x_i, y_i) \\
&\leq d(x_i, y_{i+N}) + \varepsilon \, \exp\left[(1-b)^{-1} \sum_{s=i}^{i+N-1} \lambda_s\right] \\
&\leq d + \varepsilon \, \exp\left[(1-b)^{-1}(1 + d/R)\right] \\
&< d + R.
\end{aligned}
$$

This contradiction completes the proof of Proposition 3.3.3.   □

Note that it is easy to check that $\{x_n\}$ is bounded if and only if $\{y_n\}$ is bounded if and only if the set $\{d(x_i, y_{i+n}); \ i, n \in \mathbb{N}\}$ is bounded.

As an application to both Propositions 3.3.2 and 3.3.3, we have the following approximate fixed point result.

Theorem 3.3.3. *Let (M, d) be a hyperbolic metric space and C a nonempty closed convex and bounded subset of M. Let $T : C \to C$ be a nonexpansive mapping. Fix $x_0 \in C$ and consider the Krasnoselskii-Mann iteration $\{x_n\}$ in C generated by x_0 and $(\lambda_n) \subset [0,1]$ and defined by*

$$
x_{n+1} = (1 - \lambda_n) x_n \oplus \lambda_n T(x_n), \quad \text{for any } n \in \mathbb{N}.
$$

Assume that

(i) $\displaystyle\sum_{n=1}^{+\infty} \lambda_n = +\infty$;

(ii) *there exists $b \in [0, 1)$ such that $\lambda_n \leq b$, for any $n \in \mathbb{N}$.*

Then we have

$$
\lim_{n \to +\infty} d(x_n, T(x_n)) = 0,
$$

i.e., $\{x_n\}$ is an approximate fixed point sequence of T. In particular, we have

$$
\inf\{d(x, T(x)) : x \in C\} = 0.
$$

The boundedness assumption in Theorem 3.3.3 is crucial. We may relax it if we assume that the hyperbolic metric space is uniformly convex.

Theorem 3.3.4 [27]. *Let (M, d) be a hyperbolic metric space. Assume that M is uniformly convex. Let C be a nonempty, closed and convex subset of M. Let $T : C \to C$ be a nonexpansive mapping with a nonempty fixed points set. Fix $x_0 \in C$ and consider the Krasnoselskii-Mann iteration $\{x_n\}$ in C generated by x_0 and $\lambda \in (0, 1)$ and defined* *by*

$$x_{n+1} = (1 - \lambda)x_n \oplus \lambda T(x_n), \quad \textit{for any } n \in \mathbb{N}.$$

Then we have $\lim\limits_{n \to \infty} d(Tx_n, x_n) = 0$. *Moreover* $\{x_n\}$ Δ-*converges to* $x \in C$ *which is a fixed point of* T, *i.e.,* $x \in Fix(T)$.

3.4. Approximate Fixed Points of Monotone Mappings

Following the publication of Ran and Reurings' fixed point theorem in metric spaces endowed with a partial order [52], many mathematicians became interested in investigating the fixed point problem for monotone mappings. In this section, we discuss the approximate fixed points for monotone mappings or order-preserving mappings. Let (M,d) be a metric space endowed with a partial order $\preceq$. We will say that $x, y \in M$ are *comparable* whenever $x \preceq y$ or $y \preceq x$. Next, we give the definition of monotone mappings.

Definition 3.4.1. Let $(M, d, \preceq)$ be a metric space endowed with a partial order. A map $T : M \to M$ is said to be *monotone* or *order-preserving* if

$$x \preceq y \implies T(x) \preceq T(y), \quad \text{for any } x, y \in M.$$

Next we give the definition of monotone Lipschitzian mappings.

Definition 3.4.2. Let $(M, d, \preceq)$ be a metric space endowed with a partial order. A map $T : M \to M$ is said to be *monotone Lipschitzian mapping* if T is monotone and there exists $k \geq 0$ such that

$$d(T(x), T(y)) \leq k\, d(x, y),$$

for any $x, y \in M$ such that x and y are comparable. If $k < 1$, then we say that T is a *monotone contraction mapping*. And if $k = 1$, T is called a *monotone nonexpansive mapping*.

Note that monotone Lipschitzian mappings are not necessarily continuous. The following result is the extension of the Banach contraction principle to monotone contraction mappings.

Theorem 3.4.1 [52]. *Let* $(M, d, \preceq)$ *be a complete metric space endowed with a partial order. Let* $T : M \to M$ *be a continuous monotone contraction mapping. Assume that there exists* $x_0 \in M$ *such that* x_0 *and* $T(x_0)$ *are comparable. Then the sequence* $\{T^n(x_0)\}$ *converges to a fixed point* ω *of* T. *Moreover if* $x \in M$ *is comparable to* x_0, *then* $\lim\limits_{n \to +\infty} T^n(x) = \omega$.

Proof. Without loss of generality, we assume that $x_0 \preceq T(x_0)$. Since T is monotone, we conclude that $T^n(x_0) \preceq T^{n+1}(x_0)$, for any $n \in \mathbb{N}$. Since T is a monotone contraction mapping, there exists $k \in [0,1)$ such that

$$d(T^n(x_0), T^{n+1}(x_0)) \leq k \, d(T^{n-1}(x_0), T^n(x_0)) \leq k^n \, d(x_0, T(x_0)), \quad \text{for any } n \in \mathbb{N}.$$

Hence $\sum_{n \in \mathbb{N}} d(T^n(x_0), T^{n+1}(x_0))$ is convergent because $k < 1$. This will imply that $\{T^n(x_0)\}$ is a Cauchy sequence. Using the completeness of (M,d), there exists $\omega \in M$ such that $\{T^n(x_0)\}$ converges to ω. Since T is continuous, we conclude that

$$\omega = \lim_{n \to +\infty} T^{n+1}(x_0) = T\left(\lim_{n \to +\infty} T^n(x_0) \right) = T(\omega),$$

i.e., ω is a fixed point of T. Finally, let $x \in M$ such that x and x_0 are comparable. Since T is monotone, then $T^n(x)$ and $T^n(x_0)$ are also comparable for any $n \in \mathbb{N}$. Hence we have

$$d(T^n(x), T^n(x_0)) \leq k \, d(T^{n-1}(x), T^{n-1}(x_0)) \leq k^n \, d(x, x_0), \quad \text{for any } n \in \mathbb{N}.$$

In particular, we have $\lim_{n \to +\infty} d(T^n(x), T^n(x_0)) = 0$, which implies that $\{T^n(x)\}$ also converges to ω. $\qquad\qquad\square$

In fact under suitable conditions, the fixed point may be unique, which completes the conclusion of the Banach contraction principle for monotone contraction mappings.

Theorem 3.4.2 [52]. *Let $(M, d, \preceq)$ be a metric space endowed with a partial order such that every pair $x, y \in M$ has an upper bound or a lower bound in M. Let $T : M \to M$ be a continuous monotone contraction mapping. Assume that there exists $x_0 \in M$ such that x_0 and $T(x_0)$ are comparable. Then T has a unique fixed point ω and $\lim_{n \to +\infty} T^n(x) = \omega$, for any $x \in M$.*

Proof. Using Theorem 3.4.1, we only have to prove that for any $x \in M$, we have $\lim_{n \to +\infty} T^n(x) = \omega$. In order to do this, notice that if x and z are comparable, then $T^n(x)$ and $T^n(z)$ are comparable, for any $n \in \mathbb{N}$, since T is monotone. Hence we have

$$d(T^n(x), T^n(z)) \leq k \, d(T^{n-1}(x), T^{n-1}(z)) \leq k^n \, d(x, z), \quad \text{for any } n \in \mathbb{N}.$$

Since $k < 1$, we conclude that $\lim_{n \to +\infty} d(T^n(x), T^n(z)) = 0$. Let $x \in M$. Without loss of generality, we assume that that every pair of elements in M has an upper bound. Hence there exists $z \in M$ such that $x \preceq z$ and $x_0 \preceq z$. Therefore we have

$$\lim_{n \to +\infty} T^n(x) = \lim_{n \to +\infty} T^n(z) = \lim_{n \to +\infty} T^n(x_0) = \omega,$$

which completes the proof of Theorem 3.4.2. $\qquad\qquad\qquad\qquad\qquad\qquad\square$

Remark 3.4.1. Both Theorems 3.4.1 and 3.4.2 extend easily to mappings $T : M \to M$ such that $T(x)$ and $T(y)$ are comparable whenever x and y are comparable. This is true because the contractive condition is symmetric between x and y.

The continuity assumption in Theorem 3.4.1 may be relaxed. This point was noted by Nieto and Rodríguez-López [50].

Theorem 3.4.3 [50]. *Let $(M, d, \preceq)$ be a complete metric space endowed with a partial order. Assume that $\preceq$ satisfies the following property:*

> (∗) *for any $\{x_n\}$ in M such that $x_n \preceq x_{n+1}$ (respectively, $x_{n+1} \preceq x_n$), for any $n \geq 1$, and $\lim\limits_{n \to +\infty} x_n = x$, then $x_n \preceq x$ (respectively, $x \preceq x_n$).*

Let $T : M \to M$ be a monotone contraction mapping. Assume that there exists $x_0 \in M$ such that x_0 and $T(x_0)$ are comparable. Then the sequence $\{T^n(x_0)\}$ converges to a fixed point ω of T. Moreover if $x \in M$ is comparable to x_0, then $\lim\limits_{n \to +\infty} T^n(x) = \omega$.

Remark 3.4.2. If we assume that order intervals are closed, then $\preceq$ satisfies the property (∗). Recall that an order interval is any of the subsets

$$[a, \to) = \{x \in X; a \preceq x\}, \ (\leftarrow, a] = \{x \in X; x \preceq a\}, \ [a, b] = \{x \in X; a \preceq x \preceq b\}$$

for any $a, b \in X$.

The fixed point theory for monotone nonexpansive mappings is fairly recent and follows the same approach as for nonexpansive mappings. Recall that the first results for nonexpansive mappings find their roots in the works of Browder [17], Göhde [33] and Kirk [42]. As we said in the introduction, it took four decades to achieve positive results for nonexpansive mappings. In other words, the study of the fixed point property for nonexpansive mappings is far more complicated and complex than the contraction mappings. Since the first results were discovered in Banach spaces, it was natural to ask whether these results are still valid for monotone mappings defined in Banach spaces.

Let $(X, \|.\|, \preceq)$ be a Banach space endowed with a partial order. Throughout, we will assume that order intervals are convex and closed. Let C be a convex and bounded subset of X not reduced to one point. Let $T : C \to C$ be a monotone nonexpansive mapping. Fix $x_0 \in C$ and consider the *Krasnoselskii-Mann iteration* $\{x_n\}$ in C generated by x_0 and $(\lambda_n) \subset [0,1]$ and defined by

$$x_{n+1} = (1 - \lambda_n)x_n + \lambda_n T(x_n), \quad \text{for any } n \in \mathbb{N}. \tag{KMS}$$

Assume that x_0 and $T(x_0)$ are comparable. Without loss of generality, we assume that $x_0 \preceq T(x_0)$. Since order intervals are convex, we have $x_0 \preceq x_1 \preceq T(x_0)$. Since T is monotone, we get $T(x_0) \preceq T(x_1)$. By induction, we will prove that

$$x_n \preceq x_{n+1} \preceq T(x_n) \preceq T(x_{n+1}), \quad \text{for any } n \in \mathbb{N}.$$

Let ω be a weak-cluster point of $\{x_n\}$, i.e., there exists a subsequence $\{x_{\phi(n)}\}$ of $\{x_n\}$ which converges weakly to ω. Since order intervals are closed and convex and $\{x_n\}$ is monotone increasing, we conclude that $x_n \preceq \omega$, for any $n \in \mathbb{N}$. From this we conclude that $\{x_n\}$ has at most one weak-cluster point. Indeed let z be another weak-cluster point of $\{x_n\}$. Since $x_n \preceq \omega$, for any $n \in \mathbb{N}$, and the order intervals are closed and convex, then we have $z \preceq \omega$. If we reverse the roles of z and ω, we get $\omega \preceq z$. Therefore we must have $z = \omega$. If we assume that C is weakly compact, then $\{x_n\}$ will be weakly convergent. This is surprising since this result does not hold in the general case of nonexpansive mappings. Since x_n and $T(x_n)$ are comparable and T is monotone nonexpansive, then

$$\|T(x_{n+1}) - T(x_n)\| \le \|x_{n+1} - x_n\|, \quad \text{for any } n \in \mathbb{N}.$$

Using Propositions 3.3.2 and 3.3.3, we get the following approximate fixed point result for monotone nonexpansive mappings.

Theorem 3.4.4. *Let $(X, \|.\|, \preceq)$ be a partially ordered Banach space having the above properties. Let C be a convex and bounded subset of X not reduced to one point. Let $T : C \to C$ be monotone nonexpansive mapping. Assume there exists $x_0 \in C$ such that x_0 and $T(x_0)$ are comparable. Consider the sequence $\{x_n\}$ in C generated by x_0 and $(\lambda_n) \subset [0,1]$ and defined by (KMS). Assume that*

(i) $\sum\limits_{n=1}^{+\infty} \lambda_n = +\infty$;

(ii) *there exists $b \in [0,1)$ such that $\lambda_n \le b$, for any $n \in \mathbb{N}$.*

Then $\lim\limits_{n \to +\infty} \|x_n - T(x_n)\| = 0$, i.e., $\{x_n\}$ is an approximate fixed point sequence of T. In particular, we have $\inf\{\|x - T(x)\|; x \in C\} = 0$.

If we assume in Theorem 3.4.4 that C is weakly compact, then C will be bounded and $\{x_n\}$ will be weakly convergent. A natural question to ask is: when is the weak limit of $\{x_n\}$ a fixed point of T? This is done, for example, via the Opial property [51].

Definition 3.4.3. $(X, \|.\|)$ is said to satisfy the *weak-Opial condition* if whenever any sequence $\{y_n\}$ in X weakly converges to y, we have

$$\limsup_{n \to +\infty} \|y_n - y\| < \limsup_{n \to +\infty} \|y_n - z\|, \quad \text{for any } z \in X \text{ such that } z \neq y.$$

The first fixed point result obtained for monotone nonexpansive mappings is:

Theorem 3.4.5 [7, 16]. *Let $(X, \|.\|, \preceq)$ be a partially ordered Banach space as described above. Assume X satisfies the weak-Opial condition. Let C be a weakly compact convex nonempty subset of X. Let $T : C \to C$ be a monotone nonexpansive mapping. Assume there exists $x_0 \in C$ such that x_0 and $T(x_0)$ are comparable. Then T has a fixed point.*

Proof. Without loss of generality, we may assume that $x_0 \preceq T(x_0)$. Consider the Krasnoselskii-Mann iteration $\{x_n\}$ in C generated by x_0 and $\lambda \in (0, 1)$ and defined by

$$x_{n+1} = (1 - \lambda)x_n + \lambda T(x_n), \quad \text{for any } n \in \mathbb{N}.$$

Then $\{x_n\}$ is weakly convergent to some $x \in C$ with x_n and x being comparable, for any $n \in \mathbb{N}$. Assume to the contrary that $T(x) \neq x$. Since the assumptions of Theorem 3.4.4 are satisfied, then we must have $\lim_{n \to +\infty} \|x_n - T(x_n)\| = 0$. Using the definition of the weak-Opial condition, we get

$$\begin{aligned}
\liminf_{n \to \infty} \|x_n - x\| &< \liminf_{n \to \infty} \|x_n - T(x)\| \\
&= \liminf_{n \to \infty} \|T(x_n) - T(x)\| \\
&\leq \liminf_{n \to \infty} \|x_n - x\|.
\end{aligned}$$

This is a contradiction. Hence $T(x) = x$, i.e., x is a fixed point of T. As we saw, x and x_0 are comparable, which completes the proof of Theorem 3.4.5. $\square$

Since the classical Banach spaces ℓ_p, $1 < p < +\infty$, satisfies the weak-Opial condition, the following result is a straightforward consequence of Theorem 3.4.5.

Corollary 3.4.1. *Let C be a bounded closed convex nonempty subset of ℓ_p, $1 < p < +\infty$. Consider the pointwise partial ordering in ℓ_p, i.e., $(\alpha_n) \preceq (\beta_n)$ if and only if $\alpha_n \leq \beta_n$, for any $n \geq 1$. Then any monotone nonexpansive mapping $T : C \to C$ has a fixed point provided there exists a point $x_0 \in C$ such that x_0 and $T(x_0)$ are comparable.*

Remark 3.4.3. The case of $p = 1$ is not interesting for the weak topology since l_1 is a Schur Banach space. But if we consider the weak* topology $\sigma(l_1, c_0)$ on l_1 or the pointwise convergence topology, then l_1 satisfies the Opial condition for these topologies. Note that these two topologies are Hausdorff. In this case we have a similar conclusion to Corollary 3.4.1 for l_1. Recall the definitions of l_p and c_0 spaces:

(a) $l_p = \left\{ (\alpha_n) \in \mathbb{R}^{\mathbb{N}}, \sum_n |\alpha_n|^p < +\infty \right\}$, for $1 \leq p < +\infty$;

(b) $c_0 = \left\{ (\alpha_n) \in \mathbb{R}^{\mathbb{N}}, \lim_{n \to +\infty} \alpha_n = 0 \right\}$.

Note that the conclusion of Theorem 3.4.5 holds in any Banach space that is uniformly convex [16]. This extension allows us to prove a fixed point result for monotone nonexpansive mappings in $L_p([0,1])$ spaces, $p > 1$, of functions for which the pth power of the absolute value is Lebesgue integrable despite the fact that $L_p([0,1])$ spaces fail to satisfy the weak-Opial property. The case $p = 1$ is interesting and allows for the use of a different topology.

Theorem 3.4.6 [39]. *Let $C \subset L_1([0,1])$ be nonempty, convex and compact for the convergence almost everywhere. Let $T : C \to C$ be a monotone nonexpansive mapping. Assume there exists $f_0 \in C$ such that f_0 and $T(f_0)$ are comparable. Consider the sequence $\{f_n\}$ in C generated by f_0 and $(\lambda_n) \subset [0,1]$ and defined by (KMS). Assume that*

(i) $\sum_{n=1}^{+\infty} \lambda_n = +\infty$;

(ii) *there exists $b \in [0,1)$ such that $\lambda_n \leq b$, for any $n \in \mathbb{N}$.*

Then $\{f_n\}$ converges almost everywhere to some $f \in C$ and $\lim_{n \to \infty} \| f_n - T(f_n)\| = 0$. Moreover, f is a fixed point of T and is comparable to f_n for any $n \in \mathbb{N}$.

In the next example, we give an application to integral equations which involves monotone nonexpansive mappings.

Example 3.4.1 [7]. Let us consider the following integral equation of the form:

$$x(t) = g(t) + \int_0^1 F(t,s,x(s))ds, \quad t \in [0,1], \tag{IE}$$

where

(a) g is in $L^2([0,1],\mathbb{R})$,

(b) $F : [0,1] \times [0,1] \times L^2([0,1],\mathbb{R}) \to \mathbb{R}$ is measurable and satisfies the condition:

$$0 \leq F(t,s,x) - F(t,s,y) \leq x - y, \tag{3.4.1}$$

where $t,s \in [0,1]$, and $x,y \in L^2([0,1],\mathbb{R})$ such that $y \leq x$.

Recall that for any $u,v \in L^2([0,1],\mathbb{R})$, we have

$$u \leq v \iff u(t) \leq v(t) \text{ almost everywhere } t \in [0,1].$$

Assume there exists a non-negative function $h(\cdot,\cdot) \in L^2([0,1] \times [0,1])$, and $M < \frac{1}{2}$ such that:

$$|F(t,s,x)| \leq h(t,s) + M|x|, \tag{3.4.2}$$

where $t,s \in [0,1]$, and $x \in L^2([0,1],\mathbb{R})$.

Let

$$B = \left\{ y \in L^2([0,1],\mathbb{R}), \text{ such that } \|x\|_{L^2([0,1],\mathbb{R})} \leq \rho \right\},$$

where ρ is sufficiently large, i.e., B is the closed ball of $L^2([0,1],\mathbb{R})$ centered at 0 with radius ρ. Consider the operator defined by

$$\widetilde{F}(t)(y)(s) = F(t,s,y(s)),$$

and define the operator $J : L^2([0,1],\mathbb{R}) \to L^2([0,1],\mathbb{R})$ by

$$(Jy)(t) = g(t) + \int_0^1 \widetilde{F}(t)(y)(s)ds.$$

We have $J(B) \subset B$. Indeed, let $x \in B$, then by using the Cauchy-Schwarz inequality, the condition (3.4.2) and the quadratic inequality $(a+b)^2 \leq 2a^2 + 2b^2$ for any $a,b \in \mathbb{R}$, we have:

$$
\begin{aligned}
\|Jx\|^2_{L^2([0,1],\mathbb{R})} &= \int_0^1 |Jx(t)|^2 dt. \\
&= \int_0^1 |g(t) + \int_0^1 \widetilde{F}(t)(x)(s)ds|^2 dt \\
&\leq 2\int_0^1 |g(t)|^2 dt + 2\int_0^1 \int_0^1 |\widetilde{F}(t)(x)(s)|^2 ds\,dt \\
&\leq 2\int_0^1 |g(t)|^2 dt + 2\int_0^1 \int_0^1 |h(t,s) + M|x(s)||^2 ds\,dt \\
&\leq 2\int_0^1 |g(t)|^2 dt + 4\int_0^1 \int_0^1 |h(t,s)|^2 ds\,dt + 4M^2 \int_0^1 \int_0^1 |x(s)|^2 ds\,dt \\
&= 2\int_0^1 |g(t)|^2 dt + 4\int_0^1 \int_0^1 |h(t,s)|^2 ds\,dt + 4M^2 \int_0^1 |x(s)|^2 ds \\
&= 2\int_0^1 |g(t)|^2 dt + 4\int_0^1 \int_0^1 |h(t,s)|^2 ds\,dt + 4M^2 \|x\|^2_{L^2([0,1],\mathbb{R})} \\
&\leq 2\int_0^1 |g(t)|^2 dt + 4\int_0^1 \int_0^1 h^2(t,s)ds\,dt + 4M^2\rho^2.
\end{aligned}
$$

Since $M < 1/2$, choosing ρ such that

$$\frac{2}{(1-4M^2)} \int_0^1 |g(t)|^2 dt + \frac{4}{(1-4M^2)} \int_0^1 \int_0^1 h^2(t,s)ds\,dt \leq \rho^2,$$

we will get $J(x) \in B$ as claimed. Next we prove that J is monotone nonexpansive. First from the condition (3.4.1), J is obviously monotone. Let $x,y \in L^2([0,1],\mathbb{R})$ such that

$y \leq x$. Using the Cauchy-Schwarz inequality, we have:

$$
\begin{aligned}
\|Jx - Jy\|^2_{L^2([0,1],\mathbb{R})} &= \int_0^1 (Jx(t) - Jy(t))^2 dt \\
&= \int_0^1 (\int_0^1 (\tilde{F}(t)(x)(s) - \tilde{F}(t)(y)(s))ds)^2 dt \\
&\leq \int_0^1 (\int_0^1 (x(s) - y(s))ds)^2 dt \\
&\leq \int_0^1 (x(s) - y(s))^2 ds = \|x - y\|^2_{L^2([0,1],\mathbb{R})},
\end{aligned}
$$

which implies that J is a monotone nonexpansive operator as claimed. In order to use Theorem 3.4.5, we need to check its assumptions. First note that $X = L^2([0,1],\mathbb{R})$ is a Hilbert space. Then X satisfies the weak-Opial condition. It is easy to check that order intervals are convex. In order to show that order intervals are closed, we will show that if $\{u_n\}$ is a non-negative sequence of elements in X which converges weakly to u, then u is positive. Let $a < 0$, then the set $A = \{t \in [0,1]; u(t) \leq a\}$ has measure 0. Indeed, we have

$$
\int_A u(t)dt = \lim_{n \to +\infty} \int_A u_n(t)dt,
$$

because of weak-convergence. So

$$
0 \leq \lim_{n \to +\infty} \int_A u_n(t)dt = \int_A u(t)dt \leq a\, m(A) \leq 0.
$$

Hence $m(A) = 0$. Set

$$
D = \bigcup_{n \geq 1} \left\{ t \in [0,1]; u(t) \leq -\frac{1}{n} \right\}.
$$

Then D has measure 0, which implies that $u(t) \geq 0$ for almost every $t \in [0,1]$. Using Theorem 3.4.5, we get the following result:

Theorem 3.4.7. *Under the above assumptions, we conclude that*

(a) *the integral equation (IE) has a non-negative solution provided we assume that* $g(t) + \int_0^1 F(t,s,0)ds \geq 0$ *for almost every $t \in [0,1]$ (which implies $J(0) \geq 0$);*

(b) *the integral equation (IE) has a nonpositive solution provided we assume that* $g(t) + \int_0^1 F(t,s,0)ds \leq 0$ *for almost every $t \in [0,1]$ (which implies $J(0) \leq 0$).*

3.5. Approximate Fixed Points of Nonlinear Semigroups

In this section, we discuss the common fixed points and approximate common fixed points of a semigroup of mappings $\{T_t\}_{t \geq 0}$, i.e., a family of mappings such that $T_0(x) = x$, $T_{s+t} = T_s \circ T_t$. In functional differential equations, a semigroup family of mappings is usually denoted by $\{T(t)\}$. Such a situation is quite common in mathematics and applications. For instance, in the theory of dynamical systems, the vector function space would define the state space and the mapping $(t,x) \to T_t(x)$ would represent the evolution function of a dynamical system. The question about the existence of common fixed points, and about the structure of the set of common fixed points, can be interpreted as a question whether there exist points that are fixed during the state space transformation T_t at any given point of time t, and if so, what the structure of a set of such points may look like. The existence of common fixed points for families of contractions and nonexpansive mappings in Banach spaces has been the subject of intense research since the early 1960s, as investigated by Belluce and Kirk [12, 13], Browder [17], Bruck [18], DeMarr [21], and Lim [48]. The asymptotic approach for finding common fixed points of semigroups of Lipschitzian mappings has also been investigated (see, e.g., Tan and Xu [57]).

In the setting of this section, the state space is a nonlinear hyperbolic metric space.

Definition 3.5.1. Let (M,d) be a metric space and $C \subset M$ be a nonempty subset. A one-parameter family $\mathscr{F} = \{T_t; t \geq 0\}$ of mappings from C into itself is said to be a *nonexpansive semigroup* on C if $\mathscr{F}$ satisfies the following conditions:

(i) $T_0(x) = x$, for $x \in C$;

(ii) $T_{t+s}(x) = T_t(T_s(x))$, for $x \in C$ and $t,s \in [0,\infty)$;

(iii) for each $t \geq 0$, T_t is a nonexpansive mapping.

Define the set of all common fixed points of $\mathscr{F}$ as $Fix(\mathscr{F}) = \bigcap_{t \geq 0} Fix(T_t)$, and the set of common approximate fixed points of $\mathscr{F}$ as $AFPS(\mathscr{F}) = \bigcap_{t \geq 0} AFPS(T_t)$.

The following technical result is the extension of Bruck's result [18] to metric spaces.

Lemma 3.5.1 [15]. *Let (M,d) be a hyperbolic metric space. Assume that (M,d) is strictly convex. Let C be a subset of M. Let S and T be nonexpansive mappings from C into M with a common fixed point. Then for each $\lambda \in (0,1)$, the mapping $U : C \to M$ defined by $U(x) = \lambda S(x) \oplus (1 - \lambda)T(x)$, for $x \in C$, is nonexpansive and $Fix(U) = Fix(S) \cap Fix(T)$ holds.*

Recall that a one-parameter family $\mathscr{F} = \{T_t; t \geq 0\}$ of mappings from C into M, where C is a nonempty subset of a metric space (M,d), is said to be *continuous* on C if for any $x \in C$, the mapping $t \to T_t(x)$ is continuous, i.e., for any $t_0 \geq 0$, we have

$$\lim_{t \to t_0} d(T_t(x), T_{t_0}(x)) = 0.$$

Moreover, $\mathscr{F}$ is said to be *strongly continuous* on C if for any bounded nonempty subset $K \subset C$, we have

$$\lim_{t \to t_0} \sup_{x \in K} \left(d(T_t(x), T_{t_0}(x)) \right) = 0.$$

Proposition 3.5.1. *Let (M, d) be a metric space and $C \subset M$ be nonempty and closed. Let $\mathscr{F} = \{T_t; t \geq 0\}$ be a one-parameter semigroup of mappings from C into M which is continuous on C. Let A be a dense subset of $[0, +\infty)$. Then the set of common fixed points of $\mathscr{F}$ is $\mathrm{Fix}(\mathscr{F}) = \bigcap_{a \in A} \mathrm{Fix}(T_a)$.*

The following lemma gives an interesting example of dense subsets of $[0, +\infty)$.

Lemma 3.5.2 [55]. *Let G be a nonempty additive subgroup of $\mathbb{R}$. Then G is either dense in $\mathbb{R}$ or there exists $a > 0$ such that $G = a \cdot \mathbb{Z} = \{a \, n, n \in \mathbb{Z}\}$. Therefore if α and β are two real numbers such that $\frac{\alpha}{\beta}$ is irrational, then the set*

$$G(\alpha, \beta) = \{\alpha \, n + \beta \, m; \, n, m \in \mathbb{Z}\}$$

is dense in $\mathbb{R}$. In particular, the set $G(\alpha, \beta) \cap [0, +\infty)$ is dense in $[0, +\infty)$.

Proposition 3.5.2 [15]. *Let (M, d) be a metric space and $C \subset M$ be nonempty and closed. Let $\mathscr{F} = \{T_t; t \geq 0\}$ be a one-parameter semigroup of mappings from C into M. Let α and β be any two positive real numbers. Then*

$$\bigcap_{a \in G_+(\alpha, \beta)} \mathrm{Fix}(T_a) = \mathrm{Fix}(T_\alpha) \cap \mathrm{Fix}(T_\beta),$$

where $G_+(\alpha, \beta) = \{\alpha \, n + \beta \, m; \, n, m \in \mathbb{Z}\} \cap [0, +\infty)$.

Using Propositions 3.5.1 and 3.5.2 we get the following result.

Theorem 3.5.1 [15, 56]. *Let (M, d) be a metric space and $C \subset M$ be nonempty and closed. Let $\mathscr{F} = \{T_t; t \geq 0\}$ be a one-parameter semigroup of mappings from C into M which is continuous on C. Let α and β be two positive real numbers such that $\frac{\alpha}{\beta}$ is irrational, then we have $\mathrm{Fix}(\mathscr{F}) = \mathrm{Fix}(T_\alpha) \cap \mathrm{Fix}(T_\beta)$. In particular, we have*

$$\mathrm{Fix}(\mathscr{F}) = \mathrm{Fix}(T_1) \cap \mathrm{Fix}(T_{\sqrt{2}}) = \mathrm{Fix}(T_1) \cap \mathrm{Fix}(T_\pi).$$

The main conclusion of Theorem 3.5.1 still holds when one considers approximate fixed point sequences. Indeed, we have the following result.

Theorem 3.5.2 [37]. *Let (M,d) be a metric space and $C \subset M$ be nonempty and bounded. Let $\mathscr{F} = \{T_t; t \geq 0\}$ be a one-parameter nonexpansive semigroup of mappings from C into C. Assume that $\mathscr{F}$ is strongly continuous. Let α and β be two positive real numbers such that $\frac{\alpha}{\beta}$ is irrational. Then*

$$AFPS(\mathscr{F}) = AFPS(T_\alpha) \cap AFPS(T_\beta).$$

If the metric space (M,d) is hyperbolic strictly convex, Lemma 3.5.1 allows us to get the following result:

Theorem 3.5.3. *Let (M,d) be a hyperbolic metric space. Assume that (M,d) is strictly convex and $C \subset M$ be nonempty and closed. Let $\mathscr{F} = \{T_t; t \geq 0\}$ be a one-parameter semigroup of nonexpansive mappings from C into M which is continuous on C. Let α and β be two positive real numbers such that $\frac{\alpha}{\beta}$ is irrational, then*

$$Fix(\mathscr{F}) = Fix(\lambda T_\alpha + (1 - \lambda)T_\beta), \quad \text{for any } \lambda \in (0,1).$$

Next, we use the previous results to investigate the behavior of Krasnoselskii-Mann iterates generated by two mappings. These iterations will allow us to approximate common fixed points of a continuous semigroup. Let (M,d) be a hyperbolic metric space and $C \subset M$ be a nonempty convex subset. Let $T, S : C \to C$ be two mappings. Fix $x_0 \in C$, Das and Debata [20] studied the strong convergence of the iterates $\{x_n\}$ generated by x_0 and $\alpha_n, \beta_n \in [0,1]$ defined by

$$x_{n+1} = \alpha_n S\Big(\beta_n T(x_n) \oplus (1 - \beta_n)x_n\Big) \oplus (1 - \alpha_n)x_n, \quad \text{for any } n \in \mathbb{N}. \quad \text{(DDS)}$$

Under suitable assumptions, we will show that $\{x_n\}$ is an approximate fixed point sequence of both T and S. Assume that T and S are nonexpansive and have a common fixed point $p \in C$. Then we have

$$
\begin{aligned}
d(x_{n+1}, p) &= d(\alpha_n S(y_n) \oplus (1 - \alpha_n)x_n, p) \\
&\leq \alpha_n d(S(y_n), p) + (1 - \alpha_n)d(x_n, p) \\
&\leq \alpha_n d(y_n, p) + (1 - \alpha_n)d(x_n, p) \\
&= \alpha_n d(\beta_n T(x_n) \oplus (1 - \beta_n)x_n, p) + (1 - \alpha_n)d(x_n, p) \\
&\leq \alpha_n \Big[\beta_n d(T(x_n), p) + (1 - \beta_n)d(x_n, p)\Big] + (1 - \alpha_n)d(x_n, p) \\
&\leq d(x_n, p),
\end{aligned}
$$

where $y_n = \beta_n T(x_n) \oplus (1 - \beta_n)x_n$. This proves that $\{d(x_n, p)\}$ is decreasing, which implies that $\lim\limits_{n\to\infty} d(x_n, p)$ exists. Using the above inequalities, we get

$$
\begin{aligned}
\lim_{n\to\infty} d(x_n, p) &= \lim_{n\to\infty} d(\alpha_n S y_n \oplus (1-\alpha_n) x_n, p) \\
&= \lim_{n\to\infty} \left[\alpha_n d(S y_n, p) + (1-\alpha_n) d(x_n, p) \right] \\
&= \lim_{n\to\infty} \left[\alpha_n d(y_n, p) + (1-\alpha_n) d(x_n, p) \right] \\
&= \lim_{n\to\infty} \left[\alpha_n d(\beta_n T(x_n) \oplus (1-\beta_n) x_n, p) + (1-\alpha_n) d(x_n, p) \right] \\
&= \lim_{n\to\infty} \left[\alpha_n \Big(\beta_n d(T(x_n), p) + (1-\beta_n) d(x_n, p) \Big) + (1-\alpha_n) d(x_n, p) \right].
\end{aligned}
$$

Theorem 3.5.4 [37]. *Let C be a nonempty, closed and convex subset of a complete uniformly convex hyperbolic space (M, d). Let $S, T : C \to C$ be nonexpansive mappings such that $Fix(T) \cap Fix(S) \neq \emptyset$. Fix $x_0 \in C$ and generate $\{x_n\}$ by (DDS). Assume that $\alpha_n, \beta_n \in [\alpha, \beta]$, with $0 < \alpha \leq \beta < 1$, then*

$$
\lim_{n\to\infty} d(x_n, S(x_n)) = 0, \quad \text{and} \quad \lim_{n\to\infty} d(x_n, T(x_n)) = 0,
$$

i.e., $\{x_n\} \in AFPS(T) \cap AFPS(S)$.

Remark 3.5.1 [37]. The existence of a common fixed point of T and S is crucial. If one assumes that T and S commute, i.e., $S \circ T = T \circ S$, then a common fixed point of T and S exists under the assumptions of Theorem 3.5.4 provided we assume C is bounded.

Theorem 3.5.5. *Let (M, d) be a complete hyperbolic metric space. Assume that M is uniformly convex. Let C be a nonempty, closed and convex subset of M. Let $\mathscr{F} = \{T_t; t \geq 0\}$ be a one-parameter semigroup of nonexpansive mappings from C into C. Assume that $\mathscr{F}$ is continuous and has a nonempty common fixed points set. Let α and β be two positive real numbers such that $\frac{\alpha}{\beta}$ is irrational. Fix $\lambda, \mu \in (0, 1)$ such that $\lambda + \mu < 1$. Let $x_0 \in C$ and define a sequence $\{x_n\}$ in C by*

$$
x_{n+1} = (\mu + \lambda)\Big(\mu T_\alpha(x_n) \oplus \lambda T_\beta(x_n) \Big) \oplus (1 - \mu - \lambda) x_n, \quad \text{for any } n \in \mathbb{N}.
$$

Then $\{x_n\} \in AFPS(T_\alpha) \cap AFPS(T_\beta)$ and Δ-converges to a common fixed point of the semigroup $\mathscr{F}$. Moreover if we assume that $\mathscr{F}$ is strongly continuous on C, then we have $\{x_n\} \in AFPS(\mathscr{F})$.

3.6. Approximate Fixed Points of Monotone Nonlinear Semigroups

In this section, we discuss the extension of the previous results to the case of monotone semigroups. We restrict the domain to the case of a Banach space endowed with a partial order though most of the results discussed here may be stated in hyperbolic metric spaces. Let $(X, \|.\|)$ be a Banach vector space and suppose $\preceq$ is a partial order on X. Throughout this section, we will assume that the partial order $\preceq$ and the linear structure of X obey the following convexity property:

$$a \preceq b \text{ and } c \preceq d \implies \alpha\, a + (1 - \alpha)\, c \preceq \alpha\, b + (1 - \alpha)\, d,$$

for any $\alpha \in [0,1]$, and $a,b,c,d \in X$. Moreover, we will assume that order intervals are closed (see Remark 3.4.2 for the definition of an order interval). Next we give the definition of monotone nonexpansive semigroup of mappings.

Definition 3.6.1. Let $(X, \|.\|, \preceq)$ be as above. Let C be a nonempty subset of X. A one-parameter family $\mathscr{F} = \{T_t; t \geq 0\}$ of mappings from C into C is said to be a *monotone nonexpansive semigroup* if $\mathscr{F}$ satisfies the following conditions:

(i) $T_0 x = x$, for $x \in C$;

(ii) $T_{t+s} = T_t \circ T_s$, for $t, s \in [0, \infty)$;

(iii) for each $t \geq 0$, T_t is a monotone nonexpansive mapping.

Next we give an example of such a semigroup.

Example 3.6.1 [8]. Let $(X, \| \cdot \|, \preceq)$ be as above. Let C be a nonempty closed, bounded convex subset of X. Let $J : C \to C$ be a monotone nonexpansive mapping. Let $x \in C$ be such that $x \preceq J(x)$. Consider the recurrent sequence defined by

$$\begin{cases} u_0(t) = x, \\ u_{n+1}(t) = e^{-t}x + \displaystyle\int_0^t e^{s-t} J(u_n(s))ds, \end{cases} \tag{3.6.3}$$

for any $t \in [0, A]$, where A is a fixed positive number. Then the sequence $\{u_n(t)\}$ is Cauchy for any $t \in [0, A]$. Indeed, let us suppose that $x, y : [0, A] \to X$ are continuous functions. For each $t \in [0.A]$, we have

$$\left\| e^{-t} y(t) + \int_0^t e^{s-t} x(s)ds \right\| \leq e^{-t} \|y\|_\infty + K(t) \|x\|_\infty, \tag{3.6.4}$$

where $\|u\|_\infty = \sup\{\|u(s)\|; t \in [0,A]\}$ and $K(t) = \displaystyle\int_0^t e^{s-t}ds = 1 - e^{-t}$. Indeed, let

$t \in [0,A]$ be fixed, and $\tau = \{t_i; i = 0, 1, \ldots n\}$ be any subdivision of $[0,t]$. Set

$$S_\tau = e^{-t} y(t) + \sum_{i=0}^{n-1} (t_{i+1} - t_i) e^{t_i - t} x(t_i).$$

The family $\{S_\tau\}$ is norm-convergent to:

$$S = e^{-t} y(t) + \int_0^t e^{s-t} x(s) ds,$$

when $|\tau| = \sup\{|t_{i+1} - t_i|; i = 0, 1, \ldots, (n-1)\}$ goes to 0. Since

$$\sup_{s \in [0,t]} \|x(s)\| \leq \sup_{s \in [0,A]} \|x(s)\| = \|x\|_\infty$$

and

$$\|S_\tau\| \leq e^{-t} \|y(t)\| + \sum_{i=0}^{n-1} (t_{i+1} - t_i) e^{t_i - t} \|x\|_\infty \leq e^{-t} \|y(t)\| + K(t) \|x\|_\infty,$$

where we used

$$\sum_{i=0}^{n-1} (t_{i+1} - t_i) e^{t_i - t} \leq \int_0^t e^{s-t} ds = K(t),$$

then we have $\|S\| \leq e^{-t} \|y(t)\| + K(t) \|x\|_\infty$. Therefore we have

$$\left\| e^{-t} y(t) + \int_0^t e^{s-t} x(s) ds \right\| \leq e^{-t} \|y\|_\infty + K(t) \|x\|_\infty, \quad \text{for any } t \in [0,A].$$

Next, we discuss the sequence $\{u_n(t)\}$. Since

$$u_1(t) = e^{-t} x + \int_0^t e^{s-t} J(x) ds = e^{-t} x + (1 - e^{-t}) J(x), \quad t \in [0,A],$$

we have $u_0(t) \preceq u_1(t)$, for any $t \in [0,A]$. Assume that $u_n(t) \preceq u_{n+1}(t)$, for any $t \in [0,A]$. Then we have $J(u_n(t)) \preceq J(u_{n+1}(t))$, for any $t \in [0,A]$, because J is monotone. Using the properties of the partial order $\preceq$, we get

$$u_{n+1}(t) = e^{-t} x + \int_0^t e^{s-t} J(u_n(s)) ds \preceq e^{-t} x + \int_0^t e^{s-t} J(u_{n+1}(s)) ds = u_{n+2}(t),$$

for any $t \in [0,A]$. By induction, we conclude that $\{u_n(t)\}$ is monotone increasing for any $t \in [0,A]$. Moreover, we have

$$u_{n+2}(t) - u_{n+1}(t) = \int_0^t e^{s-t} \Big(J(u_{n+1}(s)) - J(u_n(s)) \Big) ds, \quad t \in [0,A],$$

which implies by using the monotone nonexpansive behavior of J

$$\|u_{n+2}(t) - u_{n+1}(t)\| \leq \int_0^t e^{s-t} \left\|J(u_{n+1}(s)) - J(u_n(s))\right\| ds, \quad t \in [0,A],$$

or

$$\|u_{n+2}(t) - u_{n+1}(t)\| \leq \int_0^t e^{s-t} \left\|u_{n+1}(s) - u_n(s)\right\| ds, \quad t \in [0,A],$$

for any $n \geq 0$. From this inequality and by induction, we get

$$\|u_{n+1}(t) - u_n(t)\| \leq (1 - e^{-A})^n \ diam(C), \quad t \in [0,A],$$

where $diam(C) = \sup\{\|y - z\|; y, z \in C\} < \infty$, since C is bounded. Hence

$$\|u_{n+h}(t) - u_n(t)\| \leq \frac{(1 - e^{-A})^n}{1 - (1 - e^{-A})} \ diam(C) = e^A \ (1 - e^{-A})^n \ diam(C),$$

for any $t \in [0,A]$ and $n, h \in \mathbb{N}$. Clearly this implies that $\{u_n(t)\}$ is Cauchy for any $t \in [0,A]$. Since order intervals are closed, we conclude that the limit $u(t)$ of $\{u_n(t)\}$ satisfies $u_n(t) \preceq u(t)$, for any $n \geq 0$ and $t \in [0,A]$. Using the monotone nonexpansive behavior of J, we get

$$\|J(u_n(t)) - J(u(t))\| \leq \|u_n(t) - u(t)\| \leq e^A (1 - e^{-A})^n \ diam(C),$$

for any $t \in [0,A]$ and $n, h \in \mathbb{N}$. In particular, we have $\lim\limits_{n \to +\infty} J(u_n(t)) = J(u(t))$, uniformly in $[0,A]$, which implies

$$u(t) = e^{-t}x + \int_0^t e^{s-t} J(u(s)) \, ds, \quad t \in [0,A]. \tag{3.6.5}$$

Moreover, the same proof as above will show that if $x \preceq y$, then $u_n(t) \preceq U_n(t)$, for any $n, t \geq 0$, where u_n and U_n are the functions obtained by the initial values x and y respectively, by (3.6.3). This will imply that $u(t) \preceq U(t)$, for any $t \geq 0$. Define the one-parameter family $\mathscr{F} : [0,\infty) \times C \to C$ by

$$T_t(x) = u(t).$$

Then $\mathscr{F}$ defines a semigroup which is monotone nonexpansive. Indeed, we have $\|u_0(t) - U_0(t)\| = \|x - y\| = \|x - y\|_\infty$. Assume $\|u_n(t) - U_n(t)\| \leq \|x - y\|$, then by using (3.6.4), we will have

$$\|u_{n+1}(t) - U_{n+1}(t)\| \leq e^{-t}\|x - y\| + \int_0^t e^{s-t}\|u_n(s) - U_n(s)\| ds \leq \|x - y\|.$$

By induction, $\|u_n(t) - U_n(t)\| \leq \|x - y\|$ holds for any $n \in \mathbb{N}$ and $t \in [0,A]$. Hence

$$\|T_t(x) - T_t(y)\| = \|u(t) - U(t)\| \leq \|x - y\|,$$

for any $t \in [0,A]$. Moreover, we have

$$
\begin{aligned}
T_{t+s}(x) &= e^{-(t+s)}x + \int_0^{t+s} e^{-(t+s-\sigma)}J(T_\sigma(x))d\sigma \\
&= e^{-(t+s)}x + \int_0^t e^{-(t+s-\sigma)}J(T_\sigma(x))d\sigma + \int_t^{t+s} e^{-(t+s-\sigma)}J(T_\sigma(x))d\sigma, \\
&= e^{-s}\left(e^{-t}x + \int_0^t e^{-(t-\sigma)}J(T_\sigma(x))d\sigma\right) + \int_0^s e^{-(s-\sigma)}J(T_{t+\sigma}(x))d\sigma \\
&= e^{-s}T_t(x) + \int_0^s e^{-(s-\sigma)}J(T_{t+\sigma}(x))d\sigma \\
&= T_s\Big(T_t(x)\Big),
\end{aligned}
$$

for any $t \geq 0$. Note that if we started by a point $x \in C$ such that $J(x) \preceq x$, then we would have found that the sequence $\{u_n(t)\}$ is monotone decreasing, for any $t \geq 0$. In other words, the conclusion would have been the same. Moreover, if x is a fixed point of J, then we have $u_n(t) = x$, for any $n \in \mathbb{N}$ and $t \geq 0$. Hence $T_t(x) = x$, for any $t \geq 0$, i.e., $x \in \bigcap_{t \geq 0} Fix(T_t)$. On the other hand, it is easy to show that if $x \in Fix(\mathscr{F})$, then $x \in Fix(J)$, which means we have

$$
Fix(J) = \bigcap_{t \geq 0} Fix(T_t).
$$

As we have seen in the previous section the common fixed points and approximate common fixed points of a semigroup of mappings are exactly the fixed points and approximate common fixed points of two maps. Let us first discuss the case of two monotone mappings and then discuss the case of a semigroup of monotone mappings. Let $(X, \|.\|, \preceq)$ be a Banach space endowed with a partial order as described before. Let $C \subset X$ be a nonempty convex subset. Let $J, H : C \to C$ be two mappings. Fix $x_0 \in C$. Consider the sequence $\{x_n\}$ generated by x_0 and $\alpha_n, \beta_n \in [0,1]$ defined by (DDS), i.e.,

$$
x_{n+1} = \alpha_n H\Big(\beta_n J(x_n) + (1 - \beta_n)x_n\Big) + (1 - \alpha_n)x_n, \quad \text{for any } n \in \mathbb{N}. \tag{DD}
$$

As we did in the previous section, we will show that, under suitable assumptions, $\{x_n\}$ is an approximate fixed point sequence of both J and H. Assume that J and H are monotone nonexpansive. Assume that $x_0 \preceq J(x_0)$ and $x_0 \preceq H(x_0)$. We will also assume that H and J have a common fixed point $p \in C$ such that x_0 and p are comparable. Using the convexity properties of the partial order $\preceq$, we will easily show that x_n and p are comparable. Since H and J are monotone nonexpansive, we get

$\|H(x_n) - p\| \leq \|x_n - p\|$ and $\|J(x_n) - p\| \leq \|x_n - p\|$ for any $n \in \mathbb{N}$. Hence

$$
\begin{aligned}
\|x_{n+1} - p\| &= \|\alpha_n H(y_n) + (1 - \alpha_n) x_n - p\| \\
&\leq \alpha_n \|H(y_n) - p\| + (1 - \alpha_n) \|x_n - p\| \\
&\leq \alpha_n \|y_n - p\| + (1 - \alpha_n) \|x_n - p\| \\
&= \alpha_n \|\beta_n J(x_n) + (1 - \beta_n) x_n - p\| + (1 - \alpha_n) \|x_n - p\| \\
&\leq \alpha_n \Big[\beta_n \|J(x_n) - p\| + (1 - \beta_n) \|x_n - p\| \Big] + (1 - \alpha_n) \|x_n - p\| \\
&\leq \alpha_n \Big[\beta_n \|x_n - p\| + (1 - \beta_n) \|x_n - p\| \Big] + (1 - \alpha_n) \|x_n - p\| \\
&\leq \|x_n - p\|,
\end{aligned}
$$

where $y_n = \beta_n J(x_n) + (1 - \beta_n) x_n$, for any $n \geq 1$. This proves that $\{\|x_n - p\|\}$ is decreasing which implies that $\lim_{n \to \infty} \|x_n - p\|$ exists. Using the above inequalities, we get

$$
\begin{aligned}
\lim_{n \to \infty} \|x_n - p\| &= \lim_{n \to \infty} \|\alpha_n H(y_n) + (1 - \alpha_n) x_n - p\| \\
&= \lim_{n \to \infty} \Big[\alpha_n \|H(y_n) - p\| + (1 - \alpha_n) \|x_n - p\| \Big] \\
&= \lim_{n \to \infty} \Big[\alpha_n \|y_n - p\| + (1 - \alpha_n) \|x_n - p\| \Big] \\
&= \lim_{n \to \infty} \Big[\alpha_n \|\beta_n J(x_n) + (1 - \beta_n) x_n - p\| + (1 - \alpha_n) \|x_n - p\| \Big] \\
&= \lim_{n \to \infty} \Big[\alpha_n \Big(\beta_n \|J(x_n) - p\| + (1 - \beta_n) \|x_n - p\| \Big) + (1 - \alpha_n) \|x_n - p\| \Big].
\end{aligned}
$$

Recall that if X is uniformly convex, then X is reflexive [11]. Moreover, we have [27]:

$$
\|\alpha x + (1 - \alpha) y\| \leq 1 - \delta_X \Big(2 \min\{\alpha, 1 - \alpha\} \varepsilon \Big),
$$

for any $\alpha \in (0,1)$, $\varepsilon > 0$ and $x, y \in X$ such that $\|x\| \leq 1$, $\|y\| \leq 1$ and $\|x - y\| \geq \varepsilon$.

The following technical lemma is useful and will prove to be crucial whenever we deal with successive iterations that use convex combinations.

Lemma 3.6.1 [27]. *Let* $(X, \|.\|)$ *be a uniformly convex Banach space. Let* $\{x_n\}$ *and* $\{y_n\}$ *be in* X *such that* $\limsup_{n \to \infty} \|x_n\| \leq R$, $\limsup_{n \to \infty} \|y_n\| \leq R$, *and* $\lim_{n \to \infty} \|\alpha_n x_n + (1 - \alpha_n) y_n\| = R$, *where* $\alpha_n \in [a, b]$, *with* $0 < a \leq b < 1$, *and* $R \geq 0$. *Then we have* $\lim_{n \to \infty} \|x_n - y_n\| = 0$.

Next we state the main result of this section.

Theorem 3.6.1. *Let C be a nonempty, closed and convex subset of a uniformly convex Banach space $(X, \|.\|)$. Let $H, J : C \to C$ be monotone nonexpansive mappings such that H is uniformly continuous. Assume that there exists $x_0 \in C$ such that $x_0 \preceq J(x_0)$ and $x_0 \preceq H(x_0)$. We will also assume that H and J have a common fixed point $p \in C$ such that x_0 and p are comparable. Consider the sequence $\{x_n\}$ defined by x_0 and the recurrent formula (DD). Assume that $\alpha_n, \beta_n \in [\alpha, \beta]$, with $0 < \alpha \leq \beta < 1$, then*

$$\lim_{n \to \infty} \|x_n - H(x_n)\| = 0, \text{ and } \lim_{n \to \infty} \|x_n - J(x_n)\| = 0,$$

i.e., $\{x_n\} \in AFPS(J) \cap AFPS(H)$.

Proof. We have already seen that $\{\|x_n - p\|\}$ is decreasing. Set $c = \lim_{n \to \infty} \|x_n - p\|$. If $c = 0$, then all the conclusions are trivial. Therefore we will assume that $c > 0$. We have already seen that

$$\begin{aligned} \lim_{n \to \infty} \|x_n - p\| &= \lim_{n \to \infty} \|\alpha_n H(y_n) + (1 - \alpha_n) x_n - p\| \\ &= \lim_{n \to \infty} \left[\alpha_n \|H(y_n) - p\| + (1 - \alpha_n) \|x_n - p\| \right]. \end{aligned}$$

We claim that $\lim_{n \to \infty} \|H(y_n) - p\| = c$. Indeed, let $\mathscr{U}$ be a nontrivial ultrafilter over $\mathbb{N}$. Then we have $\lim_{n, \mathscr{U}} \alpha_n = \alpha_\infty \in [\alpha, \beta]$ and $\lim_{n, \mathscr{U}} \|x_n - p\| = c$. Hence

$$c = \alpha_\infty \lim_{n, \mathscr{U}} \|H(y_n) - p\| + (1 - \alpha_\infty) c.$$

Since $\alpha_\infty \neq 0$, we get $\lim_{n, \mathscr{U}} \|H(y_n) - p\| = c$. Since $\mathscr{U}$ was arbitrary, we get $\lim_{n \to \infty} \|H(y_n) - p\| = c$ as claimed. Therefore we have

$$\lim_{n \to \infty} \|x_n - p\| = \lim_{n \to \infty} \|H(y_n) - p\| = \lim_{n \to \infty} \|\alpha_n H(y_n) + (1 - \alpha_n) x_n - p\|.$$

Using Lemma 3.6.1 , we get $\lim_{n \to \infty} \|H(y_n) - x_n\| = 0$. Since we already proved

$$\lim_{n \to \infty} \|x_n - p\| = \lim_{n \to \infty} \left[\alpha_n \|y_n - p\| + (1 - \alpha_n) \|x_n - p\| \right],$$

a similar argument will show that $\lim_{n \to \infty} \|y_n - p\| = \lim_{n \to \infty} \|x_n - p\|$. Moreover, we have

$$\lim_{n \to \infty} \|x_n - p\| = \lim_{n \to \infty} \left[\alpha_n \left(\beta_n \|J(x_n) - p\| + (1 - \beta_n) \|x_n - p\| \right) + (1 - \alpha_n) \|x_n - p\| \right].$$

If we again use ultrafilters, one can easily prove that $\lim_{n \to \infty} \|J(x_n) - p\| = \lim_{n \to \infty} \|x_n - p\|$. Hence we have

$$\lim_{n \to \infty} \|x_n - p\| = \lim_{n \to \infty} \|J(x_n) - p\| = \lim_{n \to \infty} \|\beta_n J(x_n) + (1 - \beta_n) x_n - p\|.$$

Again using Lemma 3.6.1, we get $\lim\limits_{n\to\infty} \|J(x_n) - x_n\| = 0$. Since

$$\lim_{n\to\infty} \|y_n - x_n\| = \lim_{n\to\infty} \beta_n \|J(x_n) - x_n\| = 0,$$

and H is uniformly continuous, we get $\lim_{n\to\infty} \|H(x_n) - H(y_n)\| = 0$. Combined with $\lim\limits_{n\to\infty} \|H(y_n) - x_n\| = 0$, we get

$$\lim_{n\to\infty} \|H(x_n) - x_n\| = 0.$$

$\square$

Remark 3.6.1. If we assume that $\alpha_n = \alpha$ and $\beta_n = \beta$ with $\alpha, \beta \in (0,1)$, then under the assumptions of Theorem 3.6.1, we can prove that

$$x_n \preceq x_{n+1} \text{ and } y_n \preceq y_{n+1}, \quad \text{for any } n \in \mathbb{N}.$$

Since $\{\|x_n - p\|\}$ and $\{\|y_n - p\|\}$ are convergent, then both $\{x_n\}$ and $\{y_n\}$ are bounded. Therefore they have a weak-cluster point since X is reflexive. Using the convexity properties of the partial order $\preceq$ in X, we will conclude that in fact $\{x_n\}$ and $\{y_n\}$ are weakly convergent. It is not clear that the weak-limit of $\{x_n\}$ is a fixed point of H and J. This will be the case if we have a demi-closed property for monotone nonexpansive mappings. Otherwise we may assume that X satisfies the Opial condition [51]. In this case the weak-limit of $\{x_n\}$ will be a fixed point of T and S comparable to x_0. As an example of a Banach space which satisfies all of the above assumptions, one may take $X = \ell_p$, $1 < p < +\infty$.

As an application of Theorem 3.6.1, we get the following approximate fixed point result on monotone nonexpansive semigroups.

Theorem 3.6.2. *Let $(X, \|.\|)$ be a Banach space. Assume that X is uniformly convex. Let C be a nonempty, closed and convex subset of X. Let $\mathscr{F} = \{T_t ; t \geq 0\}$ be a one-parameter continuous semigroup of monotone nonexpansive mappings uniformly continuous from C into C. Assume that there exists $x_0 \in C$ such that $x_0 \preceq T_t(x_0)$ for any $t \geq 0$. We also assume that $\mathscr{F}$ have a common fixed point $p \in C$ such that x_0 and p are comparable. Let α and β be two positive real numbers such that $\frac{\alpha}{\beta}$ is irrational. Fix $\lambda, \mu \in (0,1)$ such that $\lambda + \mu < 1$. Consider the sequence $\{x_n\}$ in C defined by*

$$x_{n+1} = (\mu + \lambda)\Big(\mu T_\alpha(x_n) + \lambda T_\beta(x_n)\Big) + (1 - \mu - \lambda)x_n, \quad \text{for any } n \in \mathbb{N}.$$

Then $\{x_n\} \in AFPS(T_\alpha) \cap AFPS(T_\beta)$. Moreover, if we assume that $\mathscr{F}$ is strongly continuous on C, then we have $\{x_n\} \in AFPS(\mathscr{F})$.

REFERENCES

1. Alfuraidan MR, Fixed points of monotone nonexpansive mappings with a graph. Fixed Point Theory Appl. 2015:49. DOI: 10.1186/s13663-015-0299-0.
2. Alfuraidan MR, On monotone Ciric quasi-contraction mappings with a graph. Fixed Point Theory Appl. 2015:93. DOI: 10.1186/s13663-015-0341-2.
3. Alfuraidan MR, On monotone pointwise contractions in Banach spaces with a graph. Fixed Point Theory Appl. 2015:139. DOI: 10.1186/s13663-015-0390-6.
4. Alfuraidan MR, Bachar M, Khamsi MA, On monotone contraction mappings in modular function spaces. Fixed Point Theory Appl. 2015:28. DOI 10.1186/s13663-015-0274-9.
5. Alfuraidan MR, Khamsi MA, Fixed points of monotone nonexpansive mappings on a hyperbolic metric space with a graph. Fixed Point Theory Appl. 2015:44. DOI 10.1186/s13663-015-0294-5.
6. Aksoy AG, Khamsi MA, Nonstandard Methods in Fixed Point Theory. New York: Springer; 1990.
7. Bachar M, Khamsi MA, Fixed points of monotone mappings and application to integral equations. Fixed Point Theory Appl. 2015:110. DOI:10.1186/s13663-015-0362-x.
8. Bachar M, Khamsi MA, On common approximate fixed points of monotone nonexpansive semi-groups in Banach spaces. Fixed Point Theory Appl. 2015:160 DOI:10.1186/s13663-015-0405-3.
9. Baillon JB, Bruck RE, Reich S, On the asymptotic behavior of nonexpansive mappings and semi-groups in Banach spaces. Houston J. Math. 1978;4:1–9.
10. Banach S, Sur les opérations dans les ensembles abstraits et leurs applications. Fund. Math. 1922;3:133–181.
11. Beauzamy B, Introduction to Banach Spaces and Their Geometry. North-Holland, Amsterdam; 1985.
12. Belluce LP, Kirk WA, Fixed-point theorems for families of contraction mappings. Pacific J. Math. 1966;18:213–217.
13. Belluce LP, Kirk WA, Nonexpansive mappings and fixed-points in Banach spaces. Illinois J. Math. 1967;11:474–479.
14. Berggren JL, Episodes in the Mathematics of Medieval Islam. New York: Springer; 1986.
15. Bin Dehaish BA, Khamsi MA, Approximating common fixed points of semigroups in metric spaces. Fixed Point Theory Appl. 2015:51 DOI:10.1186/s13663-015-0291-8.
16. Bin Dehaish BA, Khamsi MA, Mann iteration process for monotone nonexpansive mappings. Fixed Point Theory Appl. 2015:177 DOI:10.1186/s13663-015-0416-0.
17. Browder FE, Nonexpansive nonlinear operators in a Banach space. Proc. Nat. Acad. Sci. U.S.A. 1965;4:1041–1044.
18. Bruck RE, A common fixed point theorem for a commuting family of nonexpansive mappings. Pacific J. Math. 1974;53:59–71.
19. Clarkson JA, Uniformly convex spaces. Trans. Am. Math. Soc. 1936;40:396–414.
20. Das G, Debata P, Fixed points of quasi-nonexpansive mappings. Indian J. Pure Appl. Math. 1986;17:1263–1269.
21. DeMarr RE, Common fixed-points for commuting contraction mappings. Pacific J. Math. 1963;13:1139–1141.
22. Diestel J, Geometry of Banach spaces – Selected Topics. Springer Lecture Notes in Math. no. 485, New York: Springer; 1957.
23. Dugundji J, Granas A, Fixed Point Theory. Polska Akademia Nauk, Instytut Matematyczny, PWN-Polish Scientific Publ., Warsaw; 1982.
24. Edelstein M, A remark on a theorem of M. A. Krasnoselski. Am. Math. Monthly 1966;73:509–510.
25. Edelstein M, O'Brien RC, Nonexpansive mappings, asymptotic regularity and successive approximations. J. London Math. Soc. 1978;17:547–554.
26. El-Sayed SM, Ran ACM, On an iteration method for solving a class of nonlinear matrix equations. SIAM J. Matrix Anal. Appl. 2002;23:632–645.
27. Fukhar-ud-din H, Khamsi MA, Approximating common fixed points in hyperbolic spaces. Fixed Point Theory Appl. 2014:113 DOI:10.1186/1687-1812-2014-113.
28. Genel A, Lindenstrauss J, An example concerning fixed points. Israel J. Math. 1975;22:81–86.
29. Goebel K, Kirk WA, Iteration processes for nonexpansive mappings. Contemp. Math. 1983;21:115–123.

30. Goebel K, Kirk WA, Topics in Metric Fixed Point Theory. Cambridge Univ. Press, Cambridge; 1990.

31. Goebel K, Reich S, Uniform convexity, hyperbolic geometry, and nonexpansive mappings. Monographs and Textbooks in Pure and Applied Mathematics, 83. New York:Marcel Dekker; 1984.

32. Goebel K, Sekowski T, Stachura A, Uniform convexity of the hyperbolic metric and fixed points of holomorphic mappings in the Hilbert ball. Nonlinear Anal. 1980;4:1011–1201.

33. Göhde D, Zum Prinzip der kontraktiven Abbildung. Math. Nachr. 1965;30:251–258.

34. Ishikawa S, Fixed points and iterations of a nonexpansive mapping in a Banach space. Proc. Am. Math. Soc. 1976;59:65–71.

35. Jachymski J, The Contraction Principle for Mappings on a Metric Space with a Graph. Proc. Am. Math. Soc. 2008;136:1359–1373.

36. Khamsi MA, On metric spaces with uniform normal structure. Proc. Am. Math. Soc. 1989;106:723–726.

37. Khamsi MA, Approximate fixed point sequences of nonlinear semigroup in metric spaces. Canadian Math. Bull. http://dx.doi.org/10.4153/CMB-2014-026-x, 9 pages.

38. Khamsi MA, Khan AR, Inequalities in metric spaces with applications. Nonlinear Anal. 2011;74:4036–4045.

39. Khamsi MA, Khan AR, On monotone nonexpansive mappings in $L_1([0,1])$. Fixed Point Theory Appl. 2015:94. DOI 10.1186/s13663-015-0346-x.

40. Khamsi MA, Kirk WA, An Introduction to Metric Spaces and Fixed Point Theory. New York: John Wiley; 2001.

41. Khamsi MA, Kozlowski WK, On asymptotic pointwise nonexpansive mappings in modular function spaces. J. Math. Anal. Appl. 2011;380:697–708.

42. Kirk WA, A fixed point theorem for mappings which do not increase distances. Am. Math. Monthly 1965;72:1004–1006.

43. Kirk WA, Krasnosel'skii iteration process in hyperbolic spaces. Numer. Funct. Anal. Optimiz. 1982;4:371–381.

44. Kohlenbach U, Leustean L, Mann iterates of directionally nonexpansive mappings in hyperbolic spaces. Abstr. Appl. Anal. 2003;8:449–477.

45. Krasnoselskii MA, Two observations about the method of successive approximations. Uspehi Mat. Nauk 1955;10:123–127.

46. Kuczumow T, An almost convergence and its applications. Ann. Univ. Mariae Curie-Skłodowska Sect. 1978;32:79–88.

47. Leustean L, A quadratic rate of asymptotic regularity for CAT(0)-spaces. J. Math. Anal. Appl. 2007;325:386–399.

48. Lim TC, A fixed point theorem for families of nonexpansive mappings. Pacific J. Math. 1974;53:487–493.

49. Nadler SB, Multi-valued contraction mappings. Maruzen, Tokyo 1950;30:475–488.

50. Nieto JJ, Rodríguez-López R, Contractive mapping theorems in partially ordered sets and applications to ordinary differential equations. Order 2005; 22:223–239.

51. Opial Z, Weak convergence of the sequence of successive approximations for nonexpansive mappings. Bull. Am. Math. Soc. 1967;73:591–597.

52. Ran ACM, Reurings MCB, A fixed point theorem in partially ordered sets and some applications to matrix equations. Proc. Am. Math. Soc. 2004;132:1435–1443.

53. Reich S, Shafrir I, Nonexpansive iterations in hyperbolic spaces. Nonlinear Anal., Theory, Methods Appl. 1990;15:537–558.

54. Reich S, Zaslavski AJ, Generic aspects of metric fixed point theory. In: Kirk WA, Sims B, editors, Handbook of Metric Fixed Point Theory, Kluwer Academic; 2001; 557–576.

55. Stewart I, Tall D, Algebraic Number Theory and Fermat's Last Theorem. Third Edition, Massachusetts, USA: A K Peters Ltd; 2001.

56. Suzuki T, The set of common fixed points of a one-parameter continuous semigroup of mappings is $F(T(1)) \cap F(T(\sqrt{2}))$. Proc. Am. Math. Soc. 2006;134:673–681.
57. Tan K-K, Xu H-K, An ergodic theorem for nonlinear semigroups of Lipschitzian mappings in Banach spaces. Nonlinear Anal. 1992;19:805–813.
58. Turinici M, Fixed points for monotone iteratively local contractions. Dem. Math. 1986;19:171–180.
59. Turinici M, Ran and Reurings theorems in ordered metric spaces. J. Indian Math. Soc. 2011;78:207–2014.
60. Zeidler E, Nonlinear Functional Analysis and Its Applications I: Fixed-Point Theorems. New York: Springer; 1986.

CHAPTER 4

Viscosity Methods for Some Applied Nonlinear Analysis Problems

Paul-Emile Maingé[*] and Abdellatif Moudafi[a†]

[*] Université des Antilles, Département Scientifique Interfacultaire, Campus de Schoelcher, Martinique, 97233
[‡] Aix Marseille Université CNRS-LSIS UMR 7296,13397, Marseille, France
[a] Corresponding: abdellatif.moudafi@univ-amu.fr

Contents

Abstract

The aim of this chapter is to present a useful viscosity approach for solving some applied nonlinear analysis problems which were solved through translations into split fixed point problems. The underlying principle is to operate a certain nonlinear quasi-nonexpansive mapping iteratively and generate a convergent sequence to its fixed point. However, such a mapping often has infinitely many fixed points, meaning that a selection from the fixed point set should be of great importance. Nevertheless, most fixed point methods can only return an unspecified solution. Therefore, based on a viscosity idea used in solving optimization problems which are not well posed (namely, to consider a family of regularized problems such that each of them is well posed, then a viscosity approximation method is applied to seek a particular solution of the original problem as a limit of the solutions of the regularized problems), we accomplish this challenging task by viscosity approximation methods and obtain a particular solution from the fixed point set that solves a variational inequality criterion. We present the mathematical idea of the proposed approach in the context of fixed point theory together with examples in linear programming, semicoercive elliptic problems and in the area of finance as well as elegant techniques of analysis to overcome the no-Fejer monotonicity of the generated sequences. Split common fixed point problems are then investigated, the convergence is studied with different step sizes, convergence results in some

http://dx.doi.org/10.1016/B978-0-12-804295-3.50004-8

specific cases when the projections onto the involved sets are not available are proposed and numerical experiments on an inverse heat equation are provided. Special attention is also given to split equilibrium problems by making a link with the hybrid steepest descent approach. We then conclude with a reference to viscosity methods for hierarchical fixed point problems which permit the design of many powerful schemes with the strong support of both viscosity approximation and the Mann iterative process.

4.1. Introduction

Viscosity methods provide an efficient approach to a large number of problems arising from different branches of applied mathematics, such as mathematical programming, variational problems, game theory, control theory, finance and ill-posed problems. A major feature of these methods is to provide, as a limit of the solutions of the well-posed approximate problems, a particular solution of the original problem which has suitable properties. In this chapter, we focus our attention on viscosity methods for fixed point problems. In a quite general setting, relying on fixed point techniques, we prove that the viscosity solution solves, among all the fixed points of the original operator, a variational inequality criterion. In so doing, we are able to explain, in a unified way, a number of viscosity approaches. It is worthwhile remembering that the first abstract formulations of the properties of the viscosity approximation were given by Tikhonov [63] in 1963 while studying ill-posed problems. The viscosity method has been successfully applied to various problems from calculus of variations as minimal surface problems and plasticity theory, and various applications can be obtained in optimal control theory, game theory and partial differential equations (see the excellent paper by Attouch [2] and references therein).

To begin with, it is worth mentioning that fixed point problems define a broad class of problems that arises in the context of optimization as well as in fields as diverse as economics, game theory, transportation science, and more recently image and signal processing. These problems can be modeled by an equation of the type $x = Tx$, where T is a nonlinear operator defined on a metric space. The solutions to this equation are called fixed points of the operator T. Their widespread applicability motivates the need to develop and study efficient solution algorithms. The literature for solving fixed points is vast. Many authors and more recently Marquez [41] have summarized and categorized many algorithms for this problem as well as the application of these methods to some special settings. A standard method for solving fixed point problems for contractive maps T, on a complete metric space (E, d), is to apply the iteration $x_{n+1} = Tx_n$. The classical Banach Fixed Point Theorem shows that for contractive maps, the sequence of Picard iterates $\{T^n x\}_{n \in \mathbb{N}}$ strongly converges, for any starting point $x \in E$, to the unique fixed point of T. When the map is not contractive, for instance nonexpansive, that is, a mapping verifying $d(Tx, Ty) \leq d(x, y)$ for any $x, y \in E$, the sequence $\{T^n x\}_{n \in \mathbb{N}}$ does not converge in general and, indeed, the map T does not have a fixed point (or it might have several). Then, we must assume additional conditions on T and/or the underlying space to ensure the existence of fixed points. Since the 1960s, the study of the class of nonexpansive mappings has been one of the major and most active research areas of applied nonlinear analysis and the development of feasible iterative methods for approximating fixed points of a nonexpansive mapping T has been of particular importance. For a nice survey about the asymptotic behavior of nonexpansive mappings in Hilbert and Banach spaces, see for instance Ref. [10]. Remember that there are two ways to solve fixed point problems: one implicit and one explicit, namely Mann and Halpern iterations. Both iterations have been studied extensively and are still the focus of most research.

The Mann iteration, initially due to Mann [39], is the averaged algorithm defined by the recursive scheme

$$x_{n+1} = (1 - \alpha_n)x_n + \alpha_n T x_n, \quad n \geq 0, \tag{4.1.1}$$

where x_0 is an arbitrary point of the domain of T and (α_n) is a sequence of positive real parameters in $(0, 1)$. One of the classical results, due to Reich [53], states that if the underlying space is uniformly convex and has a Fréchet differentiable norm, T has fixed points and $\sum_{n \geq 0} \alpha_n(1 - \alpha_n) = \infty$, then the sequence $\{x_n\}$ defined by the Mann algorithm weakly converges to a fixed point of T. Moreover, a counterexample provided by Genel and Lindenstrauss [20] shows that in infinite-dimensional spaces Mann iteration cannot have strong convergence. The Halpern iteration, first presented in Ref. [22], is generated by the recursive formula

$$x_{n+1} = \alpha_n u + (1 - \alpha_n) T x_n, \quad n \geq 0, \tag{4.1.2}$$

for real parameters $\alpha_n \in (0, 1)$ and an arbitrary $x_0, u \in C$. Unlike the Mann iteration, the Halpern algorithm can be proved to have strong convergence provided that the underlying space is smooth enough. Viscosity approximation methods are a generalization of this second method and have very attractive properties that make them particularly well suited for selecting solutions with remarkable properties. One of the sources of the relevance of the iterative methods for approximating fixed points of a nonexpansive mapping is its application in other fields via, for instance, the orthogonal projection on a closed convex set, the proximity mapping of a proper convex and lower semicontinuous function or the resolvent of a maximal monotone operator, as well as that of a monotone bifunction. Hence, besides the problem of zeros of a monotone operator, they can be applied to finding a solution of a variational inequality, a minimizer of a convex function and Nash equilibria, among other problems.

The organization of this chapter is as follows. In Section 4.2 we recall some results of viscosity methods for fixed point problems, provide some examples in linear programming, semicoercive elliptic problems and problems in the area of finance, and prove a convergence result for quasi-nonexpansive mappings. Convergence results for viscosity methods for split common fixed point problems are established in Section 4.3, some specific cases are also provided as well as numerical experiments for an inverse heat equation. In Section 4.4 a convergence result for a viscosity approximation method for split equilibrium problems is discussed and a reference to viscosity methods for hierarchical fixed point problems is made.

4.2. Viscosity Method for Fixed Point Problems

To begin with let us state some convergence results for both implicit and explicit schemes of the Halpern iteration. To this end, let H be a real Hilbert space endowed with an inner product and its induced norm denoted by $\langle .,.\rangle$ and $\|\cdot\|$ respectively. Let C be a closed convex set of H and $T : C \to C$ a nonexpansive mapping, that is, a mapping satisfying $\|Tx - Ty\| \le \|x - y\|$ for any $x, y \in C$, and P_C be the metric projection of H onto C which assigns to each $x \in H$ the unique point in C, denoted $P_C x$, such that $\|x - P_C x\| = \min_{y \in C} \|x - y\|$.

Now, for $t \in (0,1)$ and an arbitrary $u \in C$, let $x_t \in C$ denote the unique fixed point of $T_t := tu + (1-t)T$, namely

$$x_t = tu + (1-t)Tx_t. \tag{4.2.3}$$

Assume that $\mathrm{Fix}(T) = \{x \in C : Tx = x\} \ne \emptyset$. In 1967, Browder [9] proved the following result:

Theorem 4.2.1. *The net* $\{x_t\}$ *converges strongly to a fixed point* $\bar{x}$ *of* T *that is closest to* u, *in other words* $\bar{x} = P_{Fix(T)}u$.

Now, for a sequence $\{\alpha_n\} \in (0,1)$ and an arbitrary $u \in C$, let the sequence $\{x_n\}$ be iteratively defined from $x_0 \in C$ by

$$x_{n+1} = \alpha_n u + (1 - \alpha_n)Tx_n, \quad n \ge 0, \tag{4.2.4}$$

The iterative method (4.2.4) was first introduced in 1967 by Halpern [22] in the framework of Hilbert spaces. He proved the weak convergence of $\{x_n\}$ to a fixed point of T where $\alpha_n = n^{-\theta}$, $\theta \in (0,1)$. In [31], Lions improved the result of Halpern, still in Hilbert spaces, by proving the following result:

Theorem 4.2.2. *If* $\{\alpha_n\}$ *satisfies the following conditions:*

(i) $\lim_{n \to +\infty} \alpha_n = 0$;

(ii) $\sum_{n=0}^{\infty} \alpha_n = \infty$;

(iii) $\lim_{n \to +\infty} \dfrac{\|\alpha_n - \alpha_{n+1}\|}{\alpha_{n+1}^2} = 0$.

Then $\{x_n\}$ converges strongly to a fixed point $\bar{x}$ of T, that is, closest to u, in other words $\bar{x} = P_{FixT}\, u$.

In 1980, Reich [53] proved that Halpern's result remains true when E is a uniformly smooth Banach space. More precisely, he established the following:

Theorem 4.2.3. *Let E be a uniformly smooth Banach space, then both the net $\{x_t\}$ and the sequence $\{x_n\}$ defined by (4.2.3) and (4.2.4) converge strongly to the same fixed point of T. If we define*

$$Q : C \to Fix(T) \ by \ Q(u) = s - \lim_{t \to 0} x_t,$$

then Q is the sunny nonexpansive retraction from C onto $FixT$.

It is worth mentioning that both Halpern's and Lions's conditions on the parameter α_n are not valid for $\alpha_n = \frac{1}{1+n}$. This was overcome in 1992 by Wittman [66], who proved, in the Hilbert spaces context, the strong convergence of $\{x_n\}$ under (i), (ii) above and in addition the following condition:

(iv) $\sum_{n=0}^{\infty} \|\alpha_{n+1} - \alpha_n\| < \infty$.

In [53], Reich extended the result of Wittman to Banach spaces which are uniformly smooth and have weakly sequentially continuous duality mappings and the parameters $\{\alpha_n\}$ satisfy (i) and (ii) and also are decreasing (hence verifying (iv)). The conditions on the spaces exclude L_p spaces with $1 < p < \infty$, $p \neq 2$. In [55], Shioji and Takahashi extended Wittman's result to real Banach spaces with uniformly Gâteaux differentiable norms. More precisely, they obtained the following result:

Theorem 4.2.4. *Let E be a real Banach space with uniformly Gâteaux differentiable norm and let C be a closed convex subset of E. Let $T : C \to C$ be a nonexpansive mapping with $Fix(T) \neq \emptyset$. Let $\{\alpha_n\}$ satisfy (i), (ii) and in addition the following condition:*

(v) $\lim_{n \to +\infty} \|\alpha_{n+1} - \alpha_n\| = 0$.

Assume that $\{x_t\}$ (the net given by (4.2.3)) converges strongly to a fixed point $\bar{x}$ of T as $t \to 0$. Then $\{x_n\}$ (generated by (4.2.4)) converges strongly to $\bar{x}$.

In 2002, Xu [68] improved the result by Lions under a condition weaker than (iii) and in the context of uniformly smooth Banach spaces. He proved the following result:

Theorem 4.2.5. *Let E be a uniformly smooth Banach space, C be a closed convex set of E, and $T : C \to C$ be a nonexpansive mapping with a fixed point. Let $u, x_0 \in C$ and assume that the parameters $\{\alpha_n\} \in (0,1)$ satisfy (i), (ii) and in addition the following condition:*

$$\text{(vi)} \quad \lim_{n \to +\infty} \frac{\|\alpha_n - \alpha_{n+1}\|}{\alpha_{n+1}} = 0.$$

Then $\{x_n\}$ converges strongly to a fixed point of T that is closest to u, in other words $\bar{x} = P_{Fix(T)} u$.

In [46], Moudafi presented an extension of Browder's and Halpern's results in another direction (Moudafi's generalizations are called viscosity approximations and are very important, because they are applied to convex optimization, linear programming, monotone inclusions and elliptic differential equations, among other possible applications). The viscosity extension of the above results was thus first studied by Moudafi [46] and further developed by Xu [69]. More precisely, in the viscosity approximation method, the implicit and explicit schemes (4.2.3) and (4.2.4) are replaced respectively by the following viscosity schemes:

(IVS) $$x_t = t f(x_t) + (1 - t) T x_t$$

and

(EVS) $$x_{n+1} = \alpha_n f(x_n) + (1 - \alpha_n) T x_n,$$

where f is a contraction on C, $t \in (0,1)$ and α_n is a sequence in $[0,1]$.

Theorem 4.2.6 [46, 69]. *In a Hilbert space H, assume that $Fix(T)$ is nonempty and let $\{x_t\}$ be the net given by (IVS). Then $\{x_t\}$ strongly converges to the unique solution $\bar{x} \in Fix(T)$ of the variational inequality criterion*

$$\langle (I - f)\bar{x}, x - \bar{x} \rangle \geq 0, \quad \text{for all } x \in Fix(T).$$

Theorem 4.2.7 [46, 69]. *In a Hilbert space, assume that $Fix(T)$ is nonempty and let $\{x_n\}$ be the sequence generated by the algorithm (EVS). Assume conditions (i) and (ii) above, and in addition, either (iv) or (vi), hold. Then $\{x_n\}$ converges strongly to the unique solution $\bar{x} \in Fix(T)$ of the following variational inequality criterion:*

$$\langle (I - f)\bar{x}, x - \bar{x} \rangle \geq 0, \quad \text{for all } x \in Fix(T).$$

In [69], Xu generalized also Moudafi's Theorems to uniformly smooth Banach spaces. Since then, viscosity approximations have been widely studied and extended; for example, Suzuki [57] discusses Moudafi's viscosity approximations with Meir-Keeler contractions. The viscosity approximations were also modified by many authors either to weaken the conditions on the regularization parameters α_n or to improve the class of operators and / or spaces. For instance, from Hilbert or uniformly convex Banach spaces to CAT(k) spaces that form a natural extension of spherical geometry and which generalize among others complete Riemanian manifolds with curvature bounded above (see for example Ref. [51]), or from single-valued operators to multivalued ones (see for instance Ref. [50] and references therein). The usefulness of our approach was illustrated also by applying it in the literature to a large class of algorithms including the Proximal Point, Forward-Backward and the Douglas-Rachford algorithms, since in various fields of science and engineering, such as signal/image processing, many problems can be cast as solving a structured fixed point problem. We would like to emphasize that the idea we used, namely to replace the original problem by a family of regularized ones, with each having a unique solution, and a particular (viscosity) solution of the original problem being obtained as a limit of these unique solutions, has already been considered in solving some nonlinear problems, among others by Attouch [2] in the context of convex optimization. Here, we obtained its generalization to fixed point problems.

It should also be noticed that applications of this approach have been illustrated, among others, in the following specific cases:

Example 4.2.1. Specific situations are given below.

(a) **Log-barrier and exponential penalty for linear programming:** More precisely, when considering an inequality constrained program of the form

$$\min_{x \in \mathbb{R}^n} \left\{ c^t x : Ax \le b \right\}, \tag{4.2.5}$$

with a nonempty and bounded feasible set $\{x \in \mathbb{R}^n : Ax \le b\}$, the corresponding log-barrier approximation is given by

$$\min_{x \in \mathbb{R}^n} \left\{ c^t x - \varepsilon \sum_{i=1}^{n} \ln(b_i - a_i^t x) \right\}, \tag{4.2.6}$$

where a_i denotes the rows of A and an alternative penalty approach is to consider

$$\min_{x \in \mathbb{R}^n} \left\{ c^t x - \varepsilon \sum_{i=1}^{n} \exp\left(-(b_i - a_i^t x)/\varepsilon \right) \right\}. \tag{4.2.7}$$

Problem (4.2.6) (respectively (4.2.7)) has a unique solution x_ε (respectively $\tilde{x}_\varepsilon$)

which converges when $\varepsilon \to 0$ to the analytic center of the optimal set S, that is, the unique solution of

$$\min_{x \in \mathbb{R}^n} \left\{ \sum_{i \notin I_0} \ln(b_i - a_i^t x) \right\}, \quad \text{where } I_0 = \{i : a_i^t x = b_i \text{ for all } x \in S\} \quad (4.2.8)$$

(respectively to $\tilde{x} \in S$, called the *centroid*, see Ref. [18] for details).

(b) **Semicoercive elliptic problem:** Let Ω be an open bounded subset of $\mathbb{R}^n$, $n \in \mathbb{N}^*$ with regular boundary $\partial \Omega$. Given $f \in L^2(\Omega)$, we consider the following boundary value problem: find $u \in H^1(\Omega)$ such that

$$\begin{cases} -\Delta u = f, & \text{on } \Omega; \\ \frac{\partial u}{\partial n} = 0, & \text{on } \partial \Omega. \end{cases} \quad (4.2.9)$$

Here, $\partial/\partial n$ denotes the exterior normal derivative. Setting $\tilde{f} = \frac{1}{\|\Omega\|} \int f(x)dx$, it is well known that (4.2.9) admits a solution (unique upon a constant) if, and only if, $\tilde{f} = 0$ (see, for example, Brézis and Lions [8]). The sequence $\{u_\varepsilon\}$ generated by the viscosity approximation method, namely,

$$\begin{cases} -\varepsilon u_\varepsilon \Delta u_\varepsilon = f, & \text{on } \Omega; \\ \frac{\partial u_\varepsilon}{\partial n} = 0, & \text{on } \partial \Omega, \end{cases} \quad (4.2.10)$$

converges in H^1 to $\tilde{u}$ unique solution of minimal norm for (4.2.9). This is equivalent to

$$\begin{cases} -\Delta \tilde{u} = f, & \text{on } \Omega; \\ \frac{\partial \tilde{u}}{\partial n} = 0, & \text{on } \partial \Omega, \\ \int \tilde{u}(x)dx = 0. \end{cases} \quad (4.2.11)$$

(c) **A selection concept in finance:** Lehdili and Salem [30] consider a single period model and an investor who faces a capital market with n riskly financial assets. Let $\tilde{R}_i$, $i = 1, 2, \ldots, n$, be the random variable which represents the stochastic return of capital asset i. The problem of choosing an optimal portfolio for an investor is formulated as an expected utility criterion: Assume that there is an investor whose preferred structure is characterized by its utility function $U : \mathbb{R} \to \mathbb{R}$. In particular, suppose that U is a differentiable, concave, and strictly increasing function. If we denote by W_0 the investor's initial wealth and by x_i, $i = 1, 2, \ldots, n$, the proportion of this wealth invested in the financial asset i, then the terminal wealth is a random variable given by

$$\tilde{W}(X) := W_0 \left(\sum_{i=1}^n x_i \tilde{R}_i \right),$$

with

$$X = (x_1, \ldots, x_n), \quad \sum_{i=1}^{n} x_i = 1, \quad \text{and} \quad x_i \geq 0, i = 1, 2, \ldots, n.$$

Hence, the optimization problem will be expressed as maximizing the expected utility of terminal wealth, that is,

$$\max_{x \in C} E[U(\tilde{W}(X))],$$

where

$$C := \left\{ (x_1, \ldots, x_n) : \sum_{i=1}^{n} x_i = 1 \text{ and } x_i \geq 0, \ i = 1, 2, \ldots, n \right\}.$$

The investor chooses his best trading strategy by solving problem (S), but if there are several optimal solutions the "good" portfolio must be selected amongst all optimal portfolios, for example the optimal portfolio which minimizes the risk! When one uses the variance as a measure of risk, we will show that the problem is to select the optimal solution which minimizes some norm. Indeed, set

$$\Omega := [\sigma_{i,j}]_{1 \leq i,j \leq n}, \quad \sigma_{i,j} := cov(\tilde{R}_i, \tilde{R}_j), \quad i, j = 1, 2, \ldots, n,$$

in other words Ω is the $n \times n$ variance-covariance matrix. It is then easily seen that

$$var\left(\frac{\tilde{W}}{W_0}\right) = \sum_{i=}^{n} \sum_{j=1}^{n} \sigma_{i,j} x_i x_j = \langle \Omega X, X \rangle = \|X\|_{\Omega}^2.$$

Consequently, instead of directly solving the original problem, we consider the following approximate ones obtained by the combination of penalization and Tikhonov regularization:

$$\max_{x \in \mathbb{R}^n} \left\{ E[U(\tilde{W}(x))] - \frac{1}{2\varepsilon^\theta} dist^2(x, C) - \frac{\varepsilon}{2} \|x\|_{\Omega}^2 \right\},$$

with $\theta > 1$. Applying a visco-penalization analysis, we directly select the unique optimal portfolio which presents the minimal risk.

Also, we suppose that there is a riskless asset and denote R_0 as its deterministic return. Markowitz's mean variance criterion provides an optimal portfolio by solving the following approximate problems based on Tikhonov regularization:

$$\max_{X \in \mathbb{R}^n} \left\{ E[U(\tilde{W}(X))] - \frac{\varepsilon}{2} var[\tilde{W}(X)] \right\},$$

with

$$\tilde{W}(X) := W_0 \left(R_0 + \sum_{i=1}^{n} x_i \left(\tilde{R}_i - R_0 \right) \right).$$

The interesting improvement to quasi-nonexpansive mappings of these results was obtained by Maingé [35]. Before stating this result let us first recall the definition of quasi-nonexpansive operators which appear naturally when using subgradient projection operator techniques in solving some feasibility problems and also some definitions of classes of operators often used in fixed point theory and which are commonly encountered in the literature.

- An operator T belongs in the general class $\mathcal{E}_Q$ of (possibly discontinuous) *quasi-nonexpansive mappings* if

$$\text{for all } (x,y) \in H \times \text{Fix}(T), \quad \|Tx-y\| \leq \|x-y\|. \qquad (4.2.12)$$

- T belongs to the set $\mathcal{E}_N$ of *nonexpansive mappings* if

$$\text{for all } (x,y) \in H \times H, \quad \|Tx-Ty\| \leq \|x-y\|.$$

- T belongs to the set $\mathcal{E}_A$ of *attracting mappings* if

$$\text{for all } (x,y) \in H \times \text{Fix}(T), \quad \|Tx-y\| < \|x-y\|.$$

- T belongs to the set $\mathcal{E}_{SA}$ of α-*attracting mappings* if there exists $\alpha > 0$ satisfying

$$\text{for all } (x,y) \in H \times \text{Fix}(T), \quad \alpha \|x-Tx\|^2 \leq \|x-y\|^2 - \|Tx-y\|^2. \qquad (4.2.13)$$

It is easily observed that $\mathcal{E}_N \subset \mathcal{E}_Q$ and that $\mathcal{E}_{SA} \subset \mathcal{E}_A \subset \mathcal{E}_Q$.

- A quasi-nonexpansive mapping T is said to be w-*averaged* if there exists some $w \in (0,1)$, and some quasi-nonexpansive mapping N such that $T = (1-w)I + wN$.

  In this case T satisfies an obvious relation $\text{Fix}(T) = \text{Fix}(N)$. In particular, if T is $\frac{1}{2}$-averaged, T is called *firmly quasi-nonexpansive*.

This suggests that given a quasi-nonexpansive mapping N, we can always construct an α-attracting quasi-nonexpansive T that shares the same fixed point set with N.

Remark 4.2.1. It is easy to verify that given a quasi-nonexpansive mapping T and its relaxed operator $T_w := (1-w)I + wT$, with $\text{Fix}(T) \neq \emptyset$ and $w \in (0,1]$, the following statements are valid:

(a) $\text{Fix}(T) = \text{Fix}(T_w)$;

(b) T_w is quasi-nonexpansive;

(c) $\|T_w x - q\|^2 \leq \|x-q\|^2 - w(1-w)\|Tx-x\|^2$ for all $(x,q) \in C \times \text{Fix}(T)$;

(d) $\langle x - T_w x, x - q \rangle \geq w \|x - Tx\|^2$ for all $(x,q) \in C \times \text{Fix}(T)$.

From (c) of Remark 4.2.1 it is easily seen that a quasi-nonexpansive mapping T is w-averaged for some $w \in (0,1)$ if and only if T is $\frac{1-w}{w}$-attracting. From this we infer that a quasi-nonexpansive mapping T is $\frac{1}{2}$-averaged if and only if T is 1-strongly attracting. Recently Bauschke and Combettes [4] have considered a class of mappings satisfying the condition

$$\langle y - Tx, x - Tx \rangle \leq 0, \quad \text{for all } (x,y) \in H \times \text{Fix}(T).$$

It can be easily seen that the class of mappings satisfying the latter condition coincides with that of firmly quasi-nonexpansive mappings (in other words $\frac{1}{2}$-averaged quasi-nonexpansive mappings).

Maingé proved the following theorem.

Theorem 4.2.8 [35]. *Let C be a closed convex subset of a Hilbert space H. Let $T : C \to C$ be a quasi-nonexpansive mapping such that $I - T$ is demi-closed at zero, that is, $z \in \text{Fix}T$ whenever $\{z_n\}$ is a sequence in C such that z_n converges weakly to z and $z_n - Tz_n$ strongly converges to zero. Suppose that $f : C \to C$ is a contraction (with some modulus $\rho \in (0,1)$). Let $\{x_n\}$ be a sequence in C defined by, from $x_0 = x$ arbitrarily chosen,*

$$x_{n+1} = \alpha_n f(x_n) + (1 - \alpha_n) T_w x_n, \tag{4.2.14}$$

with $\omega \in (0,1)$, I is the identity mapping and α_n is a sequence in $(0,1)$ satisfying

(i) $\lim\limits_{k \to +\infty} \alpha_n = 0$;

(ii) $\sum_{k=0}^{\infty} \alpha_n = \infty$;

Then $\{x_n\}$ converges strongly to the unique element $\bar{x}$ verifying

$$\bar{x} = \left(P_{\text{Fix}(T)} \circ f \right) \bar{x}, \tag{4.2.15}$$

which equivalently solves the following variational inequality criterion:

$$\bar{x} \in \text{Fix}(T), \text{ and for all } x \in \text{Fix}(T), \quad \langle (I - f)\bar{x}, x - \bar{x} \rangle \geq 0. \tag{4.2.16}$$

Before proving Theorem 4.2.8, we give a series of preliminary results.

Lemma 4.2.1. *The sequence $\{x_n\}$ generated by (4.2.14), with T quasi-nonexpansive, $\{\alpha_n\} \subset (0,1)$, and $w \in (0,1]$, is bounded and satisfies*

$$\|x_n - q\| \leq \max \left\{ \|x_0 - q\|, \frac{\|fq - q\|}{1 - \rho} \right\}, \tag{4.2.17}$$

where q is any element in $\mathrm{Fix}(T)$.

Proof. From $q \in \mathrm{Fix}(T)$ and using (4.2.14), we have

$$\|x_{n+1} - q\| = \|\alpha_n(fx_n - fq) + \alpha_n(fq - q) + (1 - \alpha_n)(T_w x_n - q)\|,$$

hence it is evident that

$$\|x_{n+1} - q\| \leq \alpha_n\|fx_n - fq\| + \alpha_n\|fq - q\| + (1 - \alpha_n)\|T_w x_n - q\|. \qquad (4.2.18)$$

In light of Remark 4.2.1(c), we also have

$$\|T_w x_n - q\| \leq \|x_n - q\|, \qquad (4.2.19)$$

which, by (4.2.18) and $\|fx_n - fq\| \leq \rho\|x_n - q\|$ (as f is assumed to be a ρ-contraction), amounts to

$$\|x_{n+1} - q\| \leq [1 - (1 - \rho)\alpha_n]\|x_n - q\| + \alpha_n\|fq - q\|. \qquad (4.2.20)$$

Then, setting $Q_n := \max\left\{\|x_n - q\|, \frac{\|fq - q\|}{1 - \rho}\right\}$, we clearly have $|x_{n+1} - q| \leq Q_n$, hence it is obviously checked that $Q_{n+1} \leq Q_n$, so that $Q_n \leq Q_0$, which leads to (4.2.17) and proves the boundedness of $\{x_n\}$. $\qquad\square$

Lemma 4.2.2. *If $\bar{x}$ is a solution of (4.2.16) with T demi-closed and $\{y_n\} \subset C$ is a bounded sequence such that $\|Ty_n - y_n\| \to 0$, then*

$$\liminf_{n\to\infty}\langle(I - f)\bar{x}, y_n - \bar{x}\rangle \geq 0. \qquad (4.2.21)$$

Proof. Clearly, by $\|Ty_n - y_n\| \to 0$ and T being demi-closed, we know that any weak cluster point of $\{y_n\}$ belongs to $\mathrm{Fix}(T)$. It is also a simple matter to see that there exists $\bar{y}$ and a subsequence $\{y_{n_k}\}$ of $\{y_n\}$ such that $\{y_{n_k}\} \rightharpoonup \bar{y}$ weakly as $k \to \infty$ (hence $\bar{y} \in \mathrm{Fix}(T)$) and such that

$$\liminf_{n\to\infty}\langle(I - f)\bar{x}, y_n - \bar{x}\rangle = \lim_{k\to\infty}\langle(I - f)\bar{x}, y_{n_k} - \bar{x}\rangle, \qquad (4.2.22)$$

which by (4.2.16) obviously leads to

$$\liminf_{n\to\infty}\langle(I - f)\bar{x}, y_n - \bar{x}\rangle = \langle(I - f)\bar{x}, \bar{y} - \bar{x}\rangle \geq 0, \qquad (4.2.23)$$

that is the desired result. $\qquad\square$

Remark 4.2.2. A helpful result in our study is given by the following classical equality:

$$\text{for all } u, v \in H, \quad \langle u, v\rangle = -(1/2)\|u - v\|^2 + (1/2)\|u\|^2 + (1/2)\|v\|^2. \qquad (4.2.24)$$

Remark 4.2.3. Let $f : C \to C$ be a contraction of modulus $\rho \in [0,1)$. It is a simple matter to see that the operator $I - f$ is $(1 - \rho)$-strongly monotone over C, i.e.,

$$\langle (I - f)x - (I - f)y, x - y \rangle \geq (1 - \rho)|x - y|^2, \quad \text{for all } (x, y) \in C^2. \tag{4.2.25}$$

The techniques of analysis used for proving Theorem 4.2.8 are based on the above result that was established in Ref. [34].

Lemma 4.2.3 [34, Lemma 1.3]. *Let $\{\Gamma_n\}$ be a sequence of real numbers that does not decrease at infinity, in the sense that there exists a subsequence $\{\Gamma_{n_j}\}_{j \geq 0}$ of $\{\Gamma_n\}$ which satisfies*

$$\Gamma_{n_j} < \Gamma_{n_j+1}, \quad \text{for all } j \geq 0. \tag{4.2.26}$$

Also consider the sequence of integers $\{\tau(n)\}_{n \geq n_0}$ defined by

$$\tau(n) = \max\{k \leq n : \Gamma_k < \Gamma_{k+1}\}. \tag{4.2.27}$$

Then $\{\tau(n)\}_{n \geq n_0}$ is a nondecreasing sequence verifying $\lim_{n \to \infty} \tau(n) = \infty$ and, for all $n \geq n_0$, it holds that $\Gamma_{\tau(n)} \leq \Gamma_{\tau(n)+1}$ and we have

$$\Gamma_n \leq \Gamma_{\tau(n)+1}. \tag{4.2.28}$$

Proof. Clearly, we can see that $\{\tau(n)\}$ is a well-defined sequence, and the fact that it is nondecreasing is obvious as well as $\lim_{n \to \infty} \tau(n) = \infty$ and $\Gamma_{\tau(n)} \leq \Gamma_{\tau(n)+1}$. Let us prove (4.2.28). It is easily observed that $\tau(n) \leq n$. Consequently, we prove (4.2.28) by distinguishing the three cases: (c1) $\tau(n) = n$; (c2) $\tau(n) = n - 1$; (c3) $\tau(n) < n - 1$.

In the first case (i.e., $\tau(n) = n$), (4.2.28) is immediately given by $\Gamma_{\tau(n)} \leq \Gamma_{\tau(n)+1}$. In the second case (i.e., $\tau(n) = n - 1$), (4.2.28) becomes obvious. In the third case (i.e., $\tau(n) \leq n - 2$), by (4.2.27) and for any integer $n \geq n_0$, we easily observe that $\Gamma_j \geq \Gamma_{j+1}$ for $\tau(n) + 1 \leq j \leq n - 1$, namely

$$\Gamma_{\tau(n)+1} \geq \Gamma_{\tau(n)+2} \geq \cdots \geq \Gamma_{n-1} \geq \Gamma_n,$$

which entails the desired result. □

Proof. Let $\bar{x}$ be the solution of (4.2.16). From (4.2.14), we obviously have

$$x_{n+1} - x_n + \alpha_n(x_n - fx_n) = (1 - \alpha_n)(T_w x_n - x_n), \tag{4.2.29}$$

hence

$$\langle x_{n+1} - x_n + \alpha_n(I - f)x_n, x_n - \bar{x} \rangle = -(1 - \alpha_n)\langle x_n - T_w x_n, x_n - \bar{x} \rangle. \tag{4.2.30}$$

Moreover, by $\bar{x} \in \text{Fix}(T)$ and using Remark 4.2.1(d), we have

$$\langle x_n - T_w x_n, x_n - \bar{x} \rangle \geq w \| x_n - T x_n \|^2,$$

which together with (4.2.30) entails

$$\langle x_{n+1} - x_n + \alpha_n (I - f) x_n, x_n - \bar{x} \rangle \leq -w(1 - \alpha_n) \| x_n - T x_n \|^2, \tag{4.2.31}$$

or equivalently,

$$-\langle x_n - x_{n+1}, x_n - \bar{x} \rangle \leq -\alpha_n \langle (I - f) x_n, x_n - \bar{x} \rangle - w(1 - \alpha_n) \| x_n - T x_n \|^2. \tag{4.2.32}$$

Furthermore, using (4.2.24) and setting $\Gamma_n := (1/2) \| x_n - \bar{x} \|^2$, we have

$$\langle x_n - x_{n+1}, x_n - \bar{x} \rangle = -\Gamma_{n+1} + \Gamma_n + (1/2) \| x_n - x_{n+1} \|^2, \tag{4.2.33}$$

so that (4.2.32) can be equivalently rewritten as

$$\Gamma_{n+1} - \Gamma_n - (1/2) \| x_n - x_{n+1} \|^2 \leq -\alpha_n \langle (I - f) x_n, x_n - \bar{x} \rangle - w(1 - \alpha_n) \| x_n - T x_n \|^2. \tag{4.2.34}$$

Now using (4.2.29) again, we have

$$\| x_{n+1} - x_n \|^2 = \| \alpha_n (f x_n - x_n) + (1 - \alpha_n)(T_w x_n - x_n) \|^2, \tag{4.2.35}$$

hence it is a classical matter to see that

$$\| x_{n+1} - x_n \|^2 \leq 2\alpha_n^2 \| f x_n - x_n \|^2 + 2(1 - \alpha_n)^2 \| T_w x_n - x_n \|^2, \tag{4.2.36}$$

which by $\| T_w x_n - x_n \| = w \| x_n - T x_n \|$ and $(1 - \alpha_n)^2 \leq (1 - \alpha_n)$ yields

$$(1/2) \| x_{n+1} - x_n \|^2 \leq \alpha_n^2 \| f x_n - x_n \|^2 + (1 - \alpha_n) w^2 \| T x_n - x_n \|^2. \tag{4.2.37}$$

Then from (4.2.34) and (4.2.37), we obtain

$$\Gamma_{n+1} - \Gamma_n + (1 - w) w (1 - \alpha_n) \| x_n - T x_n \|^2 \leq$$
$$\alpha_n \left(\alpha_n \| f x_n - x_n \|^2 - \langle (I - f) x_n, x_n - \bar{x} \rangle \right). \tag{4.2.38}$$

The rest of the proof will be divided into two cases:

Case 1. Suppose that there exists n_0 such that $\{\Gamma_n\}_{n \geq n_0}$ is nonincreasing. In this situation, $\{\Gamma_n\}$ is then convergent because it is also nonnegative (hence it is bounded from below), so that $\lim_{n \to \infty} \{\Gamma_{n+1} - \Gamma_n\} = 0$; hence, in light of (4.2.38) together with $\{\alpha_n\} \to 0$, and the boundedness of $\{x_n\}$ (hence, because of the continuity of f, $\{f x_n\}$ is also bounded), we obtain

$$\lim_{n \to \infty} \| x_n - T x_n \| = 0. \tag{4.2.39}$$

From (4.2.38) again, we have

$$\alpha_n \left(-\alpha_n \| f x_n - x_n \|^2 + \langle (I - f) x_n, x_n - \bar{x} \rangle \right) \leq \Gamma_n - \Gamma_{n+1}. \tag{4.2.40}$$

Then, by $\sum_n \alpha_n = \infty$, we obviously deduce that

$$\liminf_{n \to \infty} \left(-\alpha_n \| fx_n - x_n \|^2 + \langle (I - f)x_n, x_n - \bar{x} \rangle \right) \leq 0, \qquad (4.2.41)$$

or equivalently (as $\alpha_n \| fx_n - x_n \|^2 \to 0$)

$$\liminf_{n \to \infty} \left(\langle (I - f)x_n, x_n - \bar{x} \rangle \right) \leq 0. \qquad (4.2.42)$$

Moreover, by Remark 4.2.3, we have

$$2(1 - \rho)\Gamma_n + \langle (I - f)\bar{x}, x_n - \bar{x} \rangle \leq \langle (I - f)x_n, x_n - \bar{x} \rangle, \qquad (4.2.43)$$

which by (4.2.42) entails

$$\liminf_{n \to \infty} \left(2(1 - \rho)\Gamma_n + \langle (I - f)\bar{x}, x_n - \bar{x} \rangle \right) \leq 0,$$

hence, recalling that $\lim_{n \to \infty} \Gamma_n$ exists, we equivalently obtain

$$2(1 - \rho) \lim_{n \to \infty} \Gamma_n + \liminf_{n \to \infty} \langle (I - f)\bar{x}, x_n - \bar{x} \rangle \leq 0,$$

namely

$$2(1 - \rho) \lim_{n \to \infty} \Gamma_n \leq -\liminf_{n \to \infty} \langle (I - f)\bar{x}, x_n - \bar{x} \rangle. \qquad (4.2.44)$$

From (4.2.39) and invoking Lemma 4.2.2, we have

$$\liminf_{n \to \infty} \langle (I - f)\bar{x}, x_n - \bar{x} \rangle \geq 0, \qquad (4.2.45)$$

which by (4.2.44) yields $\lim_{n \to \infty} \Gamma_n = 0$, so that (x_n) converges strongly to $\bar{x}$.

CASE 2. Suppose there exists a subsequence $\{\Gamma_{n_k}\}_{k \geq 0}$ of $\{\Gamma_n\}_{n \geq 0}$ such that $\Gamma_{n_k} < \Gamma_{n_k + 1}$ for all $k \geq 0$. In this situation, we consider the sequence of indexes $\{\tau(n)\}$ as defined in Lemma 4.2.3. It follows that $\Gamma_{\tau(n)+1} - \Gamma_{\tau(n)} > 0$, which by (4.2.38) amounts to

$$(1 - w)w(1 - \alpha_{\tau(n)})|x_{\tau(n)} - Tx_{\tau(n)}|^2 <$$
$$\alpha_{\tau(n)} \left(\alpha_{\tau(n)} \| fx_{\tau(n)} - x_{\tau(n)} \|^2 - \langle (I - f)x_{\tau(n)}, x_{\tau(n)} - \bar{x} \rangle \right), \quad (4.2.46)$$

hence, by the boundedness of $\{x_n\}$ and $\alpha_n \to 0$, we immediately obtain

$$\lim_{n \to \infty} \| x_{\tau(n)} - Tx_{\tau(n)} \| = 0. \qquad (4.2.47)$$

Now by (4.2.46), we clearly have

$$\langle (I - f)x_{\tau(n)}, x_{\tau(n)} - \bar{x} \rangle \leq \alpha_{\tau(n)} \| fx_{\tau(n)} - x_{\tau(n)} \|^2, \qquad (4.2.48)$$

which in light of (4.2.43) yields

$$2(1 - \rho)\Gamma_{\tau(n)} + \langle (I - f)\bar{x}, x_{\tau(n)} - \bar{x} \rangle \leq \alpha_{\tau(n)} \| fx_{\tau(n)} - x_{\tau(n)} \|^2,$$

hence (as $\alpha_{\tau(n)} \|f x_{\tau(n)} - x_{\tau(n)}\|^2 \to 0$) it follows that

$$2(1-\rho) \limsup_{n\to\infty} \Gamma_{\tau(n)} \leq -\liminf_{n\to\infty} \langle (I-f)\bar{x}, x_{\tau(n)} - \bar{x} \rangle. \qquad (4.2.49)$$

From (4.2.47) and invoking Lemma 4.2.2, we have

$$\liminf_{n\to\infty} \langle (I-f)\bar{x}, x_{\tau(n)} - \bar{x} \rangle \geq 0, \qquad (4.2.50)$$

which by (4.2.49) yields $\limsup_{n\to\infty} \Gamma_{\tau(n)} = 0$, so that $\lim_{n\to\infty} \Gamma_{\tau(n)} = 0$. Then, recalling that $\Gamma_n \leq \Gamma_{\tau(n)}$ (by Lemma 4.2.3), we get $\lim_{n\to\infty} \Gamma_n = 0$, so that $x_n \to \bar{x}$ strongly. □

To conclude this section, since our approach can be applied to analyze the convergence behavior of the iterates generated by a variety of fixed point methods, the class of viscosity approximation methods has been regularly enriched with increasingly sophisticated algorithms as the structure of problems to handle becomes more complex.

4.3. Viscosity Method for Split Common Fixed Point Problems

4.3.1. Algorithms and Related Convergence Results

Given two real Hilbert spaces $\mathcal{H}_1$ and $\mathcal{H}_2$ endowed with inner products and induced norms denoted by $\langle .,. \rangle$ and $\| \cdot \|$ respectively, we consider the following general two-operator split common fixed point problem:

$$\text{find } x^* \in \text{Fix}(T) \text{ such that } Ax^* \in \text{Fix}(S), \qquad (4.3.51)$$

where $A : \mathcal{H}_1 \to \mathcal{H}_2$ is a (nonzero) bounded linear operator, $T : \mathcal{H}_1 \to \mathcal{H}_1$ and $S : \mathcal{H}_2 \to \mathcal{H}_2$ are general operators. The solution set of (4.3.51) is denoted by Ω and we make the following standing assumption:

$$\Omega = \{ q \in \mathcal{H}_1 : q \in \text{Fix}(T) \text{ and } Aq \in \text{Fix}(S) \} \neq \emptyset. \qquad (4.3.52)$$

Real-world inverse problems can be cast into this framework by making different choices of the spaces and by choosing appropriate operators. Note that particular cases of this formulation were used for solving an inverse problem in radiation therapy treatment planning and other problems encountered in image/signal processing.

To solve (4.3.51), Censor and Segal [13] proposed and proved, in finite dimensional spaces, the convergence of the following algorithm:

$$x_{n+1} = T\left(x_n + \gamma A^t (S-I) A x_n \right), \quad n \in \mathbb{N}, \qquad (4.3.53)$$

where $\gamma \in (0, 2/\lambda)$, with λ being the largest eigenvalue of the matrix $A^t A$ (t stands for matrix transposition).

Very recently, Byrne et al. [12] have considered the following algorithm:

$$x_{n+1} = \alpha_n x_0 + (1 - \alpha_n) T \left(x_n + \gamma A^t (S - I) A x_n \right), \quad n \in \mathbb{N}, \tag{4.3.54}$$

with two special nonexpansive mappings T, S and proved that any sequence (x_n) generated by algorithm (4.3.54) strongly converges to some element $x^* \in \Omega$ as long as $\gamma \in (0, 2/\lambda)$ and $(\alpha_n) \subset (0, 1)$ satisfies $\lim\limits_{n \to +\infty} \alpha_n = 0$ and $\sum_{n=0}^{\infty} \alpha_n = +\infty$.

Inspired by their work and having in hand Theorem 4.2.8, we propose a simple and alternative proof of a result obtained in Ref. [74] concerning a strong convergence viscosity-type result for the split common fixed point problem of the wide class of quasi-nonexpansive operators in Hilbert spaces.

To begin with, we present a result which, together with Theorem 4.2.8 above, will be needed in the proof of the next Theorem. It summarizes some properties of attracting mappings (see, for example, Ref. [3]).

Proposition 4.3.1. *Let H be a real Hilbert space and suppose that $T_1 : H \to H$ is a quasi-nonexpansive mapping and that $T_2 : H \to H$ is an attracting quasi-nonexpansive mapping satisfying $Fix(T_1) \cap Fix(T_2) \neq \emptyset$. Then*

$$T_2 \circ T_1 \text{ is quasi-nonexpansive and } Fix(T_2 \circ T_1) = Fix(T_1) \cap Fix(T_2).$$

In particular, if each $T_i (i = 1, 2)$ is α_i-attracting (where $\alpha_i > 0$), then $T_2 \circ T_1$ is $\frac{\alpha_1 \alpha_2}{\alpha_1 + \alpha_2}$-attracting.

Taking into account this latter result we focus on the following viscosity-type approximation method discussed in Ref. [74] relative to (4.3.51) and that involves the relaxed operator $U_w = (1 - w)I + wU$ with $w \in (0, 1)$ and $U := T(I - \gamma A^*(I - S)A)$, where A^* is the adjoint operator of A.

Algorithm 4.3.1. *Initialization*: Let $x_0 \in \mathscr{H}_1$ be arbitrary.
Iterative step: for $n \in \mathbb{N}$, let

$$x_{n+1} = \alpha_n f(x_n) + (1 - \alpha_n) U_w(x_n), \quad n \in \mathbb{N}, \tag{4.3.55}$$

where $f : \mathscr{H}_1 \to \mathscr{H}_2$ is a strict contraction of modulus $\rho \in [0, 1)$, $\alpha_n \in (0, 1)$ and $\gamma \in \left(0, \frac{1}{\lambda}\right)$ with λ being the spectral radius of A^*A.

Firstly, we propose a simple proof based on attracting operator properties. Then we propose a modification of this algorithm and prove its strong convergence. We conclude by mentioning as an example of attracting operators the class of projected-gradient mappings which was used in a wide variety of online tasks, which span from classical linear adaptive filtering to nonlinear classification and regression tasks (see, for example, the very recent work on online learning).

Remark 4.3.1. Let us recall that the fixed point set of any quasi-nonexpansive mapping is closed and convex (see, e.g., Refs [24, Corollary 1; 72, Proposition 1]). It is then immediately deduced that Ω is a closed and convex subset of $\mathscr{H}_1$.

Now, we establish the strong convergence of the iterates given by (4.3.55) to the unique element $x_* \in \Omega$ verifying the variational inequality problem

$$\text{for all } v \in \Omega, \quad \langle (I - f)x_*, v - x_* \rangle \geq 0, \tag{4.3.56}$$

which equivalently solves the fixed point problem

$$x_* = (P_\Omega \circ f)x_*, \tag{4.3.57}$$

where P_Ω denotes the metric projection onto Ω.

Theorem 4.3.1. *Let $\{x_n\}$ be the sequence generated by the algorithm (4.3.55) with a bounded linear operator $A : \mathscr{H}_1 \to \mathscr{H}_2$, two quasi-nonexpansive operators $T : \mathscr{H}_1 \to \mathscr{H}_1$ and $S : \mathscr{H}_2 \to \mathscr{H}_2$, together with parameters $\gamma \in \left(0, \frac{1}{\lambda}\right)$, $w \in (0,1)$ and $\alpha_n \subset (0,1)$ such that $\lim_{n \to +\infty} \alpha_n = 0$ and $\sum_{n=0}^{\infty} \alpha_n = +\infty$. Assume in addition that $I - U$ is demi-closed at 0. Then $\{x_n\}$ strongly converges to the split common fixed point $x^* \in \Omega$ satisfying (4.2.16).*

Proof. First, we prove that the operator $I - \gamma A^*(I - S)A$ is α-attracting for some $\alpha > 0$. To this end, taking $y \in \Omega$, i.e., $y \in \text{Fix}(T)$; $Ay \in \text{Fix}(S)$, we can write

$$\begin{aligned} \langle A^*(I - S)Ax - A^*(I - S)Ay, x - y \rangle &= \langle (I - S)Ax - (I - S)Ay, Ax - Ay \rangle \\ &\geq \frac{1}{2}\|(I - S)Ax\|^2. \end{aligned}$$

From definition of λ follows

$$\begin{aligned} \langle A^*(I - S)Ax, A^*(I - S)Ax \rangle &= \langle (I - S)Ax, AA^*(I - S)Ax \rangle \\ &\leq \frac{1}{\lambda}\|(I - S)Ax\|^2. \end{aligned}$$

Thus

$$\langle A^*(I - S)Ax, x - y \rangle \geq \frac{\lambda}{2}\|A^*(I - S)Ax\|^2.$$

As a consequence, we obtain

$$\begin{aligned}
\|(I-\gamma A^*(I-S)A)x-y\|^2 &= \|(x-y)-\gamma A^*(I-S)Ax\|^2 \\
&= \|x-y\|^2 - 2\gamma\langle A^*(I-S)Ax, x-y\rangle + \gamma^2\|A^*(I-S)Ax\|^2 \\
&\leq \|x-y\|^2 - \gamma(\lambda-\gamma)\|A^*(I-S)Ax\|^2.
\end{aligned}$$

This ensures that $I-\gamma A^*(I-S)A$ is $\frac{\lambda-\gamma}{\gamma}$-attracting and in the light of Proposition 4.3.1, we infer that $U := T \circ (I-\gamma A^*(I-S)A)$ is quasi-nonexpansive. Therefore by Theorem 4.2.8 and taking into account the fact that $I-U_w = w(I-U)$ and that $U-I$ is assumed to be demi-closed at 0, we obtain that the sequence $\{x_n\}$ generated by algorithm (4.3.55) strongly converges to the fixed point x^* of U satisfying

$$\langle(I-f)x^*, y-x^*\rangle \geq 0, \quad \text{for all } y \in \text{Fix}(U).$$

It remains to show that $x^* \in \Omega$.

Let $z \in \text{Fix}(T)$ and $Az \in \text{Fix}(S)$, so we can write

$$\begin{aligned}
(I-\gamma A^*(I-S))Az &= z-\gamma A^*(I-S)Az \\
&= z-\gamma A^*Az + \gamma A^*SAz \\
&= z-\gamma A^*Az + \gamma A^*Az = z.
\end{aligned}$$

Hence $z \in \text{Fix}((I-\gamma A^*(I-S)A)$. Furthermore $T(I-\gamma A^*(I-SA)(z) = T(z) = z$. Therefore any $z \in \Omega$ is a fixed point of $T(I-\gamma A^*(I-S)A)$. Since $\Omega \neq \emptyset$, we get that

$$\text{Fix}(T) \cap \text{Fix}(I-\gamma A^*(I-S)A) = \text{Fix}(T \circ (I-\gamma A^*(I-S)A)).$$

Since x^* is a fixed point of $T \circ (I-\gamma A^*(I-S)A)$, we have $x^* \in \text{Fix}(S)$ and $x^* \in \text{Fix}(I-\gamma A^*(I-S)A)$. Thus $A^*(I-S)Ax^* = 0$ or equivalently $S(Ax^*) = Ax^* + w$, where $A^*w = 0$. Since $S(Az) = Az$, we obtain

$$S(Ax^*) - S(Az) = Ax^* + w - Az.$$

So,

$$\begin{aligned}
\|Ax^* - Az\|^2 &\geq \|S(Ax^*) - S(Az))\|^2 \\
&= \|Ax^* + w - Az\|^2 \\
&= \|Ax^* - Az\|^2 + 2\langle x^* - z, A^*w\rangle + \|w\|^2 \\
&= \|Ax^* - Az\|^2 + \|w\|^2.
\end{aligned}$$

Hence $w = 0$, which means that $S(Ax^*) = Ax^*$. This completes the proof.   □

It should be noticed that by taking $f \equiv x_0$, we obtain, in a general context, a Halpern-type algorithm and the corresponding strong convergence result which generalizes, in some sense, the result obtained in Ref. [12] for a class of nonexpansive mappings. The algorithm takes the following form:

$$x_{n+1} = \alpha_n x_0 + (1 - \alpha_n) U_w(x_n), \quad n \in \mathbb{N},$$

where $U := T(I - \gamma A^*(I - S)A)$.

Observe that it is enough to relax the operator T instead of relaxing the composition of T by $I - \gamma A^*(I - S)A$. In this case T_w is $\frac{1-w}{w}$-attracting and its composition by $I - \gamma A^*(I - S)A$ is still α-attracting (for some $\alpha > 0$), and thus (in the light of proposition 1) $\frac{1}{1+\alpha}$-averaged. We obtain the next algorithm:

$$x_{n+1} = \alpha_n f(x_n) + (1 - \alpha_n) T_w(I - \gamma A^*(I - S)A)(x_n), \quad n \in \mathbb{N}. \qquad (4.3.58)$$

Theorem 4.2.8 is again applicable (provided that the demi-closedness assumption at 0 is verified). So, any sequence $\{x_n\}$ generated by algorithm (4.3.58) converges strongly to a point $x^* \in \text{Fix}(T_w(I - \gamma A^*(I - S)A))$ as long as this set is nonempty. Following the proof of Theorem 4.3.1 and having in mind that $\text{Fix}(T_w) = \text{Fix}(T)$, we conclude that $x^* \in \Omega$ and satisfies the variational inequality (4.3.56).

Now, a suitable case is the so-called subgradient projection relative to a convex function. Subgradient projection techniques occur, for instance, in signal and image processing as an alternative to projection methods [4, 72]. Given a continuous convex function $\Phi : \mathscr{H} \to \mathbb{R}$ ($\mathscr{H}$ being a real Hilbert space with norm denoted by $\|.\|$), we set $\text{lev}_{\leq 0}(\Phi) := \{x \in \mathscr{H} : \Phi(x) \leq 0\} \neq \emptyset$ and we denote by $T_{(\Phi)}$ a subgradient projection operator relative to Φ.

Remark 4.3.2. An operator $T_{(\Phi)} : \mathscr{H} \to \mathscr{H}$ is a *subgradient projection* relative to Φ whenever it is of the form:

$$(\text{for all } x \in \mathscr{H}) \quad T_{(\Phi)}(x) := \begin{cases} x - \dfrac{\Phi(x)}{|\Phi'(x)|^2}\Phi'(x), & \text{if } \Phi(x) > 0, \\ x, & \text{otherwise,} \end{cases}$$

where $\Phi' : \mathscr{H} \to \mathscr{H}$ is a selection of $\partial \Phi : \mathscr{H} \to 2^{\mathscr{H}}$, the Fenchel subdifferential of Φ, namely $\Phi'(x) \in \partial \Phi(x)$ for all $x \in \mathscr{H}$, where $\partial \Phi(x) := \{x^* \in H : \Phi(y) \geq \Phi(x) + \langle x^*, y - x \rangle \text{ for all } y \in H\}$.

It is well known that $T_{(\Phi)}$ belongs to $\mathscr{E}_{FQ}$ (the class of firmly quasi-nonexpansive mappings) and that $Fix(T_{(\Phi)}) = \text{lev}_{\leq 0}(\Phi)$ (see, for instance, Ref. [3]). We also recall that $I - T_{(\Phi)}$ is demi-closed (at zero) whenever $\partial \Phi$ is bounded (that is, $\partial \Phi$ maps bounded sets to bounded sets) [33, Lemma 4.6].

It is readily checked that the relaxed operator denoted $T_{(\Phi)}^{(w)} := (1 - w)I + wT_{(\Phi)}$ with $w \in (0, 2)$ is $\frac{2-w}{w}$-attracting and that $I - T_{(\Phi)}^{(w)}$ is demi-closed at zero whenever this latter property holds for $I - T_{(\Phi)}$.

If we set $T = T_{(\Phi)}$, algorithm (4.3.55) can be used to solve the following two-operator split common fixed point problem

$$\text{find } x^* \in \text{Fix}(S) \text{ such that } Ax^* \in \text{lev}_{\leq 0}\Phi. \qquad (4.3.59)$$

If in addition we take $S = T_{(\Theta)}$, Θ being another continuous convex function $\Theta : \mathscr{H} \to \mathbb{R}$, algorithm (4.3.58) can be used to select a particular solution of the following split common level-set problem

$$\text{find } x^* \in \text{lev}_{\leq 0}\Theta \text{ such that } Ax^* \in \text{lev}_{\leq 0}\Phi. \qquad (4.3.60)$$

Another interesting particular case is obtained by taking $T = P_C$ and $S = P_Q$, C and Q being two closed convex sets. In this case S is nonexpansive and the relaxed projection operator $T_w := (1-w)I + wP_C$, with $w \in (0,2)$, is $\frac{2-w}{w}$-attracting, $I - T_w$ is demi-closed at 0, $\text{Fix}(U_w) = C$ and $\text{Fix}(S) = Q$ (see, for instance, Ref. [4]).

Algorithm (4.3.58) can be used to select a particular solution of the following split feasibility problem:

$$\text{find } x^* \in C \text{ such that } Ax^* \in Q. \qquad (4.3.61)$$

It turns out for convergence that all the above-mentioned fixed point methods are based upon the knowledge of the spectral radius of the operator A^*A. Now, we propose an alternative approach to solving the common fixed point problem (see Maingé [37]). Our study will be concerned with S and T belonging to the general class of quasi-nonexpansive operators ($\mathscr{E}_Q$), which includes commonly used classes such as firmly nonexpansive ($\mathscr{E}_{FN}$), nonexpansive ($\mathscr{E}_N$) and firmly quasi-nonexpansive (or directed, $\mathscr{E}_{FQ}$) operators. More precisely, we prove a strong convergent result regarding a variant of the previously considered algorithms, under very classical conditions, together with a range of variable step sizes that does not depend on $\|A^*A\|$. An important particular case of our less restrictive strategy is also considered, leading to Polyak-type algorithms [52].

Let us recall that the class $\mathscr{E}_Q$ of quasi-nonexpansive mappings includes the class $\mathscr{E}_{FQ}$ of firmly quasi-nonexpansive mappings. However, as can be noticed from the literature (related to (4.3.51)), the conditions on the parameters for convergence of related algorithms are somewhat different, when dealing either with element of $\mathscr{E}_Q$ or with element of $\mathscr{E}_{FQ}$. So, in order to state precise results regarding these two situations, we will use the concept of demicontractive mappings.

In the sequel, $\mathscr{H}$ will refer to any of the real Hilbert spaces $\mathscr{H}_i$ involved in (4.3.51).

Definition 4.3.1. An operator $U : \mathscr{H} \to \mathscr{H}$ is called η-*demicontractive* (for some real value $\eta \in (-\infty, 1)$) if for any $(x, q) \in \mathscr{H} \times \text{Fix}(U)$, it satisfies (d1) $\|Ux - q\|^2 \leq \|x - q\|^2 + \eta \|x - Ux\|^2,$

or equivalently,

(d2) $\langle x - Ux, x - q \rangle \geq (1/2)(1 - \eta)\|x - Ux\|^2.$

Remark 4.3.3. Let us observe that Definition 4.3.1 is satisfied for $U \in \mathcal{E}_Q$ and $U \in \mathcal{E}_{FQ}$, with $\eta = 0$ and $\eta = -1$ respectively. It is then evident that any quasi-nonexpansive or directed mapping $U : \mathcal{H} \to \mathcal{H}$ is also $(1 - k_U)$-demicontractive, with $k_U = 1$ and $k_U = 2$ respectively, namely

$$\text{for all } (x, q) \in \mathcal{H} \times \text{Fix}(U), \quad \langle x - Ux, x - q \rangle \geq (1/2)k_U |x - Ux|^2. \qquad (4.3.62)$$

Remark 4.3.4. It is worth stressing that our study can be extended to the class of demicontractive mappings. However, as discussed in Ref. [35, Remark 1.2], any demicontractive mapping can be reduced to an element of $\mathcal{E}_Q$ that has the same fixed point set, by means of a simple relaxation process. Namely, if $U : \mathcal{H} \to \mathcal{H}$ is η-demicontractive, then the relaxed operator $T_v = (1 - v)I + vU$ belongs to $\mathcal{E}_Q$ and $\text{Fix}(T_v) = \text{Fix}(T)$, whenever $v \in (0, 1 - \eta]$. So we restrict ourselves to solving (4.3.51) with quasi-nonexpansive operators T and S.

From now on, when dealing with a quasi-nonexpansive operator U, we assume that U is $(1 - k_U)$-demicontractive, with the value $k_U = 1$ (e.g., when U is a subgradient projection operator), or with $k_U = 2$ (e.g., when U is a metric projection). These considerations will be helpful for legibility, so as to provide results that can be compared with other related works (involving projections or more general operators).

In order to compute a solution of (4.3.51) we focus on the sequence (x_n) generated by the following algorithm:

$$\text{Take } \{w_n\} \subset (0, +\infty), \ \{\alpha_n\} \subset (0, 1], \text{ and select } x_0 \in \mathcal{H}_1, \qquad (4.3.63)$$

$$v_n = x_n - \beta_n A^*(I - S)Ax_n, \qquad (4.3.64)$$

$$x_{n+1} = \alpha_n f(v_n) + (1 - \alpha_n)T_{w_n}v_n, \qquad (4.3.65)$$

where $T_{w_n} := (1 - w_n)I + w_n T$, $f : \mathcal{H}_1 \to \mathcal{H}_1$ is often referred to as a "selecting operator", while $\{\beta_n\}$ is some control sequence of nonnegative real numbers.

Conditions on the operators:

(C1) $A : \mathcal{H}_1 \to \mathcal{H}_2$ is a (nonzero) bounded linear operator;

(C2) $T : \mathcal{H}_1 \to \mathcal{H}_1$ (respectively $S : \mathcal{H}_2 \to \mathcal{H}_2$) is $(1 - k_T)$-demicontractive (respectively $(1 - k_S)$-demicontractive), with $k_T = 1$, or $k_T = 2$ if T belongs to $\mathcal{E}_{FQ}$ (respectively $k_S = 1$, or $k_S = 2$ if S belongs to $\mathcal{E}_{FQ}$);

(C3) $f : \mathscr{H}_1 \to \mathscr{H}_1$ is a strict contraction, i.e., there exists some $\rho \in [0,1)$ such that $\|f(x) - f(y)\| \le \rho \|x - y\|$ for all $(x,y) \in \mathscr{H}_1^2$;

(C4) $I - T$ and $I - S$ are demi-closed operators (at zero).

Conditions on the parameters $\{\alpha_n\}$ and $\{w_n\}$:

(C5) $\{w_n\} \subset [w_a, w_b]$ for values w_a and w_b such that $0 < w_a \le w_b < k_T$;

(C6) $\{\alpha_n\} \subset (0,1]$, $\lim\limits_{n\to\infty} \alpha_n = 0$, $\sum_{n\ge 0} \alpha_n = \infty$.

Remark 4.3.5. Let us stress that condition (C5) allows over-relaxation for $T \in \mathscr{E}_{FN}$ (hence $k_T = 2$), namely the relaxation parameter (w_n) can be chosen in the range $(1,2)$. This procedure is supposed to speed up the convergence of some fixed point iterations [54].

Several possibilities are then considered (in the next section) regarding the choice of the control sequence $\{\beta_n\}$.

Roughly speaking, as can be noticed, algorithmic solutions to (4.3.51) under conditions (C1) and (C2) usually involve a (possibly constant) control sequence $\{\beta_n\}$ verifying

$$\{\beta\} \subset [\beta_a, \beta_b] \text{ with } \beta_a \text{ and } \beta_b \text{ such that } 0 < \beta_a \le \beta_b < k_S \gamma, \tag{4.3.66}$$

where $\gamma = \|A^* A\|^{-1}$.

Instead of a choice such as (4.3.66), we adopt an alternative methodology involving the set of indexes $\mathscr{J}$ and the sequence $\{\mu_n\}$ defined by

$$\mathscr{J} = \{n \in \mathbb{N} : (S - I)Ax_n \ne 0\}, \tag{4.3.67}$$

$$\mu_n = \begin{cases} \dfrac{|(S-I)Ax_n|^2}{|A^*(S-I)Ax_n|^2}, & \text{if } n \in \mathscr{J}, \\ \mu, & \text{otherwise} \quad (\mu \text{ being any nonnegative value}). \end{cases} \tag{4.3.68}$$

Remark 4.3.6. The value of μ does not influence the considered process, but it was introduced just for the sake of clarity. Furthermore, it can be checked that $\{\mu_n\}$ is well defined and satisfies

$$\mu_n \ge \gamma, \quad \text{for } n \in \mathscr{J}. \tag{4.3.69}$$

Let us emphasize that (4.3.69) is obtained provided that (C1) holds, S is $(1 - k_S)$ demicontractive for some positive value k_S (which is guaranteed by (C2)), and that $R(A)$ (the range of A) is such that $R(A) \cap \text{Fix}(S) \ne \emptyset$ (which is obviously satisfied whenever (4.3.51) has a nonempty solution set).

Denoting the cardinal of $\mathscr{J}$ by $\text{Card}(\mathscr{J})$, we establish a main (strong) convergence result (Theorem 4.3.2) relative to algorithm (4.3.63) with step sizes $\{\beta_n\}$ verifying

$$\beta_n \in (0, k_S \mu_n), \quad \text{for } n \in \mathscr{J}, \tag{4.3.70}$$

together with the following condition:

$$\lim_{n \to +\infty, \, n \in \mathscr{J}} \frac{\alpha_n}{\beta_n (k_S \mu_n - \beta_n)} = 0, \text{ whenever } \text{Card}(\mathscr{J}) = +\infty). \tag{4.3.71}$$

Remark 4.3.7. Let us clarify that condition (4.3.71) should be understood in the sense that $\lim_{n \to +\infty} \Theta_{\phi(n)} = 0$, with $\Theta_n = \frac{\alpha_n}{\beta_n (k_S \mu_n - \beta_n)}$, for any increasing mapping $\phi : \mathbb{N} \to \mathscr{J}$, where $\mathbb{N}$ denotes the set of integers.

Remark 4.3.8. Clearly, condition (4.3.70) makes sense and it is more flexible than (4.3.66), in the sense that it provides a wider range of parameters for $\{\beta_n\}$. Numerical experiments also show that (4.3.66) is far from being optimal, since the value of γ can be very small compared with the values of μ_n for $n \in \mathscr{J}$ (see Figure 4.5). It is also obvious that (4.3.70) and (4.3.71) are satisfied by (4.3.66) whenever $\alpha_n \to 0$ (as $n \to +\infty$), since $\mu_n \geq \gamma$ for $n \in \mathscr{J}$ (from Remark 4.3.6).

The following theorems were established in Ref. [37].

Theorem 4.3.2 (Convergence with general step-sizes). *Let $\{x_n\}$ be the sequence generated by (4.3.64) and (4.3.65) under conditions (C1)–(C6), together with values $\{\beta_n\}$ satisfying*

$$\beta_n \in (0, k_S \mu_n), \quad \text{for } n \in \mathscr{J}, \tag{4.3.72}$$

$$\lim_{n \to +\infty, \, n \in \mathscr{J}} \frac{\alpha_n}{\beta_n (k_S \mu_n - \beta_n)} = 0, \quad \text{whenever } \text{Card}(\mathscr{J}) = +\infty, \tag{4.3.73}$$

where $\mathscr{J}$ and $\{\mu_n\}$ are given by (4.3.67) and (4.3.68) respectively, and $\text{Card}(\mathscr{J})$ is the cardinal of $\mathscr{J}$. Then $\{x_n\}$ converges strongly to the unique solution x_* of (4.2.16).

Two consequences of Theorem 4.3.2 are also considered through Theorems 4.3.3 and 4.3.4. Specifically, Theorem 4.3.3 is concerned with the particular choice of Polyak-type step sizes, usually involved in projected subgradient methods [52] (also see Ref. [29] for a performance study), namely $\beta_n = k_S(1 - \varepsilon_n)\mu_n$, where $\{\varepsilon_n\} \subset (0,1)$. The sequence $\{\varepsilon_n\}$ is assumed to be such that $\lim_{n \to +\infty} \frac{\alpha_n}{\varepsilon_n(1 - \varepsilon_n)} = 0$.

A first application of Theorem 4.3.2 is given below.

Theorem 4.3.3 (Convergence with Polyak-type step-sizes). *Let $\{x_n\}$ be the sequence generated by (4.3.64) and (4.3.65) under conditions (C1)–(C6) and step sizes $\{\beta_n\}$ such that*

$$\beta_n = k_S(1 - \varepsilon_n)\mu_n, \tag{4.3.74}$$

where $\{\mu_n\}$ is given by (4.3.68), and $\{\varepsilon_n\} \subset (0,1)$ satisfies

$$\lim_{n \to +\infty} \frac{\alpha_n}{\varepsilon_n(1 - \varepsilon_n)} = 0. \tag{4.3.75}$$

Then $\{x_n\}$ converges strongly to the unique solution x_* of (4.2.16).

A second application of Theorem 4.3.2 is given below relative to the (limiting) case of vanishing sequences $\{\beta_n\}$. Note that such a procedure can be encountered, for instance, in projected subgradients methods [1].

Theorem 4.3.4 (Convergence with vanishing step sizes). *Let $\{x_n\}$ be the sequence generated by (4.3.64) and (4.3.65) under conditions (C1)–(C6) and together with a sequence of positive numbers $\{\beta_n\}$ verifying*

$$\lim_{n \to +\infty} \beta_n = 0 \text{ and } \lim_{n \to +\infty} \frac{\alpha_n}{\beta_n} = 0. \tag{4.3.76}$$

Then $\{x_n\}$ converges strongly to the unique solution x_* of (4.2.16).

Remark 4.3.9. Note that conditions (C6) and (4.3.76) are satisfied, for instance, when taking $\alpha_n = 1/n^{p_1}$ and $\beta_n = 1/n^{p_2}$ with positive exponents $p_1 \in (0,1]$ and $p_2 \in (0,p_1)$.

4.3.2. Convergence Results in Some Specific Cases

As an interesting instance of Theorem 4.3.3 we give a result regarding a special case of split problems, when the projections onto the involved sets are not available, e.g.: find an element belonging to some set Ω of the form

$$\Omega := \{q \in \mathcal{H}_1 : q \in Q_1 \text{ and } Aq \in Q_2\} \neq \emptyset, \tag{4.3.77}$$

where $A : \mathcal{H}_1 \to \mathcal{H}_2$ is a (nonzero) bounded linear operator, and each set Q_k $(k = 1, 2)$ is some nonempty closed and convex subset of $\mathcal{H}_k$ given by

$$Q_k = \{x \in \mathcal{H}_k : \Phi_k(x) \leq 0\}, \tag{4.3.78}$$

where Φ_k is a convex function on $\mathcal{H}_k$. We also pay some attention to the particular case of (4.3.78) when Q_k is a collection of r_k half subspaces $H_{k,i}$ $(i = 1, \ldots, r_k)$, associated with some family $(a_{k,i}, b_{k,i})_{i=1,\ldots,r_k} \subset \mathcal{H}_k \times \mathbb{R}$, namely

$$Q_k = \bigcap_{i=1,\ldots,r_k} H_{k,i}, \tag{4.3.79}$$

$$H_{k,i} = \{x \in \mathcal{H}_k : l_{k,i}(x) = \langle a_{k,i}, x \rangle + b_{k,i} \geq 0\}. \tag{4.3.80}$$

In this setting, (4.3.77) can be reformulated as a split common fixed point problem by means of subgradient projections and other appropriate types of operators.

Theorem 4.3.5. *Suppose that Ω is some nonempty set of the form (4.3.77), where $A : \mathcal{H}_1 \to \mathcal{H}_2$ is a (nonzero) bounded linear operator, and suppose that $Q_k := \{x \in \mathcal{H}_k : \Phi_k(x) \leq 0\}$ (for $k = 1,2$), where $\Phi_k : \mathcal{H}_k \to \mathbb{R}$ is a convex and continuous function with bounded subdifferential $\partial \Phi_k$. Let $\{x_n\}$ be the sequence generated from any $x_0 \in \mathcal{H}_1$ by*

$$v_n = x_n + \beta_n A^*(T_{(\Phi_2)} - I)Ax_n, \tag{4.3.81}$$

$$x_{n+1} = \alpha_n f(x_n) + (1 - \alpha_n)[(1 - w)I + wT_{(\Phi_1)}]v_n, \tag{4.3.82}$$

where $T_{(\Phi_k)} : \mathcal{H}_k \to \mathcal{H}_k$ (for $k = 1,2$) is a subgradient projection operator relative to Φ_k, $f : \mathcal{H}_1 \to \mathcal{H}_1$ is a strict contraction, $w \in (0,2)$, $\{\alpha_n\}$ is a sequence verifying (C6), and the parameter $\{\beta_n\}$ is such that

$$\beta_n = 2(1 - \varepsilon_n)\frac{|(T_{(\Phi_2)} - I)Ax_n|^2}{|A^*(T_{(\Phi_2)} - I)Ax_n|^2}, \quad \text{if } (T_{(\Phi_2)} - I)Ax_n \neq 0,$$

where $\{\varepsilon_n\} \subset 0,1)$ satisfies $\displaystyle\lim_{n \to +\infty} \frac{\alpha_n}{\varepsilon_n} = 0$. Then $\{x_n\}$ converges strongly to the unique element $u \in \Omega$ (Ω being defined in (4.3.77)) verifying $u = (P_\Omega \circ f)(u)$.

This result is deduced from Remark 4.3.2 and Theorem 4.3.3 with regard to the special situation when $T = T_{(\Phi_1)}$ and $S = T_{(\Phi_2)}$.

A more explicit operator can be introduced when facing constraints of the form (4.3.79) in (4.3.77), which obviously is included in the case (4.3.78) with $\Phi_k(x) = \max_{1 \leq i \leq r_k} -l_{k,i}(x)$.

Remark 4.3.10. Let Q be a collection of r half subspaces H_i ($i = 1,\ldots,r$), associated with a family of pairs $(a_i, b_i)_{i=1,\ldots,r} \subset \mathcal{H} \times \mathbb{R}$ (with $a_i \neq 0$ for $i = 1,\ldots,r$), namely

$$Q = \cap_{i=1,\ldots,r} H_i, \tag{4.3.83}$$

where $H_i = \{x \in \mathcal{H} : l_i(x) = \langle a_i, x \rangle + b_i \geq 0\}$. Let $P : \mathcal{H} \to \mathcal{H}$ be the mapping defined for $x \in \mathcal{H}$ by

$$Px = \begin{cases} x - l_{i_x}(x)|a_{i_x}|^{-2}a_{i_x}, & \text{if } I_x \neq \emptyset, \\ x, & \text{if } I_x = \emptyset, \end{cases} \tag{4.3.84}$$

$$\text{where } I_x := \{j \in \{1, \ldots, r\} | \mathcal{H} l_j(x) \leq 0\}, \tag{4.3.85}$$

$$i_x = \min\{i \in I_x | \mathcal{H} |l_{i_x}(x)| = \max_{j \in I_x} |l_j(x)|\}, \text{ if } I_x \neq \emptyset. \tag{4.3.86}$$

Then it can be observed that $Q = \text{Fix}(P)$ and that Px is nothing but the projection of x onto the hyperplan

$$K_{i_x} = \{z \in \mathcal{H} : l_{i_x}(z) = 0\}. \tag{4.3.87}$$

The following result was established in Ref. [37].

Lemma 4.3.1. *Let $P : \mathcal{H} \to \mathcal{H}$ be defined as in Remark 4.3.10. Then P satisfies the following statements:*

(c1) $P \in \mathscr{E}_{FQ}$;

(c2) $I - P$ is demi-closed (at zero).

Remark 4.3.11. From Remark 4.3.10 and Lemma 4.3.1, we can see that Theorem 4.3.5 still holds when the involved set of constraints Q_k (for $k = 1$ and/or $k = 2$) is given by (4.3.79) and $T_{(\Phi_k)}$ is replaced by the operator given by (4.3.84) relative to Q_k.

Remark 4.3.12. In the specific case when $f = I - \lambda \nabla \Psi$, where $\nabla \Psi : \mathcal{H}_1 \to \mathcal{H}_1$ denotes the gradient of some convex and differentiable function $\Psi : \mathcal{H}_1 \to \mathbb{R}$, and λ is some positive parameter, we stress that the solution of (4.2.16) (that is, the limit attained by (4.4.98)) is a minimizer of Ψ over Ω provided that (for some positive values L and η) $\nabla \Psi : \mathcal{H}_1 \to \mathcal{H}_1$ is L-Lipchitz continuous (i.e., $\|\nabla \Psi(x) - \nabla \Psi(y)\| \leq L\|x - y\|$, for $x, y \in \mathcal{H}_1$) and η-strongly monotone on $\mathcal{H}_1$ (i.e., $\langle \nabla \Psi(x) - \nabla \Psi(y), x - y \rangle \geq \eta \|x - y\|^2$, for $x, y \in \mathcal{H}_1$), and that $\lambda \in (0, 2\eta/L^2)$. Indeed, under these conditions, we readily obtain $\|f(x) - f(y)\|^2 \leq (1 - (2\eta - \lambda L^2)) |x - y|^2$. Hence, for $\lambda \in (0, 2\eta/L^2)$, C becomes a strict contraction. It is then a classical matter to see that (4.2.16) and the minimization problem $\min_{\Omega} \Psi$ are equivalent. In particular, when $\Psi(x) = (1/2)\|x - a\|^2$ (for some $a \in \mathcal{H}_1$), we have $\nabla \Psi(x) = x - a$. So, $\nabla \Psi$ is 1-Lipchitz continuous and 1-strongly monotone, and $f = I - \lambda \nabla \Psi$ reduces to $f \equiv a$ (for $\lambda = 1$).

4.3.3. Numerical Experiments for an Inverse Heat Equation

In this section we illustrate the effectiveness of the proposed algorithm (4.3.63) relative to an inverse heat conduction problem.

Consider the *Volterra integral equation* of the first kind: find $u \in L^2(0,1)$ such that

$$Ku = \int_0^t \kappa(t-s)u(s)ds = g^\delta(t), \quad 0 \le t \le 1, \tag{4.3.88}$$

where κ is a kernel function of convolution type defined by $\kappa(0) = 0$ and $\kappa(t) = \frac{1}{2\sqrt{\pi} t^{3/2}} e^{-1/4t}$, for $t \in (0,1]$, while the data g is some real-valued function on $[0,1]$ verifying $g(0) = 0$.

Remark 4.3.13. K is a (compact) linear operator from $L^2(0,1)$ to $L^\infty(0,1)$ (see, e.g., Ref. [15]).

Remark 4.3.14. Problem (4.3.88) corresponds to a sideways heat equation in a quarter plane, which consists of determining the value of a heat source $u = u(t)$ ($0 < t \le 1$) at position $x = 0$, by exploiting temperature measurements $g = g(t)$ ($0 \le t \le 1$) at position $x = 1$. The interested reader is referred to Refs. [15, 27, 28] for more information on this subject.

In practice, the measurement of the data g is often corrupted by some noise, so that only noisy data is available. Specifically we focus on the case of a uniformly distributed noise. Regarding this framework, Clason [15] has recently discussed some regularization-like processes based on L^∞ fitting, so as to approximate the original solution. It turns out that the L^∞ norm is more appropriate (than the L^2 norm) to measure the data misfit. The particular case of noisy data g^δ with level noise δ (for some positive value δ) was also considered, namely the (unknown) true data g satisfies $\|g^\delta - g\|_{L^\infty(0,1)} \le \delta$. In this latter situation, we seek an approximation u^δ of the exact solution u, through the following regularized problem : find $u^\delta \in L^2(0,1)$ such that

$$\|u^\delta\|_{L^2(0,1)} = \min_{v \in \Gamma} \|v\|_{L^2(0,1)}, \tag{4.3.89}$$

where $\Gamma = \{w \in L^2(0,1) : \|Kw - g^\delta\|_{L^\infty(0,1)} \le \delta\}$. Such approaches gave rise to numerical algorithms based upon methods (using judicious penalty functions) combined with semismooth Newton methods.

Remark 4.3.15. Note that the set of constraints Γ in (4.3.89) can be rewritten as

$$\Gamma = \left\{w \in L^2(0,1) : Kw \in \overline{B}_\infty(g^\delta, \delta)\right\},$$

where $\overline{B}_\infty(g^\delta, \delta) \subset L^2(0,1)$ denotes the closed ball in the L^∞ norm of center g^δ and radius δ, namely $\overline{B}_\infty(g^\delta, \delta) = \{w \in L^\infty(0,1) : \|w - g^\delta\|_{L^\infty(0,1)} \le \delta\}$.

From a practical viewpoint, we define an approximate problem for (4.3.89), setting $h = 1/N$ and denoting $t_i = ih$ (for $i = 0, \ldots, N$) and $t_{i+1/2} = t_i + h/2$ (for $i = 0, \ldots, N - 1$).

First, we first perform a discretization of problem (4.3.88) based on the composite midpoint integration rule $\int_0^{t_i} \phi(t)dt \approx h \sum_{j=0}^{i-1} \phi(t_{j+1/2})$, for any given continuous function ϕ on $[0, 1]$. By approximating the integral in (4.3.88), we are led to the midpoint method for the approximate solution of (4.3.88), given by $h \sum_{j=0}^{i-1} \kappa(t_i - t_{j+1/2})u_{j+1/2} = g(t_i)$ (for $i = 1, \ldots, N$), where $u_{j+1/2}$ is an approximation for $u(t_{j+1/2})$. Equivalently, introducing the two elements of R^N defined by

$$G = (g(t_1), \ldots, g(t_N))^T \text{ and } U = (u_{0+1/2}, \ldots, u_{(N-1)+1/2}),$$

we obtain the matrix equation

$$AU = G, \tag{4.3.90}$$

where $A = (a_{i,j})_{1 \le i,j \le N}$ is a lower-triangular $N \times N$ matrix with entries such that $a_{i,j} = h\kappa(t_i - t_{(j-1)+1/2})$ for $i = 1, \ldots, N$ and $j = 1, \ldots, i$. More precisely, it can be checked that the discretization method used generates a Toeplitz matrix A given by

$$A = \begin{pmatrix} k_1 & 0 & \cdots & 0 \\ k_2 & k_1 & \cdots & 0 \\ \vdots & \vdots & \ddots & \vdots \\ k_N & k_{N-1} & \cdots & k_1 \end{pmatrix} \tag{4.3.91}$$

with positive values $k_i = h\kappa((i - \frac{1}{2})h)$ (for $i = 1, \ldots, N$).

Next, we denote by $\|.\|_p$ the L^p norm of $\mathbb{R}^N$ (for $p = 2$ or $p = \infty$), and we extend (4.3.90) to an approximate model of (4.3.89). Precisely, we define an approximation of $u^\delta(t_j - (1/2)h)$ (for $j = 1, \ldots, N$) by the value u_j^δ obtained from the element $U^\delta = (u_1^\delta, \ldots, u_N^\delta)$ verifying

$$\left\| U^\delta \right\|_2 = \min_{X \in \Delta} \|X\|_2, \quad \text{where } \Delta = \{Z \in R^N : AZ \in \overline{D}(G^\delta, \delta)\}, \tag{4.3.92}$$

where A is the matrix defined in (4.3.91), $G^\delta = (g^\delta(t_1), \ldots, g^\delta(t_N))$ is the discrete noisy data, and $\overline{D}(G^\delta, \delta) = \{W \in R^N : \mathscr{H}\|W - G^\delta\|_\infty \le \delta\}$. Clearly, (4.3.92) enters the setting of problem (4.2.16) relative to the special case when $\mathscr{H}_1 = \mathscr{H}_2 = \mathbb{R}^N$, $f = 0$, $T = I$ (identity operator on $\mathbb{R}^N$), while $S : \mathbb{R}^N \to \mathbb{R}^N$ is any quasi-nonexpansive operator such that $\text{Fix}(S) = \overline{D}(G^\delta, \delta)$. Note indeed that the corresponding solution to (4.3.57) reduces to $x_* = P_\Delta 0$.

Remark 4.3.16. Let us observe that the discrete model (4.3.92) can be enhanced (with regard to the desired purposes) when prior information regarding the exact solution U is known. For instance, if U is known to belong to some (explicitly given) convex and closed subset Υ of $\mathbb{R}^N$, then it seems natural to consider (instead of (4.3.92)) the following model:

$$\left\| U^\delta \right\|_2 = \min_{X \in \Delta'} \|X\|_2, \tag{4.3.93}$$

where $\Delta' = \{Z \in \Upsilon : AZ \in \overline{D}(G^\delta, \delta)\}$, which is also encompassed by (4.2.16) with $\mathscr{H}_1 = \mathscr{H}_2 = \mathbb{R}^N$, $f = 0$, together with any quasi-nonexpansive operators $S, T : \mathbb{R}^N \to \mathbb{R}^N$ such that $\text{Fix}(S) = \overline{D}(G^\delta, \delta)$ and $\text{Fix}(T) = \Upsilon$.

Our goal here is to show the influence of the step sizes (β_n) on algorithm (4.3.63) relative to the discrete problem (4.3.93) with $N = 400$, $\Upsilon = \mathbb{R}^N_+$, and the other data chosen as follows.

First, we set $U = (u(t_0), \ldots, u(t_{N-1}))^T \in \mathbb{R}^N$, for some continuous and nonnegative function u on $[0,1]$, and we let $G = AU$ (hence U is the solution of (4.3.90) with data F). Next, we consider the noisy data given by $G^\delta = AU + \xi^\delta$, where the noise $\xi^\delta \in \mathbb{R}^N$ is a uniformly distributed random value of level $\delta := \|\xi^\delta\|_\infty = (1/4)\|G\|_\infty$. Finally, by using algorithm (4.4.98), we compute the solution U^δ of (4.3.93), so as to obtain a convenient reconstruction of U. The sequence generated by (4.3.63) is denoted $\{U_n^\delta\}$ and the following parameters and mappings are used: $U_0^\delta = 0$, $w_n = 0.5$, $\alpha_n = 1/n$, T is the metric projection onto $\mathbb{R}^N_+$, while S is of the type defined in Remark 4.3.10 relative to $\overline{D}(G^\delta, \delta)$.

A classical realization of noisy data is displayed in Figure 4.1. Figure 4.2 shows its corresponding reconstruction U^δ.

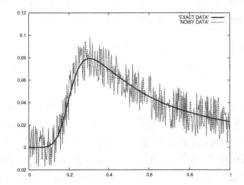

Figure 4.1 Profiles of G and G^δ (exact and noisy data).

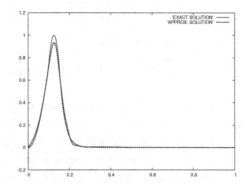

Figure 4.2 Reconstructed solution U^δ compared with exact solution U.

Figures 4.3 and 4.4 display the asymptotic behavior the error $\|U_n^\delta - U^\delta\|_2$ with respect to the number of iterations n, for different choices of the step size β_n.

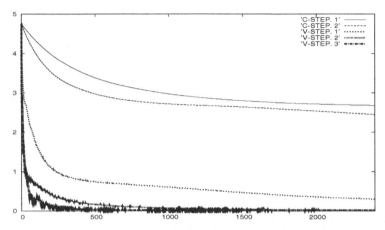

Figure 4.3 Error $\|U_n^\delta - U^\delta\|_2$ (with respect to n) for several step sizes: $\beta_n = 7$ (C-STEP. 1), $\beta_n = 15$ (C-STEP. 2), $\beta_n = 0.2\mu_n$ (V-STEP. 1), $\beta_n = \mu_n$ (V-STEP. 2) and $\beta_n = 1.8\mu_n$ (V-STEP. 3).

Regarding the spectral radius of A^*A (using the symmetric power method) we obtain $2\gamma \approx 15.8$ (where $\gamma = 1/\|A^*A\|$). The convergence of the method for constant step sizes in the range $(0, 2\gamma)$ appears to be very slow compared with Polyak-type step sizes (Figure 4.3). The convergence of the method is also illustrated with regard to some

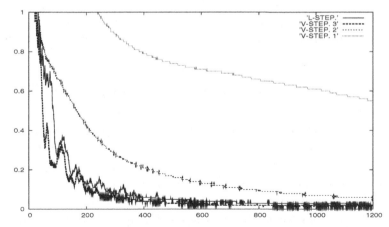

Figure 4.4 Error $\|U_n^\delta - U^\delta\|_2$ (with respect to n) for the limiting case $\beta_n = 2(1 - \varepsilon_n)\mu_n$ with $\varepsilon_n = \frac{0.1}{n^{0.5}}$ (L-STEP.) compared with the cases: $\beta_n = 0.2\mu_n$ (V-STEP. 1), $\beta_n = \mu_n$ (V-STEP. 2) and $\beta_n = 1.8\mu_n$ (V-STEP. 3).

limiting case of $\{\beta_n\}$ (Figure 4.4). It appears for step sizes of the form $\beta_n = \lambda_n \mu_n$ (with $\lambda_n \in (0,2)$) that the convergence of the algorithm is slower the greater λ_n is. Nevertheless, the convergence for the considered limiting case closely follows an optimal convergence obtained for λ_n close to 1.8.

Figure 4.5 shows how small may be the value of γ compared with the quantity μ_n involved in the Polyak-type step sizes.

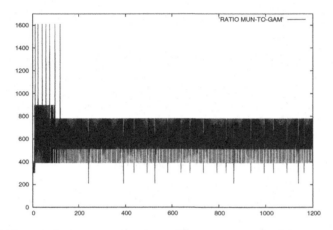

Figure 4.5 Ratio μ_n/γ ($\gamma \approx 7.9$) with respect to n in the case $\beta_n = \mu_n$.

4.4. Viscosity Method for Split Equilibrium Problems

Throughout this section, H is again a real Hilbert space with inner product $\langle .,. \rangle$ and induced norm $\| \cdot \|$. Let C be a nonempty closed convex subset of H and denote by S_F the set of solutions of the following *equilibrium problem*:

$$\text{find } u \in C \text{ such that } F(u,y) \geq 0, \quad \text{for all } y \in C, \qquad (4.4.94)$$

where $F : C \times C \to \mathbb{R}$ is a bifunction. Problem (4.4.94) is very general in the sense that it includes, as special cases, optimization problems, variational inequalities, min-max problems, Nash equilibrium problems in noncooperative games and others (see, for instance, Refs. [6, 64] and the references cited therein). In recent years, methods for solving equilibrium problems have been studied extensively. In [45], Moudafi extended the proximal method to monotone equilibrium problems and in [25] Konnov used the proximal method to solve problem (4.4.94) with weakly monotone bifunctions. Recently, Mastroeni [43] extended the so-called auxiliary problem principle to (4.4.94) involving strong monotone equilibrium problems. Other solution methods such as bundle methods and extragradient methods are extended to (1.87) in Refs. [64, 65].

Now, consider a (possibly) nonlinear mapping $T : H \to \mathcal{H}$ with a fixed point set denoted by $Fix(T) := \{x \in H : Tx = x\}$ and satisfying $Fix(T) \cap S_F \neq \emptyset$. In this section, we are interested in approximating a solution of the mixed problem:

$$\text{find } u \in S_F \cap Fix(T). \tag{4.4.95}$$

It is worth noting that numerous algorithms were proposed for solving fixed point problems for nonexpansive and even more general mappings [4, 23, 32, 36, 46, 58, 71, 72]. Other numerical methods were proposed for solving (4.4.95) in the special case when $F(x,y) = \langle Ax, y - x \rangle$, where $A : C \to H$ is a monotone and Lipschitz continuous mapping and T is nonexpansive. In this latter case, the proposed methods can be regarded as a suitable combination of the extragradient method initiated in Ref. [26] and either a Mann-type iteration [39, 48], a Halpern-type process [22, 73] or the hybrid steepest descent method [49, 67]. Very recently, a numerical approach was considered in Ref. [60] for solving the more general problem (4.4.95) where the bifunction F verifies the following usual conditions:

(A1) $F(x,x) = 0$ for all $x, y \in C$;

(A2) F is monotone, i.e., $F(x,y) + F(y,x) \leq 0$ for all $x, y \in C$;

(A3) $\lim_{t \downarrow 0} F(tz + (1-t)x, y) \leq F(x,y)$ for any $x, y, z \in C$;

(A4) for each $x \in C$, $y \to F(x,y)$ is convex and lower-semicontinuous.

Ref. [60] introduced an iterative scheme by the viscosity approximation method for finding a common element of the set of solutions of (4.4.95) and the set of fixed points of a nonexpansive mapping in a Hilbert space and proved the following strong convergence theorem:

Theorem 4.4.1 [60, Theorem 3.2]. *Let C be a nonempty closed convex subset of H. Let F be a bifunction from $C \times C$ to $\mathbb{R}$ satisfying (A1)–(A4) and let T be a nonexpansive mapping of C into H such that $S_F \cap Fix(T) \neq \emptyset$. Let f be a contraction of $\mathcal{H}$ into itself and let $\{x_n\}$ and $\{u_n\}$ be sequences generated by $x_0 \in H$ and*

$$\begin{cases} F(u_n, y) + \dfrac{1}{r_n} \langle y - u_n, u_n - x_n \rangle \geq 0, \quad \forall y \in C, \\ x_{n+1} = \alpha_n f(x_n) + (1 - \alpha_n) T u_n, \end{cases} \tag{4.4.96}$$

for all $n \in \mathbb{N}$, where $\{\alpha_n\} \subset (0,1]$ and $\{r_n\} \subset (0,\infty)$ satisfy: $\lim_{n \to \infty} \alpha_n = 0$, $\sum_n \alpha_n = \infty$, $\sum_n |\alpha_{n+1} - \alpha_n| < \infty$, $\liminf_{n \to \infty} r_n > 0$, and $\sum_n |r_{n+1} - r_n| < \infty$. Then, $\{x_n\}$ and $\{y_n\}$ converge strongly to z in $Fix(T) \cap S_F$, where $z = P_{Fix(T) \cap S_F} f(z)$.

Motivated by the above work and based upon the proximal method [26, 48, 73], we propose an alternative method for solving (4.4.95) in the more general case when T is demicontractive and demi-closed. Then, we prove a strong convergence theorem which improves or develops several corresponding results in this field.

First of all remember (see Definition 4.3.1 above) that as T is β-demicontractive this means that there exists a constant $\beta \in [0,1)$ such that $\|Tx - q\|^2 \le \|x - q\|^2 + \beta\|x - Tx\|^2$ for all $(x,q) \in H \times Fix(T)$, which is equivalent to (see Ref. [44])

$$\langle x - Tx, x - q \rangle \ge \frac{1-\beta}{2}\|x - Tx\|^2, \quad \text{for all } (x,q) \in \mathscr{H} \times Fix(T). \tag{4.4.97}$$

Let us also recall that T is called *demi-closed* (see Ref. [21]) if for any sequence $\{z_n\} \subset H$ and $z \in H$, we have:

$$z_n \to z \text{ weakly}, (I - T)(z_n) \to 0 \text{ strongly} \Rightarrow z \in Fix(T).$$

An operator satisfying (4.4.97) will be referred to as a β-demicontractive mapping. It is worth noting that the class of demicontractive maps contains important operators such as the quasi-nonexpansive maps and the strictly pseudocontractive maps with fixed points (see Ref. [23, 42, 44]). Remember again that a mapping $T : H \to H$ is called:

(a) *nonexpansive* if $\|Tx - Ty\| \le \|x - y\|$ for all $(x,y) \in H \times H$;

(b) *quasi-nonexpansive* if $\|Tx - q\| \le \|x - q\|$ for all $(x,q) \in H \times Fix(T)$;

(c) *strictly pseudocontractive* if $\|Tx - Ty\|^2 \le \|x - y\|^2 + \rho\|x - y - (Tx - Ty)\|^2$ for all $(x,y) \in H \times H$ (for some $\rho \in [0,1)$).

Observe also that the nonexpansive operators are both quasi-nonexpansive and strictly pseudocontractive maps and are well known for being demi-closed.

In view of selecting a particular solution of (4.4.95) and to make a connection with the hybrid steepest descent method investigated in Ref. [36], we consider an operator $\mathscr{F} := I - f$ where f is a contraction. It is then easily seen that $\mathscr{F}$ is strongly monotone and Lipshitz. So, in what follows we consider an operator $\mathscr{F} : C \to H$ satisfying the following two conditions:

(LC) $\mathscr{F}$ is L-Lipschitz continuous (for some $L > 0$), i.e., $\|\mathscr{F}(x) - \mathscr{F}(y)\| \le L\|x - y\|$ for all $x, y \in C$;

(SM) $\mathscr{F}$ is η-strongly monotone (for some $\eta > 0$), i.e., $\langle \mathscr{F}(x) - \mathscr{F}(y), x - y \rangle \ge \eta\|x - y\|^2$ for all $x, y \in C$, and we investigate the asymptotic behavior of the sequence $\{x_n\}$ generated, from an arbitrary x_0 in H, by the following algorithm:

$$\begin{cases} \text{compute } u_n \text{ by} \\ F(u_n, y) + \frac{1}{r_n}\langle y - u_n, u_n - x_n \rangle \ge 0, \quad \text{for all } y \in C; \\ \text{set } x_{n+1} := T_w v_n, \text{ with } v_n := (I - \alpha_n \mathscr{F})(u_n), \end{cases} \tag{4.4.98}$$

where $T_w := (1-w)I + wT$, with T assumed to be demicontractive, $I : H \to H$ stands for the identity mapping and the parameters are such that: $\{\alpha_n\} \subset [0,1)$, $\{r_n\} \subset (0,\infty)$ and $w \in (0,1)$.

More precisely, we will prove that the limit is the solution to the following variational inequality $VIP(\mathscr{F}, S_F \cap Fix(T))$: find $x^* \in Fix(T) \cap S_F$ such that

$$\langle v - x^*, \mathscr{F}(x^*) \rangle \geq 0, \quad \text{for all } v \in Fix(T) \cap S_F. \tag{4.4.99}$$

It is worth mentioning that the existence and the uniqueness of the solution of (4.4.99) are ensured by the conditions (LC), (SM) and by the fact that $S_F \cap Fix(T)$ is a nonempty closed and convex set.

We would like to emphasize that when $F \equiv 0$ and $C = H$, (4.4.98) reduces to a modified version of the hybrid steepest descent method recently investigated [36] as an algorithmic solution for solving $VIP(\mathscr{F}, Fix(T))$. The convergence in norm of the iterates generated by this scheme is obtained in the more general case when T is demicontractive. On the other hand, we would like to emphasize that the relaxation process induced by the mapping T_w in (4.4.98) was mainly suggested by the work of Suzuki [58] (see also Refs. [23, 39, 32]) and permits us to relax substantially the conditions on parameters α_n and r_n. Finally, let us notice that when $C = H$ and $F(x,y) = \max_{u \in Ax} \langle u, y - x \rangle$, where A is a maximal monotone operator, (4.4.98) amounts to finding zeros of the operator A and the sequence u_n given by (4.4.98) is nothing but the resolvent operator associated with A at x_n, namely $u_n = J_{r_n}^A x_n = (I + r_n A)^{-1} x_n$, so that the algorithm (4.4.98) reduces to $x_{n+1} := T_w v_n$ with $v_n := (I - \alpha_n \mathscr{F})(J_{r_n}^A x_n)$.

Under classical assumptions on the operators and the parameters, we will prove that the sequences $\{x_n\}$ and $\{u_n\}$ generated by the scheme (4.4.98) converge strongly to the unique solution of (4.4.99). Thus, by algorithm (4.4.98), we provide an efficient selecting method for solving the initial mixed problem (4.4.95) for a new broad class of maps. Moreover, the techniques of proofs are simple and different from the usual ones.

We begin with the following preliminary facts.

Fact 1 (see, for instance, Refs. [6, 17]). Let C be a nonempty closed convex subset of H and let F be a bifunction from $C \times C$ into $\mathbb{R}$ satisfying (A1)–(A4).

(i) Let $r > 0$ and $x \in H$. Then there exists $z \in C$ such that

$$F(z,y) + \frac{1}{r}\langle y - z, z - x \rangle \geq 0, \quad \text{for all } y \in C.$$

(ii) Let $T_r : H \to C$ be the mapping defined by

$$T_r(x) = \left\{ z \in C : F(z,y) + \frac{1}{r}\langle y - z, z - x \rangle \geq 0 \text{ for all } y \in C \right\}.$$

Then the following hold:

(a) T_r is single-valued;

(b) T_r is firmly nonexpansive, i.e.,

$$\|T_r x_1 - T_r x_2\|^2 \le \langle T_r x_1 - T_r x_2, x_1 - x_2 \rangle;$$

(c) $Fix(T_r) = S_F$ (S_F being the set of solutions of (4.4.95));

(d) S_F is closed and convex.

Fact 2 [60] Assume that $\{x_n\}$ and $\{u_n\}$ are two sequences in H verifying $r_n > 0$ and $u_n = T_{r_n} x_n$ for all $n \ge 0$. Then

$$\|u_n - u\|^2 \le \|x_n - u\|^2 - \|x_n - u_n\|^2, \quad \text{for all } n \ge 0, \qquad (4.4.100)$$

where u is any element in S_F.

Fact 3. Let $\{x_n\}$ and $\{u_n\}$ are two sequences in H verifying $u_n = T_{r_n} x_n$ for all $n \ge 0$ and assume that $r_n \in [\delta, \infty)$ for some $\delta > 0$. If, in addition, there exists a subsequence $\{u_{n_k}\}$ of $\{u_n\}$ such that

(i) $\{u_{n_k}\}$ converges weakly to some u in H;

(ii) $\|u_{n_k} - x_{n_k}\| \to 0$, it easily seen then that u belongs in S_F.

Fact 4 (See also Ref. [36]) Let T be a β-demicontractive self-mapping on H with $Fix(T) \ne \emptyset$ and set $T_w := (1-w)I + wT$ for $w \in (0,1]$. Then T_w is quasi-nonexpansive if $w \in [0, 1-\beta]$.

Indeed, for any arbitrary element $(x,q) \in H \times Fix(T)$, we have

$$\|T_w x - q\|^2 = \|x - q\|^2 - 2w\langle x - q, x - Tx \rangle + w^2 \|Tx - x\|^2,$$

which according to (4.4.97) yields

$$\|T_w x - q\|^2 \le \|x - q\|^2 - w(1 - \beta - w)\|Tx - x\|^2. \qquad (4.4.101)$$

Furthermore, we clearly have $Fix(T) = Fix(T_w)$ if $w \ne 0$. As a consequence, the operator T_w is quasi-nonexpansive for $w \in [0, 1-\beta]$ and $Fix(T)$ is then a closed convex subset of H (see Ref. [72, Proposition 1]).

The following lemma shows that the sequences $\{x_n\}$ and $\{u_n\}$ generated by (4.4.98) are bounded.

Lemma 4.4.1. *Suppose $T : H \to H$ is β-demicontractive with $S_F \cap Fix(T) \ne \emptyset$ and let $\mathscr{F} : H \to H$ be an operator satisfying (LC) and (SM). Assume, in addition, that $w \in (0, 1-\beta]$, $\{r_n\} \subset (0, \infty)$ and $\{\alpha_n\} \subset [0, \delta)$ (for some small enough $\delta > 0$). Then the sequences $\{x_n\}$ and $\{u_n\}$ generated by (4.4.98) are bounded.*

Proof. Without loss of generality, we may assume $0 < \eta < L$. Given $\mu \in (0, \infty)$ and $x, y \in H$, by using properties (SM) and (LC), we can write

$$
\begin{aligned}
&\|(\mu\mathscr{F} - I)(x) - (\mu\mathscr{F} - I)(y)\|^2 \\
&= \mu^2 \|\mathscr{F}(x) - \mathscr{F}(y)\|^2 - 2\mu \langle x - y, \mathscr{F}(x) - \mathscr{F}(y) \rangle + \|x - y\|^2 \\
&\leq \mu^2 L^2 \|x - y\|^2 - 2\mu\eta \|x - y\|^2 + \|x - y\|^2,
\end{aligned}
$$

so that

$$
\|(\mu\mathscr{F} - I)(x) - (\mu\mathscr{F} - I)(y)\| \leq \left(\sqrt{1 - 2\mu\eta + \mu^2 L^2} \right) \|x - y\|. \qquad (4.4.102)
$$

Furthermore, taking $q \in Fix(T) \cap S_F$ and recalling that $v_n = u_n - \alpha_n \mathscr{F}(u_n)$, we have

$$
\begin{aligned}
&\|v_{n+1} - (q - \alpha_{n+1}\mathscr{F}(q))\| = \|(u_{n+1} - \alpha_{n+1}\mathscr{F}(u_{n+1})) - (q - \alpha_{n+1}\mathscr{F}(q))\| \\
&= \left\| \left(1 - \frac{\alpha_{n+1}}{\mu}\right)(u_{n+1} - q) - \frac{\alpha_{n+1}}{\mu}((\mu\mathscr{F} - I)(u_{n+1}) - (\mu\mathscr{F} - I)(q)) \right\| \\
&\leq \left(1 - \frac{\alpha_{n+1}}{\mu}\right)\|u_{n+1} - q\| + \frac{\alpha_{n+1}}{\mu}\|(\mu\mathscr{F} - I)(u_{n+1}) - (\mu\mathscr{F} - I)(q)\|
\end{aligned}
$$

provided that $\{\alpha_n\} \subset [0, \mu)$, which by (4.4.102) yields

$$
\|v_{n+1} - (q - \alpha_{n+1}\mathscr{F}(q))\| \leq \left(1 - \frac{\alpha_{n+1}}{\mu}v\right)\|u_{n+1} - q\|, \qquad (4.4.103)
$$

where $v := 1 - \sqrt{1 - 2\mu\eta + \mu^2 L^2}$. Clearly, we have that $v \in (0, 1)$ when $\mu \in (0, \mu_0)$ for some small enough μ_0. Using (4.4.100) and observing that $T_w := (1 - w)I + wT$ is quasi-nonexpansive for $w \in (0, 1 - \beta]$ (because of fact 4), by (4.4.101) we additionally have

$$
\|u_{n+1} - q\| \leq \|x_{n+1} - q\| = \|T_w v_n - q\| \leq \|v_n - q\|. \qquad (4.4.104)
$$

Combining (4.4.103) and (4.4.104), we then get

$$
\|v_{n+1} - (q - \alpha_{n+1}\mathscr{F}(q))\| \leq \left(1 - \frac{\alpha_{n+1}}{\mu}v\right)\|v_n - q\|.
$$

As a consequence, we deduce

$$
\begin{aligned}
\|v_{n+1} - q\| &\leq \|v_{n+1} - (q - \alpha_{n+1}\mathscr{F}(q))\| + \|(q - \alpha_{n+1}\mathscr{F}(q)) - q\| \\
&\leq \left(1 - \frac{\alpha_{n+1}v}{\mu}\right)\|v_n - q\| + \alpha_{n+1}\|\mathscr{F}(q)\| \\
&= \left(1 - \frac{\alpha_{n+1}v}{\mu}\right)\|v_n - q\| + \left(\frac{\alpha_{n+1}v}{\mu}\right)\left(\frac{\mu\|\mathscr{F}(q)\|}{v}\right),
\end{aligned}
$$

and hence

$$
\max\left\{\|v_{n+1} - q\|, \frac{\mu\|\mathscr{F}(q)\|}{v}\right\} \leq \max\left\{\|v_n - q\|, \frac{\mu\|\mathscr{F}(q)\|}{v}\right\},
$$

so that for all $n \geq 0$,

$$\|v_n - q\| \leq \max\left\{\|v_0 - q\|, \frac{\mu\|\mathscr{F}(q)\|}{\nu}\right\}. \tag{4.4.105}$$

Thus $\{v_n\}$ is bounded, which by (4.4.104) leads to the boundedness of $\{x_n\}$ and $\{u_n\}$.

$\square$

In order to prove our main convergence theorem, we start with some preliminary results.

Lemma 4.4.2. *Suppose* $T : H \to H$ *is a* β*-demicontractive mapping with* $Fix(T) \cap S_F \neq \emptyset$ *and let* $\mathscr{F} : H \to H$ *is a given operator. Assume in addition that the following condition holds:*

(H1) $w \in \left(0, \frac{1-\beta}{2}\right]$.

Then for all $n \geq 0$ *the sequences* $\{x_n\}$ *and* $\{u_n\}$*, given by (4.4.98), satisfy the following inequality:*

$$\begin{aligned} &\|x_{n+1} - q\|^2 - \|x_n - q\|^2 \\ &+ \|x_{n+1} - u_n\|^2 + \|x_n - u_n\|^2 \leq -2\alpha_n\langle x_{n+1} - q, \mathscr{F}(u_n)\rangle, \end{aligned} \tag{4.4.106}$$

where q *is any element in* $Fix(T) \cap S_F$*.*

Proof. Let $q \in Fix(T) \cap S_F$. From (4.4.98) and (4.4.101) we obtain

$$\|x_{n+1} - q\|^2 \leq \|v_n - q\|^2 - w(1 - \beta - w)\|v_n - Tv_n\|^2 \tag{4.4.107}$$

and by virtue of (4.4.98) we also have $Tv_n - v_n = \frac{1}{w}(x_{n+1} - v_n)$. Consequently, setting $\rho := \frac{1}{w}(1 - \beta - w)$, we obtain

$$\|x_{n+1} - q\|^2 \leq \|v_n - q\|^2 - \rho\|x_{n+1} - v_n\|^2, \tag{4.4.108}$$

hence if $w \in \left(0, \frac{1-\beta}{2}\right]$ (so that $\rho \geq 1$) we get

$$\begin{aligned} \|x_{n+1} - q\|^2 &\leq \|v_n - q\|^2 - \|x_{n+1} - v_n\|^2 \\ &= \|(u_n - q) - \alpha_n\mathscr{F}(u_n)\|^2 - \|(u_n - x_{n+1}) - \alpha_n\mathscr{F}(u_n)\|^2 \\ &= \|u_n - q\|^2 - 2\alpha_n\langle x_{n+1} - q, \mathscr{F}(u_n)\rangle - \|x_{n+1} - u_n\|^2. \end{aligned} \tag{4.4.109}$$

Furthermore, from (4.4.100) we have

$$\|u_n - q\|^2 \leq \|x_n - q\|^2 - \|x_n - u_n\|^2, \tag{4.4.110}$$

which, combined with (4.4.109), gives the desired result.

$\square$

Lemma 4.4.3. *Let* $\mathscr{F} : H \to H$ *be any operator satisfying (LC). Suppose that* $T : H \to H$ *is demi-closed with* $Fix(T) \cap S_F \neq \emptyset$ *and that the following conditions on the parameters hold:*

(C1) $\{r_n\} \subset [\delta, \infty)$ *(for some* $\delta > 0$*);*

(H2) $\{\alpha_n\} \subset [0, 1)$, $\alpha_n \to 0$.

Let $\{x_n\}$, $\{u_n\}$ *be the sequences generated by (4.4.98) and assume further the existence of a subsequence* $\{u_{n_k}\}$ *of* $\{u_n\}$ *such that*

(i) $\|u_{n_k} - x_{n_k+1}\| \to 0$;

(ii) $\|u_{n_k} - x_{n_k}\| \to 0$.

Then any weak cluster point of $\{u_{n_k}\}$ *belongs to* $Fix(T) \cap S_F$. *Moreover, if in addition* $\{u_{n_k}\}$ *is bounded, then*

$$\liminf_{k \to \infty} \langle u_{n_k} - u^*, \mathscr{F}(u^*) \rangle \geq 0, \tag{4.4.111}$$

where u^* *is any solution of (4.4.99).*

Proof. Let $u \in H$ be a weak cluster point of $\{u_{n_k}\}$. Then there exists a bounded subsequence of $\{u_{n_k}\}$ (labeled $\{u_{m_n}\}$) which weakly converges to u. By (i) and (ii), we also have $\|x_{m_n+1} - u_{m_n}\| \to 0$ and $\|x_{m_n} - u_{m_n}\| \to 0$. If in addition $\alpha_n \to 0$, we easily deduce that $v_{m_n} := u_{m_n} - \alpha_{m_n} \mathscr{F}(u_{m_n})$ weakly converges to u (because $\mathscr{F}(u_{m_n})$ is bounded because of (LC)), hence $\alpha_{m_n} \|\mathscr{F}(u_{m_n})\| \to 0$), which by (4.4.98) entails

$$\|Tv_{m_n} - v_{m_n}\| = \frac{1}{w} \|x_{m_n+1} - v_{m_n}\| = \frac{1}{w} \|(x_{m_n+1} - u_{m_n}) + \alpha_{m_n} \mathscr{F}(u_{m_n})\| \to 0.$$

Now, as T is assumed to be demi-closed, we then obtain $u \in Fix(T)$. Furthermore, recalling that $\|x_{m_n} - u_{m_n}\| \to 0$ and assuming $r_n \geq \delta > 0$, by Lemma 4.4.2 we get $u \in S_F$. Consequently, the set of weak cluster points of (u_{n_k}) is included in $S_F \cap Fix(T)$. If $\{u_{n_k}\}$ is also a bounded sequence, so is the quantity $\langle u_{n_k} - u^*, \mathscr{F}(u^*) \rangle$. It is then evident that there exists a subsequence of $\{u_{n_k}\}$ (denoted $\{u_{m_n}\}$) which converges weakly to some element v in H (hence $v \in Fix(T) \cap S_F$) and such that

$$\liminf_{k \to \infty} \langle u_{n_k} - u^*, \mathscr{F}(u^*) \rangle = \lim_{n \to \infty} \langle u_{m_n} - u^*, \mathscr{F}(u^*) \rangle.$$

Thus, by the weak convergence of $\{u_{m_n}\}$ and by recalling that u^* is the solution of (4.4.99), we easily deduce

$$\liminf_{k \to \infty} \langle u_{n_k} - u^*, \mathscr{F}(u^*) \rangle = \langle v - u^*, \mathscr{F}(u^*) \rangle \geq 0.$$

This ends the proof. $\qquad\qquad\qquad\qquad\qquad\qquad\qquad\qquad\qquad\qquad\qquad\qquad\quad\square$

Lemma 4.4.4. *Assume that $T : H \to H$ is β-demicontractive, demi-closed and such that $Fix(T) \cap S_F \neq \emptyset$; $\mathscr{F} : H \to H$ satisfies (LC) and (SM) and suppose in addition that*

(C1) $\{r_n\} \subset [\delta, \infty)$ *(for some $\delta > 0$);*

(H2) $\{\alpha_n\} \subset [0,1)$, $\alpha_n \to 0$.

Let $\{x_n\}$, $\{u_n\}$ be the sequences generated by (4.4.98) and assume furthermore the existence of a bounded subsequence $\{u_{n_k}\}$ of $\{u_n\}$ such that

(i) $\|u_{n_k} - x_{n_k+1}\| \to 0$;

(ii) $\|x_{n_k} - u_{n_k}\| \to 0$;

(iii) $\langle x_{n_k+1} - x_*, \mathscr{F}(u_{n_k}) \rangle \leq 0$, *where x_* is the solution of (4.4.99).*

Then $\{x_{n_k}\}$ and $\{u_{n_k}\}$ converge strongly to x_*.

Proof. From the boundedness of $\{u_{n_k}\}$, we can extract a subsequence (again labeled $\{u_{n_k}\}$) which converges weakly to some q in H and such that (i), (ii) and (iii) still hold. From Lemma 4.4.3, we infer that $q \in Fix(T) \cap S_F$. Furthermore, by (SM) we observe that

$$\eta \|u_{n_k} - x_*\|^2 \leq \langle u_{n_k} - x_*, \mathscr{F}(u_{n_k}) - \mathscr{F}(x_*) \rangle$$
$$= \langle x_{n_k+1} - x_*, \mathscr{F}(u_{n_k}) \rangle + \langle u_{n_k} - x_{n_k+1}, \mathscr{F}(u_{n_k}) \rangle - \langle u_{n_k} - x_*, \mathscr{F}(x_*) \rangle,$$

which in the light of (iii) entails

$$\eta \|u_{n_k} - x_*\|^2 \leq \langle u_{n_k} - x_{n_k+1}, \mathscr{F}(u_{n_k}) \rangle - \langle u_{n_k} - x_*, \mathscr{F}(x_*) \rangle. \tag{4.4.112}$$

Hence by (4.4.112), (ii) and (4.4.99) we obviously have

$$\limsup_{k \to +\infty} \|u_{n_k} - x_*\|^2 \leq -(1/\eta)\langle q - x_*, \mathscr{F}(x_*) \rangle \leq 0.$$

Therefore we obtain $\lim_{k \to +\infty} \|u_{n_k} - x_*\| = 0$ and by virtue of (ii) and the uniqueness of x_* we deduce that $\{x_{n_k}\}$ also converge strongly to x_*, which completes the proof. $\square$

Lemma 4.4.5. *Let $\mathscr{F} : H \to H$ be an operator satisfying (LC) and (SM), suppose that $T : H \to H$ is β-demicontractive, demi-closed with $Fix(T) \cap S_F \neq \emptyset$ and assume that the following conditions on the parameters hold:*

(H1) $w \in (0, \frac{1-\beta}{2}]$;

(H2) $\{\alpha_n\} \subset [0,1)$, $\alpha_n \to 0$;

(C2) $\{r_n\} \subset [\delta, \infty)$ *(for $\delta > 0$);*

(SP) $\sum_{n\geq 0} \alpha_n = \infty$.

Assume furthermore that the sequences $\{x_n\}$ and $\{u_n\}$ given by (4.4.98) satisfy:

(i) $\|u_n - x_{n+1}\| \to 0$;

(ii) $\|x_n - u_n\| \to 0$;

(iii) $\lim_{n\to\infty} \|x_n - x_*\|$ *exists, x_* being the solution of (4.4.99).*

Then $\{x_n\}$ and $\{u_n\}$ converge strongly to x_*.

Proof. Using condition (SM), we obviously obtain

$$\langle x_{n+1} - x_*, \mathscr{F}(u_n) \rangle \geq \eta \|u_n - x_*\|^2 + \langle u_n - x_*, \mathscr{F}(x_*) \rangle + \langle x_{n+1} - u_n, \mathscr{F}(u_n) \rangle.$$
(4.4.113)

Set $\lim_{n\to\infty} \|x_n - x_*\| = \lambda \geq 0$. $\{x_n\}$ is thus bounded and, from (ii), so is $\{u_n\}$. Since Lemma 4.4.3 is applicable, we also have

$$\liminf_{n\to\infty} \langle u_n - x_*, \mathscr{F}(x_*) \rangle \geq 0.$$
(4.4.114)

Therefore by (4.4.113), (4.4.114) and (i), we get

$$\liminf_{n\to\infty} \langle x_{n+1} - x_*, \mathscr{F}(u_n) \rangle \geq \eta \lambda^2.$$

Hence, for $\varepsilon > 0$, from Lemma 4.4.2, we deduce that for $n \geq n_0$ (for some n_0 large enough),

$$\|x_{n+1} - x_*\|^2 - \|x_n - x_*\|^2 \leq -2\alpha_n(\eta\lambda^2 - \varepsilon).$$

This easily leads to

$$\|x_{n+1} - x_*\|^2 - \|x_{n_0} - x_*\|^2 \leq -2(\lambda^2\eta - \varepsilon) \sum_{k=n_0}^{n} \alpha_n.$$

Assuming $\sum \alpha_n = \infty$, we observe that this last inequality is absurd for $\lambda > 0$, since $\{x_n\}$ is bounded. As a straightforward consequence, we obtain $\lambda = 0$, namely $\{x_n\}$ converges strongly to x_* and according to (ii), so does $\{u_n\}$. This completes the proof. $\qquad \square$

We are now in a position to give the following main convergence theorem.

Theorem 4.4.2. *Suppose $T : H \to H$ is β-demicontractive, demi-closed with $Fix(T) \cap S_F \neq \emptyset$. Let $\mathscr{F} : H \to H$ satisfy (LC) and (SM) and assume the following conditions hold:*

(H1) $w \in (0, \frac{1-\beta}{2}]$;

(H2) $\{\alpha_n\} \subset [0,1)$, $\alpha_n \to 0$;

(H3) $\{r_n\} \subset [\delta, \infty)$ *(where $\delta > 0$);*

(SP) $\sum_{n\geq 0} \alpha_n = \infty$.

Then the sequences $\{x_n\}$ and $\{u_n\}$ generated by (4.4.98) converge strongly to x_*, the unique solution of the variational inequality (4.4.99).

Proof. The boundedness of $\{x_n\}$ and $\{u_n\}$ is deduced from Lemma 4.4.1, so that there exists a constant $C \geq 0$ such that $\|\langle x_{n+1} - x_*, \mathscr{F}(u_n)\rangle\| \leq C$ for all $n \geq 0$. Consequently, by Lemma 4.4.2, we get

$$\|x_{n+1} - x_*\|^2 - \|x_n - x_*\|^2 + \|x_{n+1} - u_n\|^2 + \|x_n - u_n\|^2 \leq 2C\alpha_n. \qquad (4.4.115)$$

The rest of the proof can be divided into two cases:

CASE 1. Assume $(\|x_n - x_*\|)$ is a monotone sequence. In other words, for n_0 large enough, $(\|x_n - x_*\|)_{n \geq n_0}$ is either nondecreasing or nonincreasing and being also bounded, $(\|x_n - x_*\|)$ is thus convergent. Clearly, we then have $\|x_{n+1} - x_*\|^2 - \|x_n - x_*\|^2 \to 0$, which by (4.4.115) yields $\|x_{n+1} - u_n\| \to 0$ and $\|x_n - u_n\| \to 0$. Consequently, by Lemma 4.4.5, we deduce that $\{x_n\}$, $\{u_n\}$ converge strongly to x_*.

CASE 2. Assume $(\|x_n - x_*\|)$ is not a monotone sequence and set $\Gamma_n := \|x_n - x_*\|^2$. Clearly, there exists a subsequence $\{\Gamma_{n_k}\}_{k \geq 0}$ of $\{\Gamma_n\}_{n\geq 0}$ such that $\Gamma_{n_k} < \Gamma_{n_k+1}$ for all $k \geq 0$. So, we consider the sequence of indexes $\{\tau(n)\}$ as defined in Lemma 4.2.3. It follows that τ is a nondecreasing sequence such that $\tau(n) \to +\infty$ (as $n \to +\infty$) and $\Gamma_{\tau(n)} \leq \Gamma_{\tau(n)+1}$ (for $n \geq n_0$), which by (4.4.115) entails

$$\|x_{\tau(n)+1} - u_{\tau(n)}\|^2 + \|x_{\tau(n)} - u_{\tau(n)}\|^2 \leq 2C\alpha_{\tau(n)} \to 0,$$

thus $\|x_{\tau(n)+1} - u_{\tau(n)}\| \to 0$ and $\|x_{\tau(n)} - u_{\tau(n)}\| \to 0$, so that $\|x_{\tau(n)+1} - x_{\tau(n)}\| \to 0$. Furthermore, by Lemma 4.4.2 we have

$$\text{for any } j \geq 0 \quad \langle x_{j+1} - x_*, \mathscr{F}(u_j)\rangle > 0 \quad \Rightarrow \quad \Gamma_{j+1} < \Gamma_j.$$

As a consequence, since $\Gamma_{\tau(n)} \leq \Gamma_{\tau(n)+1}$, we get

$$\langle x_{\tau(n)+1} - x_*, \mathscr{F}(u_{\tau(n)})\rangle \leq 0, \quad \text{for all } n \geq n_0.$$

Applying Lemma 4.4.4, we deduce that $\|x_{\tau(n)} - x_*\| \to 0$ and it is then evident that $\lim_{n\to\infty} \Gamma_{\tau(n)} = \lim_{n\to\infty} \Gamma_{\tau(n)+1} = 0$, since $\|x_{\tau(n)+1} - x_{\tau(n)}\| \to 0$. Again using Lemma 4.2.3, we have for all $n \geq n_0$,

$$0 \leq \Gamma_n \leq \Gamma_{\tau(n)+1}.$$

Hence $\lim_{n\to\infty} \Gamma_n = 0$, that is, $\{x_n\}$ converges strongly to x_*. In view of (4.4.115), we also obtain the strong converge of $\{u_n\}$ to x_*, which completes the proof. $\qquad\square$

We end this section with two important particular cases. First, when $F \equiv 0$, we have $u_n = P_C x_n$. So as a direct consequence of Theorem 4.4.2, we obtain that the sequence generated from x_0 by

$$x_{n+1} := T_w(I - \alpha_n \mathscr{F})(P_C x_n)$$

converges strongly to $x^* \in FixT$, which solves the variational inequality

$$\langle v - x^*, \mathscr{F}(x^*)\rangle \geq 0, \quad \text{for all } v \in Fix(T).$$

In the case when $T = I$, algorithm (4.4.98) generates from an arbitrary $x_0 \in H$ two sequences $\{u_n\}$ and $\{x_n\}$ by the following rule:

$$\begin{cases} \text{compute } u_n \text{ such that:} \\ F(u_n, y) + \frac{1}{r_n}\langle y - u_n, u_n - x_n \rangle \geq 0, \quad \text{for all } y \in C; \\ \text{set } x_{n+1} = (I - \alpha_n \mathscr{F})(u_n). \end{cases} \quad (4.4.116)$$

As a direct consequence of Theorem 4.4.2, we obtain that the sequences $\{x_n\}$ and $\{u_n\}$ strongly converge to $x^* \in S_F$, which solves the variational inequality

$$\langle v - x^*, \mathscr{F}(x^*)\rangle \geq 0, \quad \text{for all } v \in S_F.$$

Various applications in signal/image recovery, machine learning and problems encountered in computer science can be cast in the formulation above by particularizing the bifunction F (see, for instance, Ref. [16]). Indeed, in order to obtain, for instance, reliable solutions to inverse problems in signal/image processing, it is necessary to incorporate in the mathematical formulation of the problem the various pieces of *a priori* information that are available (spatial constraints, sparsity constraints, statistical properties of the noise, etc.). In many cases, it is possible to map the set of images possessing a certain property to the fixed point set of a nonlinear operator. The inverse problem then reduces to a minimization problem over the set of common fixed points which naturally has the variational inequality characterization above. Such formulations are used for establishing the convergence properties of algorithms including hybrid steepest descent methods, that we obtained here as a special case, and are available to solve problems such as online learning or to design robust smart antennas (see, for instance, Ref. [56] and references therein). By particularizing again the bifunction F, algorithm (4.4.98) appears to be a suitable method for a variational rationality approach of human behavior which unifies many different theories of stability (habits, routines, equilibrium, traps, etc.) and changes (creations, innovations, learning and destructions, etc.) in behavioral sciences (see, for instance, Ref. [5] and references therein).

To conclude, it is worth mentioning that we have also proposed a viscosity method for hierarchical fixed point problems in Hilbert spaces (see Refs. [38, 47]). We introduced a more general approach which consists of finding a particular part of the solution set of a given fixed point problem, i.e., fixed points which solve a variational inequality *criterion*. Our goal was to present a method for finding hierarchically a fixed point of a nonexpansive mapping T with respect to another nonexpansive mapping P, namely

Find $\tilde{x} \in Fix(T)$ such that $\langle \tilde{x} - P(\tilde{x}), x - \tilde{x} \rangle \geq 0$,   for all $x \in Fix(T)$,   (4.4.117)

i.e., $0 \in (I - P)\tilde{x} + N_{Fix(T)}(\tilde{x})$.

More precisely, we considered in Ref. [47] the viscosity method for the hierarchical fixed point problem of nonexpansive mappings (4.4.117) as follows:

Given a contraction f on C and two nonexpansive mappings P and T. Then for $s, t \in (0, 1)$, the mapping

$$sf(x) + (1 - s)(tPx + (1 - t)Tx)$$

is a contraction on C. So it has a unique fixed point, denoted $x_{s,t} \in C$; thus

$$x_{s,t} = sf(x_{s,t}) + (1 - s)(tPx_{s,t} + (1 - t)Tx_{s,t}).$$

Its convergence was studied under certain conditions [47] and the strong convergence of its explicit version was obtained under a regularity metric condition [38]. These implicit and explicit iterations were then further analyzed by many authors. Specifically efforts have been made recently to further investigate their convergence in, for example, Refs. [14, 40, 70] and the references therein.

REFERENCES

1. Albert YI, Iusem AN, Solodov MV, On the projected subgradient method for nonsmooth convex optimization in a Hilbert space. Math. Prog. 1998;81:23–35.
2. Attouch H, Viscosity approximation methods for minimization problems. SIAM J. Optim. 1996;6:769–806.
3. Bauschke HH, Borwein JM, On projection algorithms for solving convex feasibility problems. SIAM Rev. 1996;38:367–426.
4. Bauschke HH, Combettes PL, A weak-to-strong convergence principle for Fejer monotone methods in Hilbert spaces. Math. Oper. Res. 2001;26:248–264.
5. Bento GC, Cruz Neto JX, Soares PA, Soubeyran A, Behavioral traps and the equilibrium problem on Hadamard Manifolds. ArXiv 1307.7200v2 (2013).
6. Blum E, Oettli W, From optimization and variational inequalities to equilibrium problems. Math. Student 1994;63:123–145.
7. Brezis H, Operateurs Maximaux Monotones. Mathematics studies 5: North-Holland; 1973.
8. Brézis H, Lions PL, Produits infinis de resolvantes. Israel J. Math. 1978;29:329–345.
9. Browder FE, Convergence theorems for sequences of nonlinear operators in Banach spaces. Math. Zeitschr. 1967;100:201–225.
10. Bruck RE, Asymptotic behavior of nonexpansive mapping. Contemporary Math. 1983;18:1–47.
11. Byrne CH, Iterative oblique projection onto convex sets and the split feasibility problem. Inverse Problems 2002;18:441–453.
12. Byrne CH, Censor Y, Gibali A, Reich S, Weak and strong convergence of algorithms for the split common null point problem. ArXiv 1108.5953 (2011).
13. Censor Y, Segal A, The split common fixed point problem for directed operators. J. Convex Anal. 2009;16:587–600.

14. Cianciaruso F, Colao V, Muglia L, Xu HK, On a implicit hierarchical fixed point approach to variational inequalities. Bull. Aust. Math. Soc. 2009;80:117–124.
15. Clason C, L^∞ fitting for inverse problems with uniform noise. Inverse Problems 2012;28:Article ID 10.
16. Combettes PL, A block-iterative surrogate constraint splitting method for quadratic signal recovery. IEEE Trans. Signal Processing 2003;51:1771–1782.
17. Combettes PL, Hirstoaga A, Equilibrium programming in Hilbert spaces. J. Nonlinear Convex Anal. 2005;6:117–136.
18. Cominetti R, San Martin S, Asymptotical analysis of the exponential penalty trajectory in linear programming. Math. Program. 1994;67:169–187.
19. Ekeland I, Themam R, Convex Analysis and Variational Problems. Classic in Applied Mathematics 28: SIAM; 1999.
20. Genel A, Lindenstrauss J, An example concerning fixed points. Israel J. Math. 1975; 22: 81–86.
21. Goebel K, Kirk WA, Topics in metric fixed point theory. Cambridge Studies in Advanced Mathematics 28: Cambridge University Press, Cambridge; 1990.
22. Halpern B, Fixed points of nonexpanding maps. Bull. Am. Math. Soc. 1967;73:591–597.
23. Hicks TL, Kubicek JD, On the Mann iteration process in Hilbert spaces. J. Math. Anal. Appl. 1977; 59:498–504.
24. Itoh S, Takahashi W, The common fixed point theory of singlevalued mappings and multivalued mappings. Pacific J. Math. 1978;79:493–508.
25. Konnov IV, Application of the proximal method to nonmonotone equilibrium problems. J. Optim. Theory Appl. 2003; 119:317–333.
26. Korpelevich GM, The extragradient method for finding saddle points and other problems. Matecon 1976;12:747–756.
27. Lamm PK, Elden L, Numerical solution of first-kind Volterra equations by sequential Tikhonov regularization. SIAM J. Numer. Anal. 1997;34:1432–1450.
28. Lamm PK, A survey on regularization methods for first-kind Volterra equation. In: Colton D, Engl HW, Louis A, McLaughlin JR, Rundell W, editors, Surveys on Solution Methods for Inverse Problems, Vienna, New York: Springer; 2000.
29. Larsson T, Patricksson M, Stromberg AB, Conditional subgradient optimization – Theory and applications. Eur. J. Oper. Res. 1996;88:382–403.
30. Lehdili N, Ould Ahmed Salem C, Proximal methods: Decomposition and selection. In: Lecture Notes in Economics and Mathematical Systems 481, Springer Berlin; 2000;218–233.
31. Lions PL, Approximation des points fixes de contractions. C. R. Acad. Sci. Ser. A-B Paris 1977;284;1357–1359.
32. Liu L, Approximation of fixed points of a strictly pseudocontractive mapping. Proc. Am. Math. Soc. 1997;125:1363–1366.
33. Maingé PE, Inertial interative process for fixed points of certain quasi-nonexpansive mappings. Set Valued Anal. 2007; 15:67–79.
34. Maingé PE, Strong convergence of projected subgradient methods for nonsmooth and nonstrictly convex minimization. Set-Valued Anal. 2008;16:899–912.
35. Maingé PE, The viscosity approximation process for quasi-nonexpansive mappings in Hilbert spaces. Computers Math. Appl. 2010; 59:74–79.
36. Maingé PE, New approach to solving a system of variational inequalities and hierarchical problems. J. Optim. Theory Appl. 2008;138:459–477.
37. Maingé PE, A viscosity method with no spectral radius requirements for the split common fixed point problem. Eur. J. Oper. Res. 2014;235:17–27.
38. Maingé PE, Moudafi A, Strong convergence of an iterative method for hierarchical fixed point problems. Pacific J. Optim. 2007;3:529–538.
39. Mann WR, Mean value methods in iteration. Proc. Am. Math. Soc. 1953;4:506–510.
40. Marino G, Xu HK, Explicit hierarchical fixed point approach to variational inequalities. J. Optim. Theory Appl. 2011;149:61–78.

41. Marquez VM, Fixed point approximations methods for nonexpansive mappings. Ph.D. Thesis (Universidad de Sevilla); 2010.
42. Maruster S, The solution by iteration of nonlinear equations in Hilbert spaces. Proc. Am. Math. Soc. 1997;63:69–73.
43. Mastroeni G, On auxiliary principle for equilibrium problems. Publicatione del Departimento di mathematica dell'Universita di Pisa 2000;3:1244–1258.
44. Moore C, Iterative aproximation of fixed points of demicontractive maps. Scientific Report of the Abdus Salam. Intern. Centre for Theoretical Physics (Trieste, Italy) 1998; IC /98/214.
45. Moudafi A, Proximal point algorithm extended for equilibrium problems. J. Natural Geometry 1999;15:91–100.
46. Moudafi A, Viscosity approximations methods for fixed-points problems. J. Math. Anal. Appl. 2000;241:46–55.
47. Moudafi A, Maingé PE, Towards viscosity approximations of hierarchical fixed-points problems. Fixed Point Theory Appl. 2006;2006: Article ID 95453.
48. Nadezhkina N, Takahashi W, Weak convergence theorem by an extragradient method for nonexpansive mappings and monotone mappings. J. Optim. Theory Appl. 2006;128:191–201.
49. Nadezhkina N, Takahashi W, Strong convergence theorem by an hybrid method for nonexpansive mappings and Lipschitz continuous monotone mappings. SIAM J. Optim. 2006;16:1230–1241.
50. Panyanak B, Suantai S, Viscosity approximation methods for multivalued nonexpansive mappings in geodesic spaces. Fixed Point Theory Appl. 2015;2015:1-14.
51. Piatek B, Viscosity iteration in CAT (k) spaces. Numer. Funct. Anal. Optim. 2013;34:1245–1264.
52. Polyak BT, Introduction to Optimization. New York: Optimization Software; 1987.
53. Reich S, Weak convergence theorems for nonexpansive mappings in Banach spaces. J. Math. Anal. Appl. 1979;67:274–276.
54. Rockafellar RT, Monotone operators and the proximal point algorithm. SIAM J. Control. Optim. 1976;14:877–898.
55. Shioji N, Takahashi W, Strong convergence of approximated sequences for nonexpansive mappings in Banach spaces. Proc. Am. Math. Soc. (1997);125:3641–3645.
56. Slavakis K, Yamada I, The adaptive projected subgradient method constrained by families of quasi-nonexpansive mappings and its application to online learning. SIAM J. Optim. 2013;23:126–152.
57. Suzuki T, Moudafi's viscosity approximations with Meir-Keeler contractions. Math. Anal. Appl. 2007;325:342–352.
58. Suzuki T, A sufficient and necessary condition for Halpern-type strong convergence to fixed points of nonexpansive mappings. Proc. Am. Math. Soc. 2007;135:99–106.
59. Tada A, Takahashi W, Strong convergence theorem for an equilibrium problem and a nonexpansive mapping. In: Takahashi W, Tanaka T, Nonlinear Analysis and Convex Analysis, Yokohama Publishers; 2006.
60. Takahashi S, Takahashi W, Viscosity approximation methods for equilibrium problems and fixed point problems in Hilbert spaces. J. Math. Anal. Appl. 2007;331:506–515.
61. Takahashi W, Nonlinear Functional Analysis. Japan: Yokohama Publishers, Yokohama; 2000.
62. Takahashi W, Toyoda M, Weak convergence theorems for nonexpansive mappings and monotone mappings. J. Optim. Theory Appl. 2003;118:417–428.
63. Tikhonov AN, Solution of incorrectly formulated problems and regularization method. Soviet Math. Dokl. 1963;4:1035–1038.
64. Tran DQ, Muu LD, Nguyen VH, Extragradient algorithms extended to solving equilibrium problems. Rapport de recherche 2006, Université de Namur (Belgique).
65. Van Nguyen TT, Strodiot JJ, Nguyen VH, A bundle method for solving equilibrium problems, Math. Progr. Ser. B (2009);116:529–552.
66. Wittman R, Approximation of fixed points of nonexpansive mappings. Arch. Math. 1992;58:486–491.
67. Xu HK, Kim TH, Convergence of hybrid steepest descent methods for variational inequalities. J. Optim. Theory Appl. 2003;119:185–201.
68. Xu HK, Iterative algorithms for nonlinear operators. J. London Math. Soc. 2002;66:240–256.

69. Xu HK, Viscosity approximation methods for nonexpansive mappings. J. Math. Anal. Appl. 2004;298:279–291.
70. Xu HK, Viscosity methods for hierarchical fixed point approach to variational inequalities. Taiwanese J. Math. 2010;14:463–478.
71. Yamada I, The hybrid steepest descent method for the variational inequality over the intersection of fixed point sets of nonexpansive mappings. In: Butnariu D, Censor Y, Reich S (Editors), Inherently Parallel Algorithm for Feasibility and Optimization and Their Applications; Elsevier; 2001;473–504.
72. Yamada I, Ogura N, Hybrid steepest descent method for the variational inequality problem over the fixed point set of certain quasi-nonexpansive mappings. Numer. Funct. Anal. Optim. 2004;25:619–655.
73. Zeng LC, Yao JC, Strong convergence theorem by an extragradient method for fixed point problems and variational inequality problems. Taiwanese J. Math. 2006;10:1293–1303.
74. Zhao J, He S, Strong convergence of the viscosity approximation process for the split common fixed-point problem of quasi-nonexpansive mapping. J. Appl. Math. 2012;2012:Article ID 438023.

CHAPTER 5

Extragradient Methods for Some Nonlinear Problems

Qamrul Hasan Ansari[a][*][†] and D. R. Sahu[‡]

[*] Aligarh Muslim University, Department of Mathematics, Aligarh, India
[†] King Fahd University of Petroleum & Minerals, Department of Mathematics & Statistics, King Fahd University of Petroleum & Minerals, Dhahran, Saudi Arabia
[‡] Banaras Hindu University, Department of Mathematics, Varanasi, India
[a]Corresponding: qhansari@gmail.com

Contents

Abstract

In this chapter, we discuss several extragradient iterative algorithms for some nonlinear problems, namely fixed point problems, variational inequality problems, hierarchical variational inequality problems and split feasibility problems. We also present and analyze extragradient iterative algorithms for finding a common solution to a fixed point problem and the variational inequality problem.

http://dx.doi.org/10.1016/B978-0-12-804295-3.50005-X

187

5.1. Introduction

Let C be a nonempty closed convex subset of a real Hilbert space H and $\mathscr{F} : C \to H$ be a nonlinear operator. The classical *variational inequality problem* (in short, VIP) is defined as follows:

$$\text{Find } x^* \in C \text{ such that } \langle \mathscr{F}x^*, x - x^* \rangle \geq 0, \quad \text{for all } x \in C. \qquad (5.1.1)$$

It was introduced separately by Fichera [23] and Stampacchia [43] in the early 1960s. The problem (5.1.1) is denoted by $VI(C, \mathscr{F})$. The solution set of the variational inequality problem $VI(C, \mathscr{F})$ is denoted by $\Omega[VI(C, \mathscr{F})]$, i.e.,

$$\Omega[VI(C, \mathscr{F})] = \{x^* \in C : \langle \mathscr{F}x^*, x - x^* \rangle \geq 0 \text{ for all } x \in C\}.$$

It is well known that the following *convex minimization problem:*

$$\text{Find } x^* \in C \text{ such that } g(x^*) = \min\{x \in C : g(x)\},$$

where $g : H \to \mathbb{R}$ is a convex Fréchet differentiable function, can be formulated equivalently as the *variational inequality problem* $VI(C, \nabla g)$ with $\mathscr{F}x = \nabla gx$. The convex minimization problems have a great impact and influence on the development of almost all branches of pure and applied sciences (see, e.g., Refs [7, 19, 20, 40]).

During the last three decades, the theory of variational inequalities has been studied in different directions, namely, existence results, solution methods and applications. It is a powerful unified methodology to study partial differential equations, optimization problems, optimal control problems, mathematical programming problems, financial management problems, industrial management problems, equilibrium problems from traffic network, spatial price equilibrium problems, oligopolistic market equilibrium problems, financial equilibrium problems, migration equilibrium problems, environmental network problems, knowledge network problems, and so on. A large number of iterative methods for solving variational inequality problems $VI(C, \mathscr{F})$ has been studied. For further details on variational inequalities, their generalizations and applications, see Refs. [3, 5, 6, 22, 25–27, 36, 38] and the references therein.

In 1976, Korpelevich [28] introduced the following *extragradient method* for computing a solution of the variational inequality problem $VI(C, A)$:

$$\begin{cases} y_n &= P_C(x_n - \lambda_n A x_n), \\ x_{n+1} &= P_C(x_n - \lambda_n A y_n), \end{cases}$$

for all $n \in \mathbb{N}$, where $A : C \to \mathbb{R}^n$ is a monotone and L-Lipschitz continuous mapping, $0 < \lambda_n < \frac{1}{L}$ and P_C is the metric projection mapping from $\mathbb{R}^n$ onto a nonempty closed convex subset C of $\mathbb{R}^n$. She proved that the sequences generated by this method converge to a solution of the variational inequality problem $VI(C, A)$. Motivated by the idea of an extragradient method, Nadezhkina and Takahashi [35] introduced an iterative algorithm for finding a common element of the set of fixed points of a nonexpansive mapping and the set of solutions of a variational inequality problem in the Hilbert spaces setting. Extragradient methods for finding a common element of the set of solutions of some nonlinear problems (e.g., fixed point problem, split feasibility problem) and the set of solutions of a variational inequality problem in the setting of Hilbert spaces have been extensively studied in recent years (see, e.g., Refs. [11–14, 34, 39, 52], and the references therein).

In this chapter, an attempt is made to present some extensions of the extragradient method in which the conclusion is obtained under inexactness of nonexpansive mappings and which play an important role in the development of perturbation theory of nonexpansive mappings. We include some known results on the geometry of Hilbert spaces as well as some fundamental properties of the metric projection operator for approximating fixed points of nonexpansive mappings and solutions of variational inequality problems. In Section 5.3, we present a projection gradient method for finding the solutions of variational inequality problems. Section 5.4 deals with the extragradient method for finding the fixed points of nonexpansive mappings and the solutions of variational inequality problems. Section 5.5 is devoted to the modified extragradient method for finding a common solution of the fixed point problem for nonexpansive mappings and variational inequality problems. In Section 5.6, we discuss hierarchical variational inequality problems, that is, variational inequality problems defined on the set of fixed points of a mapping. We present implicit and explicit extragradient methods for finding the solutions of hierarchical variational inequality problems. In the last section, we study split feasibility problems and an extragradient method for finding a common solution of split feasibility problems and a fixed point problem for nonexpansive mappings.

5.2. Preliminaries

Throughout this chapter, we denote by $\mathbb{N}$ the set of natural numbers. We use the following notations:

- $\rightharpoonup$ for weak convergence and $\rightarrow$ for strong convergence.
- $\omega_w(\{x_n\})$ denotes the weak ω-limit set of sequence $\{x_n\}$.

Let H be a real Hilbert space whose inner product and norm are denoted by $\langle .,.\rangle$ and $\|\cdot\|$ respectively. For all $x, y, z \in H$ and $\alpha, \beta, \gamma \in [0,1]$ with $\alpha + \beta + \gamma = 1$, we have

$$\|\alpha x + \beta y + \gamma z\|^2 = \alpha \|x\|^2 + \beta \|y\|^2 + \gamma \|z\|^2 - \alpha\beta \|x-y\|^2 - \beta\gamma \|y-z\|^2 - \alpha\gamma \|x-z\|^2.$$

Let C be a nonempty subset of H. A mapping $T : C \to H$ is said to be

(a) *L-Lipschitz* if there exists a nonnegative real number L such that

$$\|Tx - Ty\| \leq L\|x - y\|, \quad \text{for all } x, y \in C;$$

(b) *k-contraction* if there exists a real number $k \in [0,1)$ such that

$$\|Tx - Ty\| \leq k\|x - y\|, \quad \text{for all } x, y \in C;$$

(c) *nonexpansive* if

$$\|Tx - Ty\| \leq \|x - y\|, \quad \text{for all } x, y \in C;$$

(d) η-*strongly monotone* if there exists a positive real number η such that

$$\langle Tx - Ty, x - y \rangle \geq \eta\|x - y\|^2, \quad \text{for all } x, y \in C;$$

(e) ν-*inverse-strongly monotone* if there exists a positive real number ν such that

$$\langle Tx - Ty, x - y \rangle \geq \nu\|Tx - Ty\|^2, \quad \text{for all } x, y \in C.$$

Let C be a nonempty closed convex subset of H and $x \in H$. An element $y_0 \in C$ is said to be a *best approximation* to x if $\|x - y_0\| = d(x,C)$, where $d(x,C) = \inf_{y \in C} \|x - y\|$. The set of all best approximations from x to C is denoted by

$$P_C(x) = \{y \in C : \|x - y\| = d(x,C)\}.$$

This defines a mapping P_C from H onto C and is called the *nearest point projection mapping* or *metric projection mapping* onto C (see Ref. [1]).

We need the following lemma for our main results.

Lemma 5.2.1 [1]. *The metric projection mapping P_C has the following properties:*

(P1) $P_C(x) \in C$ *for all* $x \in H$;

(P2) $\langle x - P_C(x), P_C(x) - y \rangle \geq 0$ *for all* $x \in H, y \in C$;

(P3) $\|P_C(x) - y\|^2 \leq \|x - y\|^2 - \|x - P_C(x)\|^2$ *for all* $x \in H, y \in C$;

(P4) $\langle P_C(x) - P_C(y), x - y \rangle \geq \|P_C(x) - P_C(y)\|^2$ *for all* $x, y \in H$;

(P5) P_C *is a nonexpansive mapping.*

Proposition 5.2.1 [45]. *Let C be a nonempty closed convex subset of a real Hilbert space H. Let $\{x_n\}$ be a sequence in H such that $\|x_{n+1} - v\| \leq \|x_n - v\|$ for all $v \in C$ and $n \in \mathbb{N}_0 := \mathbb{N} \cup \{0\}$. Then the sequence $\{P_C(x_n)\}$ converges strongly to some $z \in C$.*

Proof. Set $u_n = P_C(x_n)$. Let $m > n$ and note

$$\|u_n - x_m - (u_m - x_m)\|^2 + \|u_n - x_m + (u_m - x_m)\|^2 = 2\|u_n - x_m\|^2 + 2\|u_m - x_m\|^2.$$

Hence

$$
\begin{aligned}
\|u_n - u_m\|^2 &= 2\|u_n - x_m\|^2 + 2\|u_m - x_m\|^2 - \|u_n - x_m + (u_m - x_m)\|^2 \\
&= 2\|u_n - x_m\|^2 + 2\|u_m - x_m\|^2 - 4\|x_m - (u_m + u_n)/2\|^2 \\
&\le 2\|u_n - x_m\|^2 + 2\|u_m - x_m\|^2 - 4\|x_m - u_m\|^2 \\
&= 2\|x_m - u_n\|^2 - 2\|x_m - u_m\|^2 \\
&\le 2\|x_{m-1} - u_n\|^2 - 2\|x_m - u_m\|^2 \\
&\cdots \\
&\le 2\|x_n - u_n\|^2 - 2\|x_m - u_m\|^2, \quad\quad\quad (5.2.2)
\end{aligned}
$$

which gives that

$$
\|x_m - u_m\| \le \|x_n - u_n\|.
$$

Thus $\{\|x_n - u_n\|\}$ is monotonically decreasing and hence $\lim_{n \to \infty} \|x_n - u_n\|$ exists. Taking the limit as $m, n \to \infty$ on (5.2.2), we get $\|u_n - u_m\| \to 0$ as $m, n \to \infty$. Hence $\{u_n\}$ is a Cauchy sequence in C. Therefore $\{P_C(x_n)\}$ converges strongly to some $z \in C$.   □

Let C be a nonempty subset of a Banach space X. We denote by $\mathscr{B}(C)$ the collection of all bounded subsets of C. Let $T_1, T_2 : C \to X$ be mappings. The deviation between T_1 and T_2 on $B \in \mathscr{B}(C)$ [42], denoted by $\mathscr{D}_B(T_1, T_2)$, is defined by

$$
\mathscr{D}_B(T_1, T_2) = \sup\{\|T_1 x - T_2 x\| : x \in B\}.
$$

A Banach space X is said to satisfy the *Opial condition* (see Ref. [1]) if, for each sequence $\{x_n\}$ in X which converges weakly to a point $x \in X$, we have

$$
\liminf_{n \to \infty} \|x_n - x\| < \liminf_{n \to \infty} \|x_n - y\|, \quad \text{for all } y \in X,\, y \ne x.
$$

Note that "$\liminf_{n \to \infty}$" can be replaced by "$\limsup_{n \to \infty}$". It is well known that every Hilbert space enjoys the Opial condition (see Ref. [1]).

The following lemma is a consequence of the Opial condition.

Proposition 5.2.2. *Let C be a nonempty closed convex subset of a Hilbert space H and $T : C \to C$ be a nonexpansive mapping. Then $I - T$ is demi-closed at zero, that is, if $\{x_n\}$ is a sequence in C, converges weakly to x and $\{(I - T)x_n\}$ converges strongly to zero, then $(I - T)x = 0$.*

Proof. Let $\{x_n\}$ be a sequence in C such that it converges weakly to x and $\{(I - T)x_n\}$ converges strongly to zero. Suppose on the contrary that $x \ne Tx$. Then by the Opial condition, we have

$$
\begin{aligned}
\limsup_{n\to\infty} \|x_n - x\| \;&<\; \limsup_{n\to\infty} \|x_n - Tx\| \\
&\leq\; \limsup_{n\to\infty} (\|x_n - Tx_n\| + \|Tx_n - Tx\|) \\
&\leq\; \limsup_{n\to\infty} \|x_n - x\|,
\end{aligned}
$$

a contradiction. Thus $(I - T)x = 0$.   □

Proposition 5.2.3 [31]. *Let C be a nonempty closed convex subset of a real Hilbert space H. Let $A : C \to H$ be a monotone mapping and weakly continuous along line segments, that is, $Fx + ty \to Fx$ weakly as $t \to 0$. Then the variational inequality problem $VI(C,A)$:*

$$
\text{find } x^* \in C \text{ such that } \langle Ax^*, x - x^* \rangle \geq 0, \quad \text{for all } x \in C,
$$

is equivalent to the dual variational inequality problem $DVI(C,A)$:

$$
\text{find } x^* \in C \text{ such that } \langle Ax, x - x^* \rangle \geq 0, \quad \text{for all } x \in C.
$$

Proposition 5.2.4. *Let C be a nonempty closed convex subset of a real Hilbert space H and $A : C \to H$ be a monotone and Lipschitz continuous mapping. For $\lambda > 0$, define $T = P_C(I - \lambda A)$. Then $I - T$ is demi-closed at zero, that is, if $\{x_n\}$ is a sequence in C converges weakly to z and $\{(I - T)x_n\}$ converges strongly to zero, then $(I - T)z = 0$.*

Proof. Note A is monotone and continuous. From Proposition 5.2.3, we have

$$
x^* \in \Omega[VI(C,A)] \;\Leftrightarrow\; \langle Ax, x - x^* \rangle \geq 0, \quad \text{for all } x \in C.
$$

Let $\{x_n\}$ be a sequence in C that converges weakly to z and $\{(I - T)x_n\}$ converges strongly to zero. We now show that $(I - T)z = 0$. Let $x \in C$. Set $y_n = Tx_n$. From (P2), we have

$$
\langle x_n - \lambda Ax_n - y_n, x - y_n \rangle = \langle x_n - \lambda Ax_n - P_C(I - \lambda A)x_n, x - P_C(I - \lambda A)x_n \rangle \leq 0,
$$

for all $n \in \mathbb{N}$. Then by the monotonicity of A, we have

$$
\begin{aligned}
\langle \lambda Ax, x_n - x \rangle &= \lambda \langle Ax_n, x_n - x \rangle - \lambda \langle Ax_n - Ax, x_n - x \rangle \\
&\leq \lambda \langle Ax_n, x_n - x \rangle \\
&= \lambda \langle Ax_n, x_n - y_n \rangle + \lambda \langle Ax_n, y_n - x \rangle \\
&= \lambda \langle Ax_n, x_n - y_n \rangle - \lambda \langle Ax_n, x - y_n \rangle \\
&= \lambda \langle Ax_n, x_n - y_n \rangle + \langle x_n - \lambda Ax_n - y_n, x - y_n \rangle - \langle x_n - y_n, x - y_n \rangle \\
&\leq \lambda \langle Ax_n, x_n - y_n \rangle - \langle x_n - y_n, x - y_n \rangle \\
&\leq (\lambda \|Ax_n\| + \|y_n - x\|) \|x_n - y_n\|,
\end{aligned}
$$

which implies that

$$\langle Ax, x_n - x \rangle \leq (\|Ax_n\| + \frac{1}{\lambda}\|y_n - x\|)\|x_n - y_n\|.$$

Since $\{Ax_n\}$ is bounded, $x_n - y_n \to 0$ and $x_n \rightharpoonup z$, we have

$$\langle Ax, z - x \rangle \leq 0.$$

Hence $z \in \Omega[VI(C,A)]$. Therefore $z = P_C(I - \lambda A)z$.   □

Proposition 5.2.5. *Let C be a nonempty closed convex subset of a real Hilbert space H and $A : C \to H$ be a monotone and L-Lipschitz continuous mapping. Let $\{x_n\}$ be a bounded sequence in C and $\{y_n\}$ be a sequence in C defined by $y_n = P_C(I - \lambda_n A)x_n$, where $\{\lambda_n\}$ is a sequence in $(0, \infty)$ satisfying $0 < a \leq \lambda_n \leq b < 1/L$ for all $n \in \mathbb{N}_0$. Suppose that $\{\|x_n - y_n\|\}$ converges to zero and that there exists a subsequence $\{x_{n_i}\}$ of $\{x_n\}$ that converges weakly to a point $p \in C$. Then $p \in \Omega[VI(C,A)]$.*

Proof. Since $\{\lambda_{n_i}\}$ is in $[a,b]$, there exists a subsequence $\{\lambda_{n_{i_k}}\}$ of $\{\lambda_{n_i}\}$ such that $\lim_{k \to \infty} \lambda_{n_{i_k}} = \lambda \in [a,b]$. We may assume, without loss of generality, that $\lim_{i \to \infty} \lambda_{n_i} = \lambda$. Define $T = P_C(I - \lambda A)$. Note that

$$\|y_n - Tx_n\| = \|P_C(I - \lambda_n A)x_n - P_C(I - \lambda A)x_n\| \leq |\lambda_n - \lambda| \, \|Ax_n\|.$$

Clearly, $x_{n_i} - Tx_{n_i} \to 0$ as $i \to \infty$, and we obtain from Proposition 5.2.4 that $p \in \Omega[VI(C,A)]$.   □

The following technical results will be used to establish the results in this chapter.

Lemma 5.2.2 [1, Lemma 6.1.5]. *Let $\{a_n\}$, $\{\beta_n\}$ and $\{\delta_n\}$ be sequences of nonnegative real numbers satisfying the inequality:*

$$a_{n+1} \leq (1 + \beta_n)a_n + \delta_n, \quad \text{for all } n \in \mathbb{N}.$$

If $\sum_{n=1}^{\infty} \beta_n < \infty$ and $\sum_{n=1}^{\infty} \delta_n < \infty$, then $\lim_{n \to \infty} a_n$ exists.

Lemma 5.2.3 [1, Theorem 2.3.13]. *Let X be a uniformly convex Banach space and $\{t_n\}$ be a sequence of real numbers in $(0,1)$ bounded away from 0 and 1. Let $\{x_n\}$ and $\{y_n\}$ be sequences in X such that*

$$\limsup_{n \to \infty} \|x_n\| \leq a, \ \limsup_{n \to \infty} \|y_n\| \leq a \text{ and } \limsup_{n \to \infty} \|t_n x_n + (1 - t_n)y_n\| = a,$$

for some $a \geq 0$. Then $\lim_{n \to \infty} \|x_n - y_n\| = 0$.

Lemma 5.2.4 [44]. *Let $\{x_n\}$ and $\{y_n\}$ be bounded sequences in a Banach space X and $\{\beta_n\}$ be a sequence in $[0,1]$ with $0 < \liminf_{n\to\infty} \beta_n \leq \limsup_{n\to\infty} \beta_n < 1$. Suppose that*

$$x_{n+1} = (1 - \beta_n)x_n + \beta_n y_n, \quad \text{for all } n \in \mathbb{N}_0$$

and that $\limsup_{n\to\infty}(\|y_{n+1} - y_n\| - \|x_{n+1} - x_n\|) \leq 0$. Then $\lim_{n\to\infty} \|y_n - x_n\| = 0$.

Lemma 5.2.5 [33]. *Let $\{a_n\}$ and $\{c_n\}$ be sequences of nonnegative real numbers and $\{b_n\}$ be a sequence in $\mathbb{R}$ satisfying the following condition:*

$$a_{n+1} \leq (1 - \alpha_n)a_n + b_n + c_n, \quad \text{for all } n \in \mathbb{N},$$

where $\{\alpha_n\}$ is a sequence in $(0,1]$. Assume that $\sum_{n=1}^{\infty} c_n < \infty$. Then the following statements hold:

(a) *If $b_n \leq K\alpha_n$ for all $n \in \mathbb{N}$ and for some $K \geq 0$, then*

$$a_{n+1} \leq \delta_n a_1 + (1 - \delta_n)K + \sum_{j=1}^{n} c_j, \quad \text{for all } n \in \mathbb{N},$$

where $\delta_n = \prod_{j=1}^{n}(1 - \alpha_j)$ and hence $\{a_n\}$ is bounded.

(b) *If $\sum_{n=1}^{\infty} \alpha_n = \infty$ and $\limsup_{n\to\infty}(b_n/\alpha_n) \leq 0$, then $\{a_n\}_{n=1}^{\infty}$ converges to zero.*

5.3. Projection Gradient Method

It is important to note that fixed point theory and the theory of variational inequalities have important roles in the study of many diverse disciplines, for instance partial differential equations, optimal control, optimization, mathematical programming, mechanics, finance, etc. (see, for example, Refs. [29, 47, 51] and references therein). The relationship between $VI(\mathscr{F}, C)$ and a fixed point problem can be made through the property (P2) of the projection operator P_C as follows:

Theorem 5.3.1. *Let C be a nonempty closed convex subset of a real Hilbert space H and $\mathscr{F} : C \to H$ be a given operator. Then the following statements are equivalent:*

(a) $x^* \in \Omega[VI(C, \mathscr{F})]$.

(b) $x^* \in Fix(P_C(I - \mu\mathscr{F}))$ *for any $\mu > 0$.*

Proof. (a) $\Rightarrow$ (b). Suppose that $x^* \in C$ is a solution of $VI(C, \mathscr{F})$, i.e.,

$$\langle \mathscr{F}x^*, y - x^* \rangle \geq 0, \quad \text{for all } y \in C. \tag{5.3.3}$$

Let $\mu > 0$ such that $P_C(I - \mu\mathscr{F}) : C \to C$. Multiplying (5.3.3) by $-\mu$ and then adding

$\langle x^*, y - x^* \rangle \geq 0$ to both sides, we get

$$\langle x^* - \mu \mathscr{F} x^*, y - x^* \rangle \leq \langle x^*, y - x^* \rangle, \quad \text{for all } y \in C.$$

From (P2), we get $x^* = P_C(I - \mu \mathscr{F})x^*$.

(b) $\Rightarrow$ (a). Suppose that $\mu > 0$ and x^* is the fixed point of the mapping $P_C(I - \mu \mathscr{F})$: $C \to C$. It is easy to see from (P2) that $x^* \in C$ is a solution of $VI(C, \mathscr{F})$. $\square$

This equivalence formulation is useful for existence, uniqueness and computation of solutions of the variational inequality problem $VI(C, \mathscr{F})$. In particular, the Banach contraction principle guarantees that $VI(C, \mathscr{F})$ has a unique solution x^* and the sequence of the Picard iteration process converges strongly to x^*. In fact, we have the following.

Theorem 5.3.2 (Projected Gradient Method). [24, 51] *Let C be a nonempty closed convex subset of a real Hilbert space H and $\mathscr{F} : C \to H$ be a κ-Lipschitz continuous and η-strongly monotone mapping. Let μ be a positive constant such that $\mu < 2\eta / \kappa^2$. Then*

(a) *$P_C(I - \mu \mathscr{F}) : C \to C$ is $\sqrt{1 - 2\mu\eta + \mu^2\kappa^2}$-contraction.*

(b) *There exists a unique solution $x^* \in C$ of the variational inequality problem $VI(C, \mathscr{F})$ and the sequence $\{x_n\}$ generated by the Picard iteration process, given by*

$$x_{n+1} = P_C(I - \mu \mathscr{F})x_n, \quad n \in \mathbb{N}_0 = \mathbb{N} \cup \{0\},$$

converges strongly to x^*.

Proof. Let $x, y \in C$. Then

$$\begin{aligned}
\|P_C(I - \mu\mathscr{F})x - P_C(I - \mu\mathscr{F})y\|^2 &\leq \|(I - \mu\mathscr{F})x - (I - \mu\mathscr{F})y\|^2 \\
&= \|x - y\|^2 - 2\mu\langle \mathscr{F}x - \mathscr{F}y, x - y \rangle \\
&\quad + \mu^2\|\mathscr{F}x - \mathscr{F}y\|^2 \\
&\leq (1 - 2\mu\eta + \mu^2\kappa^2)\|x - y\|^2.
\end{aligned}$$

Part (b) follows from the Banach contraction principle. $\square$

Motivated by the nonexpensiveness of P_C in Theorem 5.3.2, Yamada [47] introduced the following *hybrid steepest descent method* for solving $VI_{Fix(T)}(H, \mathscr{F})$:

$$u_{n+1} = Tu_n - \lambda_n \mu \mathscr{F} Tu_n, \quad \text{for all } n \in \mathbb{N},$$

where $0 < \mu < 2\eta / \kappa^2$, $T : H \to H$ is a nonexpansive mapping with $Fix(T) \neq \emptyset$ and $\mathscr{F} : H \to H$ is κ-Lipschitz continuous and η-strongly monotone. Yamada proved that the hybrid steepest descent method converges strongly to a unique solution $x^* \in Fix(T)$ of the variational inequality problem $VI_{Fix(T)}(H, \mathscr{F})$. The hybrid steepest descent method (HSDM) has been extensively studied in recent years (see, e.g., Refs. [10, 41, 42], and references therein).

5.4. Extragradient Method for Nonexpansive Mappings and Variational Inequalities

Let C be a nonempty closed convex subset of a real Hilbert space H and $A : C \to H$ be a monotone operator. For $\lambda > 0$, define a mapping $E_{\lambda}^A \equiv E_{\lambda} : C \to C$ by

$$E_{\lambda}(x) = P_C(x - \lambda A P_C(x - \lambda Ax)), \quad \text{for all } x \in C. \tag{5.4.4}$$

E_{λ} is called an *extragradient mapping* associated with $\lambda > 0$ and an iterative algorithm which involves extragradient mappings will be called the *extragradient iterative method*. Thus the notion of extragradient mapping suggests immediately a Mann-like method for computation of the elements of $\Omega[VI(C,A)]$ as follows:

$$x_{n+1} = (1 - \alpha_n)x_n + \alpha_n E_{\lambda_n}(x_n), \quad n \in \mathbb{N}_0, \tag{5.4.5}$$

where $\{\alpha_n\}$ is a sequence in $(0,1)$ and $\{\lambda_n\}$ is a sequence in $(0, \infty)$ satisfying suitable conditions.

In this section, we deal with computation of elements of $Fix(S) \cap \Omega[VI(C,A)]$ by the following *extragradient iterative method:*

$$x_{n+1} = (1 - \alpha_n)x_n + \alpha_n S_n E_{\lambda_n}(x_n), \quad n \in \mathbb{N}_0, \tag{5.4.6}$$

where $S : C \to C$ is a nonexpansive mapping such that $Fix(S) \cap \Omega[VI(C,A)] \neq \emptyset$ and $\{S_n\}$ is a sequence of nonexpansive mappings from C into itself such that $\{S_n\}$ converges to S in some sense.

We begin with some properties of monotone mappings and extragradient mappings generated by monotone mappings.

Proposition 5.4.1. *Let C be a nonempty closed convex subset of a real Hilbert space H. Let $A : C \to H$ be monotone and L-Lipschitz continuous such that $\Omega[VI(C,A)] \neq \emptyset$. For $\lambda > 0$, let $E_{\lambda} : C \to H$ be an extragradient mapping defined by (5.4.4). Then the following assertions hold:*

(a) *$I - \lambda A$ is $(1 + \lambda L)$-Lipschitz continuous.*

(b) *For $x \in C$ with $y = P_C(x - \lambda Ax)$ and $v \in \Omega[VI(C,A)]$, we have*

$$\|E_{\lambda}(x) - v\|^2 \leq \|x - v\|^2 - (1 - \lambda L)\|x - y\|^2 - (1 - \lambda L)\|y - E_{\lambda}(x)\|^2.$$

Proof. (a) For $x, y \in C$, we have

$$
\begin{aligned}
\|(I - \lambda A)x - (I - \lambda A)y\|^2 &= \|x - y - \lambda(Ax - Ay)\|^2 \\
&= \|x - y\|^2 + \lambda^2\|Ax - Ay\|^2 - 2\lambda\langle Ax - Ay, x - y\rangle \\
&\leq \|x - y\|^2 + \lambda^2 L^2\|x - y\|^2 = (1 + \lambda^2 L^2)\|x - y\|^2,
\end{aligned}
$$

which implies that

$$
\|(I - \lambda A)x - (I - \lambda A)y\| \leq (1 + \lambda L)\|x - y\|.
$$

It can also be shown without the monotonicity of A.

(b) Let $x \in C$ and $v \in \Omega[VI(C, A)]$. Set $y = P_C(x - \lambda Ax)$ and $z = P_C(x - \lambda Ay)$. Then

$$
\langle Av, y - v\rangle \geq 0.
$$

By the monotonicity of A, we have

$$
\langle Ay - Av, y - v\rangle \geq 0.
$$

From Lemma 5.2.1 (P3), we have

$$
\begin{aligned}
&\|P_C(x - \lambda Ay) - v\|^2 \qquad\qquad\qquad\qquad\qquad\qquad\qquad\qquad (5.4.7)\\
&\leq \|x - \lambda Ay - v\|^2 - \|x - \lambda Ay - P_C(x - \lambda Ay)\|^2 \\
&= \|x - v - \lambda Ay\|^2 - \|x - z - \lambda Ay\|^2 \\
&\leq \|x - v\|^2 - 2\lambda\langle Ay, x - v\rangle + \lambda^2\|Ay\|^2 - \|x - z\|^2 + 2\lambda\langle Ay, x - z\rangle - \lambda^2\|Ay\|^2 \\
&= \|x - v\|^2 - \|x - z\|^2 + 2\lambda\langle Ay, v - z\rangle \\
&= \|x - v\|^2 - \|x - z\|^2 + 2\lambda[\langle Ay, v - y\rangle + \langle Ay, y - z\rangle] \\
&= \|x - v\|^2 - \|x - z\|^2 + 2\lambda[\langle Ay - Av, v - y\rangle + \langle Av, v - y\rangle + \langle Ay, y - z\rangle] \\
&\leq \|x - v\|^2 - \|x - z\|^2 + 2\lambda\langle Ay, y - z\rangle. \qquad\qquad\qquad\qquad (5.4.8)
\end{aligned}
$$

Substituting x by $x - \lambda Ax$ and y by z in (P2), we get

$$
\langle x - \lambda Ax - P_C(x - \lambda Ax), z - P_C(x - \lambda Ax)\rangle \leq 0,
$$

which gives that

$$
\langle x - \lambda Ax - y, z - y\rangle \leq 0. \qquad\qquad\qquad\qquad (5.4.9)
$$

Thus, from (5.4.9), we have

$$
\begin{aligned}
\langle x - \lambda Ay - y, z - y\rangle &= \langle x - \lambda Ax - y, z - y\rangle + \langle \lambda Ax - \lambda Ay, z - y\rangle \\
&\leq \langle \lambda Ax - \lambda Ay, z - y\rangle \\
&\leq \lambda L\|x - y\|\,\|z - y\| \\
&\leq \frac{\lambda L}{2}\left(\|x - y\|^2 + \|y - z\|^2\right). \qquad\qquad (5.4.10)
\end{aligned}
$$

Hence, from (5.4.8) and (5.4.10), we get

$$
\begin{aligned}
\|z - v\|^2 &\le \|x - v\|^2 - \|x - y + (y - z)\|^2 + 2\lambda \langle Ay, y - z \rangle \\
&= \|x - v\|^2 - \|x - y\|^2 - \|y - z\|^2 - 2\langle x - y, y - z \rangle + 2\lambda \langle Ay, y - z \rangle \\
&= \|x - v\|^2 - \|x - y\|^2 - \|y - z\|^2 - 2\langle x - \lambda Ay - y, y - z \rangle \\
&\le \|x - v\|^2 - \|x - y\|^2 - \|y - z\|^2 + \lambda L(\|x - y\|^2 + \|y - z\|^2) \\
&= \|x - v\|^2 - (1 - \lambda L)\|x - y\|^2 - (1 - \lambda L)\|y - z\|^2.
\end{aligned}
$$

$\square$

We now establish a convergence theorem for extragradient method (5.4.6) for computation of an element of $Fix(S) \cap \Omega[VI(C,A)]$.

Theorem 5.4.1. *Let C be a nonempty closed convex subset of a real Hilbert space H and $A : C \to H$ be a monotone and L-Lipschitz continuous mapping. Let $S : C \to C$ be a nonexpansive mapping such that $Fix(S) \cap \Omega[VI(C,A)] \ne \emptyset$. Let $\{S_n\}$ be a sequence of nonexpansive mappings from C into itself. For given $x_0 \in C$ arbitrarily, define a sequence $\{x_n\}$ in C iteratively by extragradient method (5.4.6), where $\{\alpha_n\}$ is a sequence in $(0,1)$ and $\{\lambda_n\}$ is a sequence in $(0,\infty)$ satisfying the following conditions:*

(i) *$0 < c \le \alpha_n \le d < 1$ for all $n \in \mathbb{N}_0$,*

(ii) *$0 < a \le \lambda_n \le b < 1/L$ for all $n \in \mathbb{N}_0$,*

(iii) *$\sum_{n=0}^{\infty} \|S_n v - S v\| < \infty$ for all $v \in Fix(S)$,*

(iv) *$\lim_{n \to \infty} \mathscr{D}_B(S_n, S) = 0$ for all $B \in \mathscr{B}(C)$.*

Then the sequence $\{x_n\}$ converges weakly to an element u in $Fix(S) \cap \Omega[VI(C,A)]$.

Proof. Let $v \in Fix(S) \cap \Omega[VI(C,A)]$. Set $y_n = P_C(x_n - \lambda_n A x_n)$ and $z_n = P_C(x_n - \lambda_n A y_n)$ for all $n \in \mathbb{N}_0$. From Proposition 5.4.1, we have

$$
\begin{aligned}
\|E_{\lambda_n}(x_n) - v\|^2 &\le \|x_n - v\|^2 - (1 - \lambda_n L)\|x_n - y_n\|^2 - (1 - \lambda_n L)\|y_n - E_{\lambda_n}(x_n)\|^2 \\
&\le \|x_n - v\|^2, \quad \text{for all } n \in \mathbb{N}_0.
\end{aligned}
$$

From (5.4.6), we have

$$
\begin{aligned}
\|x_{n+1} - v\| &\le (1 - \alpha_n)\|x_n - v\| + \alpha_n \|S_n E_{\lambda_n}(x_n) - v\| \\
&\le (1 - \alpha_n)\|x_n - v\| + \alpha_n(\|S_n E_{\lambda_n}(x_n) - S_n v\| + \|S_n v - S v\|) \\
&\le \|x_n - v\| + \|S_n v - S v\|, \quad \text{for all } n \in \mathbb{N}_0.
\end{aligned}
$$

Note $\sum_{n=0}^{\infty} \|S_n v - S v\| < \infty$. Hence, from Lemma 5.2.2, we obtain that $\lim_{n \to \infty} \|x_n - v\| =:$

δ exists and hence $\{x_n\}$ is bounded. From (5.4.6), we have

$$
\begin{aligned}
\|x_{n+1} - v\|^2 &= \|(1 - \alpha_n)x_n + \alpha_n S_n E_{\lambda_n}(x_n) - v\|^2 \\
&\leq (1 - \alpha_n)\|x_n - v\|^2 + \alpha_n \|S_n E_{\lambda_n}(x_n) - v\|^2 \\
&\leq (1 - \alpha_n)\|x_n - v\|^2 + \alpha_n (\|S_n E_{\lambda_n}(x_n) - S_n v\| + \|S_n v - Sv\|)^2 \\
&\leq (1 - \alpha_n)\|x_n - v\|^2 + \alpha_n (\|E_{\lambda_n}(x_n) - v\| + \|S_n v - Sv\|)^2 \\
&\leq (1 - \alpha_n)\|x_n - v\|^2 + \alpha_n \|E_{\lambda_n}(x_n) - v\|^2 + M_1 \|S_n v - Sv\| \\
&\leq (1 - \alpha_n)\|x_n - v\|^2 + \alpha_n [\|x_n - v\|^2 - (1 - \lambda_n L)\|x_n - y_n\|^2 \\
&\quad - (1 - \lambda_n L)\|y_n - E_{\lambda_n}(x_n)\|^2] + M_1 \|S_n v - Sv\| \\
&\leq \|x_n - v\|^2 - \alpha_n (1 - \lambda_n L)\|x_n - y_n\|^2 + M_1 \|S_n v - Sv\|,
\end{aligned}
$$

for some constant $M_1 > 0$. Thus

$$
\alpha_n (1 - \lambda_n L)\|x_n - y_n\|^2 \leq \|x_n - v\|^2 - \|x_{n+1} - v\|^2 + M_1 \|S_n v - Sv\|.
$$

Since $\lim_{n \to \infty} \|x_n - v\| = \delta$ exists, we have $\|y_n - x_n\| \to 0$ as $n \to \infty$. Note

$$
\begin{aligned}
\|y_n - z_n\| &\leq \|(x_n - \lambda_n A x_n) - (x_n - \lambda_n A y_n)\| = \lambda_n \|A x_n - A y_n\| \\
&\leq \lambda_n L \|x_n - y_n\| \to 0 \text{ as } n \to \infty,
\end{aligned}
$$

and

$$
\begin{aligned}
\|S_n z_n - v\| &\leq \|S_n z_n - S_n v\| + \|S_n v - v\| \\
&\leq \|z_n - v\| + \|S_n v - v\| \\
&\leq \|x_n - v\| + \|S_n v - v\|,
\end{aligned}
$$

and we have

$$
\limsup_{n \to \infty} \|S_n z_n - v\| \leq \delta.
$$

Again, from (5.4.6), we have

$$
\limsup_{n \to \infty} \|(1 - \alpha_n)x_n + \alpha_n S_n z_n - v\| = \limsup_{n \to \infty} \|x_{n+1} - v\| = \delta.
$$

By Lemma 5.2.3, we obtain $\lim_{n \to \infty} \|x_n - S_n z_n\| = 0$. Hence, for $B = \{z_n\}$, we have

$$
\begin{aligned}
\|x_n - S z_n\| &\leq \|x_n - S_n z_n\| + \|S_n z_n - S z_n\| \\
&\leq \|x_n - S_n z_n\| + \mathscr{D}_B(S_n, S) \to 0 \text{ as } n \to \infty,
\end{aligned}
$$

which implies that

$$
\begin{aligned}
\|x_n - S x_n\| &\leq \|x_n - S z_n\| + \|S z_n - S x_n\| \\
&\leq \|x_n - S z_n\| + \|z_n - x_n\| \to 0 \text{ as } n \to \infty.
\end{aligned}
$$

Since $\{x_n\}$ is bounded, we see that $\{x_n\}$ has a weakly convergent subsequence. We show that $\omega_w(\{x_n\})$ is singleton. Assume that $\{x_{n_i}\}$ converges weakly to a point $p \in C$ and that $\{x_{n_j}\}$ converges weakly to a point $q \in C$ with $p \neq q$. From Propositions 5.2.2 and 5.2.5, we have $p \in Fix(S) \cap \Omega[VI(C,A)]$ and $q \in Fix(S) \cap \Omega[VI(C,A)]$. Also $\lim_{n\to\infty} \|x_n - p\|$ and $\lim_{n\to\infty} \|x_n - q\|$ exist. Since H satisfies the Opial condition, we have

$$\lim_{n\to\infty} \|x_n - p\| = \lim_{i\to\infty} \|x_{n_i} - p\| < \lim_{i\to\infty} \|x_{n_i} - q\| = \lim_{n\to\infty} \|x_n - q\|,$$

$$\lim_{n\to\infty} \|x_n - q\| = \lim_{k\to\infty} \|x_{n_j} - q\| < \lim_{j\to\infty} \|x_{n_j} - p\| = \lim_{n\to\infty} \|x_n - p\|,$$

a contradiction. Hence $p = q$, so $\omega_w(\{x_n\})$ is singleton. Thus $\{x_n\}$ converges weakly to p. $\qquad\square$

Corollary 5.4.1 [35, Theorem 3.1]. *Let C be a nonempty closed convex subset of a real Hilbert space H. Let $A : C \to H$ be a monotone and L-Lipschitz continuous mapping and $S : C \to C$ be a nonexpansive mapping such that $Fix(S) \cap \Omega[VI(C,A)] \neq \emptyset$. For given $x_0 \in C$ arbitrarily, define a sequence $\{x_n\}$ in C iteratively by the extragradient method:*

$$x_{n+1} = (1 - \alpha_n)x_n + \alpha_n S E_{\lambda_n}(x_n), \quad n \in \mathbb{N}_0,$$

where $\{\alpha_n\}$ is a sequence in $(0,1)$ and $\{\lambda_n\}$ is a sequence in $(0,\infty)$ satisfying the following conditions:

(i) $0 < c \leq \alpha_n \leq d < 1$ *for all $n \in \mathbb{N}_0$,*

(ii) $0 < a \leq \lambda_n \leq b < 1/L$ *for all $n \in \mathbb{N}_0$.*

Then the sequence $\{x_n\}$ converges weakly to an element u in $Fix(S) \cap \Omega[VI(C,A)]$, where $u = \lim_{n\to\infty} P_{Fix(S)\cap\Omega[VI(C,A)]}(x_n)$.

Proof. Set $u_n = P_{Fix(S)\cap\Omega[VI(C,A)]}(x_n)$. Since $\|x_{n+1} - v\| \leq \|x_n - v\|$ for all $v \in Fix(S) \cap \Omega[VI(C,A)]$ and $n \in \mathbb{N}_0$, it follows from Proposition 5.2.1 that the sequence $\{u_n\}$ converges strongly to some $u \in C$. Note that

$$\langle p - u_n, u_n - x_n \rangle \geq 0.$$

Thus $\langle p - u, u - p \rangle \geq 0$, and hence $p = u$. Therefore $\{x_n\}$ converges weakly to an element u in $Fix(S) \cap \Omega[VI(C,A)]$, where $u = \lim_{n\to\infty} P_{Fix(S)\cap\Omega[VI(C,A)]}(x_n)$. $\qquad\square$

Remark 5.4.1. The convergence analysis of extragradient method (5.4.5) for computation of an element of $\Omega[VI(C,A)]$ can be seen in Corollary 5.4.1 when $S \equiv I$.

5.5. Modified Extragradient Method for Nonexpansive Mappings and Variational Inequalities

In this section, we deal with the computation of a unique element of $Fix(S) \cap \Omega[VI(C,A)]$ by a modified extragradient method (5.5.11).

Theorem 5.5.1. *Let C be a nonempty closed convex subset of a real Hilbert space H. Let $A : C \to H$ be a monotone and L-Lipschitz continuous mapping and $S : C \to C$ be a nonexpansive mapping such that $Fix(S) \cap \Omega[VI(C,A)] \neq \emptyset$. Let $\{S_n\}$ be a sequence of nonexpansive mappings from C into itself. For fixed $u \in C$ and given $x_0 \in C$ arbitrarily, define a sequence $\{x_n\}$ in C iteratively by*

$$x_{n+1} = \alpha_n u + \beta_n x_n + \gamma_n S_n E_{\lambda_n}(x_n), \quad n \in \mathbb{N}_0, \tag{5.5.11}$$

where $\{\alpha_n\}$, $\{\beta_n\}$ and $\{\gamma_n\}$ are sequences in $[0,1]$ and $\{\lambda_n\}$ is a sequence in $(0,\infty)$ satisfying the following conditions:

(i) $\alpha_n + \beta_n + \gamma_n = 1$ *for all* $n \in \mathbb{N}_0$;

(ii) $\lim_{n\to\infty} \alpha_n = 0$, $\sum_{n=0}^{\infty} \alpha_n = \infty$;

(iii) $0 < \liminf_{n\to\infty} \beta_n \leq \limsup_{n\to\infty} \beta_n < 1$;

(iv) $0 < a \leq \lambda_n \leq b < 1/L$ *for all* $n \in \mathbb{N}_0$;

(v) $\sum_{n=0}^{\infty} \|S_n v - Sv\| < \infty$ *for all* $v \in Fix(S)$;

(vi) $\lim_{n\to\infty} \mathscr{D}_B(S_{n+1}, S_n) = 0$ *and* $\lim_{n\to\infty} \mathscr{D}_B(S_n, S) = 0$ *for all* $B \in \mathscr{B}(C)$.

Then the sequence $\{x_n\}$ generated by (5.5.11) converges strongly to $P_{Fix(S) \cap \Omega[VI(C,A)]}(u)$.

Proof. We divide the proof into the following several steps.

STEP I. $\{x_n\}$ *is bounded.*

Let $v \in Fix(S) \cap \Omega[VI(C,A)]$. Set $y_n = P_C(x_n - \lambda_n A x_n)$ and $z_n = P_C(x_n - \lambda_n A y_n)$ for all $n \in \mathbb{N}$. From Proposition 5.4.1, we have

$$\begin{aligned}
\|E_{\lambda_n}(x_n) - v\|^2 &\leq \|x_n - v\|^2 - (1 - \lambda_n L)\|x_n - y_n\|^2 - (1 - \lambda_n L)\|y_n - E_{\lambda_n}(x_n)\|^2 \\
&\leq \|x_n - v\|^2, \quad \text{for all } n \in \mathbb{N}_0.
\end{aligned}$$

Hence

$$\begin{aligned}
\|S_n E_{\lambda_n}(x_n) - v\| &\leq \|S_n E_{\lambda_n}(x_n) - S_n v\| + \|S_n v - Sv\| \\
&\leq \|E_{\lambda_n}(x_n) - v\| + \|S_n v - Sv\| \\
&\leq \|x_n - v\| + \|S_n v - Sv\|, \quad \text{for all } n \in \mathbb{N}_0.
\end{aligned}$$

For $n \in \mathbb{N}_0$, from (5.5.11), we have

$$
\begin{aligned}
\|x_{n+1} - v\| &= \|\alpha_n u + \beta_n x_n + \gamma_n S_n E_{\lambda_n}(x_n) - v\| \\
&\leq \alpha_n \|u - v\| + \beta_n \|x_n - v\| + \gamma_n \|S_n E_{\lambda_n}(x_n) - v\| \\
&\leq \alpha_n \|u - v\| + \beta_n \|x_n - v\| + \gamma_n(\|x_n - v\| + \|S_n v - Sv\|) \\
&= \alpha_n \|u - v\| + (1 - \alpha_n)\|x_n - v\| + \gamma_n \|S_n v - Sv\| \\
&\leq \max\{\|u - v\|, \|x_n - v\|\} + \|S_n v - Sv\| \\
&\leq \max\{\|u - v\|, \|x_0 - v\|\} + \sum_{j=0}^{n} \|S_n v - Sv\|.
\end{aligned}
$$

Therefore $\{x_n\}$ is bounded.

STEP 2. $\|x_{n+1} - x_n\| \to 0$ *as* $n \to \infty$.

Since $\{x_n\}$ is bounded, it follows that $\{y_n\}$ is bounded. Set $B = \{z_n\}$ and

$$
M_2 = \max \left\{ \sup_{n \in \mathbb{N}_0} \|z_n\|, \ \sup_{n \in \mathbb{N}_0} \|Ax_n\|, \ \sup_{n \in \mathbb{N}_0} \|Ay_n\| \right\}.
$$

From Proposition 5.4.1(a), we have

$$
\begin{aligned}
\|y_{n+1} - y_n\| &\leq \|(x_{n+1} - \lambda_{n+1} Ax_{n+1}) - (x_n - \lambda_n Ax_n)\| \\
&= \|(I - \lambda_{n+1} A)x_{n+1} - (I - \lambda_{n+1} A)x_n + \lambda_n Ax_n - \lambda_{n+1} Ax_n\| \\
&\leq \|(I - \lambda_{n+1} A)x_{n+1} - (I - \lambda_{n+1} A)x_n\| + |\lambda_{n+1} - \lambda_n|\|Ax_n\| \\
&\leq (1 + \lambda_{n+1} L)\|x_{n+1} - x_n\| + |\lambda_{n+1} - \lambda_n|M_2. \quad (5.5.12)
\end{aligned}
$$

From (5.5.12), we have

$$
\begin{aligned}
\|z_{n+1} - z_n\| &= \|P_C(x_{n+1} - \lambda_{n+1} Ay_{n+1}) - P_C(x_n - \lambda_n Ay_n)\| \\
&\leq \|(x_{n+1} - \lambda_{n+1} Ay_{n+1}) - (x_n - \lambda_n Ay_n)\| \\
&\leq \|(x_{n+1} - \lambda_{n+1} Ax_{n+1}) - (x_n - \lambda_n Ax_n) \\
&\quad + \lambda_{n+1}(Ax_{n+1} - Ax_n) - \lambda_{n+1} Ay_{n+1} + \lambda_n Ay_n\| \\
&\leq \|(I - \lambda_{n+1} A)x_{n+1} - (I - \lambda_{n+1} A)x_n\| + \lambda_{n+1}(\|Ax_{n+1}\| + \|Ax_n\|) \\
&\quad + \lambda_{n+1}\|Ay_{n+1}\| + \lambda_n\|Ay_n\| \\
&\leq (1 + \lambda_{n+1} L)\|x_{n+1} - x_n\| + |\lambda_{n+1} - \lambda_n|M_2 + 3\lambda_{n+1} M_2 + \lambda_n M_2.
\end{aligned}
$$

Note

$$
\begin{aligned}
\|S_{n+1} z_{n+1} - S_n z_n\| &\leq \|S_{n+1} z_{n+1} - S_{n+1} z_n\| + \|S_{n+1} z_n - S_n z_n\| \\
&\leq \|z_{n+1} - z_n\| + \mathscr{D}_B(S_{n+1}, S_n).
\end{aligned}
$$

Set $x_{n+1} = (1 - \beta_n)w_n + \beta_n x_n$. Then

$$w_{n+1} - w_n = \frac{\alpha_{n+1}u + \gamma_{n+1}S_{n+1}z_{n+1}}{1 - \beta_{n+1}} - \frac{\alpha_n u + \gamma_n S_n z_n}{1 - \beta_n}$$

$$= \left(\frac{\alpha_{n+1}}{1 - \beta_{n+1}} - \frac{\alpha_n}{1 - \beta_n}\right)u + \frac{\gamma_{n+1}}{1 - \beta_{n+1}}S_{n+1}z_{n+1} - \frac{\gamma_{n+1}}{1 - \beta_{n+1}}S_n z_n$$

$$+ \frac{\gamma_{n+1}}{1 - \beta_{n+1}}S_n z_n - \frac{\gamma_n}{1 - \beta_n}S_n z_n.$$

Hence

$$\|w_{n+1} - w_n\| - \|x_{n+1} - x_n\|$$

$$\leq \left|\frac{\alpha_{n+1}}{1 - \beta_{n+1}} - \frac{\alpha_n}{1 - \beta_n}\right|\|u\| + \frac{\gamma_{n+1}}{1 - \beta_{n+1}}\|S_{n+1}z_{n+1} - S_n z_n\|$$

$$+ \left|\frac{\gamma_{n+1}}{1 - \beta_{n+1}} - \frac{\gamma_n}{1 - \beta_n}\right|\|S_n z_n\|$$

$$\leq \left|\frac{\alpha_{n+1}}{1 - \beta_{n+1}} - \frac{\alpha_n}{1 - \beta_n}\right|\|u\| + \frac{\gamma_{n+1}}{1 - \beta_{n+1}}[\lambda_{n+1}L\|x_{n+1} - x_n\| + |\lambda_{n+1} - \lambda_n|M_2$$

$$+ 3\lambda_{n+1}M_2 + \lambda_n M_2] + \left|\frac{\gamma_{n+1}}{1 - \beta_{n+1}} - \frac{\gamma_n}{1 - \beta_n}\right|\|S z_n\| + \frac{\gamma_{n+1}}{1 - \beta_{n+1}}\mathscr{D}_B(S_{n+1}, S_n).$$

Hence

$$\limsup_{n \to \infty}(\|w_{n+1} - w_n\| - \|x_{n+1} - x_n\|) \leq 0.$$

From Lemma 5.2.4, we obtain $w_n - x_n \to 0$ as $n \to \infty$. Consequently,

$$\|x_{n+1} - x_n\| = (1 - \beta_n)\|w_n - x_n\| \leq \|w_n - x_n\| \to 0 \text{ as } n \to \infty.$$

STEP 3. $\|x_n - Sx_n\| \to 0$ and $\|z_n - x_n\| \to 0$ as $n \to \infty$.
Observe that

$$\|S_n E_{\lambda_n}(x_n) - v\|^2 \leq (\|S_n E_{\lambda_n}(x_n) - S_n v\| + \|S_n v - Sv\|)^2$$

$$\leq \|E_{\lambda_n}(x_n) - v\|^2 + M_3\|S_n v - Sv\|,$$

for some constant $M_3 > 0$. For $n \in \mathbb{N}_0$, from (5.5.11), we have

$$
\begin{aligned}
\|x_{n+1} - v\|^2 &= \|\alpha_n u + \beta_n x_n + \gamma_n S_n E_{\lambda_n}(x_n) - v\|^2 \\
&\leq \alpha_n \|u - v\|^2 + \beta_n \|x_n - v\|^2 + \gamma_n \|S_n E_{\lambda_n}(x_n) - v\|^2 \\
&\leq \alpha_n \|u - v\|^2 + \beta_n \|x_n - v\|^2 + \gamma_n (\|E_{\lambda_n}(x_n) - v\|^2 + M_3 \|S_n v - Sv\|) \\
&\leq \alpha_n \|u - v\|^2 + \beta_n \|x_n - v\|^2 + \|E_{\lambda_n}(x_n) - v\|^2 + M_3 \|S_n v - Sv\| \\
&\leq \alpha_n \|u - v\|^2 + \beta_n \|x_n - v\|^2 + \|x_n - v\|^2 - (1 - \lambda_n L)\|x_n - y_n\|^2 \\
&\quad - (1 - \lambda_n L)\|y_n - E_{\lambda_n}(x_n)\|^2 + M_3 \|S_n v - Sv\| \\
&\leq \alpha_n \|u - v\|^2 + \|x_n - v\|^2 - (1 - \lambda_n L)\|x_n - y_n\|^2 + M_3 \|S_n v - Sv\|,
\end{aligned}
$$

which implies that

$$
\begin{aligned}
(1 - bL)\|x_n - y_n\|^2 &\leq (1 - \lambda_n L)\|x_n - y_n\|^2 \\
&\leq \alpha_n \|u - v\|^2 + \|x_n - v\|^2 - \|x_{n+1} - v\|^2 + M_3 \|S_n v - Sv\| \\
&= \alpha_n \|u - v\|^2 + \|x_{n+1} - x_n\|(\|x_n - v\| + \|x_{n+1} - v\|) \\
&\quad + M_3 \|S_n v - Sv\| \\
&\leq \alpha_n \|u - v\|^2 + 2\|x_{n+1} - x_n\| \sup_{k \in \mathbb{N}} \|x_k - v\| + M_3 \|S_n v - Sv\|.
\end{aligned}
$$

Hence $\|y_n - x_n\| \to 0$ as $n \to \infty$. Note

$$
\begin{aligned}
\|y_n - z_n\| &\leq \|(x_n - \lambda A x_n) - (x_n - \lambda A y_n)\| = \lambda_n \|A x_n - A y_n\| \\
&\leq \lambda_n L \|x_n - y_n\| \to 0 \text{ as } n \to \infty,
\end{aligned}
$$

and hence $\|x_n - z_n\| \to 0$ as $n \to \infty$. Observe that

$$
\begin{aligned}
\|x_{n+1} - S_n y_n\| &\leq \|x_{n+1} - S_n z_n\| + \|S_n z_n - S_n y_n\| \\
&\leq \alpha_n \|u - S_n z_n\| + \beta_n \|x_n - S_n z_n\| + \|y_n - z_n\| \\
&\leq \alpha_n \|u - S_n z_n\| + \beta_n (\|x_n - S_n x_n\| + \|S_n x_n - S_n z_n\|) + \|y_n - z_n\| \\
&\leq \alpha_n \|u - S_n z_n\| + \beta_n \|x_n - S_n x_n\| + \|x_n - z_n\| + \|y_n - z_n\|.
\end{aligned}
$$

It follows that

$$
\begin{aligned}
\|x_n - S_n x_n\| &\leq \|x_n - x_{n+1}\| + \|x_{n+1} - S_n y_n\| + \|S_n y_n - S_n z_n\| + \|S_n z_n - S_n x_n\| \\
&\leq \|x_n - x_{n+1}\| + \|x_{n+1} - S_n y_n\| + \|y_n - z_n\| + \|z_n - x_n\| \\
&\leq \|x_n - x_{n+1}\| + \alpha_n \|u - S z_n\| + \beta_n \|x_n - S_n x_n\| + 2\|y_n - z_n\| \\
&\quad + 2\|z_n - x_n\|,
\end{aligned}
$$

which implies that

$$
(1 - \beta_n)\|x_n - S_n x_n\| \leq \|x_n - x_{n+1}\| + \alpha_n \|u - S z_n\| + 2\|y_n - z_n\| + 2\|z_n - x_n\|.
$$

Hence $\|x_n - S_n x_n\| \to 0$ *as* $n \to \infty$. Thus

$$\|x_n - S x_n\| \le \|x_n - S_n x_n\| + \|S_n x_n - S x_n\| \to 0 \text{ as } n \to \infty.$$

STEP 4. $\{x_n\}$ *converges strongly to* x^*, *where* $x^* = P_{Fix(S) \cap \Omega[VI(C,A)]}(u)$.
First, we show that

$$\limsup_{n \to \infty} \langle u - x^*, x_n - x^* \rangle \le 0.$$

We choose a subsequence $\{z_{n_i}\}$ of $\{z_n\}$ such that

$$\limsup_{n \to \infty} \langle u - x^*, z_n - x^* \rangle = \lim_{i \to \infty} \langle u - x^*, z_{n_i} - x^* \rangle.$$

We may assume that $\{x_{n_i}\}$ converges weakly to a point $z \in C$. Since $\|x_n - S x_n\| \to 0$ as $n \to \infty$, we obtain from Proposition 5.2.2 that $z \in Fix(S)$. Since A is monotone and Lipschitz continuous, we obtain from Proposition 5.2.5 that $z \in \Omega[VI(C,A)]$. Thus $z \in Fix(S) \cap \Omega[VI(C,A)]$. Hence, from Lemma 5.2.1 (P2), we have

$$\limsup_{n \to \infty} \langle u - x^*, z_n - x^* \rangle = \lim_{i \to \infty} \langle u - x^*, z_{n_i} - x^* \rangle = \langle u - x^*, z - x^* \rangle \le 0.$$

It follows that $\limsup_{n \to \infty} \langle u - x^*, x_n - x^* \rangle \le 0$. For all $n \in \mathbb{N}_0$, we have

$$
\begin{aligned}
&\|x_{n+1} - x^*\|^2 \\
&= \langle \alpha_n u + \beta_n x_n + \gamma_n S_n E_{\lambda_n}(x_n) - x^*, x_{n+1} - x^* \rangle \\
&= \alpha_n \langle u - x^*, x_{n+1} - x^* \rangle + \beta_n \langle x_n - x^*, x_{n+1} - x^* \rangle + \gamma_n \langle S_n E_{\lambda_n}(x_n) - x^*, x_{n+1} - x^* \rangle \\
&= \frac{\beta_n}{2}(\|x_n - x^*\|^2 + \|x_{n+1} - x^*\|^2) + \frac{\gamma_n}{2}(\|S_n E_{\lambda_n}(x_n) - x^*\|^2 + \|x_{n+1} - x^*\|^2) \\
&\quad + \alpha_n \langle u - x, x_{n+1} - x^* \rangle \\
&\le \frac{\beta_n}{2}(\|x_n - x^*\|^2 + \|x_{n+1} - x^*\|^2) + \frac{\gamma_n}{2}(\|E_{\lambda_n}(x_n) - x^*\|^2 + M_4 \|S_n x^* - S x^*\| \\
&\quad + \|x_{n+1} - x^*\|^2) + \alpha_n \langle u - x, x_{n+1} - x^* \rangle \\
&\le \frac{1 - \alpha_n}{2}(\|x_n - x^*\|^2 + \|x_{n+1} - x^*\|^2) + \alpha_n \langle u - x^*, x_{n+1} - x^* \rangle + \frac{M_4}{2} \|S_n x^* - S x^*\|,
\end{aligned}
$$

for some constant $M_4 > 0$. It follows that

$$\|x_{n+1} - x^*\|^2 \le \|x_n - x^*\|^2 + 2\alpha_n \langle u - x^*, x_{n+1} - x^* \rangle + M_4 \|S_n x^* - S x^*\|.$$

Therefore, by Lemma 5.2.5, $\{x_n\}$ converges strongly to x^*.   □

Corollary 5.5.1 [49, Theorem 3.1]. *Let C be a nonempty closed convex subset of a real Hilbert space H. Let $A : C \to H$ be a monotone and L-Lipschitz continuous mapping and $S : C \to C$ be a nonexpansive mapping such that $Fix(S) \cap \Omega[VI(C,A)] \ne \emptyset$.*

For fixed $u \in C$ and given $x_0 \in C$ arbitrarily, define a sequence $\{x_n\}$ in C iteratively by

$$x_{n+1} = \alpha_n u + \beta_n x_n + \gamma_n SE_{\lambda_n}(x_n), \quad n \in \mathbb{N}_0, \qquad (5.5.13)$$

where $\{\alpha_n\}$, $\{\beta_n\}$ and $\{\gamma_n\}$ are sequences in $[0,1]$ and $\{\lambda_n\}$ is a sequence in $(0,\infty)$ satisfying the following conditions:

(i) *$\alpha_n + \beta_n + \gamma_n = 1$ for all $n \in \mathbb{N}_0$;*

(ii) *$\lim\limits_{n \to \infty} \alpha_n = 0$, $\sum_{n=0}^{\infty} \alpha_n = \infty$;*

(iii) *$0 < \liminf\limits_{n \to \infty} \beta_n \leq \limsup\limits_{n \to \infty} \beta_n < 1$;*

(iv) *$0 < a \leq \lambda_n \leq b < 1/L$.*

Then the sequence $\{x_n\}$ generated by (5.5.13) converges strongly to $P_{Fix(S) \cap \Omega[VI(C,A)]}(u)$.

5.6. Extragradient Method for Hierarchical Variational Inequalities

Let D be a nonempty closed convex subset of C. Then the variational inequality problem over the set D is a problem which searches $x^* \in D$ such that

$$\langle \mathscr{F}x^*, y - x^* \rangle \geq 0, \quad \text{for all } y \in D. \qquad (5.6.14)$$

We denote by $VI_D(C,\mathscr{F})$ the variational inequality problem (5.6.14) and by $\Omega[VI_D(C,\mathscr{F})]$ the set of solutions of variational inequality problem (5.6.14). For $C = D$, we use $VI(C,\mathscr{F}) = VI_D(C,\mathscr{F})$ and $\Omega[VI(C,\mathscr{F})] = \Omega[VI_D(C,\mathscr{F})]$.

In this section, we are interested in solving the following *hierarchical variational inequality problem $HVI(C,A,f)$:*

Find $z \in VI(C,A)$ such that $\langle (I - f)z, x - z \rangle \geq 0, \quad$ for all $x \in VI(C,A)$, (5.6.15)

where C is a closed convex subset of a real Hilbert space H and $A, f : C \to H$ are nonlinear operators. One can see that the hierarchical variational inequality problem (5.6.15) can be represented in terms of a variational inequality problem as follows:

$$HVI(C,A,f) = VI_{\Omega[VI(C,A)]}(C, I - f).$$

Proposition 5.6.1. *Let C be a closed convex subset of a real Hilbert space H and D be a nonempty closed convex subset of C. Let $f : C \to H$ be a contraction mapping. Then* *there exists $x^* \in D$, which is a unique solution to the following variational inequality problem:*

Find $z \in D$ such that $\langle (I - f)z, x - z \rangle \geq 0$,   for all $x \in D$.

Proof. Note $P_D f : C \to D$ is a contraction mapping. Hence there exists $x^* \in D$, which is a unique solution of the fixed point problem:

Find $z \in D$ such that $P_D fz = z$.

Since $P_D fx^* = x^*$, it follows from Theorem 5.3.1 that

$$x^* = P_D(I - (I - f))x^* \iff x^* \in \Omega[VI(C, I - f)].$$

$\square$

5.6.1. Implicit Extragradient Method for Hierarchical Variational Inequalities

First, we discuss some important properties of inverse strongly monotone operators.

Proposition 5.6.2. *Let C be a nonempty subset of a real Hilbert space H and $\mathscr{F} : C \to H$ a $\frac{1}{\kappa^2}$-inverse strongly monotone. Let t be a positive real number. Then:*

(a) *$\mathscr{F} + tI : C \to H$ is $\frac{1}{t + \kappa^2}$-inverse strongly monotone.*

(b) *$\mathscr{F} + tI$ is $(t + \kappa^2)$-Lipschitz continuous and t-strongly monotone.*

(c) *If $\mu \in (0, \frac{2t}{t + \kappa^2})$, then $P_C(I - \mu\mathscr{F}) : C \to C$ is a $\sqrt{1 - 2\mu\eta + \mu^2\kappa^2}$-contraction. In particular, if $\lambda \in (0, \frac{t}{t + \kappa^2}]$, then $P_C(I - \mu\mathscr{F})$ is a $(1 - \lambda t)$-contraction.*

Proof. (a) Let $x, y \in C$. Since $\mathscr{F}$ is $\frac{1}{\kappa^2}$-inverse strongly monotone, we have

$$\langle \mathscr{F}x - \mathscr{F}y, x - y \rangle \geq \frac{1}{\kappa^2} \|\mathscr{F}x - \mathscr{F}y\|^2.$$

Thus

$$\|\mathscr{F}x - \mathscr{F}y\| \leq \kappa^2 \|x - y\|$$

and

$$\langle \mathscr{F}x - \mathscr{F}y, x - y \rangle \leq \|\mathscr{F}x - \mathscr{F}y\| \, \|x - y\| \leq \kappa^2 \|x - y\|^2.$$

Then

$$
\begin{aligned}
(t + \kappa^2) & \langle (\mathscr{F}+tI)x - (\mathscr{F}+tI)y, x-y \rangle \\
&= (t+\kappa^2)[\langle \mathscr{F}x - \mathscr{F}y, x-y \rangle + t\|x-y\|^2] \\
&= t^2\|x-y\|^2 + t\langle \mathscr{F}x - \mathscr{F}y, x-y \rangle + \kappa^2 \langle \mathscr{F}x - \mathscr{F}y, x-y \rangle + t\kappa^2\|x-y\|^2 \\
&\geq t^2\|x-y\|^2 + t\langle \mathscr{F}x - \mathscr{F}y, x-y \rangle + \|\mathscr{F}x - \mathscr{F}y\|^2 + t(\kappa\|x-y\|)^2 \\
&\geq t^2\|x-y\|^2 + t\langle \mathscr{F}x - \mathscr{F}y, x-y \rangle + \|\mathscr{F}x - \mathscr{F}y\|^2 + t\langle \mathscr{F}x - \mathscr{F}y, x-y \rangle \\
&= t^2\|x-y\|^2 + 2t\langle \mathscr{F}x - \mathscr{F}y, x-y \rangle + \|\mathscr{F}x - \mathscr{F}y\|^2 \\
&= \|t(x-y) + \mathscr{F}x - \mathscr{F}y\|^2 \\
&= \|(\mathscr{F}+tI)x - (\mathscr{F}+tI)y\|^2.
\end{aligned}
$$

(b) Let $x, y \in C$. Since $\mathscr{F}+tI : C \to H$ is $\frac{1}{t+\kappa^2}$-inverse strongly monotone. It follows that $\mathscr{F}+tI$ is $(t+\kappa^2)$-Lipschitz continuous. So, it remains to show that $\mathscr{F}+tI$ is t-strongly monotone. Let $x, y \in C$. Note that $\mathscr{F}$ is monotone, i.e.,

$$\langle \mathscr{F}x - \mathscr{F}y, x-y \rangle \geq 0.$$

Hence

$$
\begin{aligned}
\langle (\mathscr{F}+tI)x - (\mathscr{F}+tI)y, x-y \rangle &= \langle \mathscr{F}x - \mathscr{F}y, x-y \rangle + t\|x-y\|^2 \\
&\geq t\|x-y\|^2.
\end{aligned}
$$

Therefore $\mathscr{F}+tI$ is t-strongly monotone.

(c) Note $\mathscr{F}+tI$ is t-strongly monotone and $(t+\kappa^2)$-Lipschitz continuous. Let μ be a positive constant such that $\mu < 2t/(t+\kappa^2)^2$. Then, from Theorem 5.3.2, $P_C(I - \mu\mathscr{F}) : C \to C$ is a $\sqrt{1 - 2\mu t + \mu^2(t+\kappa^2)^2}$-contraction. Let $\lambda \in (0, \frac{t}{t+\kappa^2}]$. Note $\lambda \leq \frac{t}{t+\kappa^2}$ and hence

$$1 - 2\lambda t + \lambda^2(t+\kappa^2)^2 \leq 1 - 2\lambda t + \lambda^2 t^2 = (1 - \lambda t)^2.$$

Therefore $P_C(I - \mu\mathscr{F})$ is a $(1 - \lambda t)$-contraction.   □

Proposition 5.6.3. *Let C be a nonempty subset of a real Hilbert space H and $A : C \to H$ be an α-inverse strongly monotone operator. Let λ be a positive constant such that $\lambda < 2\alpha$. Then the following assertions hold:*

(a) *For all $x, y \in C$, we have*

$$\|(I - \lambda A)x - (I - \lambda A)y\|^2 \leq \|x-y\|^2 - \lambda(2\alpha - \lambda)\|Ax - Ay\|^2. \tag{5.6.16}$$

(b) *If $T : C \to H$ is an operator defined by*

$$T = I - \lambda A, \tag{5.6.17}$$

then T is nonexpansive.

Proof. (a) Let $x, y \in C$. Then

$$
\begin{aligned}
\|(I - \lambda A)x - (I - \lambda A)y\|^2 &= \|x - y - \lambda(Ax - Ay)\|^2 \\
&= \|x - y\|^2 + \lambda^2 \|Ax - Ay\|^2 - 2\lambda \langle Ax - Ay, x - y \rangle \\
&\leq \|x - y\|^2 - \lambda(2\alpha - \lambda)\|Ax - Ay\|^2.
\end{aligned}
$$

(b) This follows from part (a). $\square$

Proposition 5.6.4. *Let C be a nonempty closed convex subset of a real Hilbert space H. Let $f : C \to H$ be a k-contraction mapping and $A : C \to H$ be an α-inverse strongly monotone operator such that $\Omega[VI(C,A)] \neq \emptyset$. For $0 < \lambda < 2\alpha$, let $T : C \to H$ be an operator defined by (5.6.17). For $t \in (0,1)$, define $G_t : C \to C$ by*

$$
G_t x = P_C[T] H_t x, \quad x \in C,
$$

where $H_t = P_C[(1-t)T + tf]$. Then the following assertions hold:

(a) *G_t is a $(1 - (1-k)t)$-contraction and has a unique fixed point x_t satisfying*

$$
x_t = P_C[T] y_t, \tag{5.6.18}
$$

where $y_t = P_C[(1-t)T + tf] x_t$.

(b) *$v = P_C[tv + (1-t)Tv]$ for all $t \in (0,1)$ and $v \in \Omega[VI(C,A)]$.*

(c) *$\|G_t x - v\| \leq (1 - (1-k)t)\|x - v\| + t\|fv - v\|$ for all $x \in C, v \in \Omega[VI(C,A)]$ and $t \in (0,1)$.*

(d) *For $x \in C$ and $v \in \Omega[VI(C,A)]$, we have*

$$
\begin{aligned}
\|x - H_t x\|^2 \leq {}& \|x - v\|^2 - \|G_t x - v\|^2 + t\|fx - Tv\|^2 \\
&+ 2(t\|fx - Tx\| + \lambda\|Ax - Av\|)\|x - H_t x\|.
\end{aligned}
$$

Proof. (a) Let $x, y \in C$. Then by nonexpansivity of $P_C[T]$, we have

$$
\begin{aligned}
\|G_t x - G_t y\| &= \|P_C[T] H_t x - P_C[T] H_t y\| \\
&\leq \|H_t x - H_t y\| = \|P_C[(1-t)Tx + tfx] - P_C[(1-t)Ty + tfy]\| \\
&\leq \|(1-t)Tx + tfx - [(1-t)Ty + tfy]\| \\
&\leq (1-t)\|Tx - Ty\| + t\|fx - fy\| \\
&\leq (1 - (1-k)t)\|x - y\|.
\end{aligned}
$$

Thus, G_t is a $(1 - (1-k)t)$-contraction. By the Banach contraction principle, there exists a unique point $x_t \in C$ such that $x_t = G_t x_t$.

(b) Let $v \in \Omega[VI(C,A)]$. From Theorem 5.3.1, we have $v = P_C[v - cAv]$ for any $c > 0$. Hence, for $t \in (0,1)$, we have

$$
\begin{aligned}
v &= P_C[v - \lambda(1-t)Av] \\
&= P_C[tv + (1-t)v - \lambda(1-t)Av] \\
&= P_C[tv + (1-t)(v - \lambda Av)] \\
&= P_C[tv + (1-t)Tv].
\end{aligned}
$$

(c) For $x \in C$, $v \in \Omega[VI(C,A)]$ and $t \in (0,1)$, we have

$$
\begin{aligned}
\|G_t x - v\| &= \|P_C[T]H_t x - P_C[T]v\| \\
&\leq \|H_t x - v\| \tag{5.6.19} \\
&= \|P_C[(1-t)Tx + tfx] - P_C[tv + (1-t)Tv]\| \\
&\leq \|(1-t)(Tx - Tv) + t(fx - v)\| \tag{5.6.20} \\
&\leq (1-t)\|Tx - Tv\| + t(\|fx - fv\| + \|fv - v\|) \\
&\leq (1-(1-k)t)\|x - v\| + t\|fv - v\|.
\end{aligned}
$$

(d) For $x \in C$ and $v \in \Omega[VI(C,A)]$, we have

$$
\begin{aligned}
\|H_t x - v\|^2 &= \|P_C[(1-t)Tx + tfx] - P_C[Tv]\|^2 \\
&\leq \langle H_t x - v, (1-t)Tx + tfx - Tv \rangle \\
&\leq \frac{1}{2}[\|(1-t)Tx + tfx - Tv\|^2 + \|H_t x - v\|^2 \\
&\quad - \|(1-t)Tx + tfx - Tv - (H_t x - v)\|^2] \\
&\leq \frac{1}{2}[(1-t)\|Tx - Tv\|^2 + t\|fx - Tv\|^2 + \|H_t x - v\|^2 \\
&\quad - \|Tx - Tv - (H_t x - v) + t(fx - Tx)\|^2] \\
&= \frac{1}{2}[(1-t)\|Tx - Tv\|^2 + t\|fx - Tv\|^2 + \|H_t x - v\|^2 \\
&\quad - \|x - H_t x - \lambda(Ax - Av) + t(fx - Tx)\|^2] \\
&\leq \frac{1}{2}[(1-t)\|x - v\|^2 + t\|fx - Tv\|^2 + \|H_t x - v\|^2 - (\|x - H_t x\|^2 \\
&\quad + \|t(fx - Tx) - \lambda(Ax - Av)\|^2 + 2t\langle x - H_t x, fx - Tx \rangle \\
&\quad - 2\lambda\langle x - H_t x, Ax - Av \rangle)],
\end{aligned}
$$

and it follows that

$$
\begin{aligned}
\|H_t x - v\|^2 \ &\leq \ \|x-v\|^2 + t\|fx - Tv\|^2 - \|x - H_t x\|^2 - \|t(fx - Tx) - \lambda(Ax - Av)\|^2 \\
&\quad + 2t\|x - H_t x\|\,\|fx - Tx\| + 2\lambda\|x - H_t x\|\,\|Ax - Av\| \\
&\leq \ \|x-v\|^2 + t\|fx - Tv\|^2 - \|x - H_t x\|^2 \\
&\quad + 2(t\|fx - Tx\| + \lambda\|Ax - Av\|)\|x - H_t x\|.
\end{aligned}
$$

From (5.6.19), we have

$$
\begin{aligned}
\|G_t x - v\|^2 \ &\leq \ \|H_t x - v\|^2 \\
&\leq \ \|x-v\|^2 + t\|fx - Tv\|^2 - \|x - H_t x\|^2 \\
&\quad + 2(t\|fx - Tx\| + \lambda\|Ax - Av\|)\|x - H_t x\|.
\end{aligned}
$$

Therefore

$$
\begin{aligned}
\|x - H_t x\|^2 &\leq \|x-v\|^2 - \|G_t x - v\|^2 + t\|fx - Tv\|^2 \\
&\quad + 2(t\|fx - Tx\| + \lambda\|Ax - Av\|)\|x - H_t x\|. \qquad \square
\end{aligned}
$$

Theorem 5.6.1 [48]. *Let C be a nonempty closed convex subset of a real Hilbert space H. Let $f : C \to H$ be a k-contraction mapping and $A : C \to H$ be an α-inverse strongly monotone operator such that $\Omega[VI(C,A)] \neq \emptyset$. For $0 < \lambda < 2\alpha$, let $T : C \to H$ be an operator defined by (5.6.17). Then the net $\{x_t\}$ generated by the implicit extragradient method (5.6.18) converges in norm, as $t \to 0^+$, to a unique solution x^* of the hierarchical variational inequality problem $HVI(C,A,f)$.*

Proof. Let $v \in \Omega[VI(C,A)]$. From Proposition 5.6.4(c), we have

$$
\|x_t - v\| = \|G_t x_t - v\| \leq (1 - (1-k)t)\|x_t - v\| + t\|fv - v\|, \quad \text{for all } t \in (0,1),
$$

and hence

$$
\|x_t - v\| \leq \frac{1}{1-k}\|fv - v\|, \quad \text{for all } t \in (0,1).
$$

Therefore $\{x_t\}$ is bounded and so are $\{fx_t\}$, $\{y_t\}$. From (5.6.16) and (5.6.20), we have

$$
\begin{aligned}
\|x_t - v\|^2 \ &\leq \ \|(1-t)(Tx_t - Tv) + t(fx_t - v)\|^2 \\
&\leq \ (1-t)\|Tx_t - Tv\|^2 + t\|fx_t - v\|^2 \\
&\leq \ (1-t)(\|x_t - v\|^2 - \lambda(2\alpha - \lambda)\|Ax_t - Av\|^2) + t\|fx_t - v\|^2 \\
&\leq \ \|x_t - v\|^2 - (1-t)\lambda(2\alpha - \lambda)\|Ax_t - Av\|^2) + t\|fx_t - v\|^2,
\end{aligned}
$$

which implies that

$$(1-t)\lambda(2\alpha-\lambda)||Ax_t - Av||^2 \le t\,||fx_t - v||^2.$$

Hence $||Ax_t - Av|| \to 0$ as $t \to 0^+$. Since $x_t = G_t x_t$, from Proposition 5.6.4(d), we have

$$
\begin{aligned}
||x_t - H_t x_t||^2 &\le t\,||fx_t - Tv||^2 + 2(t||fx_t - Tx_t|| + \lambda||Ax_t - Av||)||x_t - H_t x_t|| \\
&\le tM_5 + 2(tM_5 + \lambda||Ax_t - Av||)M_5
\end{aligned}
$$

for some constant $M_5 > 0$. Since $||Ax_t - Av|| \to 0$ as $t \to 0^+$, we have $||x_t - H_t x_t|| \to 0$ as $t \to 0^+$.

We now show that the net $\{x_t\}$ is relatively norm-compact as $t \to 0^+$. Assume that $\{t_n\}$ is a sequence in $(0,1)$ such that $t_n \to 0$ as $n \to \infty$. Put $x_n := x_{t_n}$ and $y_n = H_{t_n} x_{t_n}$. By the property (P4) of metric projection P_C and (5.6.19), we have

$$
\begin{aligned}
||y_n - v||^2 &= ||P_C[(1-t_n)Tx_n + t_n fx_n] - P_C[t_n v + (1-t_n)Tv]||^2 \\
&\le \langle (1-t_n)Tx_n + t_n fx_n - [t_n v + (1-t_n)Tv], y_n - v \rangle \\
&\le (1-t_n)\langle Tx_n - Tv, y_n - v \rangle + t_n \langle fx_n - fv + fv - v, y_n - v \rangle \\
&\le ((1-t_n)||Tx_n - Tv|| + t_n ||fx_n - fv||)\,||y_n - v|| + t_n \langle fv - v, y_n - v \rangle \\
&\le (1 - (1-k)t_n)||x_n - v||\,||y_n - v|| + t_n \langle fv - v, y_n - v \rangle \\
&\le (1 - (1-k)t_n)||y_n - v||^2 + t_n \langle fv - v, y_n - v \rangle.
\end{aligned}
$$

It follows that

$$||y_n - v||^2 \le \frac{1}{1-k}\langle fv - v, y_n - v \rangle, \quad \text{for all } v \in \Omega[VI(C,A)]. \tag{5.6.21}$$

Since $\{x_n\}$ is bounded, without loss of generality, we may assume that $\{x_n\}$ converges weakly to a point $z \in C$. Since $||x_t - H_t x_t|| \to 0$, we have $||x_n - y_n|| \to 0$. Hence $\{y_n\}$ also converges weakly to the same point z. Note $x_n = P_C[T]y_n$ and $||x_n - y_n|| \to 0$, and we obtain from Proposition 5.2.4 that $z \in \Omega[VI(C,A)]$. In particular, from (5.6.21), we have

$$||y_n - z||^2 \le \frac{1}{1-k}\langle fz - z, y_n - z \rangle. \tag{5.6.22}$$

Since $||x_n - y_n|| \to 0$ and $\{y_n\}$ converges weakly to a point $z \in C$, we obtain from (5.6.22) that $\{x_n\}$ converges strongly to z. This means that the net $\{x_t\}$ is relatively norm-compact as $t \to 0^+$. Thus, from (5.6.21), we have

$$||z - v||^2 \le \frac{1}{1-k}\langle fv - v, z - v \rangle, \quad \text{for all } v \in \Omega[VI(C,A)].$$

This shows that z solves the following variational inequality problem:

$$z \in \Omega[VI(C,A)] \text{ such that } \langle (I-f)v, v-z \rangle \geq 0, \quad \text{for all } v \in \Omega[VI(C,A)].$$

By Proposition 5.2.3, we obtain

$$z \in \Omega[VI(C,A)], \langle (I-f)z, v-z \rangle \geq 0, \quad \text{for all } v \in \Omega[VI(C,A)].$$

Thus $z = P_{\Omega[VI(C,A)]}fz$. Proposition 5.6.1 shows that $z = x^*$ is the unique solution of $VI_{\Omega[VI(C,A)]}(C, I-f)]$. Therefore the net $\{x_t\}$ generated by the implicit extragradient method (5.6.18) converges in norm, as $t \to 0^+$, to a unique solution x^* of the hierarchical variational inequality problem $HVI(C,A,f)$. $\qquad\qquad\square$

5.6.2. Explicit Extragradient Method for Hierarchical Variational Inequalities

Let C be a nonempty closed convex subset of a real Hilbert space H. Let $f : C \to H$ be a k-contraction mapping and $A : C \to H$ be an α-inverse strongly monotone operator such that $\Omega[VI(C,A)] \neq \emptyset$. For given $x_0 \in C$ arbitrarily, define a sequence $\{x_n\}$ in C iteratively by

$$\begin{cases} y_n = P_C[(1-\alpha_n)(x_n - \lambda A x_n) + \alpha_n f x_n], \\ x_{n+1} = P_C(y_n - \lambda A y_n), \quad n \in \mathbb{N}_0, \end{cases} \tag{5.6.23}$$

where $\{\alpha_n\}$ is a sequence in $(0,1)$ and $\lambda \in (0, 2\alpha)$. If $fx = 0$, then algorithm (5.6.23) reduces to an algorithm studied by Yao et al. [50].

If T is a mapping defined by (5.6.17), then algorithm (5.6.23) reduces to

$$\begin{cases} y_n = P_C[(1-\alpha_n)T x_n + \alpha_n f x_n], \\ x_{n+1} = P_C[T y_n], \quad n \in \mathbb{N}_0. \end{cases} \tag{5.6.24}$$

We now establish strong convergence of extragradient-like method (5.6.24).

Theorem 5.6.2. *Let C be a nonempty closed convex subset of a real Hilbert space H. Let $f : C \to H$ be a k-contraction mapping and $A : C \to H$ be an α-inverse strongly monotone operator such that $\Omega[VI(C,A)] \neq \emptyset$. For $0 < \lambda < 2\alpha$ and given $x_0 \in C$ arbitrarily, define a sequence $\{x_n\}$ in C iteratively by (5.6.24), where T is a mapping defined by (5.6.17) and $\{\alpha_n\}$ is a sequence in $(0,1)$ such that*

$$\lim_{n \to \infty} \alpha_n = 0, \quad \lim_{n \to \infty} \frac{\alpha_{n+1}}{\alpha_n} = 1 \text{ and } \sum_{n=1}^{\infty} \alpha_n = \infty.$$

Then the sequence $\{x_n\}$ generated by (5.6.24) converges strongly to a unique solution x^* of the hierarchical variational inequality problem $HVI(C,A,f)$.

Proof. Let $x^* = P_{\Omega[VI(C,A)]}fx^*$. Set $S = P_CT$. Thus $\text{Fix}(S) = \Omega[VI(C,A)]$. We now divide the proof into the following several steps.

STEP I. $\{x_n\}$ *is bounded.*
Note $x^* = P_C[x^* - rAx^*]$ for all $r > 0$. Thus

$$
\begin{aligned}
x^* &= P_C[x^* - \lambda(1 - \alpha_n)\,Ax^*]\\
&= P_C[\,\alpha_n x^* + (1 - \alpha_n)x^* - \lambda(1 - \alpha_n)\,Ax^*]\\
&= P_C[\,\alpha_n x^* + (1 - \alpha_n)(x^* - \lambda Ax^*)]\\
&= P_C[\,\alpha_n x^* + (1 - \alpha_n)Tx^*].
\end{aligned}
$$

Note

$$
\begin{aligned}
\|y_n - x^*\| &= \|P_C[(1 - \alpha_n)Tx_n + \alpha_n fx_n] - P_C[\,\alpha_n x^* + (1 - \alpha_n)Tx^*]\|\\
&\le \|(1 - \alpha_n)Tx_n + \alpha_n fx_n - [\,\alpha_n x^* + (1 - \alpha_n)Tx^*\|\\
&= \|(1 - \alpha_n)(Tx_n - Tx^*) + \alpha_n(fx_n - x^*)\| \qquad (5.6.25)\\
&\le (1 - \alpha_n)\|Tx_n - Tx^*\| + \alpha_n(\|fx_n - fx^*\| + \|fx^* - x^*\|)\\
&\le (1 - (1 - k)\alpha_n)\|x_n - x^*\| + \alpha_n\|fx^* - x^*\|. \qquad (5.6.26)
\end{aligned}
$$

From nonexpansiveness of T and (5.6.26), we have

$$
\begin{aligned}
\|x_{n+1} - x^*\| &= \|P_C(Ty_n) - P_C(T)x^*\|\\
&\le \|y_n - x^*\| \qquad\qquad (5.6.27)\\
&\le (1 - \alpha_n)\|x_n - x^*\| + \alpha_n\|fx^* - x^*\|\\
&\le \max\{\|x_n - x^*\|, \|fx^* - x^*\|\}\\
&\ \ \vdots\\
&\le \max\{\|x_0 - x^*\|, \|fx^* - x^*\|\}.
\end{aligned}
$$

Therefore $\{x_n\}$ is bounded.

STEP 2. $\|x_{n+1} - x_n\| \to 0$ *as $n \to \infty$.*
Since $\{x_n\}$ is bounded, it follows that $\{y_n\}$, $\{Ax_n\}$ and $\{Ay_n\}$ are bounded. From

(5.6.24), we have

$$
\begin{aligned}
\|x_{n+1} - x_n\| &= \|P_C[Ty_n] - P_C[Ty_{n-1}]\| \\
&\leq \|y_n - y_{n-1}\| \\
&= \|P_C[(1-\alpha_n)Tx_n + \alpha_n f x_n] - P_C[(1-\alpha_{n-1})Tx_{n-1} + \alpha_{n-1} f x_{n-1}]\| \\
&\leq \|[(1-\alpha_n)Tx_n + \alpha_n f x_n] - [(1-\alpha_{n-1})Tx_{n-1} + \alpha_{n-1} f x_{n-1}]\| \\
&\leq \|(1-\alpha_n)Tx_n - (1-\alpha_{n-1})Tx_{n-1}\| + \|\alpha_n f x_n - \alpha_{n-1} f x_{n-1}\| \\
&= \|(1-\alpha_n)Tx_n - (1-\alpha_n)Tx_{n-1} + (1-\alpha_n)Tx_{n-1} \\
&\qquad -(1-\alpha_{n-1})Tx_{n-1}\| + \|\alpha_n f x_n - \alpha_n f x_{n-1} + \alpha_n f x_{n-1} - \alpha_{n-1} f x_{n-1}\| \\
&\leq (1-\alpha_n)\|Tx_n - Tx_{n-1}\| + \alpha_n \|f x_n - f x_{n-1}\| \\
&\qquad + |\alpha_n - \alpha_{n-1}|(\|Tx_{n-1}\| + \|f x_{n-1}\|) \\
&\leq (1-(1-k)\alpha_n)\|x_n - x_{n-1}\| + |\alpha_n - \alpha_{n-1}|M_6,
\end{aligned}
$$

for some constant $M_6 > 0$. By Lemma 5.2.5, we obtain that $\|x_{n+1} - x_n\| \to 0$ as $n \to \infty$. From (5.6.25), (5.6.27) and the convexity of the norm, we get

$$
\begin{aligned}
\|x_{n+1} - x^*\|^2 &\leq \|y_n - x^*\|^2 \\
&\leq \|(1-\alpha_n)(Tx_n - Tx^*) + \alpha_n(f x_n - x^*)\|^2 \\
&\leq (1-\alpha_n)\|Tx_n - Tx^*\|^2 + \alpha_n \|f x_n - x^*\|^2,
\end{aligned}
$$

and it follows that

$$
\|x_{n+1} - x^*\|^2 \leq \|x_n - x^*\|^2 - \frac{(2\alpha - \lambda)}{\lambda}\|x_n - x^* - (Tx_n - Tx^*)\|^2 + \alpha_n \|f x_n - x^*\|^2.
$$

Thus

$$
\frac{(2\alpha - \lambda)}{\lambda}\|x_n - x^* - (Tx_n - Tx^*)\|^2 \leq \|x_n - x^*\|^2 - \|x_{n+1} - x^*\|^2 + \alpha_n \|f x_n - x^*\|^2,
$$

which implies that

$$
\|x_n - x^* - (Tx_n - Tx^*)\| \to 0 \text{ as } n \to \infty. \tag{5.6.28}
$$

Note

$$\begin{aligned}
\|y_n - x^*\|^2 &= \|P_C[(1 - \alpha_n)Tx_n + \alpha_n fx_n] - P_C[Tx^*]\|^2 \\
&\leq \langle y_n - x^*, (1 - \alpha_n)Tx_n + \alpha_n fx_n - Tx^* \rangle \\
&\leq \frac{1}{2}[\|(1 - \alpha_n)Tx_n + \alpha_n fx_n - Tx^*\|^2 + \|y_n - x^*\|^2 \\
&\quad - \|(1 - \alpha_n)Tx_n + \alpha_n fx_n - Tx^* - (y_n - x^*)\|^2] \\
&= \frac{1}{2}[(1 - \alpha_n)\|Tx_n - Tx^*\|^2 + \alpha_n \|fx_n - Tx^*\|^2 + \|y_n - x^*\|^2 \\
&\quad - \|Tx_n - Tx^* + \alpha_n(fx_n - Tx_n) - (y_n - x^*)\|^2] \\
&\leq \frac{1}{2}[(1 - \alpha_n)\|Tx_n - Tx^*\|^2 + \alpha_n \|fx_n - Tx^*\|^2 + \|y_n - x^*\|^2 \\
&\quad - \|Tx_n - Tx^* - (y_n - x^*) + \alpha_n(fx_n - Tx_n)\|^2] \\
&\leq \frac{1}{2}[(1 - \alpha_n)\|x_n - x^*\|^2 + \alpha_n(\|fx_n - Tx^*\|^2 + \|fx_n - Tx_n\|^2) \\
&\quad + \|y_n - x^*\|^2 - (\|Tx_n - Tx^* - (y_n - x^*)\|^2 \\
&\quad + 2\alpha_n \langle Tx_n - Tx^* - (y_n - x^*), fx_n - Tx_n \rangle)],
\end{aligned}$$

which implies that

$$\begin{aligned}
\|y_n - x^*\|^2 &\leq (1 - \alpha_n)\|x_n - x^*\|^2 + \alpha_n M_6 - \|Tx_n - Tx^* - (y_n - x^*)\|^2 \\
&\quad + 2\alpha_n \|Tx_n - Tx^* - (y_n - x^*)\| \|fx_n - Tx_n\|,
\end{aligned}$$

for some $M_7 > 0$. Hence

$$\begin{aligned}
\|x_{n+1} - x^*\|^2 &\leq \|y_n - x^*\|^2 \\
&\leq \|x_n - x^*\|^2 - \|Tx_n - Tx^* - (y_n - x^*)\|^2 + \alpha_n M_7 \\
&\quad + 2\alpha_n \|Tx_n - Tx^* - (y_n - x^*)\| \|fx_n - Tx_n\|,
\end{aligned}$$

which implies that

$$\begin{aligned}
\|Tx_n - Tx^* - (y_n - x^*)\|^2 &\leq \|x_n - x^*\|^2 - \|x_{n+1} - x^*\|^2 + \alpha_n M_7 \\
&\quad + 2\alpha_n \|Tx_n - Tx^* - (y_n - x^*)\| \|fx_n - Tx_n\|.
\end{aligned}$$

Thus we have

$$\|Tx_n - Tx^* - (y_n - x^*)\| \to 0 \text{ as } n \to \infty. \tag{5.6.29}$$

From (5.6.28) and (5.6.29) we see that

$$\|x_n - y_n\| \to 0 \text{ as } n \to \infty. \tag{5.6.30}$$

Next, we show that

$$\limsup_{n \to \infty} \langle x^* - fx^*, x^* - y_n \rangle \leq 0.$$

We choose a subsequence $\{y_{n_i}\}$ of $\{y_n\}$ such that

$$\limsup_{n \to \infty} \langle x^* - fx^*, x^* - y_n \rangle = \lim_{i \to \infty} \langle x^* - fx^*, x^* - y_{n_i} \rangle.$$

We may assume that $\{y_{n_i}\}$ converges weakly to a point $z \in C$. Since A is monotone and Lipschitz continuous, we obtain from (5.6.30) and Proposition 5.2.5 that $z \in \Omega[VI(C,A)]$. Hence

$$\limsup_{n \to \infty} \langle x^* - fx^*, x^* - y_n \rangle = \lim_{i \to \infty} \langle x^* - fx^*, x^* - y_{n_i} \rangle = \langle x^* - fx^*, x^* - z \rangle \leq 0.$$

By the property (P2) of metric projection P_C, we have

$$
\begin{aligned}
\|y_n - x^*\|^2 &= \|P_C[(1 - \alpha_n)Tx_n + \alpha_n fx_n] - P_C[\alpha_n x^* + (1 - \alpha_n)Tx^*]\|^2 \\
&\leq \langle (1 - \alpha_n)(Tx_n - Tx^*) + \alpha_n(fx_n - fx^* + fx^* - x^*), y_n - x^* \rangle \\
&\leq ((1 - \alpha_n)\|Tx_n - Tx^*\| + \alpha_n\|fx_n - fx^*\|)\|y_n - x^*\| \\
&\quad + \alpha_n \langle fx^* - x^*, y_n - x^* \rangle \\
&\leq (1 - (1-k)\alpha_n)\|x_n - x^*\|\|y_n - x^*\| + \alpha_n \langle fx^* - x^*, y_n - x^* \rangle \\
&\leq \frac{1 - (1-k)\alpha_n}{2}(\|x_n - x^*\|^2 + \|y_n - x^*\|^2) + \alpha_n \langle fx^* - x^*, y_n - x^* \rangle.
\end{aligned}
$$

It follows that

$$\|y_n - x^*\|^2 \leq (1 - (1-k)\alpha_n)\|x_n - x^*\|^2 + 2\alpha_n \langle fx^* - x^*, y_n - x^* \rangle.$$

Thus

$$\|x_{n+1} - x^*\|^2 \leq \|y_n - x^*\|^2 \leq (1 - (1-k)\alpha_n)\|x_n - x^*\|^2 + 2\alpha_n \langle fx^* - x^*, y_n - x^* \rangle.$$

Therefore, by Lemma 5.2.5, $\{x_n\}$ converges strongly to x^*.   □

Corollary 5.6.1 [50, Theorem 3.1]. *Let C be a nonempty closed convex subset of a real Hilbert space H. Let $A : C \to H$ be an α-inverse strongly monotone operator such that $\Omega[VI(C,A)] \neq \emptyset$. For $0 < \lambda < 2\alpha$ and given $x_0 \in C$ arbitrarily, define a sequence $\{x_n\}$ in C iteratively by*

$$
\begin{cases}
y_n = P_C[(1 - \alpha_n)(x_n - \lambda Ax_n)], \\
x_{n+1} = P_C(y_n - \lambda Ay_n), \quad n \in \mathbb{N}_0,
\end{cases}
\tag{5.6.31}
$$

where $\{\alpha_n\}$ is a sequence in $(0,1)$ such that

$$\lim_{n\to\infty} \alpha_n = 0, \lim_{n\to\infty} \frac{\alpha_{n+1}}{\alpha_n} = 1 \ and \ \sum_{n=1}^{\infty} \alpha_n = \infty.$$

Then the sequence $\{x_n\}$ generated by (5.6.31) converges strongly to $P_{\Omega[VI(C,A)]}(0)$ which is the minimum-norm element in $\Omega[VI(C,A)]$.

5.7. Extragradient Methods for Split Feasibility Problems

The split feasibility problem (SFP) was first introduced by Censor and Elfving [16] in the finite-dimensional Hilbert spaces. It is very useful for modeling inverse problems which arise from phase retrievals and in medical image reconstruction. A number of image reconstruction problems can be formulated as SFP (see, for example, Refs. [4, 8] and references therein). Recently, SFP has been applied to study intensity-modulated radiation therapy (see, for example, Refs. [4, 15, 17, 18, 46] and references therein).

Let H_1 and H_2 be real Hilbert spaces and $\mathscr{B}(H_1, H_2)$ denote the family of all bounded linear operators from H_1 into H_2. Let C and Q be nonempty closed convex subsets of H_1 and H_2 respectively, and $A \in \mathscr{B}(H_1, H_2)$. The *split feasibility problem* is formulated to find a point $x^* \in H_1$ with the property

$$x^* \in C \text{ such that } Ax^* \in Q. \tag{5.7.32}$$

Let Γ denote the solution set of the SFP (5.7.32). A special case of SFP is the following convex constrained linear inverse problem [21] of finding an element $x^* \in H_1$ such that

$$x^* \in C \text{ such that } Ax^* = b. \tag{5.7.33}$$

The problem (5.7.33) has been extensively investigated in the literature using the projected Landweber iterative method [30].

Throughout this section, we assume that the SFP is consistent, that is, the solution set Γ of the SFP is nonempty. Let P_C and P_Q be the metric projections onto C and Q respectively, and let A^* be the adjoint of A. Let $f : H_1 \to \mathbb{R}$ be a continuously differentiable function defined by

$$f(x) := \frac{1}{2}\|Ax - P_Q Ax\|^2, \quad \text{for all } x \in H_1.$$

The gradient of f is $\nabla f = A^*(I - P_Q)A$. Then the minimization problem:

$$\min_{x \in C} f(x)$$

is ill-posed. Following the idea of Xu [46], we now consider the Tikhonov regularized problem:

$$\min_{x \in C} f_t(x), \tag{5.7.34}$$

where $f_t(x) = f(x) + \frac{t}{2}||x||^2$ and $t > 0$ is the regularization parameter. We observe that the gradient of f_t is

$$\nabla f_t = \nabla f + tI = A^*(I - P_Q)A + tI.$$

The regularized minimization (5.7.34) has a unique solution which is denoted by x_t.

Proposition 5.7.1 [12, Proposition 3.1]. *If the SFP is consistent, then the strong $\lim_{t \to 0} x_t$ exists and is the minimum-norm solution x_{min} of the SFP.*

The minimum-norm solution $x_{min} \in \Gamma$ of the SFP has the property

$$||x_{min}|| = \min\{||x^*|| : x^* \in \Gamma\}.$$

It is easy to see that $\nabla f = A^*(I - P_Q)A$ is a $1/||A||^2$-ism. It is worth noting, from Proposition 5.6.2(c) that x_t is a fixed point of the mapping $P_C(I - \lambda \nabla f_t)$ for any $\lambda \in (0, \frac{t}{t+\kappa^2}]$. In fact, it can be obtained through the limit as $n \to \infty$ of the sequence of Picard iterates:

$$x_t^{(n+1)} = P_C(I - \lambda \nabla f_t)x_t^{(n)}, \quad \text{for all } n \in \mathbb{N}_0.$$

The fixed point algorithms are important tools for solving SFP (5.7.32). We will apply the extragradient method to solve the SFP on the basis of the following observations.

Proposition 5.7.2. *Given $x^* \in H_1$, the following statements are equivalent:*

(a) $x^* \in \Gamma$.

(b) $x^* \in Fix(P_C(I - \lambda \nabla f))$ *for any $\lambda > 0$.*

(c) $x^* \in \Omega[VI(H_1, \nabla f)]$.

Proof. (a) $\Rightarrow$ (b). Let $x^* \in \Gamma$ and $\lambda > 0$. Then

$$
\begin{aligned}
Ax^* \in Q \quad &\Rightarrow \quad (I - P_Q)Ax^* = 0 \\
&\Rightarrow \quad \lambda A^*(I - P_Q)Ax^* = 0 \\
&\Rightarrow \quad (I - \lambda A^*(I - P_Q)A)x^* = x^* \\
&\Rightarrow \quad x^* \in Fix(P_C(I - \lambda \nabla f)).
\end{aligned}
$$

(b) $\Rightarrow$ (a). Suppose that $x^* \in Fix(P_C(I - \lambda \nabla f))$. Note Γ is nonempty. Then there exists $x_0 \in C$ such that $(I - P_Q)Ax_0 = 0$. Let $\lambda > 0$. Then

$$\langle (I - \lambda A^*(I - P_Q)A)x^* - x^*, y - x^* \rangle \leq 0 \text{ for all } y \in C$$
$$\Rightarrow \quad -\lambda \langle (A^*(I - P_Q)A)x^*, y - x^* \rangle \leq 0 \text{ for all } y \in C$$
$$\Rightarrow \quad \langle (A^*(I - P_Q)A)x^*, x^* - y \rangle \leq 0 \text{ for all } y \in C$$
$$\Rightarrow \quad \langle Ax^* - P_Q Ax^*, Ax^* - Ay \rangle \leq 0 \text{ for all } y \in C$$
$$\Rightarrow \quad \langle Ax^* - P_Q Ax^*, Ax^* - Ax_0 \rangle \leq 0. \tag{5.7.35}$$

For $x \in H_2$ and $z \in Q$, we have

$$z = P_Q x \quad \Leftrightarrow \quad \langle x - P_Q x, v - P_Q x \rangle = \langle x - z, v - z \rangle \leq 0, \quad \text{for all } v \in Q.$$

Note $Ax^* \in H_2$. Let $w = P_Q Ax^*$. Then

$$\langle Ax^* - P_Q Ax^*, v - P_Q Ax^* \rangle = \langle Ax^* - w, v - w \rangle \leq 0, \quad \text{for all } v \in Q.$$

Note $Ax_0 = P_Q Ax_0 \in Q$. Then, for $v = Ax_0$, we have

$$\langle Ax^* - P_Q Ax^*, Ax_0 - P_Q Ax^* \rangle \leq 0. \tag{5.7.36}$$

Adding up (5.7.35) and (5.7.36) we find that

$$\|Ax^* - P_Q Ax^*\|^2 = \langle Ax^* - P_Q Ax^*, P_Q Ax^* - Ax^* \rangle \leq 0.$$

Hence $Ax^* - P_Q Ax^* = 0$. Thus $x^* \in \Gamma$.

(b) $\Leftrightarrow$ (c). This follows from Theorem 5.3.1. □

It is shown in Refs. [8, 9, 46] that for $\lambda \in (0, 2/\|A\|^2)$, the sequence $\{x_n\}$ generated by

$$x_{n+1} = P_C(x_n - \lambda \nabla f x_n), \quad n \in \mathbb{N}_0, \tag{5.7.37}$$

converges weakly to an element of Γ. The method with gradient ∇f given as in (5.7.37) is referred to as the CQ algorithm.

5.7.1. An Extragradient Method for Nonexpansive Mappings and Split Feasibility Problems

In this subsection, we propose an extragradient method and prove that the sequence generated by the proposed method converges weakly to an element of $Fix(S) \cap \Gamma$, where $S : C \to C$ is a nonexpansive mapping.

Theorem 5.7.1. *Let C and Q be nonempty closed convex subsets of real Hilbert spaces H_1 and H_2 respectively. Let $S : C \to C$ be a nonexpansive mapping such that $Fix(S) \cap \Gamma \neq \emptyset$. Let $\{S_n\}$ be a sequence of nonexpansive mappings from C into itself. Let $\{x_n\}$*

be the sequence in C generated by the following extragradient algorithm:

$$\begin{cases} x_0 \in C \text{ chosen arbitrarily}, \\ y_n = P_C(x_n - \lambda_n \nabla f_{t_n} x_n), \\ x_{n+1} = (1 - \alpha_n)x_n + \alpha_n S_n P_C(x_n - \lambda_n \nabla f_{t_n} y_n), \quad n \in \mathbb{N}_0, \end{cases} \qquad (5.7.38)$$

where $\{\alpha_n\}$, $\{\lambda_n\}$ and $\{t_n\}$ are sequences of positive real numbers satisfying the following conditions:

(i) $0 < c \le \alpha_n \le d < 1$ *for all* $n \in \mathbb{N}_0$,

(ii) $0 < a \le \lambda_n \le b < 1/||A||^2$ *for all* $n \in \mathbb{N}_0$,

(iii) $\sum_{n=0}^{\infty} t_n < \infty$,

(iv) $\sum_{n=0}^{\infty} ||S_n v - S v|| < \infty$ *for all* $v \in Fix(S) \cap \Gamma$ *and* $\lim_{n \to \infty} \mathscr{D}_B(S_n, S) = 0$ *for all* $B \in \mathscr{B}(C)$.

Then the sequence $\{x_n\}$ converges weakly to an element of $Fix(S) \cap \Gamma$.

Proof. It is easy to see that $\nabla f = A^*(I - P_Q)A$ is a $1/||A||^2$-ism. Hence, from Proposition 5.6.2, $F_n := \nabla f + t_n I$ is a $\frac{1}{t_n + ||A||^2}$-ism. Also F_n is $t_n + ||A||^2$-Lipschitz continuous.

From Proposition 5.6.3, we have that $I - \lambda_n F_n$ is nonexpansive for $\lambda_n \in \left(0, \frac{1}{t_n + ||A||^2}\right)$. Put $z_n = P_C(x_n - \lambda_n F_n y_n)$. Let $v \in \Gamma$. Note $P_C(I - \lambda \nabla f)v = v$ for all $\lambda \in \left(0, \frac{2}{||A||^2}\right)$. Hence $\langle \nabla f(v), v - y \rangle \le 0$ for all $y \in C$. It follows that

$$\langle F_n v, v - y_n \rangle = \langle \nabla f(v), v - y_n \rangle + t_n \langle v, v - y_n \rangle \le t_n \langle v, v - y_n \rangle, \quad \text{for all } n \in \mathbb{N}_0.$$

From (5.7.38), we have

$$\begin{aligned} ||y_n - v|| &= ||P_C(x_n - \lambda_n F_n x_n) - v|| \\ &\le ||P_C(x_n - \lambda_n F_n x_n) - P_C(v - \lambda_n F_n v)|| + ||P_C(v - \lambda_n F_n v) - v|| \\ &\le ||x_n - v|| + ||P_C(I - \lambda_n(\nabla f + t_n I))v - P_C(I - \lambda_n \nabla f)v|| \\ &\le ||x_n - v|| + \lambda_n t_n ||v||, \end{aligned}$$

and hence

$$\begin{aligned} 2\lambda_n t_n ||y_n - v|| \, ||v|| &\le t_n[||y_n - v||^2 + \lambda_n^2 ||v||^2] \\ &\le t[(||x_n - v|| + \lambda_n t_n ||v||)^2 + \lambda_n^2 ||v||^2] \\ &\le t_n[2||x_n - v||^2 + 2\lambda_n^2 t_n^2 ||v||^2 + \lambda_n^2 ||v||^2] \\ &= 2t_n ||x_n - v||^2 + \lambda_n^2(1 + 2t_n^2)t_n ||v||^2. \qquad (5.7.39) \end{aligned}$$

Note $z_n = P_C(x_n - \lambda_n \nabla f_{t_n} y_n)$. Then, by (P3), we have

$$
\begin{aligned}
\|z_n - v\|^2 &\leq \|x_n - \lambda_n F_n y_n - v\|^2 - \|x_n - \lambda_n F_n y_n - z_n\|^2 \\
&= \|x_n - v\|^2 + \|\lambda_n F_n y_n\|^2 - 2\lambda_n \langle F_n y_n, x_n - v \rangle \\
&\quad - (\|x_n - z_n\|^2 + \|\lambda_n F_n y_n\|^2 - 2\lambda_n \langle F_n y_n, x_n - z_n \rangle) \\
&= \|x_n - v\|^2 - \|x_n - z_n\|^2 + 2\lambda_n \langle F_n y_n, v - z_n \rangle \\
&= \|x_n - v\|^2 - \|x_n - z_n\|^2 + 2\lambda_n [\langle F_n y_n, v - y_n \rangle + \langle F_n y_n, y_n - z_n \rangle] \\
&= \|x_n - v\|^2 - \|x_n - z_n\|^2 + +2\lambda_n [\langle F_n y_n - F_n v, v - y_n \rangle + \langle F_n v, v - y_n \rangle \\
&\quad + \langle F_n y_n, y_n - z_n \rangle] \\
&\leq \|x_n - v\|^2 - \|x_n - z_n\|^2 + 2\lambda_n [\langle F_n v, v - y_n \rangle + \langle F_n y_n, y_n - z_n \rangle] \\
&\leq \|x_n - v\|^2 - \|x_n - y_n + y_n - z_n\|^2 + 2\lambda_n [t_n \langle v, v - y_n \rangle + \langle F_n y_n, y_n - z_n \rangle] \\
&= \|x_n - v\|^2 - \|x_n - y_n\|^2 - \|y_n - z_n\|^2 - 2\langle x_n - y_n, y_n - z_n \rangle \\
&\quad + 2\lambda_n [t_n \langle v, v - y_n \rangle + \langle F_n y_n, y_n - z_n \rangle] \\
&= \|x_n - v\|^2 - \|x_n - y_n\|^2 - \|y_n - z_n\|^2 + 2\langle y_n - (x_n - \lambda_n F_n y_n), y_n - z_n \rangle \\
&\quad + 2\lambda_n t_n \langle v, v - y_n \rangle .
\end{aligned}
$$

$$
\begin{aligned}
\langle x_n - \lambda_n F_n x_n - y_n, z_n - y_n \rangle \\
= \langle x_n - \lambda_n F_n x_n - P_C(x_n - \lambda_n F_n x_n), z_n - P_C(x_n - \lambda_n F_n x_n) \rangle \leq 0.
\end{aligned}
$$

Thus

$$
\begin{aligned}
\langle x_n - \lambda_n F_n y_n - y_n, z_n - y_n \rangle &= \langle x_n - \lambda_n F_n x_n - y_n, z_n - y_n \rangle + \lambda_n \langle F_n x_n - F_n y_n, z_n - y_n \rangle \\
&\leq \lambda_n \langle F_n x_n - F_n y_n, z_n - y_n \rangle \\
&\leq \lambda_n \|F_n x_n - F_n y_n\| \, \|z_n - y_n\| \\
&\leq \lambda_n (t_n + \|A\|^2) \|x_n - y_n\| \, \|z_n - y_n\|.
\end{aligned}
$$

Since $\lambda_n(t_n + \|A\|^2) < 1$ for all $n \in \mathbb{N}_0$, we have

$$
\begin{aligned}
\|z_n - v\|^2 &\leq \|x_n - v\|^2 - \|x_n - y_n\|^2 - \|y_n - z_n\|^2 \\
&\quad + 2\lambda_n(t_n + \|A\|^2)\|x_n - y_n\| \, \|z_n - y_n\| + 2\lambda_n t_n \langle v, v - y_n \rangle \\
&\leq \|x_n - v\|^2 - \|x_n - y_n\|^2 - \|y_n - z_n\|^2 + \lambda_n^2(t_n + \|A\|^2)^2 \|x_n - y_n\|^2 \\
&\quad + \|z_n - y_n\|^2 + 2\lambda_n t_n \langle v, v - y_n \rangle \\
&= \|x_n - v\|^2 - (1 - \lambda_n^2(t_n + \|A\|^2)^2)\|x_n - y_n\|^2 + 2\lambda_n t_n \langle v, v - y_n \rangle \\
&\leq \|x_n - v\|^2 + 2\lambda_n t_n \langle v, v - y_n \rangle \\
&\leq \|x_n - v\|^2 + 2\lambda_n t_n \|y_n - v\| \, \|v\|.
\end{aligned}
$$

Observe that

$$\begin{aligned}
\|S_n z_n - v\|^2 &\leq (\|S_n z_n - S_n v\| + \|S_n v - Sv\|)^2 \\
&\leq \|z_n - v\|^2 + M_9 \|S_n v - Sv\|,
\end{aligned}$$

for some constant $M_9 > 0$. Hence, from (5.7.38) and (5.7.39), we get

$$\begin{aligned}
&\|x_{n+1} - v\|^2 \\
&= \|(1-\alpha_n)x_n + \alpha_n S_n z_n - v\|^2 \\
&= (1-\alpha_n)\|x_n - v\|^2 + \alpha_n \|S_n z_n - v\|^2 - \alpha_n(1-\alpha_n)\|x_n - S_n z_n\|^2 \\
&\leq (1-\alpha_n)\|x_n - v\|^2 + \alpha_n(\|z_n - v\|^2 + M_9\|S_n v - Sv\|) - \alpha_n(1-\alpha_n)\|x_n - S_n z_n\|^2 \\
&\leq (1-\alpha_n)\|x_n - v\|^2 + \alpha_n[\|x_n - v\|^2 + 2\lambda_n t_n \|y_n - v\| \|v\| \\
&\quad -(1 - \lambda_n^2(t_n + \|A\|^2)^2)\|x_n - y_n\|^2] + \alpha_n M_9\|S_n v - Sv\| - \alpha_n(1-\alpha_n)\|x_n - S_n z_n\|^2 \\
&\leq \|x_n - v\|^2 + 2\lambda_n t_n \|y_n - v\| \|v\| + \alpha_n M_9\|S_n v - Sv\| \\
&\quad -\alpha_n(1 - \lambda_n^2(t_n + \|A\|^2)^2)\|x_n - y_n\|^2 - \alpha_n(1-\alpha_n)\|x_n - S_n z_n\|^2 \\
&\leq (1+2t_n)\|x_n - v\|^2 + M_9\|S_n v - Sv\| + \lambda_n^2 t_n(1 + 2t_n^2)\|v\|^2 \\
&\quad -\alpha_n(1 - \lambda_n^2(t_n + \|A\|^2)^2)\|x_n - y_n\|^2 - \alpha_n(1-\alpha_n)\|x_n - S_n z_n\|^2 \\
&= (1+\beta_n)\|x_n - v\|^2 + \delta_n - \alpha_n(1 - \lambda_n^2(t_n + \|A\|^2)^2)\|x_n - y_n\|^2 \\
&\quad -\alpha_n(1-\alpha_n)\|x_n - S_n z_n\|^2,
\end{aligned}$$

where $\beta_n = 2t_n$ and $\delta_n = M_9\|S_n v - Sv\| + \lambda_n^2 t_n(1 + 2t_n^2)\|v\|^2$. Since $0 < a \leq \lambda_n \leq b < 1/\|A\|^2$ for all $n \in \mathbb{N}_0$ and $\sum_{n=0}^{\infty} t_n < \infty$, we obtain that $\sum_{n=0}^{\infty} \beta_n < \infty$ and $\sum_{n=0}^{\infty} \delta_n < \infty$. Hence, from Lemma 5.2.2, we obtain that $\lim_{n\to\infty} \|x_n - v\|$ exists. From the last inequality, we have

$$\begin{aligned}
&c(1 - b^2(t_n + \|A\|^2)^2)\|x_n - y_n\|^2 + c(1-d)\|x_n - S_n z_n\|^2 \\
&\leq \alpha_n(1 - \lambda_n^2(t_n + \|A\|^2)^2)\|x_n - y_n\|^2 + \alpha_n(1-\alpha_n)\|x_n - S_n z_n\|^2 \\
&\leq (1+\beta_n)\|x_n - v\|^2 - \|x_{n+1} - v\|^2 + \delta_n.
\end{aligned}$$

Since $\lim_{n\to\infty} \|x_n - v\|$ exists and $\lim_{n\to\infty} t_n = 0$, we have $\lim_{n\to\infty} \|x_n - y_n\| = 0$ and $\lim_{n\to\infty} \|x_n - S_n z_n\| = 0$. Moreover,

$$\begin{aligned}
\|y_n - z_n\| &= \|P_C(x_n - \lambda_n F_n x_n) - P_C(x_n - \lambda_n F_n y_n)\| \\
&\leq \|(x_n - \lambda_n F_n x_n) - (x_n - \lambda_n F_n y_n)\| = \lambda_n \|F_n x_n - F_n y_n\| \\
&\leq \lambda_n(t_n + \|A\|^2)\|x_n - y_n\| \to 0 \text{ as } n \to \infty,
\end{aligned}$$

and hence

$$\|x_n - z_n\| \leq \|x_n - y_n\| + \|y_n - z_n\| \to 0 \text{ as } n \to \infty.$$

Observe that

$$
\begin{aligned}
\|y_n - P_C(x_n - \lambda_n \nabla f(x_n))\| &= \|P_C(x_n - \lambda_n F_n x_n) - P_C(x_n - \lambda_n \nabla f(x_n))\| \\
&\leq \lambda_n \|F_n(x_n) - \nabla f(x_n)\| = \lambda_n t_n \|x_n\|,
\end{aligned}
$$

which gives that $\lim_{n\to\infty} \|y_n - P_C(x_n - \lambda_n \nabla f(x_n))\| = 0$. On the other hand, for $B = \{z_n\}$, we have

$$
\begin{aligned}
\|x_n - Sz_n\| &\leq \|x_n - S_n z_n\| + \|S_n z_n - Sz_n\| \\
&\leq \|x_n - S_n z_n\| + \mathscr{D}_B(S_n, S) \to 0 \text{ as } n \to \infty,
\end{aligned}
$$

which implies that

$$
\begin{aligned}
\|x_n - Sx_n\| &\leq \|x_n - Sz_n\| + \|Sz_n - Sx_n\| \\
&\leq \|x_n - Sz_n\| + \|z_n - x_n\| \to 0 \text{ as } n \to \infty.
\end{aligned}
$$

Since $\{x_n\}$ is bounded, we see that $\{x_n\}$ has a weakly convergent subsequence. We show that $\omega_w(\{x_n\})$ is singleton. Assume that $\{x_{n_i}\}$ converges weakly to a point $p \in C$ and that $\{x_{n_j}\}$ converges weakly to a point $q \in C$ with $p \neq q$. From Proposition 5.2.5, we have $p \in \Omega[VI(C, \nabla f)]$. Similarly, $q \in \Omega[VI(C, \nabla f)]$. It follows that $\lim_{n\to\infty} \|x_n - p\|$ and $\lim_{n\to\infty} \|x_n - q\|$ exist. Since H_1 satisfies the Opial condition, we have

$$
\lim_{n\to\infty} \|x_n - p\| = \lim_{i\to\infty} \|x_{n_i} - p\| < \lim_{i\to\infty} \|x_{n_i} - q\| = \lim_{n\to\infty} \|x_n - q\|,
$$

$$
\lim_{n\to\infty} \|x_n - q\| = \lim_{k\to\infty} \|x_{n_j} - q\| < \lim_{j\to\infty} \|x_{n_j} - p\| = \lim_{n\to\infty} \|x_n - p\|,
$$

a contradiction. Hence $p = q$, so $\omega_w(\{x_n\})$ is singleton. Therefore $\{x_n\}$ converges weakly to p. $\qquad\square$

Corollary 5.7.1. *Let C and Q be nonempty closed convex subsets of real Hilbert spaces H_1 and H_2 respectively. Let $S : C \to C$ be a nonexpansive mapping such that $\mathrm{Fix}(S) \cap \Gamma \neq \emptyset$. Let $\{x_n\}$ be the sequence in C generated by the following extragradient algorithm:*

$$
\begin{cases}
x_0 \in C \text{ chosen arbitrarily,} \\
y_n = P_C(x_n - \lambda_n \nabla f_{t_n} x_n), \\
x_{n+1} = (1 - \alpha_n) x_n + \alpha_n S P_C(x_n - \lambda_n \nabla f_{t_n} y_n), \quad n \in \mathbb{N}_0,
\end{cases}
$$

where $\{\alpha_n\}$, $\{\lambda_n\}$ and $\{t_n\}$ are sequences of positive real numbers satisfying the following conditions:

(i) $0 < c \leq \alpha_n \leq d < 1$ for all $n \in \mathbb{N}_0$,

(ii) $0 < a \leq \lambda_n \leq b < 1/\|A\|^2$ for all $n \in \mathbb{N}_0$,

(iii) $\sum_{n=0}^{\infty} t_n < \infty$.

Then the sequence $\{x_n\}$ converges weakly to an element of $Fix(S) \cap \Gamma$.

Corollary 5.7.2. *Let C and Q be nonempty closed convex subsets of real Hilbert spaces H_1 and H_2 respectively. Let $\{x_n\}$ be the sequence in C generated by the following extragradient algorithm:*

$$
\begin{cases}
x_0 \in C \text{ chosen arbitrarily,} \\
y_n = P_C(x_n - \lambda_n \nabla f_{t_n} x_n), \\
x_{n+1} = (1 - \alpha_n)x_n + \alpha_n P_C(x_n - \lambda_n \nabla f_{t_n} y_n), \quad n \in \mathbb{N}_0,
\end{cases}
$$

where $\{\alpha_n\}$, $\{\lambda_n\}$ and $\{t_n\}$ are sequences of positive real numbers satisfying the following conditions:

(i) $0 < c \leq \alpha_n \leq d < 1$ *for all* $n \in \mathbb{N}_0$,

(ii) $0 < a \leq \lambda_n \leq b < 1/||A||^2$ *for all* $n \in \mathbb{N}_0$,

(iii) $\sum_{n=0}^{\infty} t_n < \infty$.

Then the sequence $\{x_n\}$ converges weakly to an element of Γ.

5.7.2. Strong Convergence of a Relaxed Extragradient Method for Split Feasibility Problems

Theorem 5.7.2 [12, Theorem 3.1]. *Let C and Q be nonempty closed convex subsets of real Hilbert spaces H_1 and H_2 respectively. Let $\{x_n\}$ be the sequence in C generated by the following Mann-type extragradient algorithm:*

$$
\begin{cases}
x_0 \in C \text{ chosen arbitrarily,} \\
y_n = P_C(x_n - \lambda_n \nabla f_{t_n} x_n), \\
x_{n+1} = \alpha_n x_n + \beta_n y_n + \gamma_n P_C(x_n - \lambda_n \nabla f_{t_n} y_n), \quad n \in \mathbb{N}_0,
\end{cases}
\tag{5.7.40}
$$

where $\{\alpha_n\}$, $\{\beta_n\}$, $\{\gamma_n\}$, $\{\lambda_n\}$ and $\{t_n\}$ are sequences of positive real numbers satisfying the following conditions:

(i) $\alpha_n + \beta_n + \gamma_n = 1$ *for all* $n \in \mathbb{N}_0$,

(ii) $0 < \lambda_n \leq \frac{t_n}{(t_n + ||A||^2)^2}$ *for all* $n \in \mathbb{N}_0$,

(iii) $t_n \to 0$,

(iv) $\sum_{n=0}^{\infty} \gamma_n \lambda_n t_n^2 = \infty$,

(v) $\frac{|\lambda_n - \lambda_{n+1}| + \lambda_n |t_n - t_{n+1}|}{\lambda_{n+1}^2 \gamma_{n+1} t_{n+1}^3} \to 0$,

(vi) $\frac{2\gamma_n}{t_n + ||A||^2} \leq \beta_n \lambda_n$ *for all* $n \in \mathbb{N}_0$.

Then $\{x_n\}$ *converges in norm to the minimum-norm solution of the SFP.*

Proof. Set $F_n := \nabla f + t_n I$ and $T_n := P_C(I - \lambda_n F_n)$. Note that T_n is a contraction with Lipschitz constant $1 - \frac{\lambda_n t_n}{2}$. In fact, F_n is $(t_n + ||A||^2)$-Lipschitzian and t_n-strongly monotone and hence T_n is a contraction with Lipschitz constant

$$\sqrt{1 - 2\lambda_n t_n + \lambda_n^2(t_n + ||A||^2)^2} \leq \sqrt{1 - \lambda_n t_n} \leq 1 - \frac{\lambda_n t_n}{2}.$$

For each $n \in \mathbb{N}_0$, let z_n be a unique fixed point of the contraction $T_n := P_C(I - \lambda_n F_n)$. Then $z_n = x_{t_n}$, and so $z_n \to x_{\min}$ in norm. So, it is sufficient to prove that $||x_{n+1} - z_n|| \to 0$ as $n \to \infty$. Hence

$$||y_n - z_n|| = ||T_n x_n - T_n z_n|| \leq \left(1 - \frac{\lambda_n t_n}{2}\right) ||x_n - z_n||.$$

Note

$$
\begin{aligned}
||z_{n+1} - z_n|| &= ||T_{n+1} z_{n+1} - T_n z_n|| \\
&\leq ||T_{n+1} z_{n+1} - T_{n+1} z_n|| + ||T_{n+1} z_n - T_n z_n|| \\
&\leq \left(1 - \frac{\lambda_{n+1} t_{n+1}}{2}\right) ||z_{n+1} - z_n|| + ||T_{n+1} z_n - T_n z_n||,
\end{aligned}
$$

which gives us

$$||z_{n+1} - z_n|| \leq \frac{2}{\lambda_{n+1} t_{n+1}} ||T_{n+1} z_n - T_n z_n||. \tag{5.7.41}$$

Since ∇f is Lipschitzian and $\{z_n\}$ is bounded, we have

$$
\begin{aligned}
||T_{n+1} z_n - T_n z_n|| &= ||P_C(I - \lambda_{n+1} F_{n+1}) z_n - P_C(I - \lambda_n F_n) z_n|| \\
&\leq ||(I - \lambda_{n+1} F_{n+1}) z_n - (I - \lambda_n F_n) z_n|| \\
&= ||\lambda_{n+1}(\nabla f + t_{n+1} I) z_n - \lambda_n(\nabla f + t_n I) z_n|| \\
&\leq |\lambda_{n+1} - \lambda_n| \, ||\nabla f z_n|| + |\lambda_{n+1} t_{n+1} - \lambda_n t_n| \, ||z_n|| \\
&\leq |\lambda_{n+1} - \lambda_n| \, ||\nabla f z_n|| \\
&\quad + (|\lambda_{n+1} t_{n+1} - \lambda_n t_{n+1}| + |\lambda_n t_{n+1} - \lambda_n t_n|) ||z_n|| \\
&\leq |\lambda_{n+1} - \lambda_n| \, ||\nabla f z_n|| + (|\lambda_{n+1} - \lambda_n| t_{n+1} + \lambda_n |t_{n+1} - t_n|) ||z_n|| \\
&\leq (|\lambda_{n+1} - \lambda_n| + \lambda_n |t_{n+1} - t_n|) M_{10},
\end{aligned}
$$

for some constant $M_{10} > 0$. Hence, from (5.7.40), we get

$$
\begin{aligned}
&\|x_{n+1} - z_n\| \\
&= \|\alpha_n x_n + \beta_n y_n + \gamma_n P_C(x_n - \lambda_n F_n y_n) - z_n\| \\
&\leq \alpha_n \|x_n - z_n\| + \beta_n \|y_n - z_n\| + \gamma_n \|P_C(x_n - \lambda_n F_n y_n) - P_C(z_n - \lambda_n F_n z_n)\| \\
&\leq \alpha_n \|x_n - z_n\| + \beta_n \|y_n - z_n\| + \gamma_n \|x_n - \lambda_n F_n y_n - (z_n - \lambda_n F_n z_n)\| \\
&\leq \alpha_n \|x_n - z_n\| + \beta_n \|y_n - z_n\| + \gamma_n (\|x_n - z_n\| + \lambda_n \|F_n y_n - F_n z_n\|) \\
&\leq \alpha_n \|x_n - z_n\| + \beta_n \|y_n - z_n\| + \gamma_n (\|x_n - z_n\| + \lambda_n (t_n + \|A\|^2) \|y_n - z_n\|) \\
&= (1 - \beta_n) \|x_n - z_n\| + [\beta_n + \gamma_n \lambda_n (t_n + \|A\|^2)] \|y_n - z_n\| \\
&\leq (1 - \beta_n) \|x_n - z_n\| + \left[\beta_n + \gamma_n \frac{t_n}{(t_n + \|A\|^2)^2} (t_n + \|A\|^2) \right] \|y_n - z_n\| \\
&= (1 - \beta_n) \|x_n - z_n\| + \left[\beta_n + \frac{\gamma_n t_n}{t_n + \|A\|^2} \right] \|y_n - z_n\| \\
&\leq (1 - \beta_n) \|x_n - z_n\| + \left[\beta_n + \frac{\gamma_n t_n}{t_n + \|A\|^2} \right] \left(1 - \frac{\lambda_n t_n}{2} \right) \|x_n - z_n\| \\
&= \left[1 - \beta_n + \beta_n + \frac{\gamma_n t_n}{t_n + \|A\|^2} - \left(\beta_n + \frac{\gamma_n t_n}{t_n + \|A\|^2} \right) \frac{\lambda_n t_n}{2} \right] \|x_n - z_n\| \\
&= \left[1 + \frac{\gamma_n t_n}{t_n + \|A\|^2} - \frac{\beta_n \lambda_n t_n}{2} - \frac{\gamma_n \lambda_n t_n^2}{2(t_n + \|A\|^2)} \right] \|x_n - z_n\| \\
&\leq \left[1 - \frac{\gamma_n \lambda_n t_n^2}{2(t_n + \|A\|^2)} \right] \|x_n - z_n\|.
\end{aligned}
$$

It follows from (5.7.41) that

$$
\begin{aligned}
\|x_{n+1} - z_n\| &\leq \left[1 - \frac{\gamma_n \lambda_n t_n^2}{2(t_n + \|A\|^2)} \right] (\|x_n - z_{n-1}\| + \|z_{n-1} - z_n\|) \\
&\leq \left[1 - \frac{\gamma_n \lambda_n t_n^2}{2(t_n + \|A\|^2)} \right] \|x_n - z_{n-1}\| + \frac{2}{\lambda_n t_n} \|T_n z_{n-1} - T_{n-1} z_{n-1}\| \\
&\leq \left[1 - \frac{\gamma_n \lambda_n t_n^2}{2(t_n + \|A\|^2)} \right] \|x_n - z_{n-1}\| + \frac{2M}{\lambda_n t_n} (|\lambda_n - \lambda_{n-1}| \\
&\quad + \lambda_{n-1} |t_n - t_{n-1}|).
\end{aligned}
$$

By applying Lemma 5.2.5, we conclude that $\|x_{n+1} - z_n\| \to 0$ as $n \to \infty$. Therefore $x_n \to x_{\min}$ in norm. $\quad\square$

ACKNOWLEDGMENTS

This chapter was partially written during the visit of the first author to King Fahd University of Petroleum & Minerals (KFUPM), Dhahran, Saudi Arabia. Qamrul Hasan Ansari is grateful to KFUPM for providing excellent research facilities during his visit to carry out this work.

REFERENCES

1. Agarwal RP, O'Regan D, Sahu DR, Fixed Point Theory for Lipschitzian-type Mappings with Applications. New York: Springer; 2009.
2. Anh PN, A new extragradient iteration algorithm for bilevel variational inequalities. Acta Math. Vietnamica 2012;37:95–107.
3. Ansari QH, Lalitha CS, Mehta M, Generalized Convexity, Nonsmooth Variational Inequalities, and Nonsmooth Optimization. Boca Raton: CRC Press, Taylor & Francis Group; 2014.
4. Ansari QH, Rehan A, Split feasibility and fixed point problems. In: Nonlinear Analysis: Approximation Theory, Optimization and Applications, Edited by Q.H. Ansari, Birkhäuser, Springer, New Delhi, pp. 281–322, 2014.
5. Auslender A, Teboulle M, Asymptotic Cones and Functions in Optimization and Variational Inequalities. New York: Springer; 2003.
6. Baiocchi C, Capelo A, Variational and Quasivariational Inequalities: Applications to Free Boundary Problems. Chichester: Wiley; 1984.
7. Bauschke HH, Borwein JM, On projection algorithms for solving convex feasibility problems. SIAM Review 1996;38:367–426.
8. Byrne C, Iterative oblique projection onto convex subsets and the split feasibility problem. Inverse Prob. 2002;18:441–453.
9. Byrne C, A unified treatment of some iterative algorithms in signal processing and image reconstruction. Inverse Prob. 2004;20:103–20.
10. Ceng LC, Ansari QH, Yao JC, Mann-type steepest-descent and modified hybrid steepest-descent methods for variational inequalities in Banach spaces. Numer. Funct. Anal. Optim. 2008;29:987–1033.
11. Ceng LC, Ansari QH, Schaible S, Hybrid extragradient-like methods for generalized mixed equilibrium problems, systems of generalized equilibrium problems and optimization problems. J. Global Optim. 2012;53:69–96.
12. Ceng LC, Ansari QH, Yao JC, Relaxed extragradient methods for finding minimum-norm solutions of the split feasibility problem. Nonlinear Anal. 2012;75:2116–2125.
13. Ceng LC, Ansari QH, Wong NC, Yao JC, An extragradient-like approximation method for variational inequalities and fixed point problems. Fixed Point Theory Appl. 2011;2011:Article ID 22.
14. Ceng LC, Yao JC, A relaxed extragradient-like method for a generalized mixed equilibrium problem, a general system of generalized equilibria and a fixed point problem. Nonlinear Anal. 2010;72:1922–1937.
15. Censor Y, Bortfeld T, Martin B, Trofimov A, A unified approach for inversion problems in intensity-modulated radiation therapy. Phys. Med. Biol. 2006;51:2353–2365.
16. Censor Y, Elfving T, A multiprojection algorithm using Bregman projections in a product space. Numer. Algorithms 1994;8:221–239.
17. Censor Y, Elfving T, Kopf N, Bortfeld T, The multiple-sets split feasibility problem and its applications for inverse problems. Inverse Prob. 2005;21:2071–2084.
18. Censor Y, Motova A, Segal A, Perturbed projections and subgradient projections for the multiple-sets split feasibility problem. J. Math. Anal. Appl. 2007;327:1244–1256.
19. Combettes PL, Hilbertian convex feasibility problem: Convergence of projection method. Appl. Math. Optim. 1997;35:311–330.
20. Eckstein J, Bertsekas DP, On the Douglas-Rachford splitting method and the proximal point algorithm for maximal monotone operators. Math. Program. 1992;55:293–318.
21. Eicke B, Iteration methods for convexly constrained ill-posed problems in Hilbert spaces. Numer. Funct. Anal. Optim. 1992;13:413–429.
22. Facchinei F, Pang J-S, Finite-Dimensional Variational Inequalities and Complementarity Problems, Vol. I, II. New York: Springer; 2003.
23. Fichera F, Problemi elettrostatici con vincoli unilaterali il problema di Signorini con ambigue condizioni al contorno. Atti Acad. Naz. Lincei. Mem. Cl. Sci. Fis. Mat. Nat. Sez. 1964;17:91–140.

24. Goldstein AA, Convex programming in Hilbert space. Bull. Am. Math. Soc. 1964;70:709–710.
25. Gwinner J, Stability of monotone variational inequalities. In: Giannessi F, Maugeri A, editors, Variational Inequalities and Network Equilibrium Problems, New York: Plenum Press; 1995;123–142.
26. Glowinski R, Lions JL, Trémolières R, Numerical Analysis of Variational Inequalities. Amsterdam: North-Holland; 1981.
27. Goh CJ, Yang XQ, Duality in Optimization and Variational Inequalities. London: Taylor & Francis; 2002.
28. Korpelevich GM, An extragradient method for finding saddle points and for other problems. Ekonomika Mat. Metody 1976;12:747–756.
29. Kinderlehrer D, Stampacchia G, An Introduction to Variational Inequalities and Their Applications. New York: Academic Press; 1980.
30. Landweber L, An iterative formula for Fredholm integral equations of the first kind. Am. J. Math. 1951;73:615–624.
31. Lu X, Xu HK, Yin X, Hybrid methods for a class of monotone variational inequalities. Nonlinear Anal. 2009;71:1032–1041.
32. Luo ZQ, Pang JS, Ralph D, Mathematical Programs with Equilibrium Constraints. New York: Cambridge University Press; 1996.
33. Maing PE, Approximation method for common fixed points of nonexpansive mappings in Hilbert spaces. J. Math. Anal. Appl. 2007;325:469–479.
34. Noor MA, A modified extragradient method for general monotone variational inequalities. Comput. Math. Appl. 1999;38:19–24.
35. Nadezhkina N, Takahashi W, Weak convergence theorem by an extragradient method for nonexpansive mappings and monotone mappings. J. Optim. Appl. 2006;128:191–201.
36. Nagurney A, Network Economics: A Variational Inequality Approach, vol. 1, Kluwer Academic Publishers, Boston, Massachusetts, 1993.
37. Outrata J, Kocvara M, Zowe J, Nonsmooth Approach to Optimization Problems with Equilibrium Constraints. Dordrecht: Kluwer Academic Publishers; 1998.
38. Patriksson M, Nonlinear Programming and Variational Inequality Problems: A Unified Approach. Dordrecht: Kluwer Academic Publishers; 1999.
39. Petrusel A, Yao JC, An extragradient iterative scheme by viscosity approximation methods for fixed point problems and variational inequality problems. Cent. Eur. J. Math. 2009;7(2):335–347.
40. Rockafellar RT, Monotone operators and the proximal point algorithm. SIAM J. Control Optim. 1976;14:877–898.
41. Sahu DR, Wong NC, Yao JC, A unified hybrid iterative method for solving variational inequalities involving generalized pseudo-contractive mappings. SIAM J. Control Optim. 2012;50:2335–2354.
42. Sahu DR, Wong NC, Yao JC, A generalized hybrid steepest-descent method for variational inequalities in Banach spaces. Fixed Point Theory Appl. 2011;2011:754–702.
43. Stampacchia G, Formes bilineaires coercivities sur les ensembles convexes. C. R. Acad. Sci. Paris 1964;258:4413–4416.
44. Suzuki T, Strong convergence theorems for infinite families of nonexpansive mappings in general Banach spaces. Fixed Point Theory Appl. 2005;2005:103–123.
45. Takahashi W, Toyoda M, Weak convergence theorems for nonexpansive mappings and monotone mappings. J. Optim. Theory Appl. 2003;118:417-428.
46. Xu HK, Iterative methods for the split feasibility problem in infinite-dimensional Hilbert spaces. Inverse Prob. 2010;26:1–17.
47. Yamada I, The hybrid steepest descent method for the variational inequality over the intersection of fixed point sets of nonexpansive mappings. In: Butnariu D, Censor Y, Reich S, editors, Inherently Parallel Algorithms in Feasibility and Optimization and Their Applications. Amsterdam: Elsevier, pp. 473–504 (2001).
48. Yao Y, Liou YC, An implicit extragradient method for hierarchical variational inequalities. Fixed Point Theory Appl. 2011;2011:Article ID 697248.
49. Yao Y, Liou YC, Yao JC, An extragradient method for fixed point problems and variational inequality problems. J. Ineq. Appl. 2007;2007:Article ID 38752.

50. Yao Y, Noor MA, Liou YC, Strong convergence of a modified extragradient method to the minimum-norm solution of variational inequalities. Abstr. Appl. Anal. 2012;2012:Article ID 817436.
51. Zeidler E, Nonlinear Functional Analysis and Its Applications, III: Variational Methods and Applications. New York: Springer; 1985.
52. Zeng LC, Yao JC, Strong convergence theorem by an extragradient method for fixed point problems and variational inequality problems. Taiwanese J. Math. 2006;10:1293–1303.

CHAPTER 6

Iterative Methods for Nonexpansive Type Mappings

Abdul Rahim Khan[a]* and Hafiz Fukhar-ud-din*[†]
* King Fahd University of Petroleum and Minerals, Department of Mathematics and Statistics,Dhahran, Saudi Arabia
‡ The Islamia University of Bahawalpur, Department of Mathematics,Bahawalpur, Pakistan
[a] Corresponding: arahim@kfupm.edu.sa

Contents

Abstract

The classical fixed point theorem of Goebel and Kirk for a nonexpansive mapping on a uniformly convex Banach space and a CAT(0) space is presented. The exact value of a fixed point for certain mappings cannot be found analytically. In order to find its approximate value, the iterative construction of fixed points becomes essential. In this chapter, fixed point (common fixed point) problems for nonexpansive (asymptotically quasi-nonexpansive) mappings in Banach spaces and some important classes of a metric space are studied; in particular, results on weak convergence, $\triangle$-convergence and strong convergence of explicit and multistep schemes of nonexpansive type mappings to a common fixed point in the context of uniformly convex Banach spaces, CAT(0) spaces, hyperbolic spaces and convex metric spaces are presented.

6.1. Introduction and Preliminaries

Many important problems of mathematics including boundary value problems for nonlinear ordinary or partial differential equations can be translated in terms of a fixed point equation $Tx = x$ for a given mapping T on a Banach space.

http://dx.doi.org/10.1016/B978-0-12-804295-3.50006-1
231

The class of nonexpansive mappings contains contractions as a subclass and its study has remained a popular area of research ever since its introduction. The iterative construction of fixed points of these mappings is a fascinating field of research. The fixed point problem for one or a family of nonexpansive (asymptotically nonexpansive) mappings has been studied in Banach spaces and metric spaces [4, 20, 21, 29, 35, 39, 44, 47, 61–63].

Let C be a nonempty convex subset of a real Banach space E and $\mathbb{R}$ be the set of real numbers. A mapping $T : C \rightarrow C$ is called:

(a) *nonexpansive* if $\|Tx - Ty\| \leq \|x - y\|$ for all $x, y \in C$;

(b) *quasi-nonexpansive* if the set $F(T) = \{x \in C : T(x) = x\}$ of fixed points of T is nonempty and $\|Tx - y\| \leq \|x - y\|$ for all $x \in C$ and $y \in F(T)$;

(c) *asymptotically nonexpansive* if there exists a sequence $\{k_n\} \subset [1, \infty)$ with $\lim\limits_{n \to \infty} k_n = 1$ such that $\|T^n x - T^n y\| \leq k_n \|x - y\|$ for all $x, y \in C$.

Example 6.1.1. (a) The usual mappings sin, cos, norm and isometry are nonexpansive.

(b) $T : \mathbb{R} \rightarrow \mathbb{R}$ given by $Tx = \dfrac{x}{2} \sin \dfrac{1}{x}$ and $T(0) = 0$ is quasi-nonexpansive with fixed point 0.

(c) $T : \mathbb{R} \rightarrow \mathbb{R}$ defined by $T(x) = x^2$ has two fixed points 0 and 1.

(d) The translation of a mapping $T : \mathbb{R} \rightarrow \mathbb{R}$ given by $T(x) = x + a$ has no fixed point.

(e) $T : \mathbb{R}^2 \rightarrow \mathbb{R}^2$ defined by $T(x, y) = (x, 0)$ has infinitely many fixed points, i.e., the entire set $\mathbb{R}$.

(f) The nonself mapping $T : [-1, 1] \rightarrow \mathbb{R}$ given by $T(x) = 1 - x$ has a fixed point $\frac{1}{2}$.

(g) Define $T : B \rightarrow B$ by $T(x_1, x_2, x_3, \ldots) = (0, x_1^2, a_2 x_2, a_3 x_3, \ldots)$ where B is the unit ball in the Hilbert space ℓ^2 and $\{a_i\}$ is a sequence of numbers such that $0 < a_i < 1$ and $\prod_{i=2}^{\infty} a_i = \frac{1}{2}$. Then

$$\|T^n x - T^n y\| \leq 2 \prod_{i=2}^{n} a_i \|x - y\| \text{ for } n \geq 2 \text{ and } \|Tx - Ty\| \leq 2 \|x - y\|$$

provide that T is asymptotically nonexpansive but not nonexpansive (cf. Ref. [21]).

(h) Take $C = [0, 1] \subset \mathbb{R}$ and $\frac{1}{2} < k < 1$. For each $x \in [0, 1]$, we define $T : C \rightarrow C$ by

$$T(x) = \begin{cases} kx, & \text{if } 0 \leq x \leq \frac{1}{2}, \\ -\frac{k}{2k-1}(x-k), & \text{if } \frac{1}{2} \leq x \leq k, \\ 0, & \text{if } k \leq x \leq 1. \end{cases}$$

Then T is asymptotically nonexpansive but not nonexpansive (see Ref. [34]).

For sequences, the symbol $\to$ (respectively $\rightharpoonup$) indicates norm (respectively, weak) convergence. We denote by $\omega_w(x_n)$ the weak ω-limit set of the sequence $\{x_n\}$, that is, $\omega_w(x_n) = \{x : \exists\, x_{n_k} \rightharpoonup x\}$. A mapping $T : C \to E$ is *demi-closed* at $y \in E$ if, for each sequence $\{x_n\}$ in C and each $x \in E$, $x_n \rightharpoonup x$ and $Tx_n \to y$ imply that $x \in C$ and $Tx = y$.

Picard [51] introduced the following iteration formula:

$$x_1 \in C, \quad x_{n+1} = Tx_n. \tag{6.1.1}$$

It is well-known that Picard iterations of some nonexpansive mappings fail to converge even on a Banach space. For this, consider an anticlockwise rotation of the unit disc of $\mathbb{R}^2$ about the origin through an angle of $\frac{\pi}{4}$. This mapping is nonexpansive but its Picard sequence fails to converge. Krasnoselskii [43] showed that the following iterations formula:

$$x_1 \in C, \quad x_{n+1} = \frac{1}{2}(x_n + Tx_n), \tag{6.1.2}$$

converges to the fixed point of any nonexpansive mapping T. However, the more general iterative formula for approximation of fixed points of nonexpansive mappings was introduced by Mann [46] as follows:

$$x_1 \in C, \quad x_{n+1} = (1 - \alpha_n)x_n + \alpha_n Tx_n, \tag{6.1.3}$$

where $0 \le \alpha_n \le 1$.

Apart from being an obvious generalization of the contraction mappings, nonexpansive mappings are important, as has been observed by Bruck [6], mainly for the following two reasons: (a) Nonexpansive mappings are intimately connected with the monotonicity methods developed in the early 1960s and constitute the first class of nonlinear mappings for which fixed point theorems were obtained by using the fine geometric properties of the underlying Banach spaces instead of compactness properties. (b) Nonexpansive mappings appear in applications as the transition operators for initial valued problems of differential inclusions of the form $0 \in \frac{du}{dt} + T(t)u$, where the operators $\{T(t)\}$ are, in general, set-valued and are accretive or dissipative and minimally continuous.

Construction of fixed points of nonexpansive mappings is an important subject in nonlinear mapping theory and has applications in image recovery and signal processing (see, e.g., Refs. [10, 52, 67]). For the past 30 years or so, the study of Krasnoselskii–Mann iterative procedures for the approximation of fixed points of nonexpansive mappings and some of their generalizations and approximation of zeros of accretive-type operators have been a flourishing area of research. For example, the reader can consult the recent monographs of Berinde [1] and Chidume [12]. Mann iterates are not adequate for the approximation of fixed points of pseudocontractive mappings and this led to the introduction of the Ishikawa iterative sequence [25]:

$$x_1 \in C, \ x_{n+1} = (1 - \alpha_n)x_n + \alpha_n T\left((1 - \beta_n)x_n + \beta_n Tx_n\right), \tag{6.1.4}$$

where $0 \leq \alpha_n, \beta_n \leq 1$.

Note that iterative sequences (6.1.1)–(6.1.3) are special cases of (6.1.4) (through suitable choices of α_n, β_n for all $n \geq 1$).

A Banach space E is said to be uniformly convex if, for each $\varepsilon \in (0,2]$, the modulus of convexity $\delta : (0,2] \to (0,1]$ of E given by

$$\delta(\varepsilon) = \inf\left\{1 - \frac{1}{2}\|x+y\| : \|x\| \leq 1, \|y\| \leq 1, \|x-y\| \geq \varepsilon\right\}$$

satisfies the inequality $\delta(\varepsilon) > 0$ for all $\varepsilon > 0$. It is obvious that $\lim\limits_{\varepsilon \to 0} \delta(\varepsilon) = 0$ and $\delta(2) = 1$. We denote the inverse of δ by η and observe that $\lim\limits_{y \to 0} \eta(y) = 0$. Moreover, it has been shown in [15] that δ is nondeacreasing, continuous and

$$\|\lambda x + (1-\lambda)y\| \leq \max(\|x\|, \|y\|)\left[1 - 2\lambda(1-\lambda)\delta\left(\frac{\|x-y\|}{\max(\|x\|, \|y\|)}\right)\right] \tag{6.1.5}$$

for all $x, y \in E \setminus \{0\}$ and $\lambda \in [0,1]$.

Let $S = \{x \in E : \|x\| = 1\}$ and let E^* be the dual of E, that is, the space of all continuous linear functionals f on E.

The space E has:

(a) *Gâteaux differentiable norm* [65] if

$$\lim_{t \to 0} \frac{\|x+ty\| - \|x\|}{t}$$

exists for each x and y in S;

(b) *Fréchet differentiable norm* [65] if for each x in S, the above limit exists and is attained uniformly for y in S and in this case, it has been shown in [65] that

$$\langle h, J(x)\rangle + \frac{1}{2}\|x\|^2 \leq \frac{1}{2}\|x+h\|^2 \leq \langle h, J(x)\rangle + \frac{1}{2}\|x\|^2 + b(\|h\|), \tag{6.1.6}$$

for all x, h in E, where J is the Fréchet derivative of $\frac{1}{2}\|.\|^2$ at $x \in E$, $\langle ., .\rangle$ is the pairing between E and E^*, and b is a function defined on $[0, \infty)$ such that $\lim\limits_{t \to 0^+} \frac{b(t)}{t} = 0$;

(c) *Opial's property* [50] if for any sequence $\{x_n\}$ in E, $x_n \rightharpoonup x$ implies that

$$\limsup_{n \to \infty} \|x_n - x\| < \limsup_{n \to \infty} \|x_n - y\|, \quad \text{for all } y \in E \text{ with } y \neq x;$$

(d) *Kadec-Klee property* if for every sequence $\{x_n\}$ in E, $x_n \rightharpoonup x$ and $\|x_n\| \to \|x\|$ together imply $x_n \to x$ as $n \to \infty$.

We state some useful lemmas for later use.

Lemma 6.1.1 (cf. Ref. [30]). *If $\{r_n\}$, $\{s_n\}$ and $\{t_n\}$ are non-negative real sequences satisfying*

$$r_{n+1} \leq (1+s_n)r_n + t_n \text{ for all } n \geq 1, \ \sum_{n=1}^{\infty} s_n < \infty \text{ and } \sum_{n=1}^{\infty} t_n < \infty,$$

then $\lim_{n\to\infty} r_n$ exists.

Lemma 6.1.2 [7]. *Let C be a nonempty bounded, closed and convex subset of a uniformly convex Banach space E. Then there is a strictly increasing and continuous convex function $g : [0,\infty) \to [0,\infty)$ with $g(0) = 0$ such that for a nonexpansive mapping $T : C \to E$ and for all $x, y \in C$ and $t \in [0,1]$, the following inequality holds:*

$$\|T(tx + (1-t)y) - (tTx + (1-t)Ty)\| \leq g^{-1}(\|x - y\| - \|Tx - Ty\|).$$

Lemma 6.1.3 [26]. *Let E be a reflexive Banach space such that E^* has the Kadec-Klee property. Let $\{x_n\}$ be a bounded sequence in E and $x^*, y^* \in \omega_w(x_n)$. Suppose $\lim_{n\to\infty} \|tx_n + (1-t)x^* - y^*\|$ exists for all $t \in [0,1]$. Then $x^* = y^*$.*

Lemma 6.1.4 [4]. *Let C be a nonempty closed and convex subset of a uniformly convex Banach space E and $T : C \to C$ be a nonexpansive mapping. Then $I - T$ is demi-closed at 0.*

6.2. Nonexpansive Mappings in Uniformly Convex Banach Spaces

Browder [4] and Kirk [35] showed, independently, that a nonexpansive mapping of a nonempty bounded, closed and convex subset of a uniformly convex Banach space has a fixed point. Their proofs are similar and both are based on Zorn's Lemma and other nonelementary theorems of functional analysis. Goebel [20] gave an elementary proof of this fixed point theorem using only the definition of uniform convexity and some basic theorems of topology and analysis. We prefer his method of proof due to its simplicity and present it here in detail.

The following elementary lemma is needed to prove his theorem.

Lemma 6.2.1. *If u, v, w are elements of a uniformly convex Banach space E such that $\|u - w\| \leq R$, $\|v - w\| \leq R$ and $\left\|w - \frac{u+v}{2}\right\| \geq r > 0$, then $\|u - v\| \leq R\eta\left(\frac{R-r}{R}\right)$ where η is the inverse of δ, the modulus of convexity of E.*

Proof. Note that $\left\| \frac{(w-u)+(w-v)}{2} \right\| = \left\| w - \frac{u+v}{2} \right\| \geq r = \left(1 - \frac{R-r}{R} \right) R$ and so the result follows from the definition of uniform convexity. $\qquad \square$

Here is Goebel [20] fixed point theorem for nonexpansive mappings.

Theorem 6.2.1. *Let C be a nonempty bounded, closed and convex subset of a uniformly convex Banach space E. Then every nonexpansive mapping $T : C \to C$ has at least one fixed point.*

Proof. Let $d(C)$ denote the diameter of C. For $\varepsilon \in (0,1]$, set $T_\varepsilon = (1-\varepsilon)T$. Obviously, T_ε is a contraction of C and by the well-known Banach fixed point theorem, there exists for every $\varepsilon \in (0,1]$, an x_ε such that $T_\varepsilon x_\varepsilon = x_\varepsilon$. Now we have the following relation:

$$
\begin{aligned}
\|x_\varepsilon - Tx_\varepsilon\| &= \|T_\varepsilon x_\varepsilon - Tx_\varepsilon\| \\
&= \|Tx_\varepsilon - \varepsilon Tx_\varepsilon - Tx_\varepsilon\| \\
&= \varepsilon \|Tx_\varepsilon\| \leq \varepsilon d(C);
\end{aligned}
$$

hence $\inf_{x \in C} \|x - Tx\| = 0$.

Set $C_\varepsilon = \{x : \|x - Tx\| \leq \varepsilon\}$ and $D_\varepsilon = \{x \in C_\varepsilon : \|x\| \leq a + \varepsilon\}$ where $a = \lim\limits_{\varepsilon \to 0} \left(\inf\limits_{x \in C_\varepsilon} \|x\| \right)$.

It suffices to prove that the intersection of all sets C_ε is nonempty.

If this were false, it would follow that $a > 0$ because each C_ε is closed. Choose elements u_1, u_2 in C_ε. Observe that for $i = 1, 2$,

$$
\left\| u_i - T \left(\frac{u_1 + u_2}{2} \right) \right\| \leq \|u_i - Tu_i\| \left\| Tu_i - T \left(\frac{u_1 + u_2}{2} \right) \right\| \tag{6.2.7}
$$

$$
\leq \varepsilon + \frac{1}{2} \|u_1 - u_2\|
$$

and

$$
\left\| u_i - \frac{u_1 + u_2}{2} \right\| < \varepsilon + \frac{1}{2} \|u_1 - u_2\|. \tag{6.2.8}
$$

Moreover, in view of the relation

$$
\begin{aligned}
\|u_1 - u_2\| &\leq \left\| u_1 - \frac{1}{2} \left(\frac{u_1 + u_2}{2} + T \left(\frac{u_1 + u_2}{2} \right) \right) \right\| \\
&\quad + \left\| u_1 - \frac{1}{2} \left(\frac{u_1 + u_2}{2} + T \left(\frac{u_1 + u_2}{2} \right) \right) \right\|,
\end{aligned}
$$

the inequality

$$\left\| u_i - \frac{1}{2}\left(\frac{u_1 + u_2}{2} + T\left(\frac{u_1 + u_2}{2} \right) \right) \right\| \geq \frac{1}{2}\|u_1 - u_2\| \qquad (6.2.9)$$

holds for at least one of the values of $i = 1,2$.

Now Lemma 6.2.1 and inequalities (6.2.7), (6.2.8) and (6.2.9) imply

$$
\begin{aligned}
\left\| \frac{u_1 + u_2}{2} - T\left(\frac{u_1 + u_2}{2} \right) \right\| &\leq \left(\varepsilon + \frac{1}{2}\|u_1 - u_2\| \right) \eta \left(\frac{\varepsilon}{\varepsilon + \frac{1}{2}\|u_1 - u_2\|} \right) \\
&\leq \sup_{0 < \xi \leq \frac{d(C)}{2}} (\varepsilon + \xi)\, \eta \left(\frac{\varepsilon}{\varepsilon + \xi} \right) \\
&\leq \max \left\{ \begin{array}{l} \sup_{0 < \xi \leq \sqrt{\varepsilon} - \varepsilon} (\varepsilon + \xi)\, \eta \left(\frac{\varepsilon}{\varepsilon + \xi} \right), \\ \sup_{\sqrt{\varepsilon} - \varepsilon < \xi \leq \frac{d(C)}{2}} (\varepsilon + \xi)\, \eta \left(\frac{\varepsilon}{\varepsilon + \xi} \right) \end{array} \right\} \\
&\leq \max \left\{ 2\sqrt{\varepsilon},\ \left(\frac{d(C)}{2} + \varepsilon \right) \eta\left(\sqrt{\varepsilon} \right) \right\}.
\end{aligned}
$$

Denoting the last term by $\phi(\varepsilon)$, one can say that if $u_1, u_2 \in C_\varepsilon$, then $\frac{u_1+u_2}{2} \in C_{\phi(\varepsilon)}$. Obviously, $\lim_{\varepsilon \to 0} \phi(\varepsilon) = 0$. Now for $u_1, u_2 \in D_\varepsilon$, we have the inequalities $\|u_1\| \leq a + \varepsilon$ and $\|u_2\| \leq a + \varepsilon$. Since $\frac{u_1+u_2}{2} \in C_{\phi(\varepsilon)}$, therefore the inequality $\left\| \frac{u_1+u_2}{2} \right\| \geq a\left(C_{\phi(\varepsilon)} \right)$ follows. Using once again Lemma 6.2.1, we obtain the inequality

$$
\begin{aligned}
d(D_\varepsilon) &= \sup_{u_1, u_2 \in D_\varepsilon} \|u_1 - u_2\| \\
&\leq (a + \varepsilon)\, \eta \left(\frac{a + \varepsilon - a\left(C_{\phi(\varepsilon)} \right)}{(a + \varepsilon)} \right),
\end{aligned}
$$

and hence $\lim_{\varepsilon \to 0} d(D_\varepsilon) = 0$. By Cantor's intersection theorem, the intersection of all D_ε is nonempty, and our result is proved.   □

The rest of this section is devoted to the work of Tan and Xu [65] on iterative approximation of fixed points of nonexpansive mappings in uniformly convex Banach spaces.

Lemma 6.2.2. *Let C be a nonempty bounded, closed and convex subset of a uniformly convex Banach space E and $T : C \to C$ be a nonexpansive mapping. For the sequence $\{x_n\}$ defined by (6.1.4), $\lim_{n \to \infty} \|x_n - p\|$ exists for each $p \in F(T)$.*

Proof. For $x \in C$ and each integer $n \geq 1$, we write

$$T_n(x) = (1 - \alpha_n)x + \alpha_n T[(1 - \beta_n)x + \beta_n Tx].$$

Obviously, T_n is nonexpansive and the sequence $\{x_n\}$ defined by (6.1.4) can be written as

$$x_{n+1} = T_n x_n, \quad n \geq 1.$$

It is easy to verify that $F(T_n) \supseteq F(T)$ for $n \geq 1$.

Further,

$$\|x_{n+1} - p\| = \|T_n x_n - T_n p\| \leq \|x_n - p\|.$$

That is, $\{\|x_n - p\|\}$ is a decreasing sequence of non-negative real numbers and therefore $\lim\limits_{n \to \infty} \|x_n - p\|$ exists for each $p \in F(T)$. $\qquad\square$

Lemma 6.2.3. *Let C be a nonempty bounded, closed and convex subset of a uniformly convex Banach space E and $T : C \to C$ be a nonexpansive mapping. If the sequence $\{x_n\}$ defined by (6.1.4) satisfies $\sum_{n=1}^{\infty} \alpha_n(1 - \alpha_n) = \infty$, $\sum_{n=1}^{\infty} \beta_n(1 - \alpha_n) < \infty$ and $\limsup\limits_{n \to \infty} \beta_n < 1$, then we have that $\lim\limits_{n \to \infty} \|x_n - Tx_n\| = 0$.*

Proof. We may assume that $\lim\limits_{n \to \infty} \|x_n - p\| \neq 0$ for each $p \in F(T)$.

In view of

$$\|y_n - p\| \leq \|x_n - p\|,$$

we obtain by (6.1.5) that

$$
\begin{aligned}
\|x_{n+1} - p\| &= \|\alpha_n(Ty_n - p) + (1 - \alpha_n)(x_n - p)\| \\
&\leq \|x_n - p\|\left[1 - 2\alpha_n(1 - \alpha_n)\delta\left(\frac{\|Ty_n - x_n\|}{\|x_n - p\|}\right)\right], \quad (6.2.10)
\end{aligned}
$$

where δ is the modulus of uniform convexity of E.

Now it is readily seen from (6.2.10) that $\sum_{n=1}^{\infty} \alpha_n(1 - \alpha_n)\delta\left(\frac{\|Ty_n - x_n\|}{\|x_n - p\|}\right)$ converges.

As $\sum_{n=1}^{\infty} \alpha_n(1 - \alpha_n)$ diverges, so we have $\lim\limits_{n \to \infty} \delta\left(\frac{\|Ty_n - x_n\|}{\|x_n - p\|}\right) = 0$ and thus

$$\lim_{n \to \infty} \|Ty_n - x_n\| = 0, \quad (6.2.11)$$

because δ is strictly increasing and continuous and $\lim\limits_{n \to \infty} \|x_n - p\| > 0$.

Note that

$$
\begin{aligned}
\|x_n - Tx_n\| &\leq \|Tx_n - Ty_n\| + \|Ty_n - x_n\| \\
&\leq \|x_n - y_n\| + \|Ty_n - x_n\| \\
&= \beta_n \|x_n - Tx_n\| + \|Ty_n - x_n\|,
\end{aligned}
$$

that is,

$$
\|x_n - Tx_n\| \leq \frac{1}{1 - \beta_n} \|Ty_n - x_n\|.
$$

So we have from (6.2.11) that

$$
\liminf_{n \to \infty} \|x_n - Tx_n\| = 0. \tag{6.2.12}
$$

From

$$
\begin{aligned}
\|Tx_{n+1} - x_{n+1}\| &\leq \alpha_n \|Tx_{n+1} - Ty_n\| + (1 - \alpha_n) \|Tx_{n+1} - x_n\| \\
&\leq \alpha_n \|x_{n+1} - y_n\| + (1 - \alpha_n)(\|Tx_{n+1} - x_{n+1}\| + \|x_{n+1} - x_n\|) \\
&\leq \alpha_n (\alpha_n \|Ty_n - y_n\| + (1 - \alpha_n) \|x_n - y_n\|) \\
&\quad + (1 - \alpha_n)(\|Tx_{n+1} - x_{n+1}\| + \alpha_n \|Ty_n - x_n\|),
\end{aligned}
$$

we have

$$
\begin{aligned}
\|Tx_{n+1} - x_{n+1}\| &\leq \alpha_n \|Ty_n - y_n\| + (1 - \alpha_n) \|Ty_n - x_n\| + \|x_n - y_n\| \\
&\leq \alpha_n (\beta_n \|Ty_n - Tx_n\|) + ((1 - \beta_n) \|Ty_n - x_n\|) \\
&\quad + (1 - \alpha_n)(\|Ty_n - x_n\| + \|x_n - y_n\|) \\
&\leq (1 + \alpha_n \beta_n - \alpha_n) \|x_n - y_n\| + (1 - \alpha_n \beta_n) \|Ty_n - x_n\| \\
&\leq \beta_n (1 + \alpha_n \beta_n - \alpha_n) \|x_n - Tx_n\| \\
&\quad + (1 - \alpha_n \beta_n)(\|Ty_n - Tx_n\| + \|x_n - Tx_n\|) \\
&\leq (\beta_n (1 + \alpha_n \beta_n - \alpha_n) + (1 - \alpha_n \beta_n)(1 + \beta_n)) \|x_n - Tx_n\|.
\end{aligned}
$$

Since $\sum_{n=1}^{\infty} \beta_n (1 - \alpha_n) < \infty$ and $\{\|x_n - Tx_n\|\}$ is bounded, therefore it follows from Lemma 6.1.1 that $\lim_{n \to \infty} \|x_n - Tx_n\|$ exists and equals 0 by (6.2.12). □

Lemma 6.2.4. *Let C be a nonempty closed and convex subset of a uniformly convex Banach space E and T be a nonexpansive mapping on C. Let $\{x_n\}$ be the sequence defined by (6.1.4) with $F(T) \neq \phi$. Then, for any $p_1, p_2 \in F(T)$, $\lim_{n \to \infty} \|tx_n + (1 - t)p_1 - p_2\|$ exists for any $t \in [0, 1]$.*

Proof. By Lemma 6.2.2, $\lim_{n \to \infty} \|x_n - p\|$ exists for any $p \in F(T)$ and therefore $\{x_n\}$ is bounded. Hence there exists a ball $B_r(0) = \{x \in E : \|x\| \leq r\}$ for some $r > 0$ such that

$\{x_n\} \subset C = B_r(0) \cap C$. Thus C is a nonempty bounded, closed and convex subset of E. Let $a_n(t) = \|tx_n + (1-t)p_1 - p_2\|$. Then $\lim\limits_{n\to\infty} a_n(0) = \|p_1 - p_2\|$ and $\lim\limits_{n\to\infty} a_n(1) = \lim\limits_{n\to\infty} \|x_n - p_2\|$ exists as proved in Lemma 6.2.2.

Set

$$R_{n,m} = T_{n+m-1}T_{n+m-2}\ldots T_n, \, m \geq 1 \text{ and}$$

$$b_{n,m} = \|R_{n,m}(tx_n + (1-t)p_1) - (tR_{n,m}x_n + (1-t)p_1)\|.$$

Then

$$\|R_{n,m}x - R_{n,m}y\| \leq \|x - y\| \text{ and } R_{n,m}x_n = x_{n+m}.$$

We first show that for any $p \in F(T)$, $\|R_{n,m}p - p\| \to 0$ as $n \to \infty$ and for all $m \geq 1$. Consider

$$
\begin{aligned}
\|R_{n,m}p - p\| &\leq \|T_{n+m-1}T_{n+m-2}\ldots T_n p - T_{n+m-1}T_{n+m-2}\ldots T_{n+1}p\| \\
&\quad + \|T_{n+m-1}T_{n+m-2}\ldots T_{n+1}p - T_{n+m-1}T_{n+m-2}\ldots T_{n+2}p\| \\
&\quad + \cdots + \|T_{n+m-1}p - p\| \\
&\leq \|T_n p - p\| + \|T_{n+1}p - p\| + \cdots + \|T_{n+m-1}p - p\| = 0.
\end{aligned}
$$

By Lemma 6.1.2, there exists a strictly increasing continuous function $g : [0,\infty) \to [0,\infty)$ with $g(0) = 0$ such that

$$
\begin{aligned}
g(b_{n,m}) &\leq \|x_n - p_1\| - \|R_{n,m}x_n - R_{n,m}p_1\| \\
&= \|x_n - p_1\| - \|(R_{n,m}x_n - p_1) + (p_1 - R_{n,m}p_1)\| \\
&\leq \|x_n - p_1\| + \|p_1 - R_{n,m}p_1\| - \|R_{n,m}x_n - p_1\| \\
&= \|x_n - p_1\| - \|x_{n+m} - p_1\| + \|p_1 - R_{n,m}p_1\| \to 0 \text{ as } n \to \infty.
\end{aligned}
$$

Hence $b_{n,m} \to 0$ as $n \to \infty$ and for all $m \geq 1$.

Finally, from the inequality

$$
\begin{aligned}
a_{n+m}(t) &= \|tx_{n+m} + (1-t)p_1 - p_2\| \\
&\leq b_{n,m} + \|R_{n,m}(tx_n + (1-t)p_1) - p_2\| \\
&\leq b_{n,m} + a_n(t) + \|R_{n,m}p_2 - p_2\|,
\end{aligned}
$$

it follows that

$$\limsup_{m\to\infty} a_{n+m}(t) \leq \limsup_{m\to\infty} b_{n,m} + a_n(t) + \limsup_{m\to\infty} \|R_{n,m}p_2 - p_2\|,$$

that is,

$$\limsup_{m\to\infty} a_m(t) \leq \liminf_{n\to\infty} a_n(t).$$

Hence $\lim\limits_{n\to\infty} \|tx_n + (1-t)p_1 - p_2\|$ exists for any $t \in [0,1]$. $\qquad\square$

Lemma 6.2.5. *Let C be a nonempty bounded, closed and convex subset of a uniformly convex Banach space E and $T : C \to C$ be a nonexpansive mapping. Let $\{x_n\}$ be the sequence defined by (6.1.4) with $\sum_{n=1}^{\infty} \alpha_n (1 - \alpha_n) = \infty$, $\sum_{n=1}^{\infty} \beta_n (1 - \alpha_n) < \infty$ and $\limsup_{n \to \infty} \beta_n < 1$. Then, for any $p_1, p_2 \in F(T)$, $\lim_{n \to \infty} \langle x_n, J(p_1 - p_2) \rangle$ exists; in particular, $\langle p - q, J(p_1 - p_2) \rangle = 0$ for all $p, q \in \omega_w(x_n)$.*

Proof. Take $x = p_1 - p_2$ with $p_1 \neq p_2$ and $h = t(x_n - p_1)$ in the inequality (6.1.6) to get

$$\frac{1}{2} \|p_1 - p_2\|^2 + t \langle x_n - p_1, J(p_1 - p_2) \rangle \quad \leq \quad \frac{1}{2} \|t x_n + (1 - t)p_1 - p_2\|^2$$

$$\leq \quad \frac{1}{2} \|p_1 - p_2\|^2 + t \langle x_n - p_1, J(p_1 - p_2) \rangle$$
$$+ b(t \|x_n - p_1\|).$$

As $\sup_{n \geq 1} \|x_n - p_1\| \leq M$ for some $M > 0$, so it follows that

$$\frac{1}{2} \|p_1 - p_2\|^2 + t \limsup_{n \to \infty} \langle x_n - p_1, J(p_1 - p_2) \rangle \quad \leq \quad \frac{1}{2} \lim_{n \to \infty} \|t x_n + (1 - t)p_1 - p_2\|^2$$

$$\leq \quad \frac{1}{2} \|p_1 - p_2\|^2 + b(tM)$$
$$+ t \liminf_{n \to \infty} \langle x_n - p_1, J(p_1 - p_2) \rangle,$$

that is,

$$\limsup_{n \to \infty} \langle x_n - p_1, J(p_1 - p_2) \rangle \leq \liminf_{n \to \infty} \langle x_n - p_1, J(p_1 - p_2) \rangle + \frac{b(tM)}{tM} M.$$

If $t \to 0$, then $\lim_{n \to \infty} \langle x_n - p_1, J(p_1 - p_2) \rangle$ exists for all $p_1, p_2 \in F(T)$; in particular, we have $\langle p - q, J(p_1 - p_2) \rangle = 0$ for all $p, q \in \omega_w(x_n)$. $\square$

Theorem 6.2.2. *Let C be a nonempty bounded, closed and convex subset of a uniformly convex Banach space E and $T : C \to C$ be a nonexpansive mapping. Let $\{x_n\}$ be the sequence defined by (6.1.4) with $\sum_{n=1}^{\infty} \alpha_n (1 - \alpha_n) = \infty$, $\sum_{n=1}^{\infty} \beta_n (1 - \alpha_n) < \infty$ and $\limsup_{n \to \infty} \beta_n < 1$. Assume that one of the following conditions holds:*

(i) *E satisfies the Opial property;*

(ii) *E has a Fréchet differentiable norm;*

(iii) *E^* has the Kadec-Klee property.*

Then $\{x_n\}$ converges weakly to some $p \in F(T)$.

Proof. Lemma 6.2.2 tells us that $\lim_{n \to \infty} \|x_n - p\|$ exists for any $p \in F(T)$. As E is uniformly convex and hence reflexive, so there exists a subsequence $\{x_{n_i}\}$ of $\{x_n\}$ converging weakly to some $z_1 \in C$. By Lemmas 6.2.3 and 6.1.4, $\lim_{n \to \infty} \|x_n - Tx_n\| = 0$ and $I - T$ is demi-closed at 0. Therefore we obtain $Tz_1 = z_1$, that is, $z_1 \in F(T)$. In order to show that $\{x_n\}$ converges weakly to z_1, take another subsequence $\{x_{n_j}\}$ of $\{x_n\}$ converging weakly to some $z_2 \in C$. Again, as before, we can prove that $z_2 \in F(T)$. Next, we prove that $z_1 = z_2$. Assume (i) is given and suppose that $z_1 \neq z_2$. Then by the Opial property, we obtain

$$
\begin{aligned}
\lim_{n \to \infty} \|x_n - z_1\| &= \lim_{n_i \to \infty} \|x_{n_i} - z_1\| \\
&< \lim_{n_i \to \infty} \|x_{n_i} - z_2\| \\
&= \lim_{n \to \infty} \|x_n - z_2\| \\
&= \lim_{n_j \to \infty} \|x_{n_j} - z_2\| \\
&< \lim_{n_j \to \infty} \|x_{n_j} - z_1\| \\
&= \lim_{n \to \infty} \|x_n - z_1\|.
\end{aligned}
$$

This contradiction implies that $z_1 = z_2$. Next suppose that (ii) is satisfied. By Lemma 6.2.5, we have that $\langle p - q, J(p_1 - p_2)\rangle = 0$ for all $p, q \in \omega_w(x_n)$. Now $\|z_1 - z_2\|^2 = \langle z_1 - z_2, J(z_1 - z_2)\rangle = 0$ gives that $z_1 = z_2$.

Finally, let (iii) be given. Since $\lim_{n \to \infty} \|tx_n + (1-t)z_1 - z_2\|$ exists, therefore by Lemma 6.2.5, we obtain $z_1 = z_2$. Hence $x_n \rightharpoonup p \in F(T)$. $\square$

In order to prove the corresponding strong convergence theorem, we need the following: A mapping T is said to satisfy condition (A) [56] if there exists a nondecreasing function $f : [0, \infty) \to [0, \infty)$ with $f(0) = 0$, $f(r) > 0$ for all $r \in (0, \infty)$ such that $\|x - Tx\| \geq f(d(x, F(T)))$ for all $x \in C$, where $d(x, F(T)) = \inf\{\|x - p\| : p \in F(T)\}$.

Theorem 6.2.3. *Let C be a nonempty bounded, closed and convex subset of a uniformly convex Banach space E and T be a nonexpansive mapping on C satisfying the condition (A). Then $\{x_n\}$ in (6.1.4) with $\sum_{n=1}^{\infty} \alpha_n(1 - \alpha_n) = \infty$, $\sum_{n=1}^{\infty} \beta_n(1 - \alpha_n) < \infty$ and $\limsup_{n \to \infty} \beta_n < 1$, converges strongly to a fixed point of T.*

Proof. As in Lemma 6.2.2, we have

$$\|x_{n+1} - p\| \leq \|x_n - p\|,$$

which further gives that

$$\inf_{p \in F(T)} \|x_{n+1} - p\| \leq \inf_{p \in F(T)} \|x_n - p\|.$$

That is,

$$d(x_{n+1}, F) \leq d(x_n, F(T)).$$

As $\{d(x_n, F(T))\}$ is decreasing and bounded, so $\lim_{n \to \infty} d(x_n, F(T))$ exists. By Lemma 6.2.3, we have

$$\lim_{n \to \infty} \|Tx_n - x_n\| = 0.$$

Hence, by the condition (A), $\lim_{n \to \infty} f(d(x_n, F(T))) = 0$. Since f is nondecreasing and $f(0) = 0$, therefore we get $\lim_{n \to \infty} d(x_n, F(T)) = 0$.

Next, we prove that $\{x_n\}$ is a Cauchy sequence. Since $\lim_{n \to \infty} d(x_n, F(T)) = 0$, there exists an integer n_0 such that for all $n \geq n_0$,

$$d(x_n, F(T)) < \frac{\varepsilon}{2}.$$

In particular,

$$d(x_{n_0}, F(T)) < \frac{\varepsilon}{2},$$

that is,

$$\inf_{p \in F(T)} \|x_{n_0} - p\| < \frac{\varepsilon}{2}.$$

Thus there must exist $p^* \in F(T)$ such that

$$\|x_{n_0} - p^*\| < \frac{\varepsilon}{3}.$$

Now, for $n \geq n_0$, we have from the above inequality that

$$\begin{aligned}
\|x_{n+m} - x_n\| &\leq \|x_{n+m} - p^*\| + \|x_n - p^*\| \\
&\leq 2\|x_{n_0} - p^*\| \\
&< 2\left(\frac{\varepsilon}{3} + \frac{\varepsilon}{3}\right) < \varepsilon.
\end{aligned}$$

Hence $\{x_n\}$ is a Cauchy sequence in C and it must converge to a point of C. Let $\lim_{n \to \infty} x_n = q$ (say). Since $\lim_{n \to \infty} d(x_n, F(T)) = 0$ and $F(T)$ is closed, therefore $q \in F(T)$.

$\square$

6.3. Nonexpansive Mappings in CAT(0) Spaces

Most of the real world problems are nonlinear in nature and are thus, of keen interest to scientists and mathematicians. Therefore they are always striving to find methods for their solution. So translating a linear version of known problems into its equivalent nonlinear version is of great importance.

Let (X, d) be a metric space and $x, y \in X$ with $l = d(x, y)$. A *geodesic path* from x to y is a mapping $c : [0, l] \to X$ such that $c(0) = x$, $c(l) = y$, and $d(c(t), c(t')) = |t - t'|$ for all $t, t' \in [0, l]$. The image of a geodesic path is called a *geodesic segment*. A metric space X is uniquely geodesic if every two points of X are joined by a unique geodesic segment.

A geodesic triangle $\Delta(x_1, x_2, x_3)$ in a geodesic metric space X consists of three points x_1, x_2 and x_3 in X and a geodesic segment between each pair of these points. A comparison triangle for geodesic triangle $\Delta(x_1, x_2, x_3)$ in X is a triangle $\overline{\Delta}(x_1, x_2, x_3) :=$ $\Delta(\bar{x}_1, \bar{x}_2, \bar{x}_3)$ in $\mathbb{R}^2$ such that $d_{\mathbb{R}^2}(\bar{x}_i, \bar{x}_j) = d(x_i, x_j)$ for all $i, j = 1, 2, 3$.

A geodesic space X is a Cartan, Alexandrov and Toponogov (0) space, in short, CAT(0) *space* [2, p. 159], if for each Δ in X and $\overline{\Delta}$ in $\mathbb{R}^2$, the inequality

$$d(x, y) \leq d_{\mathbb{R}^2}(\bar{x}, \bar{y})$$

holds for all $x, y \in \Delta$ and $\bar{x}, \bar{y} \in \overline{\Delta}$ and is known as the CAT(0) inequality.

The open unit ball B in complex Hilbert space with respect to the Poincare metric (also called *Poincare distance*)

$$d_B(x, y) = \arg\tanh\left|\frac{x - y}{1 - x\bar{y}}\right| = \arg\tanh\left(1 - \sigma(x, y)\right)^{\frac{1}{2}},$$

where

$$\sigma(x, y) = \frac{\left(1 - |x|^2\right)\left(1 - |y|^2\right)}{|1 - x\bar{y}|^2} \quad \text{for all } x, y \in B,$$

is a nontrivial example of a CAT(0) space [41]. It is worth mentioning that fixed point theorems in CAT(0) spaces (especially in $\mathbb{R}$-trees) can be applied to graph theory (see e.g., Refs. [16, 38]). A thorough discussion of these spaces and their important role in various branches of mathematics can be found in Refs. [2, 9].

In this section, we write $(1 - \alpha)x \oplus \alpha y$ for the unique point z on the geodesic segment from x to y such that

$$d(z, x) = \alpha d(x, y) \text{ and } d(z, y) = (1 - \alpha)d(x, y).$$

We denote by $[x, y]$, the geodesic segment joining x to y, that is,

$$[x, y] = \{(1 - \alpha)x \oplus \alpha y : \alpha \in [0, 1]\}.$$

A subset C of a CAT(0) space is convex if $[x,y] \subset C$ for all $x,y \in C$.

Let $\{x_n\}$ be a bounded sequence in a metric space X and $x \in X$. Set

$$r(x,\{x_n\}) = \limsup_{n \to \infty} d(x_n,x).$$

The *asymptotic radius* $r(\{x_n\})$ of $\{x_n\}$ is given by

$$r(\{x_n\}) = \inf\{r(x,\{x_n\}) : x \in X\};$$

the *asymptotic center* $A(\{x_n\})$ of $\{x_n\}$ is the set

$$A(\{x_n\}) = \{x \in X \mid r(x,\{x_n\}) = r(\{x_n\})\}$$

and $\{x_n\}$ is *regular* if and only if $r(\{x_n\}) = r(\{u_n\})$ for every subsequence $\{u_n\}$ of $\{x_n\}$.

It is known that in a CAT(0) space, $A(\{x_n\})$ consists of exactly one point (see, e.g., Ref. [13]).

A sequence $\{x_n\}$ in X is said to be Δ-*convergent* to $x \in X$ if x is the unique asymptotic center of $\{u_n\}$ for every subsequence $\{u_n\}$ of $\{x_n\}$. We write $\Delta - \lim_n x_n = x$ and call x as Δ-limit of $\{x_n\}$.

It has been shown in a CAT(0) space [41]: for a given $\{x_n\} \subset X$ such that $\{x_n\}$ is Δ-convergent to x and given $y \in X$ with $y \neq x$,

$$\limsup_{n \to \infty} d(x_n,x) < \limsup_{n \to \infty} d(x_n,y)$$

holds. Thus a CAT(0) space satisfies the Opial property of Banach space theory.

From now onwards, in this section, we present the work of Dhompongsa and Panyanak [14] and Nanjaras and Panyanak [49] on iterative approximation of fixed points of nonexpansive mappings in CAT(0) spaces. For more material on the same lines, the interested reader should consult Refs. [55, 59, 68].

Lemma 6.3.1. *Let X be a CAT (0) space. Then*

(a) *X is uniquely geodesic.*

(b) *Let p,x,y be points of X and $\alpha \in [0,1]$. If m_1 and m_2 denote, respectively, the points of $[p,x]$ and $[p,y]$ satisfying*

$$d(p,m_1) = \alpha d(p,x) \text{ and } d(p,m_2) = \alpha d(p,y),$$

then

$$d(m_1,m_2) \leq \alpha d(x,y).$$

(c) *Let $x,y \in X$, $x \neq y$ and $z,w \in [x,y]$ such that $d(x,z) = d(x,w)$. Then $z = w$.*

(d) *Let $x, y \in X$. For each $t \in [0,1]$, there exists a unique point $z \in [x,y]$ such that*

$$d(x,z) = td(x,y) \text{ and } d(y,z) = (1-t)d(x,y).$$

Proof. (a) See Ref. [2, p.160].

(b) See Lemma 3 in Ref. [37].

(c) Let $c : [0, d(x,y)] \to X$ be the geodesic path joining x and y. As $z, w \in [x,y]$, we have $t_1, t_2 \in [0, d(x,y)]$ such that $c(t_1) = z$ and $c(t_2) = w$. Thus $d(x,z) = d(c(0), c(t_1)) = |0 - t_1| = t_1$. Similarly, $d(x,w) = t_2$. Since $d(x,z) = d(x,w)$, therefore we have $t_1 = t_2$. That is, $z = w$.

(d) If $x = y$, then the conclusion is obvious. Suppose that $x \neq y$. Take $z = c(td(x,y))$. Thus $z \in [x,y]$, $d(x,z) = d(c(0), c(td(x,y))) = |0 - td(x,y)| = td(x,y)$ and

$$d(y,z) = d(c(d(x,y)), c(td(x,y))) = |d(x,y) - td(x,y)| = (1-t)d(x,y).$$

The uniqueness of z follows from part (c).   □

Remark 6.3.1. Let (X,d) be a $CAT(0)$ space and let $x, y \in X$ such that $x \neq y$ and $s, t \in [0,1]$. Then

$$(1-t)x \oplus ty = (1-s)x \oplus sy \quad \text{if and only if} \quad s = t.$$

Lemma 6.3.2. *Let X be a $CAT(0)$ space and let $x, y \in X$ such that $x \neq y$. Then*

(a) $[x,y] = \{(1-t)x \oplus ty : t \in [0,1]\}$;

(b) $d(x,z) + d(y,z) = d(x,y)$ *if and only if $z \in [x,y]$*;

(c) *The mapping $f : [0,1] \to [x,y]$ given by $f(t) = (1-t)x \oplus ty$ is continuous and bijective.*

Proof. (a) The inclusion $\supseteq$ follows from the definition. For the converse inclusion $\subseteq$, let $z \in [x,y]$ and take $t = \frac{d(x,z)}{d(x,y)}$. Then $z, (1-t)x \oplus ty \in [x,y]$ are such that

$$d(x, (1-t)x \oplus ty) = d(x,z).$$

Therefore $z = (1-t)x \oplus ty$.

(b) Let $\Delta(\bar{x}, \bar{y}, \bar{z})$ be the comparison triangle (up to isomorphism) in $\mathbb{R}^2$ of the geodesic triangle $\Delta(x,y,z)$, and $w \in [x,y]$ be such that $d(x,w) = d(x,z)$. By the above assumption, it follows that $[\bar{x}, \bar{y}]$ is simply a straight line with the point $\bar{z}$ in between; i.e., $\bar{z} \in [\bar{x}, \bar{y}]$. Moreover, since $d(x,w) = d(x,z)$, we must have $d(\bar{x}, \bar{w}) = d(\bar{x}, \bar{z})$ and hence $\bar{z} = \bar{w}$. Then, by the CAT(0) inequality, we have $d(w,z) \leq d(\bar{x}, \bar{w}) = 0$, which implies $z = w \in [x, y]$. For the converse, let $z \in [x, y]$. By part (a), there exists $t \in [0, 1]$ such that $z = (1 - t)x \oplus ty$. Consequently, $d(x, z) + d(y, z) = td(x, y) + (1 - t)d(x, y) = d(x, y)$.

(c) Let $t_1, t_2 \in [0, 1]$ be such that $f(t_1) = f(t_2)$. That is, $(1 - t_1)x \oplus t_1 y = (1 - t_2)x \oplus t_2 y$. By Remark 6.3.1, $t_1 = t_2$. Hence f is bijective. Since the geodesic path $c : [0, d(x, y)] \to [x, y]$ and the mapping $h : [0, 1] \to [0, d(x, y)]$ defined by $h(t) = td(x, y)$ are continuous, it follows that $g : [0, 1] \to [x, y]$ given by $g(t) = c(h(t)) = c(td(x, y))$ is continuous. By Lemma 6.3.1(d), $f = g$. Therefore f is continuous.   □

Lemma 6.3.3. *Let X be a CAT (0) space. Then*

$$d((1 - t)x \oplus ty, z) \leq (1 - t)d(x, z) + td(y, z), \tag{6.3.13}$$

for all $t \in [0, 1]$ and $x, y, z \in X$.

Proof. Let $x, y, z \in X$ and $t \in [0, 1]$. Suppose that $d(z, y) \leq d(z, x)$. Let $u = (1 - t)x \oplus ty$ and let x_0 be the point of $[z, x]$ such that $d(z, x_0) = d(z, y)$. Put $v = (1 - t)x_0 \oplus ty$ and $w = (1 - t)x_0 \oplus tz$. By Lemma 6.3.1(b)

$$
\begin{aligned}
d(z, v) &\leq d(z, w) + d(w, v) \\
&\leq (1 - t)d(x_0, z) + td(z, y) \\
&= d(z, y).
\end{aligned}
$$

Suppose that (6.3.13) does not hold. Then

$$
\begin{aligned}
(1 - t)d(z, x) + td(z, y) &< d(z, u) \\
&\leq d(z, v) + d(v, u) \\
&\leq d(z, y) + d(v, u)
\end{aligned}
$$

yields

$$
\begin{aligned}
d(v, u) &> (1 - t)[d(z, x) - d(z, y)] \\
&= (1 - t)d(x, x_0).
\end{aligned}
$$

This contradicts the conclusion of Lemma 6.3.1(b).   □

If x, y_1, y_2 are points in a CAT(0) space and if y_0 is the midpoint of the segment $[y_1, y_2]$, then the CAT(0) inequality implies

$$d(x, y_0)^2 \leq \frac{1}{2}d(x, y_1)^2 + \frac{1}{2}d(x, y_2)^2 - \frac{1}{4}d(y_1, y_2)^2.$$

This inequality is known as (CN) inequality of Bruhat and Tits [8]. In fact, a geodesic space is a CAT(0) space if and only if it satisfies the (CN) inequality (see Ref. [2, p. 163]).

Using this inequality, we prove the following important fact.

Lemma 6.3.4. *Let X be a CAT* (0) *space. Then*

$$d(z, tx \oplus (1-t)y)^2 \leq (1-t)d(z,x)^2 + td(z,y)^2 - t(1-t)d(x,y)^2, \qquad (6.3.14)$$

for all $t \in [0,1]$ and $x,y,z \in X$.

Proof. We first prove the inequality (6.3.14) for $t = \frac{k}{2^n}$ where $k,n \in \mathbb{N}$ are such that $k \leq 2^n$. We use induction on n. If $n = 0$, then $\frac{k}{2^n} = k$ and $k \in \{0,1\}$, the inequality obviously holds. Suppose that (6.3.14) is true for $t = \frac{k}{2^m}$ where $k,m \in \mathbb{N}$ are such that $k \leq 2^m$, i.e.,

$$d\left(z, \left(1-\frac{k}{2^m}\right)x \oplus \frac{k}{2^m}y\right)^2 \leq \left(1-\frac{k}{2^m}\right)d(z,x)^2 + \frac{k}{2^m}d(z,y)^2 - \frac{k}{2^m}\left(1-\frac{k}{2^m}\right)d(x,y)^2.$$

Now we prove (6.3.14) for $t = \frac{k}{2^{m+1}}$ where $k,m \in \mathbb{N}$ are such that $k \leq 2^{m+1}$. Put $u = \frac{k}{2^{m+1}}x \oplus (1 - \frac{k}{2^{m+}})y$. Then we have to prove that

$$d(z,u)^2 \leq \left(1 - \frac{k}{2^{m+1}}\right)d(z,x)^2 + \frac{k}{2^{m+1}}d(z,y)^2 - \frac{k}{2^{m+1}}\left(1 - \frac{k}{2^{m+1}}\right)d(x,y)^2.$$
$$(6.3.15)$$

First we show that (6.3.15) holds for $k \leq 2^m$, i.e., $\frac{k}{2^m} \in [0,1]$. Let $\alpha = (1 - \frac{k}{2^m})x \oplus \frac{k}{2^m}y$ and $\beta = \frac{1}{2}x \oplus \frac{1}{2}\alpha$. Then

$$d(x,\beta) = \frac{1}{2}d(x,\alpha) = \frac{k}{2^{m+1}}d(x,y) = d(x,u).$$

Since $\alpha \in [x,y]$ and $\beta \in [x,\alpha]$, therefore $\beta \in [x,y]$. As $u \in [x,y]$ and $d(x,\beta) = d(x,u)$, so $u = \beta$ by Lemma 6.3.1(c). Now applying (CN) inequality and the induction hypothesis, it follows that

$$
\begin{aligned}
d(u,z)^2 &= d\left(\frac{1}{2}x \oplus \frac{1}{2}\alpha, z\right)^2 \\
&\leq \frac{1}{2}d(x,z)^2 + \frac{1}{2}d(\alpha,z)^2 - \frac{1}{4}d(x,\alpha)^2 \\
&\leq \frac{1}{2}d(x,z)^2 + \frac{1}{2}\left(\left(1-\frac{k}{2^m}\right)d(z,x)^2 + \frac{k}{2^m}d(z,y)^2 - \frac{k}{2^m}\left(1-\frac{k}{2^m}\right)d(x,y)^2\right) \\
&\quad - \frac{1}{4}\left(\frac{k}{2^m}d(x,y)\right)^2 \\
&= \left(1-\frac{k}{2^{m+1}}\right)d(z,x)^2 + \frac{k}{2^{m+1}}d(z,y)^2 - \frac{k}{2^{m+1}}\left(1-\frac{k}{2^{m+1}}\right)d(x,y)^2.
\end{aligned}
$$

Now suppose that $2^m < k \leq 2^{m+1}$ and let $p = 2^{m+1} - k$. Then $p \leq 2^m$. So by (6.3.15), we obtain

$$
\begin{aligned}
d(u,z)^2 &= d\left(\frac{p}{2^{m+1}}x \oplus \left(1 - \frac{p}{2^{m+1}}\right)y, z\right) \\
&= d\left(\left(1 - \frac{p}{2^{m+1}}\right)y \oplus \frac{p}{2^{m+1}}x, z\right)^2 \\
&\leq \left(1 - \frac{p}{2^{m+1}}\right)d(y,z)^2 + \frac{p}{2^{m+1}}d(y,z)^2 - \frac{p}{2^{m+1}}\left(1 - \frac{p}{2^{m+1}}\right)d(x,y)^2 \\
&= \left(1 - \frac{k}{2^{m+1}}\right)d(y,z)^2 + \frac{k}{2^{m+1}}d(y,z)^2 - \frac{k}{2^{m+1}}\left(1 - \frac{k}{2^{m+1}}\right)d(x,y)^2.
\end{aligned}
$$

Next we use the fact that the set $D = \left\{\frac{k}{2^n} : k,n \in \mathbb{N}, k \leq 2^n\right\}$ is dense in $[0,1]$. For $t \in [0,1]$, there exists a sequence $\{t_k\}$ in D such that $\lim_{k \to \infty} t_k = t$. Now, we have

$$
d(z, t_k x \oplus (1-t_k)y)^2 \leq (1-t_k)d(z,x)^2 + t_k d(z,y)^2 - t_k(1-t_k)d(x,y)^2.
$$

Letting $k \to \infty$, using Lemma 6.3.2(c) and the fact that d is continuous, we get (6.3.14).

$\square$

Lemma 6.3.5. *If $\{x_n\}$ is a bounded sequence in a CAT(0) space X with $A(\{x_n\}) = \{x\}$ and $\{u_n\}$ is a subsequence of $\{x_n\}$ with $A(\{u_n\}) = \{u\}$ and the sequence $\{d(x_n, u)\}$ converges, then $x = u$.*

Proof. Suppose that $x \neq u$. By the uniqueness of asymptotic centers,

$$
\begin{aligned}
\limsup_{n \to \infty} d(u_n, u) &< \limsup_{n \to \infty} d(u_n, x) \\
&\leq \limsup_{n \to \infty} d(x_n, x) \\
&< \limsup_{n \to \infty} d(x_n, u) \\
&= \limsup_{n \to \infty} d(u_n, u),
\end{aligned}
$$

a contradiction, and hence $x = u$.

$\square$

The existence of fixed points for nonexpansive mappings in a CAT(0) space has been proved by Kirk [39] as follows:

Theorem 6.3.1. *Let C be a nonempty bounded, closed and convex subset of a complete CAT(0) space X. Then any nonexpansive mapping $T : C \to C$ has a fixed point.*

We collect some basic properties of Δ-convergence of a sequence in the following result.

Lemma 6.3.6. *Let X be a CAT (0) space.*

(a) *Every bounded sequence in X has a Δ-convergent subsequence.*

(b) *If C is a closed and convex subset of X and $\{x_n\}$ is a bounded sequence in C, then the asymptotic center of $\{x_n\}$ is in C.*

(c) *If C is a closed and convex subset of X and if $T : C \to C$ is nonexpansive, then the conditions: $\{x_n\}$ Δ-converges to x and $\lim_{n\to\infty} d(x_n, Tx_n) = 0$ imply $x \in C$ and $Tx = x$.*

Using the concept of unique point $z = (1 - \alpha)x \oplus \alpha y$ on the geodesic segment from x to y, the Mann iterative sequence (6.1.3) in a CAT(0) space is defined by

$$x_1 \in C, x_{n+1} = (1 - \alpha_n)x_n \oplus \alpha_n Tx_n, \tag{6.3.16}$$

where $0 \le \alpha_n \le 1$.

Similarly, the Ishikawa iterative sequence (6.1.4) in a CAT(0) space becomes

$$x_1 \in C, x_{n+1} = (1 - \alpha_n)x_n \oplus \alpha_n T((1 - \beta_n)x_n \oplus \beta_n Tx_n), \tag{6.3.17}$$

where $0 \le \alpha_n, \beta_n \le 1$.

Lemma 6.3.7. *Let C be a nonempty closed and convex subset of a complete CAT (0) space X and let $T : C \to C$ be a nonexpansive mapping. If $\{x_n\}$ is a bounded sequence in C such that $\lim_{n\to\infty} d(Tx_n, x_n) = 0$ and $\{d(x_n, v)\}$ converges for all $v \in F(T)$, then $\omega(x_n) \subset F(T)$; here $\omega(x_n) := \bigcup A\{u_n\}$, where the union is taken over all subsequences $\{u_n\}$ of $\{x_n\}$. Moreover, $\omega(x_n)$ consists of exactly one point.*

Proof. Let $u \in \omega(x_n)$. Then there exists a subsequence $\{u_n\}$ of $\{x_n\}$ such that $A(\{u_n\}) = \{u\}$. By Lemma 6.3.6(a, b), there exists a subsequence $\{v_n\}$ of $\{u_n\}$ such that $\Delta\text{-}\lim_n v_n = v \in C$. By Lemma 6.3.6(c), $v \in F(T)$. By Lemma 6.3.5, $u = v$. This shows that $\omega(x_n) \subset F(T)$. Next, we show that $\omega(x_n)$ consists of exactly one point. Let $\{u_n\}$ be a subsequence of $\{x_n\}$ with $A(\{u_n\}) = \{u\}$ and let $A(\{x_n\}) = \{x\}$. Now $u \in \omega(x_n) \subset F(T)$ and $\{d(x_n, u)\}$ converges, so by Lemma 6.3.5, $x = u$. $\qquad\square$

The following lemmas are analogs of the corresponding ones in Section 6.2 and so their proofs are omitted.

Lemma 6.3.8. *Let C be a nonempty bounded, closed and convex subset of a CAT (0) space X and $T : C \to C$ be a nonexpansive mapping. Then for the sequence $\{x_n\}$ defined by (6.3.17), $\lim_{n\to\infty} d(x_n, p)$ exists for each $p \in F(T)$.*

Lemma 6.3.9. *Let C be a nonempty bounded, closed and convex subset of a CAT (0) space $X, T : C \to C$ be nonexpansive mapping and $\{x_n\}$ be the sequence in (6.3.17) with $\{\alpha_n\}$ and $\{\beta_n\}$ in $[0,1]$ such that $\sum_{n=1}^{\infty} \alpha_n (1 - \alpha_n) = \infty$, $\sum_{n=1}^{\infty} \beta_n (1 - \alpha_n) < \infty$ and $\limsup_{n \to \infty} \beta_n < 1$. Then $\lim_{n \to \infty} d (x_n, T x_n) = 0$.*

Now, we give an analog of a result of Opial [50]. Recall that a mapping T of a metric space (X, d) into itself is said to be asymptotically regular [5] if, for any $x \in X$, the sequence $\{d(T^{n+1}(x), T^n(x))\}$ converges to 0 as $n \to \infty$.

Theorem 6.3.2. *Let C be a nonempty bounded, closed and convex subset of a CAT (0) space X and let $T : C \to C$ be a nonexpansive asymptotically regular mapping. Then, for any $x_0 \in C$, the Picard sequence $\{T^n x_0\}$ is $\triangle$-convergent to an element of $F(T)$.*

Proof. For each $n \geq 1$, let $x_n = T^n x_0$. By the asymptotic regularity of T, it follows that

$$\lim_{n \to \infty} d (x_n, T x_n) = \lim_{n \to \infty} d(T^{n+1}(x_0), T^n(x_0)) = 0.$$

As T is nonexpansive, $\{d(x_n, v)\}$ is decreasing for each $v \in F(T)$, so it is convergent. By Lemma 6.3.7, $\omega(x_n)$ consists of exactly one point and is contained in $F(T)$. This shows that $\{x_n\}$ $\triangle$-converges to an element of $F(T)$. $\qquad\square$

Now, we prove $\triangle$-convergence of the Mann iterative sequence in a CAT(0) space.

Theorem 6.3.3. *Let C be a nonempty bounded, closed and convex subset of a CAT(0) space X and $T : C \to C$ be a nonexpansive mapping. Then for any initial point x_1 in C, the Mann iterative sequence $\{x_n\}$ defined by (6.3.16), with the restrictions that $\sum_{n=1}^{\infty} \alpha_n = \infty$ and $\limsup_{n \to \infty} \alpha_n < 1$, $\triangle$-converges to a fixed point of T.*

Proof. By Lemma 6.3.9, $\lim_{n \to \infty} d (x_n, T x_n) = 0$. As in the proof of Theorem 6.3.2, apply the fact that $\{d(x_n, v)\}$ is convergent for each $v \in F(T)$. By Lemma 6.3.7, we conclude that $\{x_n\}$ $\triangle$-converges to an element of $F(T)$. $\qquad\square$

In the same way, we can extend Theorem 6.3.3 for the Ishikawa iterative sequence in the following result to obtain an analog of Theorem 6.2.2 valid for uniformly convex Banach spaces.

Theorem 6.3.4. *Let C be a nonempty bounded, closed and convex subset of a CAT (0) space X, $T : C \to C$ a nonexpansive mapping and $\{x_n\}$ the sequence defined by (6.3.17)*

with sequences $\{\alpha_n\}$ and $\{\beta_n\}$ in $[0,1]$ such that $\sum_{n=1}^{\infty} \alpha_n (1-\alpha_n) = \infty$, $\sum_{n=1}^{\infty} \beta_n (1-\alpha_n) <$
∞ and $\limsup_{n \to \infty} \beta_n < 1$. Then $\{x_n\}$ $\triangle$-converges to a fixed point of T.

The next two results deal with strong convergence of the Ishikawa iterative sequence in the CAT(0) space setting.

Theorem 6.3.5. *Suppose that $X, C, T, \{x_n\}$ are as in Theorem 6.3.4. Suppose that C is a compact subset of X. Then $\{x_n\}$ converges strongly to a fixed point of T.*

Proof. By compactness of C, $\{x_n\}$ has a strongly convergent subsequence $\{x_{n_k}\}$ with limit z. By Lemma 6.3.9 and nonexpansiveness of T, we have

$$d(z,Tz) \leq d(z,x_{n_k}) + d(x_{n_k}, Tx_{n_k}) + d(Tx_{n_k}, Tz)$$
$$\leq 2d(x_{n_k}, z) + d(x_{n_k}, Tx_{n_k}) \to 0 \text{ as } k \to \infty.$$

Therefore $z \in F(T)$. By Lemma 6.3.8, $\lim_{n \to \infty} d(x_n, z)$ exists. Thus z is the strong limit of the sequence $\{x_n\}$ itself. $\qquad\square$

Theorem 6.3.6. *Suppose that $X, C, T, \{x_n\}$ are as in Theorem 6.3.4. If T satisfies condition (A), then $\{x_n\}$ converges strongly to a fixed point of T.*

Proof. Similar to that of Theorem 6.2.3 by replacing $\|.\|$ with d. $\qquad\square$

Let $\{x_n\}$ be a bounded sequence in a $CAT(0)$ space X and let C be a closed and convex subset of X which contains $\{x_n\}$. We adopt the notation

$$\{x_n\} \rightharpoonup w \text{ if and only if } \Phi(w) = \inf_{x \in C} \Phi(x),$$

where $\Phi(x) = \limsup_{n \to \infty} d(x_n, x)$.

We now give a connection between this kind of convergence and $\triangle$-convergence.

Lemma 6.3.10. *Let $\{x_n\}$ be a bounded sequence in a CAT(0) space X and let C be a closed and convex subset of X which contains $\{x_n\}$. Then $\triangle\text{-}\lim_n x_n = x$ implies that $\{x_n\} \rightharpoonup x$. The converse is true if $\{x_n\}$ is regular.*

Proof. Suppose that $\triangle\text{-}\lim_n x_n = x$. Then $x \in C$ by Lemma 6.3.6(b). As $A(\{x_n\}) = \{x\}$, so we have $r(\{x_n\}) = r(x, \{x_n\})$. This implies that $\Phi(x) = \inf_{y \in C} \Phi(y)$. Therefore $\{x_n\} \rightharpoonup x$.

Suppose that $\{x_n\}$ is regular and $\{x_n\} \rightharpoonup x$. We note that $\{x_n\} \rightharpoonup x$ if and only if $A(\{x_n\}) = \{x\}$. Suppose that $A(\{x_n\}) = \{y\}$; again by Lemma 6.3.6(b), we have $y \in C$. Therefore $x = y$, and hence $A(\{x_n\}) = \{x\}$. By the regularity of $\{x_n\}$, we have $A(\{x_n\}) \subset A(\{u_n\})$ for each subsequence $\{u_n\}$ of $\{x_n\}$. As the asymptotic center of any bounded sequence in X is singleton, so $\Delta\text{-}\lim_n x_n = x$. □

The following example shows that regularity in the above lemma is necessary.

Example 6.3.1. Let d be the usual metric on $\mathbb{R}$, $C = [-1, 1]$, $\{x_n\} = \{1, -1, 1, -1, \ldots\}$, $\{u_n\} = \{-1, -1, -1, \ldots\}$, and $\{v_n\} = \{1, 1, 1, \ldots\}$. Then $A(\{x_n\}) = A_C(\{x_n\}) = \{0\}$, $A(\{u_n\}) = \{-1\}$ and $A(\{v_n\}) = \{1\}$. This means that $\{x_n\} \rightharpoonup 0$ but it does not have a Δ-limit.

6.4. An Algorithm of Asymptotically Nonexpansive Mappings

A mapping $W : X^2 \times J \to X$ is a convex structure in a metric space (X, d) [60] if

$$d(u, W(x, y, \lambda)) \leq \lambda d(u, x) + (1 - \lambda) d(u, y),$$

for all $x, y, u \in X$ and $\lambda \in J = [0, 1]$. A nonempty subset C of a convex metric space is convex if $W(x, y, \lambda) \in C$ for all $x, y \in C$ and $\lambda \in J$.

Recently, Kohlenbach [42] enriched the concept of convex metric space as "hyperbolic space" by including the following additional conditions in the definition of a convex metric space:

(1) $d(W(x, y, \lambda_1), W(x, y, \lambda_2)) = |\lambda_1 - \lambda_2| d(x, y)$
(2) $W(x, y, \lambda) = W(y, x, 1 - \lambda)$
(3) $d(W(x, z, \lambda), W(y, w, \lambda)) \leq \lambda d(x, y) + (1 - \lambda) d(z, w),$

for all $x, y, z, w \in X$ and $\lambda, \lambda_1, \lambda_2 \in J$.

All normed spaces and their subsets are hyperbolic spaces as well as convex metric spaces. The class of hyperbolic spaces is properly contained in the class of convex metric spaces (see Refs. [36, 42]).

Different notions of a hyperbolic space can be found in the literature (see Refs. [23, 27, 42, 45, 36, 53] for a comparasion). The hyperbolic space introduced by Kohlenbach [42] is slightly restrictive than the space of hyperbolic type by Goebel and Kirk [22] but general than the hyperbolic space of Reich and Shafrir [53]. Moreover, this class of hyperbolic spaces also contains Hadamard manifolds, Hilbert balls equipped with the hyperbolic metric [23], $\mathbb{R}$-trees and Cartesian products of Hilbert balls as special cases.

It is remarked that every CAT(0) space is a hyperbolic space [17] (see also Refs. [24, 45]). The open unit ball in a complex Hilbert space with Poincare distance cannot be imbedded in any Banach space. Hence the class of hyperbolic spaces is wider than the class of Banach spaces and CAT(0) spaces.

A hyperbolic space X is uniformly convex [57] if for all $u,x,y \in X$, $r > 0$ and $\varepsilon \in (0,2]$, there exists a $\delta \in (0,1]$ such that $d\left(W(x,y,\frac{1}{2}),u\right) \leq (1-\delta)r$, whenever $d(x,u) \leq r, d(y,u) \leq r$ and $d(x,y) \geq \varepsilon r$.

A mapping $\eta : (0,\infty) \times (0,2] \to (0,1]$ which provides such a $\delta = \eta(r,\varepsilon)$ for $u,x,y \in X$, $r > 0$ and $\varepsilon \in (0,2]$, is called the modulus of uniform convexity of X. We say that η is monotone if it decreases with r (for a fixed ε).

A sequence $\{x_n\}$ in a metric space (X,d) has: (1) limit existence property for T if $\lim_{n\to\infty} d(x_n,p)$ exists for any $p \in F(T)$ and (2) approximate fixed point property for T if $\lim_{n\to\infty} d(x_n,Tx_n) = 0$.

Based on some geometrical properties of a convex metric space [64], we extend the algorithm (6.1.4) in a convex metric space as follows:

Let C be a nonempty convex subset of a convex metric space X and $T : C \to C$ be an asymptotically nonexpansive mapping. Then for a given $x_1 \in C$, define the sequences $\{x_n\}, \{y_n\}$ and $\{z_n\}$ by

$$z_n = W\left(T^n x_n, x_n, a_n\right),$$
$$y_n = W\left(T^n z_n, W\left(T^n x_n, x_n, \frac{c_n}{1-b_n}\right), b_n\right), \qquad (6.4.18)$$
$$x_{n+1} = W\left(T^n y_n, W\left(T^n z_n, x_n, \frac{\beta_n}{1-\alpha_n}\right), \alpha_n\right), \quad n \geq 1,$$

where $0 \leq a_n, b_n, c_n, \alpha_n, \beta_n, b_n+c_n, \alpha_n+\beta_n \leq 1$ (see also Ref. [58]).

Using "$W(x,y,0) = y$ for any x,y in X [64, Proposition 1.2(a)]", the general algorithm (6.4.18) reduces in a convex metric space to the following algorithm [66] (with $c_n = \beta_n = 0$):

$$z_n = W\left(T^n x_n, x_n, a_n\right)$$
$$y_n = W\left(T^n z_n, x_n, b_n\right), \qquad (6.4.19)$$
$$x_{n+1} = W\left(T^n y_n, x_n, \alpha_n\right), \quad n \geq 1.$$

The Ishikawa-type algorithm (with $a_n = c_n = \beta_n = 0$) is expressed as

$$y_n = W\left(T^n x_n, x_n, b_n\right),$$
$$x_{n+1} = W\left(T^n y_n, x_n, \alpha_n\right), \quad n \geq 1, \qquad (6.4.20)$$

and the Mann-type algorithm (with $a_n = b_n = c_n = \beta_n = 0$) is given by

$$x_{n+1} = W\left(T^n x_n, x_n, \alpha_n\right). \qquad (6.4.21)$$

The following lemmas will be needed.

Lemma 6.4.1 [17]. *Let C be a nonempty closed and convex subset of a uniformly convex hyperbolic space X and $\{u_n\}$ be a bounded sequence in C such that $A(\{u_n\}) =$*

$\{u\}$. If $\{v_m\}$ is any other sequence in C such that $\lim\limits_{m\to\infty} r(v_m, \{u_n\}) = r(u, \{u_n\})$, then $\lim\limits_{m\to\infty} v_m = u$.

Lemma 6.4.2 [17]. *Let X be a uniformly convex hyperbolic space with monotone modulus of uniform convexity η and $x \in X$. Let $\{\lambda_n\} \in [b,c]$ for some $b,c \in (0,1)$. If $\{u_n\}$ and $\{v_n\}$ are sequences in X such that $\limsup\limits_{n\to\infty} d(u_n, x) \leq r$, $\limsup\limits_{n\to\infty} d(v_n, x) \leq r$ and $\lim\limits_{n\to\infty} d(W(u_n, v_n, \lambda_n), x) = r$ for some $r \geq 0$, then $\lim\limits_{n\to\infty} d(u_n, v_n) = 0$.*

The following originates from the work of Khan et al. [31].

Lemma 6.4.3. *Let X be a hyperbolic space and T be an asymptotically nonexpansive mapping on a nonempty, closed and convex subset C of X with sequence $\{k_n \geq 1\}$ such that $\sum_{n=1}^{\infty}(k_n - 1) < \infty$. Then the sequence $\{x_n\}$ in (6.4.18) has the limit existence property for the mapping T.*

Proof. By the definition of convex structure, applied to the sequences $\{z_n\}$, $\{y_n\}$ and $\{x_n\}$ in (6.4.18), we obtain the inequality

$$d(x_{n+1}, p) \leq k_n^3 d(x_n, p). \tag{6.4.22}$$

Applying Lemma 6.1.1 to (6.4.22) (with $r_n = d(x_n, p)$, $s_n = k_n^3 - 1$, $t_n = 0$), we observe that $\lim\limits_{n\to\infty} d(x_n, p)$ exists for each $p \in F(T)$. Hence $\{x_n\}$ has the limit existence property for the mapping T. $\qquad\square$

Lemma 6.4.4. *Let C be a nonempty closed subset of a hyperbolic space X and $T : C \to C$ be an asymptotically nonexpansive mapping. Then $F(T)$ is closed.*

Proof. This is based on a routine argument. $\qquad\square$

Lemma 6.4.5. *Let X be a uniformly convex hyperbolic space and T be an asymptotically nonexpansive mapping on a nonempty closed and convex subset C of X with sequence $\{k_n \geq 1\}$ such that $\sum_{n=1}^{\infty}(k_n - 1) < \infty$. Let $\{a_n\}$, $\{b_n\}$, $\{c_n\}$, $\{\alpha_n\}$, $\{\beta_n\}$, $\{b_n + c_n\}$ and $\{\alpha_n + \beta_n\}$ be real sequences in $[0,1]$. Then, for the sequences $\{x_n\}$, $\{y_n\}$ and $\{z_n\}$ in (6.4.18), we have the following:*

(a) If $0 < \liminf\limits_{n\to\infty} \alpha_n \leq \limsup\limits_{n\to\infty}(\alpha_n + \beta_n) < 1$, then

$$\lim\limits_{n\to\infty} d\left(T^n y_n, W\left(T^n z_n, x_n, \frac{\beta_n}{1-\alpha_n}\right)\right) = 0.$$

(b) *If* $0 < \lim\inf_{n\to\infty} b_n \le \lim\sup_{n\to\infty} (b_n + c_n) < 1$, *then*

$$\lim_{n\to\infty} d\left(T^n z_n, W\left(T^n x_n, x_n, \frac{c_n}{1-b_n}\right)\right) = 0$$

(c) *If* $0 < \lim\inf_{n\to\infty} b_n \le \lim\sup_{n\to\infty}(b_n + c_n) < 1$ *and* $0 < \lim\inf_{n\to\infty} a_n \le \lim\sup_{n\to\infty} a_n < 1$, *then* $\lim_{n\to\infty} d\left(T^n x_n, x_n\right) = 0.$

Proof. As in Lemma 6.4.3, $\lim_{n\to\infty} d(x_n, p)$ exists for each $p \in F(T)$. Let $\lim_{n\to\infty} d(x_n, p) = c$ for some $c \ge 0$. The case $c = 0$ is trivial. We prove the result for $c > 0$.

(a) Assume that $0 < \lim\inf_{n\to\infty} \alpha_n \le \lim\sup_{n\to\infty}(\alpha_n + \beta_n) < 1$. Then there exist $\eta_1, \eta_2 \in (0,1)$ such that $0 < \eta_1 \le \alpha_n \le \alpha_n + \beta_n \le \eta_2 < 1$ for all $n \ge 1$.

By the definition of convex structure, applied to the sequences $\{z_n\}$ and $\{y_n\}$ in the general algorithm (6.4.18), we get that

$$d(y_n, p) \le k_n^2 d(x_n, p).$$

Therefore

$$\lim\sup_{n\to\infty} d(T^n y_n, p) \le c. \qquad (6.4.23)$$

Further, the inequality

$$
\begin{aligned}
d\left(W\left(T^n y_n, x_n, \frac{\beta_n}{1-\alpha_n}\right), p\right)
&\le \frac{\beta_n}{1-\alpha_n} d(T^n y_n, p) + \left(1 - \frac{\beta_n}{1-\alpha_n}\right) d(x_n, p) \\
&\le \frac{\beta_n}{1-\alpha_n} k_n d(y_n, p) + \left(1 - \frac{\beta_n}{1-\alpha_n}\right) d(x_n, p) \\
&\le \frac{\beta_n}{1-\alpha_n} k_n^2 d(x_n, p) + k_n\left(1 - \frac{\beta_n}{1-\alpha_n}\right) d(x_n, p) \\
&\le k_n^2 d(x_n, p)
\end{aligned}
$$

gives that

$$\lim\sup_{n\to\infty} d\left(W\left(T^n y_n, x_n, \frac{c_n}{1-b_n}\right), p\right) \le c. \qquad (6.4.24)$$

As $\lim_{n\to\infty} d(x_{n+1}, p) = c$, so we have

$$\lim_{n\to\infty} d\left(W\left(T^n y_n, W\left(T^n z_n, x_n, \frac{\beta_n}{1-\alpha_n}\right), \alpha_n\right), p\right) = c. \qquad (6.4.25)$$

The sequences in (6.4.23), (6.4.24) and (6.4.25) satisfy the hypotheses of Lemma 6.4.2, therefore it follows $\left(\text{with } x = p, \ \lambda_n = \alpha_n, \ u_n = T^n y_n, \ v_n = W\left(T^n z_n, x_n, \frac{\beta_n}{1-\alpha_n}\right)\right)$ that

$$\lim_{n\to\infty} d\left(T^n y_n, W\left(T^n z_n, x_n, \frac{\beta_n}{1-\alpha_n}\right)\right) = 0. \qquad (6.4.26)$$

(b) Suppose that $0 < \liminf_{n\to\infty} b_n \le \limsup_{n\to\infty}(b_n + c_n) < 1$. Then there exist $\delta_1, \delta_2 \in (0,1)$ such that $0 < \delta_1 \le a_n \le a_n + b_n \le \delta_2 < 1$ for all $n \ge 1$. Obviously,

$$\limsup_{n\to\infty} d\,(T^n z_n, p) \le c. \qquad (6.4.27)$$

As

$$
\begin{aligned}
d\left(W\left(T^n x_n, x_n, \frac{c_n}{1-b_n}\right), p\right) &\le \frac{c_n}{1-b_n} d\,(T^n x_n, p) + \left(1 - \frac{c_n}{1-b_n}\right) d\,(x_n, p) \\
&\le \frac{c_n}{1-b_n} k_n d\,(x_n, p) + \left(1 - \frac{c_n}{1-b_n}\right) d\,(x_n, p) \\
&\le \frac{c_n}{1-b_n} k_n d\,(x_n, p) + k_n\left(1 - \frac{c_n}{1-b_n}\right) d\,(x_n, p) \\
&\le k_n d\,(x_n, p),
\end{aligned}
$$

so we have

$$\limsup_{n\to\infty} d\left(W\left(T^n x_n, x_n, \frac{c_n}{1-b_n}\right), p\right) \le c. \qquad (6.4.28)$$

Next we calculate

$$
\begin{aligned}
d\,(x_{n+1}, p) &= d\left(W\left(T^n y_n, W\left(T^n z_n, x_n, \frac{\beta_n}{1-\alpha_n}\right), \alpha_n\right), p\right) \\
&\le \alpha_n d\,(T^n y_n, p) + (1-\alpha_n) d\left(W\left(T^n z_n, x_n, \frac{\beta_n}{1-\alpha_n}\right), p\right) \\
&\le \alpha_n d\,(T^n y_n, p) + (1-\alpha_n) d\left(W\left(T^n z_n, x_n, \frac{\beta_n}{1-\alpha_n}\right), p\right) \\
&\le \alpha_n d\,(T^n y_n, p) + (1-\alpha_n) d\left(W\left(T^n z_n, x_n, \frac{\beta_n}{1-\alpha_n}\right), T^n y_n\right) \\
&\quad + (1-\alpha_n) d\,(T^n y_n, p) \\
&\le k_n d\,(y_n, p) + (1-\alpha_n) d\left(W\left(T^n z_n, x_n, \frac{\beta_n}{1-\alpha_n}\right), T^n y_n\right).
\end{aligned}
$$

That is,

$$d\,(x_{n+1}, p) \le k_n d\,(y_n, p) + (1-a) d\left(W\left(T^n z_n, x_n, \frac{\beta_n}{1-\alpha_n}\right), T^n y_n\right).$$

Applying lim inf to the above inequality, we have

$$c \leq \liminf_{n \to \infty} d(y_n, p) \leq \limsup_{n \to \infty} d(y_n, p) \leq c.$$

That is,

$$\lim_{n \to \infty} d\left(W\left(T^n z_n, W\left(T^n x_n, x_n, \frac{c_n}{1 - b_n}\right), b_n\right), p\right) = c \qquad (6.4.29)$$

Again the sequences in (6.4.26)–(6.4.29) satisfy the hypotheses of Lemma 6.4.2, so it follows by its conclusion $\left(\text{with } x = p, \lambda_n = b_n, u_n = T^n z_n, v_n = W\left(T^n x_n, x_n, \frac{c_n}{1-b_n}\right)\right)$ that

$$\lim_{n \to \infty} d\left(T^n z_n, W\left(T^n x_n, x_n, \frac{c_n}{1 - b_n}\right)\right) = 0. \qquad (6.4.30)$$

(c) Let $0 < \liminf_{n \to \infty} b_n \leq \limsup_{n \to \infty}(b_n + c_n) < 1$ and $0 < \liminf_{n \to \infty} a_n \leq \limsup_{n \to \infty} a_n < 1$.
We show that $\lim_{n \to \infty} d(T^n x_n, x_n) = 0$.

As $0 < \liminf_{n \to \infty} b_n \leq \limsup_{n \to \infty}(b_n + c_n) < 1$, so as in part (b), there exist $\delta_1, \delta_2 \in (0,1)$ such that $0 < \delta_1 \leq a_n \leq a_n + b_n \leq \delta_2 < 1$ for all $n \geq 1$. Also $0 < \liminf_{n \to \infty} a_n \leq \limsup_{n \to \infty} a_n < 1$ gives that there exist $\tau_1, \tau_2 \in (0,1)$ such that $0 < \tau_1 \leq a_n \leq \tau_2 < 1$ for all $n \geq 1$. Taking lim inf in the inequality,

$$
\begin{aligned}
d(y_n, p) &= d\left(W\left(T^n z_n, W\left(T^n x_n, x_n, \frac{c_n}{1 - b_n}\right), b_n\right), p\right) \\
&\leq b_n d(T^n z_n, p) + (1 - b_n) d\left(W\left(T^n x_n, x_n, \frac{c_n}{1 - b_n}\right), p\right) \\
&\leq b_n d(T^n z_n, p) + (1 - b_n) d\left(W\left(T^n x_n, x_n, \frac{c_n}{1 - b_n}\right), T^n z_n\right) \\
&\quad + (1 - b_n) d(T^n z_n, p) \\
&\leq \delta_2 d(z_n, p) + (1 - \delta_1) d\left(W\left(T^n x_n, x_n, \frac{c_n}{1 - b_n}\right), T^n z_n\right),
\end{aligned}
$$

we have

$$c < \liminf_{n \to \infty} d(z_n, p) \leq \limsup_{n \to \infty} d(z_n, p) \leq c,$$

that is,

$$\lim_{n \to \infty} d(W(T^n x_n, x_n, a_n), p) = c.$$

Applying Lemma 6.4.2 (with $x = p$, $\lambda_n = a_n$, $u_n = T^n x_n$, $v_n = x_n$), we have

$$\lim_{n \to \infty} d\left(T^n x_n, x_n\right) = 0. \tag{6.4.31}$$

$\square$

Lemma 6.4.6. *Let X be a uniformly convex hyperbolic space and T be an asymptotically nonexpansive mapping on a nonempty closed and convex subset C of X with sequence $\{k_n \geq 1\}$ such that $\sum_{n=1}^{\infty} (k_n - 1) < \infty$. Let $\{a_n\}$, $\{b_n\}$, $\{c_n\}$, $\{\alpha_n\}$ and $\{\beta_n\}$ be real sequences in $[0,1]$ such that $b_n + c_n \in [0,1]$ and $\alpha_n + \beta_n \in [0,1]$ for all $n \geq 1$ and satisfy*

(i) $0 < \liminf\limits_{n \to \infty} \alpha_n \leq \limsup\limits_{n \to \infty} (\alpha_n + \beta_n) < 1;$

(ii) $0 < \liminf\limits_{n \to \infty} b_n \leq \limsup\limits_{n \to \infty} (b_n + c_n) < 1;$

(iii) $0 < \liminf\limits_{n \to \infty} a_n \leq \limsup\limits_{n \to \infty} a_n < 1.$

Then $\{x_n\}$ in (6.4.18), has approximate fixed point property for T.

Proof. In the proof of Lemma 6.4.5, we showed that (i), (ii) and a combination of (ii) and (iii) imply (6.4.26), (6.4.30) and (6.4.31), respectively. Now we proceed further to show that $\lim\limits_{n \to \infty} d\left(x_n, T x_n\right) = 0$.

First of all, the inequality

$$
\begin{aligned}
d\left(y_n, x_n\right) &= d\left(W\left(T^n z_n, W\left(T^n x_n, x_n, \frac{c_n}{1 - b_n}\right), b_n\right), x_n\right) \\
&\leq b_n d\left(T^n z_n, x_n\right) + (1 - b_n) d\left(W\left(T^n x_n, x_n, \frac{c_n}{1 - b_n}\right), x_n\right) \\
&\leq b_n d\left(T^n z_n, T^n x_n\right) + b_n d\left(T^n x_n, x_n\right) + c_n d\left(T^n x_n, x_n\right) \\
&\leq b_n k_n d\left(z_n, x_n\right) + b_n d\left(T^n x_n, x_n\right) + c_n d\left(T^n x_n, x_n\right) \\
&\leq a_n b_n k_n^2 d\left(T^n x_n, x_n\right) + b_n d\left(T^n x_n, x_n\right) + c_n d\left(T^n x_n, x_n\right) \\
&\leq \left(\tau_2 \delta_2 k_n^2 + 2\delta_2\right) d\left(T^n x_n, x_n\right),
\end{aligned}
$$

together with (6.4.31), gives that

$$\lim_{n \to \infty} d\left(x_n, y_n\right) = 0. \tag{6.4.32}$$

Further,

$$
\begin{aligned}
d\left(x_n, T^n z_n\right) &\leq d\left(x_n, T^n x_n\right) + d\left(T^n x_n, T^n z_n\right) \\
&\leq d\left(x_n, T^n x_n\right) + k_n d\left(x_n, z_n\right) \\
&\leq d\left(x_n, T^n x_n\right) + \delta_2 k_n^2 d\left(T^n x_n, x_n\right) \\
&= \left(1 + \delta_2 k_n^2\right) d\left(x_n, T^n x_n\right)
\end{aligned}
$$

implies that

$$
\lim_{n \to \infty} d\left(x_n, T^n z_n\right) = 0. \tag{6.4.33}
$$

Also,

$$
\begin{aligned}
d\left(x_n, T^n y_n\right) &\leq d\left(x_n, T^n x_n\right) + d\left(T^n x_n, T^n y_n\right) \\
&\leq d\left(x_n, T^n x_n\right) + k_n d\left(x_n, y_n\right),
\end{aligned}
$$

together with (6.4.31) and (6.4.32), gives that

$$
\lim_{n \to \infty} d\left(x_n, T^n y_n\right) = 0. \tag{6.4.34}
$$

Taking $\limsup$ on both sides in the inequality

$$
\begin{aligned}
d\left(x_{n+1}, x_n\right) &\leq d\left(W\left(T^n y_n, W\left(T^n z_n, x_n, \frac{\beta_n}{1 - \alpha_n}\right), \alpha_n\right), x_n\right) \\
&\leq \alpha_n d\left(T^n y_n, x_n\right) + (1 - \alpha_n) d\left(W\left(T^n z_n, x_n, \frac{\beta_n}{1 - \alpha_n}\right), x_n\right) \\
&\leq \eta_2 d\left(T^n y_n, x_n\right) + \eta_2 d\left(T^n z_n, x_n\right)
\end{aligned}
$$

and then using (6.4.33) and (6.4.34), we obtain

$$
\lim_{n \to \infty} d\left(x_{n+1}, x_n\right) = 0.
$$

Finally, the inequality

$$
\begin{aligned}
d\left(x_{n+1}, T x_{n+1}\right) &\leq d\left(x_{n+1}, T^{n+1} x_{n+1}\right) + d\left(T x_{n+1}, T^{n+1} x_{n+1}\right) \\
&\leq k_1 d\left(x_{n+1}, T^n x_{n+1}\right) + d\left(x_{n+1}, T^{n+1} x_{n+1}\right) \\
&\leq d\left(x_{n+1}, T^{n+1} x_{n+1}\right) + k_1 \left\{(1 + k_n) d\left(x_{n+1}, x_n\right) + d\left(x_n, T^n x_n\right)\right\},
\end{aligned}
$$

together with (6.4.31) and (6.4.32), gives that

$$
\lim_{n \to \infty} d\left(x_n, T x_n\right) = 0. \tag{6.4.35}
$$

This proves that $\{x_n\}$ has approximate fixed point property for T. $\square$

We need some concepts to proceed further.

Let X be a metric space and $T : C \rightarrow C$ be a mapping. Then

(a) T is *completely continuous* if, for a bounded sequence $\{x_n\}$ in C, the sequence $\{Tx_n\}$ has a convergent subsequence in C

(b) T is *semi-compact* if every bounded sequence $\{x_n\}$ in C with $d(x_n, Tx_n) \rightarrow 0$ has a convergent subsequence in C.

(c) T^m is *compact* if $\{T^m x_n\}$ has a convergent subsequence.

Remark 6.4.1. If $\{x_n\}$ has approximate fixed point property for T, then we can prove strong convergence of $\{x_n\}$ to a fixed point of T under any condition from (a) to (c).

We will prove the result only under condition (a).

Theorem 6.4.1. *Let X be a uniformly convex hyperbolic space and T be an asymptotically nonexpansive mapping on a nonempty closed and convex subset C of X with sequence $\{k_n \geq 1\}$ such that $\sum_{n=1}^{\infty} (k_n - 1) < \infty$. Let $\{a_n\}, \{b_n\}, \{c_n\}, \{\alpha_n\}$ and $\{\beta_n\}$ be real sequences in $[0,1]$ such that $b_n + c_n \in [0,1]$ and $\alpha_n + \beta_n \in [0,1]$ for all $n \geq 1$ and satisfy*

(i) $0 < \liminf_{n \to \infty} \alpha_n \leq \limsup_{n \to \infty} (\alpha_n + \beta_n) < 1,$

(ii) $0 < \liminf_{n \to \infty} b_n \leq \limsup_{n \to \infty} (b_n + c_n) < 1,$

(iii) $0 < \liminf_{n \to \infty} a_n \leq \limsup_{n \to \infty} a_n < 1.$

If T is completely continuous, then $\{x_n\}$ in (6.4.18), converges strongly to a fixed point of T.

Proof. As T is completely continuous and $\{x_n\} \subseteq C$ is bounded, so there exists a subsequence $\{x_{n_k}\}$ of $\{x_n\}$ such that $\{Tx_{n_k}\}$ converges. Therefore by (6.4.35), $\{x_{n_k}\}$ converges. Let $\lim_{k \to \infty} x_{n_k} = q$. By continuity of T and (6.4.35), we have that $Tq = q$. By Lemma 6.4.3, $\lim_{n \to \infty} d(x_n, q)$ exists and so $\{x_n\}$ converges strongly to q. $\square$

As an application of Lemma 6.4.5, we prove the following strong convergence result which generalizes the theorems of Refs. [66, Theorem 2.1] and [28, Theorem 3.6] to uniformly convex hyperbolic spaces.

Theorem 6.4.2. *Let X be a uniformly convex hyperbolic space and T be an asymptotically nonexpansive mapping on a nonempty closed and convex subset C of X with* *sequence $\{k_n \geq 1\}$ such that $\sum_{n=1}^{\infty} (k_n - 1) < \infty$. Let $\{a_n\}, \{b_n\}$ and $\{\alpha_n\}$ be real sequences in $[0,1]$ for all $n \geq 1$ and satisfy*

(i) $0 < \liminf\limits_{n \to \infty} \alpha_n \leq \limsup\limits_{n \to \infty} \alpha_n < 1,$

(ii) $0 < \liminf\limits_{n \to \infty} b_n \leq \limsup\limits_{n \to \infty} b_n < 1.$

If T is completely continuous, then $\{x_n\}$ in (6.4.19), converges strongly to a fixed point of T.

Proof. Set $c_n = \beta_n = 0$ in Lemma 6.4.5 and observe that:
(1) Algorithm (6.4.18) reduces to algorithm (6.4.19).
(2) The condition (i) implies $\lim\limits_{n \to \infty} d(T^n y_n, x_n) = 0$ and condition (ii) implies $\lim\limits_{n \to \infty} d(T^n z_n, x_n) = 0.$ Then

$$
\begin{aligned}
d(T^n x_n, x_n) &\leq d(T^n x_n, T^n y_n) + d(T^n y_n, x_n) \\
&\leq k_n d(x_n, y_n) + d(T^n y_n, x_n) \\
&= k_n d(x_n, W(T^n z_n, x_n, b_n)) + d(T^n y_n, x_n) \\
&\leq k_n b_n d(T^n z_n, x_n) + d(T^n y_n, x_n) \to 0 \text{ as } n \to \infty
\end{aligned}
$$

implies

$$
\begin{aligned}
d(x_{n+1}, T^n x_{n+1}) &\leq d(x_{n+1}, x_n) + d(x_n, T^n x_n) + d(T^n x_n, T^n x_{n+1}) \\
&\leq (1 + k_n)\alpha_n d(T^n y_n, x_n) + d(x_n, T^n x_n) \to 0 \text{ as } n \to \infty.
\end{aligned}
$$

Thus

$$
\begin{aligned}
d(x_{n+1}, T x_{n+1}) &\leq d(x_{n+1}, T^{n+1} x_{n+1}) + d(T^{n+1} x_{n+1}, T x_{n+1}) \\
&\leq d(x_{n+1}, T^{n+1} x_{n+1}) + k_1 d(T^n x_{n+1}, x_{n+1}) \to 0 \text{ as } n \to \infty,
\end{aligned}
$$

that is,

$$
\lim_{n \to \infty} d(x_n, T x_n) = 0.
$$

The rest of the proof is the same as the proof of Theorem 6.4.1. $\qquad\square$

The following Ishikawa-type convergence result generalizes the result of Ref. [54, Theorem 3].

Theorem 6.4.3. *Let X be a uniformly convex hyperbolic space and T be an asymptotically nonexpansive mapping on a nonempty closed and convex subset C of X with sequence $\{k_n \geq 1\}$ such that $\sum_{n=1}^{\infty} (k_n - 1) < \infty$. Let $\{\alpha_n\}$ and $\{b_n\}$ be real sequences in $[0,1]$ for all $n \geq 1$ and satisfy*

(i) $0 < \liminf\limits_{n\to\infty} \alpha_n \leq \limsup\limits_{n\to\infty} \alpha_n < 1$,

(ii) $0 < \liminf\limits_{n\to\infty} b_n \leq \limsup\limits_{n\to\infty} b_n < 1$.

If T is completely continuous, then $\{x_n\}$ in (6.4.20), converges strongly to a fixed point of T.

Proof. Choose $a_n = 0$ in Theorem 6.4.2.   □

Next we obtain $\triangle$-convergence results for the algorithms in (6.4.18), (6.4.19) and (6.4.20) for an asymptotically nonexpansive mapping in a uniformly convex hyperbolic space which are analogs of weak convergence results in uniformly convex Banach spaces.

Theorem 6.4.4. *Let X be a uniformly convex hyperbolic space and T be an asymptotically nonexpansive mapping on a nonempty closed and convex subset C of X with sequence $\{k_n \geq 1\}$ such that $\sum_{n=1}^{\infty}(k_n - 1) < \infty$. Let $\{a_n\}$, $\{b_n\}$, $\{c_n\}$, $\{\alpha_n\}$ and $\{\beta_n\}$ be real sequences in $[0,1]$ such that $b_n + c_n \in [0,1]$ and $\alpha_n + \beta_n \in [0,1]$ for all $n \geq 1$ and satisfy*

(i) $0 < \liminf\limits_{n\to\infty} \alpha_n \leq \limsup\limits_{n\to\infty}(\alpha_n + \beta_n) < 1$,

(ii) $0 < \liminf\limits_{n\to\infty} b_n \leq \limsup\limits_{n\to\infty}(b_n + c_n) < 1$,

(iii) $0 < \liminf\limits_{n\to\infty} a_n \leq \limsup\limits_{n\to\infty} a_n < 1$.

Then $\{x_n\}$ in (6.4.18), $\triangle$-converges to a fixed point of T .

Proof. The sequence $\{x_n\}$ is bounded by Lemma 6.4.3 and so it has a unique asymptotic center. That is, $A(\{x_n\}) = \{x\}$. Let $\{u_n\}$ be any subsequence of $\{x_n\}$ such that $A(\{u_n\}) = \{u\}$. Then by Lemma 6.4.6, we have $\lim\limits_{n\to\infty} d(u_n, Tu_n) = 0$. We claim that u is a fixed point of T.

In fact, for any $m, n \geq 1$,

$$
\begin{aligned}
d(T^m u, u_n) &\leq d(T^m u, T^m u_n) \\
&\quad + d(T^m u_n, T^{m-1} u_n) + \ldots + d(T^2 u_n, T u_n) + d(T u, u_n) \\
&\leq k_m d(u, u_n) + \sum_{i=1}^{m-1} k_i d(T u_n, u_n).
\end{aligned}
$$

By taking $\limsup$ on both sides with respect to n, we have

$$
\limsup_{n\to\infty} d(T^m u, u_n) \leq k_m r(u, \{u_n\}).
$$

Therefore

$$\limsup_{m \to \infty} r\left(T^m u, \{u_n\}\right) \leq r\left(u, \{u_n\}\right).$$

By the definition of the asymptotic center $A(\{u_n\})$ of a bounded sequence $\{u_n\}$ with respect to C, we have

$$r\left(u, \{u_n\}\right) \leq r\left(x, \{u_n\}\right) \text{ for all } x \in C.$$

This implies that

$$r\left(u, \{u_n\}\right) \leq \liminf_{m \to \infty} r\left(T^m u, \{u_n\}\right),$$

that is,

$$\lim_{m \to \infty} r\left(T^m u, \{u_n\}\right) = r\left(u, \{u_n\}\right).$$

By Lemma 6.4.1, $\lim\limits_{m \to \infty} T^m u = u$. Since T is continuous, therefore

$$Tu = T\left(\lim_{m \to \infty} T^m u\right) = \lim_{m \to \infty} T^{m+1} u = u.$$

This proves that $u \in F(T)$.

Further, $\lim\limits_{n \to \infty} d(x_n, u)$ exists by Lemma 6.4.3. Suppose $x \neq u$. Then by the uniqueness of asymptotic centers, we have

$$
\begin{aligned}
\limsup_{n \to \infty} d(u_n, u) \quad &< \quad \limsup_{n \to \infty} d(u_n, x) \\
&\leq \quad \limsup_{n \to \infty} d(x_n, x) \\
&< \quad \limsup_{n \to \infty} d(x_n, u) \\
&= \quad \limsup_{n \to \infty} d(u_n, u),
\end{aligned}
$$

which gives a contradiction. Hence $x = u$. Therefore, $A(\{u_n\}) = \{u\}$ for all subsequences $\{u_n\}$ of $\{x_n\}$. This proves that $\{x_n\}$ $\triangle$-converges to a fixed point of T. $\square$

As a special case of Theorem 6.4.2, we obtain the following result.

Corollary 6.4.1. *Let X be a CAT(0) space and T be an asymptotically nonexpansive mapping on a nonempty closed and convex subset C of X with sequence $\{k_n \geq 1\}$ such that $\sum_{n=1}^{\infty} (k_n - 1) < \infty$. Let $\{a_n\}$, $\{b_n\}$ and $\{\alpha_n\}$ be real sequences in $[0,1]$ such that*

(i) $0 < \liminf\limits_{n \to \infty} \alpha_n \leq \limsup\limits_{n \to \infty} \alpha_n < 1,$

(ii) $0 < \liminf\limits_{n \to \infty} b_n \leq \limsup\limits_{n \to \infty} b_n < 1.$

For a given $x_1 \in C$, define $\{x_n\}$ by

$$
\begin{aligned}
z_n &= a_n T^n x_n \oplus (1-a_n)x_n, \\
y_n &= b_n T^n z_n \oplus (1-b_n)x_n, \\
x_{n+1} &= \alpha_n T^n y_n \oplus (1-\alpha_n)x_n, \quad n \geq 1.
\end{aligned}
$$

Then $\{x_n\}$ $\triangle$-converges to a fixed point of T.

Proof. As every CAT(0) space is a uniformly convex hyperbolic space (take $W(x,y,\lambda) = \lambda x \oplus (1-\lambda)y$; geodesic path between the points x and y in X), so the conclusion follows from Theorem 6.4.2. □

Remark 6.4.2.   (a) Theorem 6.4.2 sets an analog to Theorem 3.1 of Ref. [32] in hyperbolic spaces.

(b) We can replace the condition $\sum_{n=1}^{\infty}(k_n - 1) < \infty$ in all the above results by $\sum_{n=1}^{\infty}(k_n^p - 1) < \infty$ for some $p > 1$ as follows:
Since $\{k_n\}$ is a bounded sequence, therefore $k_n \in [1,M]$ for all $n \geq 1$. For $p > 1$, define $f : [1,M] \to \mathbb{R}$ by $f(x) = x^p - 1 - pM^{p-1}(x-1)$. Now it follows from elementary calculus that f is a decreasing function. So $f(x) \leq f(1)$ for all $x \in [1,M]$, that is, $k_n - 1 \leq k_n^p - 1 \leq pM^{p-1}(k_n - 1)$. Therefore, $\sum_{n=1}^{\infty}(k_n - 1) < \infty$ if and only if $\sum_{n=1}^{\infty}(k_n^p - 1) < \infty$.

6.5. Existence and Approximation of Fixed Points

A metric space X is convex [40] if, for each $x,y \in X$ with $x \neq y$, there exists $z \in X, x \neq z \neq y$ such that

$$d(x,y) = d(x,z) + d(z,y).$$

Clearly, a metric space with convex structure W due to Takahashi [60], introduced in Section 6.3, is a convex metric space. For such convex metric spaces, we define $f : X \to \mathbb{R}$ as a convex function if

$$f(W(x,y,\lambda)) \leq \lambda f(x) + (1-\lambda)f(y)$$

for all $x,y \in X$ and $\lambda \in J$.

In this section, we shall utilize the convex structure W of Takahashi [60].

Let T be a continuous mapping on a nonempty, bounded, closed and convex subset C of a uniformly convex metric space X. Recently, Fukhar-ud-din et al. [18] have established the existence of a fixed point for T satisfying, for all $x,y \in C$,

$$
\begin{aligned}
d(Tx,Ty) &\leq ad(x,y) + b\{d(Tx,x) + d(Ty,y)\} \\
&\quad + c\{d(Tx,y) + d(Ty,x)\},
\end{aligned}
\tag{6.5.36}
$$

where $a+2b+2c \leq 1$. Their result applies to mappings satisfying the condition (6.5.36) as well as nonexpansive mappings.

For a a continuous mapping T on a nonempty bounded, closed and convex subset C of a uniformly convex Banach space X, Bose and Laskar [3] proved a fixed point theorem for T which satisfies

$$d\left(T^n x, T^n y\right) \leq a_n d\left(x, y\right) + b_n\left[d\left(x, T^n x\right) + d\left(y, T^n y\right)\right]$$
$$+ c_n\left[d\left(x, T^n y\right) + d\left(y, T^n x\right)\right] \tag{6.5.37}$$

for $x, y \in X$, where $\{a_n\}$, $\{b_n\}$ and $\{c_n\}$ are non-negative real sequences with $b_n + c_n < 1$ for $n \geq 1$ and $\lim\limits_{n \to \infty}\left(\dfrac{a_n + 3b_n + c_n}{1 - b_n - c_n}\right) = 1$. A mapping satisfying (6.5.37) is called a *generalized nonexpansive mapping* [3].

Actually, they proved the following result.

Theorem 6.5.1 [3]. *Let X be a uniformly convex Banach space, C a nonempty bounded, closed and convex subset of X and $T : C \to C$ a continuous mapping which satisfies (6.5.37). Then T has a fixed point. Moreover, this fixed point is unique if $a_n + 2c_n < 1$ for at least one n.*

Hereafter, we present the work of the second author published in Ref. [17].

First, we show that bounded sequences in uniformly convex metric spaces have a unique asymptotic center with respect to closed and convex subsets of the space in the following result.

Lemma 6.5.1. *Let C be a nonempty closed and convex subset of a complete uniformly convex metric space X. Then every bounded sequence $\{x_n\}$ in X has a unique asymptotic center with respect to C.*

Proof. In order to establish the result, it is enough to show that the function $r(\cdot, \{x_n\}) : C \to [0, \infty)$ attains its minimum exactly at one point of C.

Let c be the inf of $r(\cdot, \{x_n\})$ on C and define

$$C_m = \left\{x \in C : r(x, \{x_n\}) \leq c + \frac{1}{m}\right\}.$$

Obviously, $\{C_m\}$ is a decreasing sequence of subsets of a complete uniformly convex metric space X. Since

$$|r(x, \{x_n\}) - r(x_0, \{x_n\})| = \limsup_{n \to \infty} |d(x, x_n) - d(x_0, x_n)|$$
$$\leq d(x, x_0) \to 0 \text{ for any } x_0 \in C$$

and

$$r\left(W\left(x,y,\lambda\right),\{x_n\}\right) \leq \lambda \limsup d(x,x_n)+(1-\lambda)\limsup d(y,x_n)$$
$$= \lambda r(x,\{x_n\})+(1-\lambda)r(y,\{x_n\}) \text{ for any } x,y\in C,$$

therefore $r(.,\{x_n\})$ is continuous and convex on C. Moreover, $r(x,\{x_n\})\to\infty$ whenever $d(x,a)\to\infty$ for some $a\in X$. Therefore, each C_m is closed, convex and bounded. By Cantor's intersection theorem, we get that $\cap_{m=1}^{\infty}C_m\neq\phi$. Fix an element $z\in\cap_{m=1}^{\infty}C_m$. We claim that $r(z,\{x_n\})\leq c$. If $r(z,\{x_n\})>c$, then by the Archimedean property of real numbers, there exists a natural number m^* such that $m^*(r(z,\{x_n\})-c)>1$, that is, $r(z,\{x_n\})>c+\frac{1}{m^*}$. This gives that $z\notin\cap_{m=1}^{\infty}C_m$, a contradiction. Therefore, $r(z,\{x_n\})\leq c$. Since $c=\inf\{r(x,\{x_n\}) : x\in C\}$, therefore $r(z,\{x_n\})=c$.

Next we show that $r(.,\{x_n\})$ attains its minimum at exactly one point. Suppose it attains two minimum values $r(z_1,\{x_n\})$ and $r(z_2,\{x_n\})$ corresponding to two distinct points $z_1,z_2\in C$.

Let $M=\max\{r(z_1,\{x_n\}),r(z_2,\{x_n\})\}$ and $r=d(z_1,z_2)>0$. Then for $\varepsilon\in(0,1]$, there exists $n_0\geq 1$ such that $d(z_1,x_n)\leq M+\varepsilon$ and $d(z_2,x_n)\leq M+\varepsilon$ for all $n\geq n_0$. As $d(z_1,z_2)=\left(\frac{r}{M+\varepsilon}\right)\cdot(M+\varepsilon)\geq\left(\frac{r}{M+1}\right)\cdot(M+\varepsilon)$, so by uniform convexity of X, we have that

$$d\left(W\left(z_1,z_2,\frac{1}{2}\right),\{x_n\}\right)\leq(1-\alpha)(M+\varepsilon)<M+\varepsilon.$$

Hence

$$r\left(W\left(z_1,z_2,\frac{1}{2}\right),\{x_n\}\right)<M+\varepsilon.$$

When $\varepsilon\to 0$, we get that

$$r\left(W\left(z_1,z_2,\frac{1}{2}\right),\{x_n\}\right)<M,$$

a contradiction to the fact that $r(.,\{x_n\})$ attains its minimum at exactly one point. This proves the result. □

We now prove a metric version of a result of Bose and Laskar [3] which will play a crucial role in proving a fixed point theorem.

Lemma 6.5.2. *Let C be a nonempty closed and convex subset of a uniformly convex metric space X. Let ρ be the asymptotic radius of a bounded sequence $\{x_n\}$ in C and $A(\{x_n\})=\{y\}$. If $\{y_m\}$ is another sequence in C such that $\lim_{m\to\infty}r(y_m,\{x_n\})=\rho$, then $\lim_{m\to\infty}y_m=y$.*

Proof. If $y_m \nrightarrow y$, then there exists a subsequence $\{y_{m_i}\}$ of $\{y_m\}$ and $M > 0$ such that

$$d(y_{m_i}, y) \geq \frac{M}{2}, \quad \text{for all } i \geq 1.$$

As $A(\{x_n\}) = \{y\}$ and $\varepsilon \in (0, 1]$, so there exists an integer N_1 such that $d(y, x_n) \leq \rho + \varepsilon$ for all $n \geq N_1$. Also, $\lim\limits_{m \to \infty} r(y_m, \{x_n\}) = \rho = \lim\limits_{i \to \infty} r(y_{m_i}, \{x_n\})$, so there exists an integer i^* such that $r(y_m, \{x_n\}) \leq \rho + \frac{\varepsilon}{2}$ for all $i \geq i^*$. Hence there exists an integer N_2 such that $d(y_{m_i}, x_n) \leq \rho + \varepsilon$ for all $n \geq N_2$, that is,

$$d(y, x_n) \leq \rho + \varepsilon \quad \text{and} \quad d(y_{m_i}, x_n) \leq \rho + \varepsilon \text{ for all } n \geq N = \max\{N_1, N_2\}.$$

As $d(y_{m_i}, y) \geq \frac{M}{2} = \frac{M}{2(\rho + \varepsilon)} \cdot (\rho + \varepsilon) \geq \frac{M}{2(\rho + 1)} \cdot (\rho + \varepsilon)$, so by uniform convexity of X, we have

$$d\left(W\left(y, y_{m_j}, \frac{1}{2}\right), x_n\right) \leq (1 - \alpha)(\rho + \varepsilon) < \rho + \varepsilon.$$

Hence on letting $n \to \infty$, we have

$$r\left(W(y, y_{m_i}, \frac{1}{2}), \{x_n\}\right) \leq (1 - \alpha)(\rho + \varepsilon) < \rho + \varepsilon.$$

Now let $\varepsilon \to 0$ to conclude that $r\left(W\left(y, y_{m_i}, \frac{1}{2}\right), \{x_n\}\right) < \rho$, which contradicts the fact that ρ is the asymptotic radius of $\{x_n\}$. Hence $\lim\limits_{m \to \infty} y_m = y$. $\square$

Using Lemmas 6.5.1 and 6.5.2, we obtain the following generalization of Theorem 6.5.1.

Theorem 6.5.2. *Let C be a nonempty bounded, closed and convex subset of a complete uniformly convex metric space X and $T : C \to C$ a continuous mapping which satisfies (6.5.37). Then T has a fixed point. Moreover, this fixed point is unique if $a_n + 2c_n < 1$ for at least one value of n.*

Proof. Let $x_0 \in C$ and $x_n = T^n x_0$ for $n \geq 1$. Let ξ_0 and ρ be asymptotic center and asymptotic radius of $\{x_n\}$ in C, respectively. Let $\xi_j = T^j \xi_0$ for $j \geq 1$. We show that $r(\xi_j, \{x_n\}) = \limsup\limits_{n \to \infty} d(\xi_j, x_n) \to \rho$ as $j \to \infty$.

For two integers j, n $(j < n)$, we have by (6.5.37):

$$\begin{aligned}
d\left(\xi_j, x_n\right) &= d\left(T^j\xi_0, T^j x_{n-j}\right) \\
&\leq a_j d\left(\xi_0, x_{n-j}\right) + b_j\left[d\left(\xi_0, T^j\xi_0\right) + d\left(x_{n-j}, T^j x_{n-j}\right)\right] \\
&\quad + c_j\left[d\left(\xi_0, T^j x_{n-j}\right) + d\left(x_{n-j}, T^j\xi_0\right)\right] \\
&= a_j d\left(\xi_0, x_{n-j}\right) + b_j\left[d\left(\xi_0, \xi_j\right) + d\left(x_{n-j}, x_n\right)\right] \\
&\quad + c_j\left[d\left(\xi_0, x_n\right) + d\left(x_{n-j}, \xi_j\right)\right] \\
&\leq a_j d\left(\xi_0, x_{n-j}\right) + b_j\left[2d\left(\xi_0, x_n\right) + d\left(\xi_0, x_{n-j}\right) + d\left(\xi_j, x_n\right)\right] \\
&\quad + c_j\left[d\left(\xi_0, x_n\right) + d\left(x_{n-j}, \xi_j\right)\right].
\end{aligned}$$

Hence

$$\begin{aligned}
\left(1 - b_j\right) d\left(\xi_j, x_n\right) &\leq \left(a_j + b_j\right) d\left(\xi_0, x_{n-j}\right) + \left(2b_j + c_j\right) d\left(\xi_0, x_n\right) \\
&\quad + c_j d\left(x_{n-j}, \xi_j\right).
\end{aligned} \tag{6.5.38}$$

Let $\varepsilon > 0$. By the definition of ξ_0 and ρ, there exists an integer n_0 such that

$$d\left(\xi_0, x_n\right) < \rho + \frac{\varepsilon}{2}, \quad \text{for all } n \geq n_0. \tag{6.5.39}$$

For a fixed $j \geq 1$, (6.5.39) can be written as

$$d\left(\xi_0, x_{n-j}\right) < \rho + \frac{\varepsilon}{2}, \quad \text{for all } n \geq n_0 + j. \tag{6.5.40}$$

Applying (6.5.39) and (6.5.40) in (6.5.38), we have

$$\begin{aligned}
\left(1 - b_j\right) d\left(\xi_j, x_n\right) &\leq \left(a_j + b_j\right) d\left(\xi_0, x_{n-j}\right) + \left(2b_j + c_j\right) d\left(\xi_0, x_n\right) \\
&\quad + c_j d\left(x_{n-j}, \xi_j\right) \\
&\leq \left(a_j + 3b_j + c_j\right)\left(\rho + \frac{\varepsilon}{2}\right) + c_j d\left(x_{n-j}, \xi_j\right), \quad \text{for all } n \geq n_0 + j.
\end{aligned}$$

Taking lim sup on both sides in the above inequality, we obtain

$$\left(1 - b_j\right) r\left(\xi_j, \{x_n\}\right) \leq \left(a_j + 3b_j + c_j\right)\left(\rho + \frac{\varepsilon}{2}\right) + c_j r\left(\xi_j, \{x_n\}\right),$$

that is,

$$r\left(\xi_j, \{x_n\}\right) \leq \left(\frac{a_j + 3b_j + c_j}{1 - b_j - c_j}\right)\left(\rho + \frac{\varepsilon}{2}\right), \quad \text{for } j \geq n_1.$$

Since $k_j = \frac{a_j + 3b_j + c_j}{1 - b_j - c_j} \to 1$ as $j \to \infty$, therefore there is an integer n_2 such that

$$k_j\left(\rho + \frac{\varepsilon}{2}\right) < \rho + \varepsilon, \quad \text{for } j \geq n_2.$$

On choosing $n_3 \geq \max\left\{n_1, n_2\right\}$, we have that

$$r\left(\xi_j, \{x_n\}\right) - \rho < \varepsilon, \quad \text{for all } j \geq n_3,$$

that is, $r(\xi_j, \{x_n\}) \to \rho$ as $j \to \infty$. By Lemma 6.5.2, $\xi_j \to \xi_0$ as $j \to \infty$. As T is continuous, so $T\xi_j \to T\xi_0$ as $j \to \infty$. Finally, by the metric triangle inequality, we obtain

$$
\begin{aligned}
d(\xi_0, T\xi_0) &\leq d(\xi_0, T\xi_j) + d(T\xi_j, T\xi_0) \\
&= d(\xi_0, \xi_{j+1}) + d(T\xi_j, T\xi_0).
\end{aligned}
$$

This implies $\xi_0 = T\xi_0$ as desired.

For the uniqueness of the fixed point of T, assume that p and q are two distinct fixed points of T.

By inequality (6.5.37) with $x = p$, $y = q$, we have that

$$
d(p, q) = d(T^n p, T^n q) \leq (a_n + 2c_n) d(p, q),
$$

which gives that $1 \leq a_n + 2c_n$ for all $n \geq 1$, a contradiction to what is given.   □

The following example illustrates Theorem 6.5.2.

Example 6.5.1. Take $X = \mathbb{R}$, $C = [0, 1]$, $d(x, y) = |x - y|$, $T(x) = \frac{x}{3}$, $a_n = \frac{1}{3^n}$, $b_n = \frac{3n+4}{12n}$, $c_n = \frac{1}{4n}$. Then

(a) $b_n + c_n < 1$ for all $n \geq 1$, $a_n + 2c_n < 1$ for at least one $n \geq 1$,

(b) $d(T^n x, T^n y) \leq \frac{1}{3^n} d(x, y) + \left(\frac{3n+4}{12n}\right) \{d(x, T^n x) + d(y, T^n y)\}$
$\quad + \frac{1}{4n} \{d(x, T^n y) d(y, T^n x)\}$,

(c) $\lim\limits_{n \to \infty} \dfrac{a_n + 3b_n + c_n}{1 - b_n - c_n} = \lim\limits_{n \to \infty} \dfrac{\frac{1}{3^n} + 3\left(\frac{3n+4}{12n}\right) + \frac{1}{4n}}{1 - \frac{3n+4}{12n} - \frac{1}{4n}} = 1.$

By Theorem 6.5.2, T has a unique fixed point 0.

A metric version of the fundamental fixed point theorem of Goebel and Kirk [21] is obtained in the following result.

Corollary 6.5.1. *Let C be a nonempty bounded, closed and convex subset of a complete uniformly convex metric space X and $T : C \to C$ an asymptotically nonexpansive mapping with sequence $\{k_n\} \subseteq [1, \infty)$ such that $\lim\limits_{n \to \infty} k_n = 1$. Then T has a fixed point.*

Proof. The conclusion is immediate from Theorem 6.5.2 on taking $b_n = 0 = c_n$ and $k_n = a_n \geq 1$.   □

The following fixed point result is more general than the result given in Corollary 6.5.1.

Corollary 6.5.2. *Let C be a nonempty bounded, closed and convex subset of a complete uniformly convex metric space X and $T : C \to C$ be a mapping satisfying $d(T^n x, T^n y) \leq k_n d(x,y)$ where $\{k_n\} \subseteq (0,\infty)$ such that $\lim\limits_{n \to \infty} k_n = 1$. Then T has a fixed point.*

Proof. The conclusion follows on taking $b_n = 0 = c_n$ and $k_n = a_n$ in Theorem 6.5.2.   □

Corollary 6.5.1 is valid for the choice $k_n = 1 + \frac{1}{n+1} \in (1,\infty)$ while Corollary 6.5.2 includes both the cases $k_n = 1 + \frac{1}{n+1} \in (1,\infty)$ and $k_n = 1 - \frac{1}{n+1} \in (0,1)$.

Theorem 6.5.3. *Let C be a nonempty bounded, closed and convex subset of a complete uniformly convex metric space X. Let T be an equi-continuous mapping of C onto a compact subset of C satisfying (6.5.37) for all $x,y \in C$, where $a_n + 4b_n + 2c_n \leq 1$ and a_n, b_n, $c_n \geq 0$ for all $n \geq 1$. Then the average sequence $\{x_n\}$, defined by $x_n = W\left(x_n, T^n x_n, \frac{1}{2}\right)$, converges to a unique fixed point of T.*

Proof. By Theorem 6.5.2, T has a unique fixed point p (say). Now we approximate p by the sequence $x_n = W\left(x_n, T^n x_n, \frac{1}{2}\right)$. Note that

$$
\begin{aligned}
d(x_{n+1}, p) &= d\left(W\left(x_n, T^n x_n, \frac{1}{2}\right), p\right) \\
&\leq \frac{1}{2} d(x_n, p) + \frac{1}{2} d(T^n x_n, p) \qquad (6.5.41) \\
&= \frac{1}{2} d(x_n, p) + \frac{1}{2} d(T^n x_n, p).
\end{aligned}
$$

Now

$$
\begin{aligned}
d(T^n x_n, p) &\leq a_n d(x_n, p) + b_n d(x_n, T^n x_n) \\
&\quad + c_n \{d(x_n, p) + d(p, T^n x_n)\} \\
&\leq a_n d(x_n, p) + b_n \{d(x_n, p) + d(T^n x_n, p)\} \\
&\quad + c_n \{d(x_n, p) + d(T^n x_n, p)\}
\end{aligned}
$$

gives that

$$
\begin{aligned}
(1 - b_n - c_n) d(T^n x_n, p) &\leq (a_n + b_n + c_n) d(x_n, p) \\
&\leq (a_n + 3b_n + c_n) d(x_n, p).
\end{aligned}
$$

As $\frac{a_n + 3b_n + c_n}{1 - b_n - c_n} \leq 1$, so we get

$$
d(T^n x_n, p) \leq \frac{a_n + 3b_n + c_n}{1 - b_n - c_n} d(x_n, p) \leq d(x_n, p). \qquad (6.5.42)
$$

Substituting (6.5.42) in (6.5.41), we obtain

$$d(x_{n+1}, p) \leq d(x_n, p), \quad \text{for all } n \geq 1.$$

CASE 1. Suppose that there exists $\varepsilon > 0$ and a positive integer n_0 such that $d(x_n, T^n x_n) \geq \varepsilon$ for all $n \geq n_0$. Since $d(T^n x_n, p) \leq d(x_n, p)$ and

$$d(x_n, T^n x_n) \geq \frac{\varepsilon}{d(x_1, p)} d(x_n, p), \quad \text{for all } n \geq n_0,$$

therefore by uniform convexity of X, we have

$$
\begin{aligned}
d(x_{n+1}, p) &= d\left(p, W\left(x_n, T^n x_n, \frac{1}{2}\right)\right) \\
&\leq (1 - \alpha) d(x_n, p).
\end{aligned}
$$

By successive application of uniform convexity of X, we have that

$$d(x_{n+1}, p) \leq \xi^{n-n_0} d(x_{n_0}, p),$$

where $\xi = 1 - \alpha < 1$. Hence $\lim\limits_{n \to \infty} d(x_n, p) = 0$, i.e., $x_n \to p$ as $n \to \infty$ and the proof is complete.

CASE 2.   Suppose that for each $\varepsilon > 0$, there exists an integer n_0 such that $d(x_n, T^n x_n) < \varepsilon$ for all $n \geq n_0$. Then there exists a subsequence $\{x_{n_k}\}$ of $\{x_n\}$ such that $\lim\limits_{k \to \infty} d(x_{n_k}, T^{n_k} x_{n_k}) = 0$. Since T is equi-continuous, therefore $\lim\limits_{k \to \infty} d(x_{n_k}, T x_{n_k}) = 0$. As T maps C into a compact subset of C, so there exists a subsequence $\{x_{n_{k_l}}\}$ of $\{x_{n_k}\}$ such that $\{T x_{n_{k_l}}\}$ converges to a point u in C as $l \to \infty$. From $\lim\limits_{l \to \infty} d(x_{n_{k_l}}, T x_{n_{k_l}}) = 0$, we obtain that $x_{n_{k_l}} \to u$ as $l \to \infty$. Since T is continuous and $\lim\limits_{l \to \infty} T x_{n_{k_l}} = u$, therefore $Tu = u$. That is, $u = p$. From (6.5.42) and $\lim\limits_{l \to \infty} d(x_{n_{k_l}}, p) = 0$, we conclude that $x_n \to p$. $\qquad\square$

6.6. Viscosity Method for Generalized Asymptotically Nonexpansive Mappings

In this section, we employ the viscosity method due to Moudafi [48] to improve some of our earlier established results.

Let C be a convex subset of a normed space. Khan et al. [30] introduced the following multistep iterative method:

$$x_1 \in C,$$
$$x_{n+1} = (1 - a_{rn})x_n + a_{rn}T_r^n \, y_{(r-1)n},$$
$$y_{(r-1)n} = (1 - a_{(r-1)n})x_n + a_{(r-1)n}T_{r-1}^n \, y_{(r-2)n},$$
$$y_{(r-2)n} = (1 - a_{(r-2)n})x_n + a_{(r-2)n}T_{r-2}^n \, y_{(r-3)n},$$

$$\cdot \qquad\qquad\qquad\qquad\qquad\qquad\qquad (6.6.43)$$

$$\cdot$$

$$\cdot$$

$$y_{2n} = (1 - a_{2n})x_n + a_{2n}T_2^n \, y_{1n},$$
$$y_{1n} = (1 - a_{1n})x_n + a_{1n}T_1^n \, y_{0n},$$

where $\{T_i : i \in I\}$ is a family of mappings of C, $0 \le a_{in} \le 1$, $y_{0n} = x_n$ for all n.

Now we assume that $F = \bigcap_{i \in I} F(T_i) \ne \phi$.

The iterative method (6.6.43) extends the Mann iterative method (6.1.3), the Ishikawa iterative method (6.1.4), Khan and Takahashi iterative method [33] and the three-step iterative method of Xu and Noor [66], simultaneously.

Recently, Chang et al. [11] introduced and studied the following viscosity iterative method:

$$x_{n+1} = (1 - \alpha_n)f(x_n) + \alpha_n T^n \, y_n$$
$$y_n = (1 - \beta_n)x_n + \beta_n T^n x_n, \quad n \ge 1, \qquad (6.6.44)$$

where T is an asymptotically nonexpansive mapping and f is a fixed contraction.

We define S_n-mapping generated by a family $\{T_i : i \in I\}$ of generalized asymptotically quasi-nonexpansive mappings on C as:

$$S_n x = U_{rn}x, \qquad (6.6.45)$$

where $U_{0n} = I$ (the identity mapping),

$$U_{1n}x = W(T_1^n U_{0n}x, x, a_{1n}),$$
$$U_{2n}x = W(T_2^n U_{1n}x, x, a_{2n}),$$

$$\vdots$$

$$U_{rn}x = W(T_r^n U_{(r-1)n}x, x, a_{rn}).$$

For $\{\alpha_n\} \subset J$, a fixed contraction f on C and S_n given by (6.6.45), we define $\{x_n\}$ as follows:

$$x_1 \in C, x_{n+1} = W(f(x_n), S_n x_n, \alpha_n), \qquad (6.6.46)$$

and call it a *general viscosity iterative method* in a convex metric space.

We now set out to give details of the work of Fukhar-ud-din et al. [19].

Lemma 6.6.1. *Let C be a nonempty closed and convex subset of a convex metric space X and $\{T_i : i \in I\}$ be a family of generalized asymptotically quasi-nonexpansive mappings of C, i.e., $d\left(T_i^n x, p_i\right) \leq (1 + u_{in})d\left(x, p_i\right) + c_{in}$ for all $x \in C$ and $p_i \in F(T_i)$, $i \in I$ where $\{u_{in}\}$ and $\{c_{in}\}$ are sequences in $[0, \infty)$ with $\sum_{n=1}^{\infty} u_{in} < \infty$, $\sum_{n=1}^{\infty} c_{in} < \infty$ for each i. Then, for the sequence $\{x_n\}$ in (6.6.46) with $\sum_{n=1}^{\infty} \alpha_n < \infty$, there are sequences $\{v_n\}$ and $\{\xi_n\}$ in $[0, \infty)$ satisfying $\sum_{n=1}^{\infty} v_n < \infty, \sum_{n=1}^{\infty} \xi_n < \infty$ such that*

(a) $d\left(x_{n+1}, p\right) \leq (1 + v_n)^r d\left(x_n, p\right) + \xi_n$, *for all $p \in F$ and all $n \geq 1$;*

(b) $d\left(x_{n+m}, p\right) \leq M_1 \left(d\left(x_n, p\right) + \sum_{n=1}^{\infty} \xi_n\right)$, *for all $p \in F$ and $n \geq 1$, $m \geq 1$, $M_1 > 0$.*

Proof. (a) Let $p \in F$ and $v_n = \max_{i \in I} u_{in}$ for all $n \geq 1$. Since $\sum_{n=1}^{\infty} u_{in} < \infty$ for each i, therefore $\sum_{n=1}^{\infty} v_n < \infty$. Now we have

$$
\begin{aligned}
d\left(U_{1n}x_n, p\right) &= d\left(W\left(T_1^n U_{0n}x_n, x_n, a_{1n}\right), p\right) \\
&\leq (1 - \alpha_{1n})d\left(x_n, p\right) + \alpha_{1n}\left(T_1^n x_n, p\right) \\
&\leq (1 - \alpha_{1n})d\left(x_n, p\right) + \alpha_{1n}\left[(1 + u_{1n})d\left(x_n, p\right) + c_{1n}\right] \\
&\leq (1 + u_{1n})d\left(x_n, p\right) + c_{1n} \\
&= (1 + v_n)^1 d\left(x_n, p\right) + c_{1n}.
\end{aligned}
$$

Assume that $d\left(U_{kn}x_n, p\right) \leq (1 + v_n)^k d\left(x_n, p\right) + (1 + v_n)^{k-1}\sum_{i=1}^{k} c_{in}$ holds for some

$k > 1$. Consider

$$
\begin{aligned}
d\left(U_{(k+1)n}x_n, p\right) &= d\left(W(T^n_{k+1}U_{kn}x_n, x_n, a_{(k+1)n}), p\right) \\
&\leq (1 - a_{(k+1)n})d\left(x_n, p\right) + a_{(k+1)n}d\left(T^n_{k+1}U_{kn}x_n, p\right) \\
&\leq (1 - a_{(k+1)n})d\left(x_n, p\right) + a_{(k+1)n}(1 + u_{(k+1)n})d\left(U_{kn}x_n, p\right) \\
&\quad + a_{(k+1)n}c_{k+1n} \\
&\leq (1 - a_{(k+1)n})d\left(x_n, p\right) + a_{(k+1)n}c_{(k+1)n} \\
&\quad + a_{(k+1)n}(1 + u_{(k+1)n})d\left(U_{kn}x_n, p\right) \\
&\leq (1 - a_{(k+1)n})d\left(x_n, p\right) + a_{(k+1)n}c_{(k+1)n} \\
&\quad + a_{(k+1)n}(1 + v_n)\left[(1 + v_n)^k d\left(x_n, p\right) + (1 + v_n)^{k-1}\sum_{i=1}^{k} c_{in}\right] \\
&\leq (1 - a_{(k+1)n})(1 + v_n)^{k+1}d\left(x_n, p\right) + a_{(k+1)n}(1 + v_n)c_{(k+1)n} \\
&\quad + a_{(k+1)n}(1 + v_n)^{k+1}d\left(x_n, p\right) + a_{(k+1)n}(1 + v_n)^{k-1}\sum_{i=1}^{k+1} c_{in} \\
&\leq (1 + v_n)^{k+1}d\left(x_n, p\right) + (1 + v_n)^k\sum_{i=1}^{k+1} c_{in}.
\end{aligned}
$$

By mathematical induction, we have

$$
d\left(U_{jn}x_n, p\right) \leq (1 + v_n)^j d\left(x_n, p\right) + (1 + v_n)^{j-1}\sum_{i=1}^{j} c_{in}, \quad 1 \leq j \leq r. \tag{6.6.47}
$$

Hence

$$
d\left(S_n x_n, p\right) = d\left(U_{rn}x_n, p\right) \leq (1 + v_n)^r d\left(x_n, p\right) + (1 + v_n)^{r-1}\sum_{i=1}^{r} c_{in}. \tag{6.6.48}
$$

Now, by (6.6.46) and (6.6.48), we obtain

$$
\begin{aligned}
d\left(x_{n+1}, p\right) &= d\left(W\left(f\left(x_{n}\right), S_{n}, \alpha_{n}\right), p\right) \\
&\leq \alpha_{n} d\left(f\left(x_{n}\right), p\right)+\left(1-\alpha_{n}\right) d\left(S_{n} x_{n}, p\right) \\
&\leq \alpha_{n} \alpha d\left(x_{n}, p\right)+\alpha_{n} d\left(f(p), p\right) \\
&\quad +\left(1-\alpha_{n}\right)\left(\left(1+v_{n}\right)^{r} d\left(x_{n}, p\right)+\left(1+v_{n}\right)^{r-1} \sum_{i=1}^{r} c_{in}\right) \\
&\leq \left(1+v_{n}\right)^{r} d\left(x_{n}, p\right)+\left(1-\alpha_{n}\right)\left(1+v_{n}\right)^{r-1} \sum_{i=1}^{r} c_{in} \\
&\quad +\alpha_{n} d\left(f(p), p\right) \\
&\leq \left(1+v_{n}\right)^{r} d\left(x_{n}, p\right)+\alpha_{n} d\left(f(p), p\right)+\left(1+v_{n}\right)^{r-1} \sum_{i=1}^{r} c_{in}.
\end{aligned}
$$

Setting $\max \left\{d\left(f(p), p\right), \sup \left(1+v_{n}\right)^{r-1}\right\}=M$, we get

$$
d\left(x_{n+1}, p\right) \leq\left(1+v_{n}\right)^{r} d\left(x_{n}, p\right)+M\left(\alpha_{n}+\sum_{i=1}^{r} c_{in}\right),
$$

that is,

$$
d\left(x_{n+1}, p\right) \leq\left(1+v_{n}\right)^{r} d\left(x_{n}, p\right)+\xi_{n},
$$

where $\xi_{n}=M\left(\alpha_{n}+\sum_{i=1}^{r} c_{in}\right)$ and $\sum_{n=1}^{\infty} \xi_{n}<\infty$.

(b) We know that $1+t \leq e^{t}$ for $t \geq 0$. Thus, by part (a), we have

$$
\begin{aligned}
d\left(x_{n+m}, p\right) &\leq \left(1+v_{n+m-1}\right)^{r} d\left(x_{n+m-1}, p\right)+\xi_{n+m-1} \\
&\leq e^{r v_{n+m-1}} d\left(x_{n+m-1}, p\right)+\xi_{n+m-1} \\
&\leq e^{r\left(v_{n+m-1}+v_{n+m-2}\right)} d\left(x_{n+m-2}, p\right)+\xi_{n+m-1}+\xi_{n+m-2} \\
&\quad \vdots \\
&\leq e^{r \sum_{i=n}^{n+m-1} v_{i}} d\left(x_{n}, p\right)+\sum_{i=n+1}^{n+m-1} v_{i} \sum_{i=n}^{n+m-1} \xi_{i} \\
&\leq e^{r \sum_{i=1}^{\infty} v_{i}}\left(d\left(x_{n}, p\right)+\sum_{i=1}^{\infty} \xi_{i}\right) \\
&= M_{1}\left(d\left(x_{n}, p\right)+\sum_{i=1}^{\infty} \xi_{i}\right), \text { where } M_{1}=e^{r \sum_{i=1}^{\infty} v_{i}}.
\end{aligned}
$$

$\square$

We follow the arguments of Khan et al. [30, Theorem 2.2] to find a necessary and sufficient condition for the convergence of $\{x_n\}$ in (6.6.46) to a point of F.

Theorem 6.6.1. *Let C be a nonempty closed and convex subset of a complete convex metric space X and $\{T_i : i \in I\}$ a family of generalized asymptotically quasi-nonexpansive mappings of C, i.e., $d\left(T_i^n x, p_i\right) \leq (1 + u_{in})d\left(x, p_i\right) + c_{in}$ for all $x \in C$ and $p_i \in F(T_i)$, $i \in I$, where $\{u_{in}\}$ and $\{c_{in}\}$ are sequences in $[0, \infty)$ with $\sum_{n=1}^{\infty} u_{in} < \infty$, $\sum_{n=1}^{\infty} c_{in} < \infty$ for all i. Then, for the sequence $\{x_n\}$ in (6.6.46) with $\sum_{n=1}^{\infty} \alpha_n < \infty$, $\{x_n\}$ converges strongly to a point in F if and only if $\liminf_{n \to \infty} d(x_n, F) = 0$.*

Proof. The necessity is obvious; we only prove the sufficiency. By Lemma 6.6.1(a), we have

$$d\left(x_{n+1}, p\right) \leq (1 + v_n)^r d\left(x_n, p\right) + \xi_n \text{ for all } p \in F \text{ and } n \geq 1.$$

Therefore

$$
\begin{aligned}
d(x_{n+1}, F) &\leq (1 + v_n)^r d(x_n, F) + \xi_n, \\
&= \left(1 + \sum_{k=1}^{r} \frac{r(r-1)\cdots(r-k+1)}{k!} v_n^k\right) d(x_n, F) + \xi_n.
\end{aligned}
$$

As $\sum_{n=1}^{\infty} v_n < +\infty$, so $\sum_{n=1}^{\infty} \sum_{k=1}^{r} \frac{r(r-1)\cdots(r-k+1)}{k!} v_n^k < \infty$. Now $\sum_{n=1}^{\infty} \xi_n < \infty$ in Lemma 6.6.1(a), so by Lemma 6.1.1 and $\liminf_{n \to \infty} d(x_n, F) = 0$, we get that $\lim_{n \to \infty} d(x_n, F) = 0$.

Next, we prove that $\{x_n\}$ is a Cauchy sequence in X. Let $\varepsilon > 0$. From the proof of Lemma 6.6.1(b), we have

$$d\left(x_{n+m}, x_n\right) \leq d\left(x_{n+m}, F\right) + d\left(x_n, F\right) \leq (1 + M_1) d\left(x_n, F\right) + M_1 \sum_{i=n}^{\infty} \xi_i. \qquad (6.6.49)$$

By $\lim_{n \to \infty} d(x_n, F) = 0$ and $\sum_{i=1}^{\infty} \xi_i < \infty$, there exists a natural number n_0 such that

$$d(x_n, F) \leq \frac{\varepsilon}{2(1 + M_1)} \text{ and } \sum_{i=n}^{\infty} \xi_i < \frac{\varepsilon}{2M_1}, \quad \text{for all } n \geq n_0.$$

So for all integers $n \geq n_0$, $m \geq 1$, we obtain from (6.6.49) that

$$d\left(x_{n+m}, x_n\right) < (M_1 + 1) \frac{\varepsilon}{2(1 + M_1)} + M_1 \frac{\varepsilon}{2M_1} = \varepsilon.$$

Thus $\{x_n\}$ is a Cauchy sequence in X and so converges to $q \in X$. Finally, we show that $q \in F$. For any $\bar{\varepsilon} > 0$, there exists a natural number n_1 such that

$$d(x_n, F) = \inf_{p \in F} d\left(x_n, p\right) < \frac{\bar{\varepsilon}}{3} \text{ and } d\left(x_n, q\right) < \frac{\bar{\varepsilon}}{2}, \text{ for all } n \geq n_1.$$

There must exist $p^* \in F$ such that $d(x_n, p^*) < \frac{\bar{\varepsilon}}{2}$ for all $n \geq n_1$; in particular, $d(x_{n_1}, p^*) < \frac{\bar{\varepsilon}}{2}$ and $d(x_{n_1}, q) < \frac{\bar{\varepsilon}}{2}$. Hence

$$d(p^*, q) \leq d(x_{n_1}, p^*) + d(x_{n_1}, q) < \bar{\varepsilon}.$$

Since $\bar{\varepsilon}$ is arbitrary, therefore $d(p^*, q) = 0$, that is, $q = p^* \in F$.   □

A generalized asymptotically nonexpansive mapping is a generalized asymptotically quasi-nonexpansive, so we have the following important new results.

Corollary 6.6.1. *Let C be a nonempty closed and convex subset of a complete convex metric space X and $\{T_i : i \in I\}$ a family of generalized asymsptotically nonexpansive mappings of C , i.e., $d(T_i^n x, T_i^n y) \leq (1 + u_{in})d(x, y) + c_{in}$, for all $x, y \in C$ and $i \in I$, where $\{u_{in}\}$ and $\{c_{in}\}$ are sequences in $[0, \infty)$ with $\sum_{n=1}^{\infty} u_{in} < \infty$ and $\sum_{n=1}^{\infty} c_{in} < \infty$ for all i. Then the sequence $\{x_n\}$ in (6.6.46), converges strongly to a point $p \in F$ if and only if $\liminf_{n \to \infty} d(x_n, F) = 0$.*

Corollary 6.6.2. *Let C, $\{T_i : i \in I\}$, F and $\{u_{in}\}, \{c_{in}\}$ be as in Theorem 6.6.1. Then the sequence $\{x_n\}$ in (6.6.46), converges strongly to a point $p \in F$ if and only if there exists a subsequence $\{x_i\}$ of $\{x_n\}$ which converges strongly to p.*

Theorem 6.6.2. *Let C be a nonempty closed and convex subset of a complete convex metric space X, and $\{T_i : i \in I\}$ a family of generalized asymptotically nonexpansive mappings of C, i.e., $d(T_i^n x, T_i^n y) \leq (1 + u_{in})d(x, y) + c_{in}$, for all $x, y \in C$ and $i \in I$, where $\{u_{in}\}$ and $\{c_{in}\}$ are sequences in $[0, \infty)$ with $\sum_{n=1}^{\infty} u_{in} < \infty$ and $\sum_{n=1}^{\infty} c_{in} < \infty$ for all i. If $\lim_{n \to \infty} d(x_n, T_i x_n) = 0$ for the sequence $\{x_n\}$ in (6.6.46), $i \in I$ and one of the mappings is semi-compact, then $\{x_n\}$ converges strongly to $p \in F$.*

Proof. Let T_ℓ be semi-compact for some $1 \leq \ell \leq r$. Then there exists a subsequence $\{x_i\}$ of $\{x_n\}$ such that $x_i \to p \in C$. Hence

$$d(p, T_\ell p) = \lim_{i \to \infty} d(x_i, T_\ell x_i) = 0.$$

Thus $p \in F$ and so, by Corollary 6.6.1, $\{x_n\}$ converges strongly to a common fixed point of the family of mappings.   □

Theorem 6.6.3. *Let C, $\{T_i : i \in I\}$, F, $\{u_{in}\}$ and $\{c_{in}\}$ be as in Theorem 6.6.2. Suppose that there exists a mapping T_ℓ which satisfies the following conditions:*

(i) $\lim_{n \to \infty} d(x_n, T_\ell x_n) = 0$;

(ii) *there exists a constant M such that $d(x_n, T_\ell x_n) \geq M d(x_n, F)$, for all $n \geq 1$.*

Then the sequence $\{x_n\}$ in (6.6.46), converges strongly to a point $p \in F$.

Proof. From (i) and (ii), it follows that $\lim_{n \to \infty} d(x_n, F) = 0$. By Theorem 6.6.2, $\{x_n\}$ converges strongly to a common fixed point of the family of mappings. $\square$

Next, we establish some convergence results for the iterative method (6.6.46) of generalized asymptotically quasi-nonexpansive mappings on a uniformly convex metric space.

Lemma 6.6.2. *Let C be a nonempty closed and convex subset of a uniformly convex complete metric space X and $\{T_i : i \in I\}$ be a family of uniformly Hölder continuous and generalized asymptotically quasi-nonexpansive mappings of C, i.e., $d(T_i^n x, p_i) \leq (1 + u_{in})(x, p_i) + c_{in}$ for all $x \in C$ and $p_i \in F(T_i)$, where $\{u_{in}\}$ and $\{c_{in}\}$ are sequences in $[0, \infty)$ with $\sum_{n=1}^{\infty} u_{in} < \infty$ and $\sum_{n=1}^{\infty} c_{in} < \infty$, respectively, for each $i \in I$. Then, for the sequence $\{x_n\}$ in (6.6.46) with $a_{in} \in [\delta, 1 - \delta]$ for some $\delta \in (0, \frac{1}{2})$ and $\sum_{n=1}^{\infty} \alpha_n < \infty$, we have*

(a) $\lim_{n \to \infty} d(x_n, p)$ *exists for all $p \in F$;*

(b) $\lim_{n \to \infty} d(x_n, T_j x_n) = 0$ *for each $j \in I$.*

Proof. (a) Let $p \in F$ and $v_n = \max_{i \in I} u_{in}$ for all $n \geq 1$. By Lemmas 6.1.1 and 6.6.1(a), it follows that $\lim_{n \to \infty} d(x_n, p)$ exists for all $p \in F$. Assume that

$$\lim_{n \to \infty} d(x_n, p) = c. \tag{6.6.50}$$

(b) The inequality (6.6.47) together with (6.6.50) gives that

$$\limsup_{n \to \infty} d(U_{jn} x_n, p) \leq c, 1 \leq j \leq r. \tag{6.6.51}$$

By (6.6.46), we have

$$\begin{aligned}
d(x_{n+1}, p) &= d(W(f(x_n), S_n, \alpha_n), p) \\
&\leq \alpha_n d(f(x_n), p) + (1 - \alpha_n) d(S_n x_n, p) \\
&\leq \alpha_n d(f(x_n), p) + \alpha_n d(f(p), p) + (1 - \alpha_n) d(U_{rn} x_n, p),
\end{aligned}$$

and hence

$$c \leq \liminf_{n \to \infty} d(U_{rn} x_n, p). \tag{6.6.52}$$

Combining (6.6.51) and (6.6.52), we get

$$\lim_{n\to\infty} d\left(U_{rn}x_n,p\right) = c.$$

Note that

$$
\begin{aligned}
d\left(U_{rn}x_n,p\right) &= d\left(W(T_r^n U_{(r-1)n}x_n,x_n,a_{rn}),p\right) \\
&\leq a_{rn}d\left(T_r^n U_{(r-1)n}x_n,p\right) + (1-a_{rn})d\left(x_n,p\right) \\
&\leq a_{rn}\left[(1+v_n)d\left(U_{(r-1)n}x_n,p\right) + c_{rn}\right] + (1-a_{rn})d\left(x_n,p\right) \\
&= a_{rn}(1+v_n)d\left(W(T_{r-1}^n U_{(r-2)n}x_n,x_n,a_{(r-1)n})x_n,p\right) \\
&\quad + a_{rn}c_{rn} + (1-a_{rn})d\left(x_n,p\right) \\
&\leq a_{rn}(1+v_n)\left[a_{(r-1)n}d\left(T_{r-1}^n U_{(r-2)n}x_n,p\right) + \left(1-a_{(r-1)n}\right)d\left(x_n,p\right)\right] \\
&\quad + a_{rn}c_{rn} + (1-a_{rn})d\left(x_n,p\right) \\
&\leq a_{rn}a_{(r-1)n}(1+v_n)^2 d\left(U_{(r-2)n}x_n,p\right) + \left(1-a_{rn}a_{(r-1)n}\right)(1+v_n)^2 d\left(x_n,p\right) \\
&\quad + a_{rn}a_{(r-1)n}(1+v_n)^2 c_{(r-1)n} + a_{rn}(1+v_n)^2 c_{rn} \\
&\ \ \vdots \\
&\leq \prod_{i=j+1}^{r} a_{in}(1+v_n)^{r-j}d\left(U_{jn}x_n,p\right) + \left(1 - \prod_{i=j+1}^{r} a_{in}\right)(1+v_n)^{r-j}d\left(x_n,p\right) \\
&\quad + \prod_{i=j+1}^{r} a_{in}(1+v_n)^{r-j}c_{j+1n} + \prod_{i=j+2}^{r} a_{in}(1+v_n)^{r-j}c_{jn} \\
&\quad + \cdots + a_{rn}(1+v_n)^{r-j}c_{rn},
\end{aligned}
$$

and therefore we have

$$
\begin{aligned}
d\left(x_n,p\right) &\leq \frac{d\left(x_n,p\right)}{\delta^{r-j}} - \frac{d\left(U_{rn}x_n,p\right)}{\delta^{r-j}(1+v_n)^{r-j}} + d\left(U_{jn}x_n,p\right) \\
&\quad + c_{j+1n} + \frac{c_{jn}}{\delta} + \cdots + \frac{c_{rn}}{\delta^{r-j+1}}.
\end{aligned}
$$

Hence

$$c \leq \liminf_{n\to\infty} d\left(U_{jn}x_n,p\right), 1 \leq j \leq r. \tag{6.6.53}$$

Using (6.6.51) and (6.6.53), we have

$$\lim_{n\to\infty} d\left(U_{jn}x_n,p\right) = c,$$

that is,

$$\lim_{n\to\infty} d\left(W(T_j^n U_{(j-1)n}x_n,x_n,a_{jn}),p\right) = c, \quad \text{for } 1 \leq j \leq r.$$

This, together with (6.6.50) and (6.6.51), gives that

$$\lim_{n \to \infty} d\left(T_j^n U_{(j-1)n} x_n, x_n, \right) = 0, \quad \text{for } 1 \leq j \leq r. \tag{6.6.54}$$

If $j = 1$, we have by (6.6.54),

$$\lim_{n \to \infty} d\left(T_1^n x_n, x_n\right) = 0.$$

For $j \in \{2, 3, 4, \ldots, r\}$, we observe that

$$\begin{aligned} d\left(x_n, U_{(j-1)n} x_n\right) &= d\left(x_n, W\left(T_{j-1}^n U_{(j-2)n} x_n, x_n, a_{(j-1)n}\right)\right) \\ &\leq a_{(j-1)n} d\left(T_{j-1}^n U_{(j-2)n} x_n, x_n\right) \to 0. \end{aligned} \tag{6.6.55}$$

Since T_j is uniformly Hölder continuous, therefore the inequality

$$\begin{aligned} d\left(T_j^n x_n, x_n\right) &\leq d\left(T_j^n x_n, T_j^n U_{(j-1)n} x_n\right) + d\left(T_j^n U_{(j-1)n} x_n, x_n\right) \\ &\leq L d\left(x_n, U_{(j-1)n} x_n\right)^\gamma + d\left(T_j^n U_{(j-1)n} x_n, x_n\right), \end{aligned}$$

together with (6.6.54) and (6.6.55), gives that

$$\lim_{n \to \infty} d\left(T_j^n x_n, x_n\right) = 0.$$

Hence

$$d\left(T_j^n x_n, x_n\right) \to 0 \text{ as } n \to \infty, \quad \text{for } 1 \leq j \leq r. \tag{6.6.56}$$

Since

$$\begin{aligned} d\left(x_n, x_{n+1}\right) &= d\left(x_n, W\left(f\left(x_n\right), S_n x_n, \alpha_n\right)\right) \\ &\leq \alpha_n d\left(x_n, f\left(x_n\right)\right) + \left(1 - \alpha_n\right) d\left(x_n, S_n x_n\right) \\ &\leq \alpha_n \left[d\left(x_n, p\right) + d\left(p, f\left(p\right)\right) + d\left(f\left(p\right), f\left(x_n\right)\right)\right] \\ &\quad + \left(1 - \alpha_n\right) a_{rn} d\left(x_n, T_r^n U_{(r-1)n} x_n\right) \\ &\leq \alpha_n \left(1 + \alpha\right) d\left(x_n, p\right) + \alpha_n d\left(p, f\left(p\right)\right) \\ &\quad + \left(1 - \alpha_n\right) a_{rn} d\left(x_n, T_r^n U_{(r-1)n} x_n\right), \end{aligned}$$

therefore

$$\lim_{n \to \infty} d\left(x_n, x_{n+1}\right) = 0. \tag{6.6.57}$$

Let us observe that

$$
\begin{aligned}
d\left(x_n, T_j x_n\right) \;\leq\; & d\left(x_n, x_{n+1}\right) + d\left(x_{n+1}, T_j^{n+1} x_{n+1}\right) \\
& + d\left(T_j^{n+1} x_{n+1}, T_j^{n+1} x_n\right) + d\left(T_j^{n+1} x_n, T_j x_n\right) \\
\leq\; & d\left(x_n, x_{n+1}\right) + d\left(x_{n+1}, T_j^{n+1} x_{n+1}\right) \\
& + L d\left(x_{n+1}, x_n\right)^{\gamma} + L d\left(T_j^n x_n, x_n\right)^{\gamma}.
\end{aligned}
$$

By uniform Hölder continuity of T_j, (6.6.56) and (6.6.57), we get

$$
\lim_{n \to \infty} d\left(x_n, T_j x_n\right) = 0,\; 1 \leq j \leq r. \tag{6.6.58}
$$

$\square$

Theorem 6.6.4. *Under the hypotheses of Lemma 6.6.2, assume, for some $1 \leq j \leq r$, T_j^m is semi-compact for some positive integer m. Then $\{x_n\}$ in (6.6.46), converges strongly to a point in F.*

Proof. Fix $j \in I$ and suppose that T_j^m is semi-compact for some $m \geq 1$. By (6.6.58), we obtain

$$
\begin{aligned}
d\left(T_j^m x_n, x_n\right) \;\leq\; & d\left(T_j^m x_n, T_j^{m-1} x_n\right) + d\left(T_j^{m-1} x_n, T_j^{m-2} x_n\right) \\
& + \cdots + d\left(T_j^2 x_n, T_j x_n\right) + d\left(T_j x_n, x_n\right) \\
\leq\; & d\left(T_j x_n, x_n\right) + (m-1) L d\left(T_j x_n, x_n\right)^{\gamma} \to 0.
\end{aligned}
$$

Since $\{x_n\}$ is bounded and T_j^m is semi-compact, therefore $\{x_n\}$ has a convergent subsequence $\{x_{n_i}\}$ such that $x_{n_i} \to q \in C$. Hence, by (6.6.58), we have

$$
d\left(q, T_i q\right) = \lim_{n \to \infty} d\left(x_{n_j}, T_i x_{n_j}\right) = 0, \quad i \in I.
$$

Thus $q \in F$ and so, by Corollary 6.2.2, $\{x_n\}$ converges strongly to a common fixed point q of the family $\{T_i : i \in I\}$. $\square$

An immediate consequence of Lemma 6.6.2 and Theorem 6.6.3 is the following strong convergence result in a uniformly convex metric space.

Theorem 6.6.5. *Let C, $\{T_i : i \in I\}$, F, $\{u_{in}\}$ and $\{c_{in}\}$ be as in Lemma 6.6.2. If there exists a constant M such that $d\left(x_n, T_j x_n\right) \geq M d(x_n, F)$, for all $n \geq 1$, then the sequence $\{x_n\}$ in (6.6.46), converges strongly to a point in F.*

Remark 6.6.1.   (a) Theorems 6.2.1 and 6.3.1 are special cases of Theorem 6.5.2.

(b) The viscosity iterative approximation results in Section 6.6 carry over the corresponding results in Sections 6.2 and 6.3 onto a very general nonlinear domain "uniformly convex metric space".

ACKNOWLEDGMENTS

The authors would like to acknowledge the support provided by the Deanship of Scientific Research(DSR) at King Fahd University of Petroleum & Minerals (KFUPM) for funding this work through project No. IN141047.

REFERENCES

1. Berinde V, Iterative Approximation of Fixed Points. Lecture Notes in Mathematics 1912, Berlin: Springer; 2007.
2. Bridson M , Haeiger A, Metric Spaces of Non-positive Curvature. Berlin: Springer; 1999.
3. Bose SC, Laskar SK, Fixed point theorems for certain class of mappings. J. Math. Phys. Sci. 1985;19:503–509.
4. Browder FE, Nonexpansive nonlinear operators in a Banach space. Proc. Nat. Acad. Sci. U.S.A. 1965;54:1041–1044.
5. Browder FE, Petryshyn WV, The solution by iteration of nonlinear functional equations in Banach spaces. Bull. Am. Math. Soc. 1966;72:571–575.
6. Bruck RE, Fixed Points and Nonexpansive Mappings. AMS: Providence; 1980.
7. Bruck RE, A simple proof of the mean ergodic theorem for nonlinear contractions in Banach spaces. Israel J. Math. 1979;32:107–116.
8. Bruhat F, Tits J, Groupes réductifs sur un corps local. I. Données radicielles valuées, Inst. Hautes Études Sci. Publ. Math. 1972;41:5–251.
9. Burago D, Burago Y, Ivanov S, A Course in Metric Geometry. Vol. 33 of Graduate Studies in Mathematics, American Mathematical Society: Providence, RI; 2001.
10. Byrne C, Unified treatment of some algorithms in signal processing and image construction. Inverse Probl. 2004;20:103–120.
11. Chang SS, Lee HWJ, Chan CK, Kim JK, Approximating solutions of variational inequalities for asymptotically nonexpansive mappings. Appl. Math. Comput. 2009;212:51–59.
12. Chidume CE, Geometric Properties of Banach Spaces and Nonlinear Iterations. Verlag Series: Springer; 2009.
13. Dhompongsa S, Kirk WA, Sims B, Fixed points of uniformly Lipschitzian mappings. Nonlinear Anal. 2006;65:762–772.
14. Dhompongsa S, Panyanak B, On $\triangle$-convergence theorems in CAT(0) spaces. Comput. Math. Appl. 2008;56:2572–2579.
15. Diestel J, Geometry of Banach Spaces – Selected Topics. Berlin:New York; 1975.
16. Espinola R, Kirk WA, Fixed point theorems in R-trees with applications to graph theory, Topol. Appl. 2006;153:1046–1055.
17. Fukhar-ud-din H, Existence and approximation of fixed points in convex metric spaces, Carpathian J. Math. 2014;30:175–185.
18. Fukhar-ud-din H, Khan AR, Akhtar Z, Fixed point results for a generalized nonexpansive map in uniformly convex metric spaces. Nonlinear Anal. 2012;75:4747–4760.
19. Fukhar-ud-din H, Khamsi MA, Khan AR, Viscosity iterative method for a finite family of generalized asymptotically quasi-nonexpansive mappings in convex metric spaces, J. Nonlinear Convex Anal. 2015;16:47–58.
20. Goebel K, An elementary proof of the fixed-point theorem of Browder and Kirk. Michigan Math. J. 1969;16:381–383.
21. Goebel K, Kirk WA, A fixed point theorem for asymptotically nonexpansive mappings, Proc. Am. Math. Soc. 1972;35:171–174.

22. Goebel K, Kirk WA, Iteration processes for nonexpansive maps. In: SP Singh, S Thomeier, B Watson (eds.) Topological Methods in Nonlinear Functional Analysis: Toronto; 1982.
23. Goebel K, Reich S, Uniform Convexity, Hyperbolic Geometry, and Nonexpansive Mappings. Marcel Dekker: New York; 1984.
24. Ibn Dehaish BA, Khamsi MA, Khan AR, Mann iteration process for asymptotic pointwise nonexpansive mappings in metric spaces. J. Math. Anal. Appl. 2013;397:861–868.
25. Ishikawa S, Fixed points by a new iteration method. Proc. Am. Math. Soc. 1974;44:147–150.
26. Kaczor W, Weak convergence of almost orbits of asymptotically nonexpansive commutative semigroups. J. Math. Anal. Appl. 2002;272:565–574.
27. Khamsi MA, Khan AR, Inequalities in metric spaces with applications. Nonlinear Anal. 2011;74:4036–4045.
28. Khan AR, On modified Noor iterations for asymptotically nonexpansive mappings. Bull. Belg. Math. Soc. – Simon Stevin, 2010;17:127–140.
29. Khan AR, Common fixed point and solution of nonlinear functional equations. Fixed Point Theory Appl. 2013;2013:Article ID 290.
30. Khan AR , Domlo AA , Fukhar-ud-din H, Common fixed points Noor iteration for a finite family of asymptotically quasi-nonexpansive mappings in Banach space. J. Math. Anal. Appl. 2008;341:1–11.
31. Khan AR , Fukhar-ud-din H, Kalsoom A, Lee BS, Convergence of a general algorithm of asymptotically nonexpansive maps in uniformly convex hyperbolic spaces. Appl. Math. Comput. 2014;238:547–556.
32. Khan AR, Khamsi MA, Fukhar-ud-din H, Strong convergence of a general iteration scheme in CAT(0) spaces. Nonlinear Anal. 2011;74:783–791.
33. Khan SH, Takahashi W, Approximating common fixed points of two asymptotically nonexpansive mappings. Sci. Math. Japon. 2001;53:133–138.
34. Kim TH, Kim DH, Demiclosedness principle for total asymptotically nonexpansive mappings. Math. Anal. Lab. Sci. 2013;1821:90–106.
35. Kirk WA, A fixed point theorem for mappings which do not increase distances. Am. Math. Monthly 1965;72:1004–1006.
36. Kirk WA, Krasnoselskii iteration process in hyperbolic spaces. Num. Funct. Anal. Optim. 1982;4:371–381.
37. Kirk WA, Geodesic geometry and fixed point theory II. In: International Conference on Fixed Point Theory and Applications, Yokohama Publ.,Yokohama, 2004;113–142.
38. Kirk WA, A fixed point theorems in $CAT(0)$ spaces and $\mathbb{R}$-trees. Fixed Point Theory Appl. 2004;4:309-316.
39. Kirk WA, Geodesic geometry and fixed point theory. In: Seminar of Mathematical Analysis (Malaga/Seville, 2002/2003), in: Colecc. Abierta, vol. 64, Univ. Sevilla Secr. Publ., Seville, 2003; 195–225.
40. Kirk WA, Ray WO, A note on Lipschitzian mappings in convex metric spaces. Can. Math. Bull. 1977;20:463–466.
41. Kirk WA, Panyanak B, A concept of convergence in geodesic spaces. Nonlinear Anal. 2008;68:3689–3696.
42. Kohlenbach U, Some logical metatheorems with applications in functional analysis, Trans. Am. Math. Soc. 2005;357:89–128.
43. Krasnoselskii MA, Two remarks on the method of successive approximations. UspehiMat. Nauk 1955;10:123–127.
44. Kuhfittig PKF, Common fixed points of nonexpansive mappings by iteration. Pacific J. Math. 1981;97:137–139 .
45. Leustean I, Nonexpansive iterations in uniformly convex W-hyperbolic spaces. In: Proc. Contemporary Mathematics (IMCP'08) (Haifa, Israel), 2008;1:193–209.
46. Mann WR, Mean value methods in iterations. Proc. Am. Math. Soc. 1953;4:506–510.
47. Marr De R, Common fixed points for commuting contraction mappings. Pacific J. Math. 1963;13:1139–1141.

48. Moudafi A, Viscosity approximation methods for fixed-points problems. J. Math. Anal. Appl. 2000;41:46–55.
49. Nanjaras B, Panyanak B, Demiclosed principle for asymptotically nonexpansive mappings in CAT(0) spaces. Fixed Point Theory Appl. 2010;2010:Article ID 268780.
50. Opial Z, Weak convergence of the sequence of successive approximations for nonexpansive mappings. Bull. Am. Math. Soc. 1967;73:591–597.
51. Picard E, Memoire sur la theorie des equations aux derives partielles et la methode des approximations successives. J. Math. Pures et Appl. 1890;6:145–210.
52. Podilchuk CI, Mammone RJ, Image recovery by convex projections using a least-squares constraint, J. Opt. Soc. Am. 1990;A7:517–521.
53. Reich S, Shafrir I, Nonexpansive iterations in hyperbolic spaces. Nonlinear Anal. 1990;15:537–338.
54. Rhoades BE, Comments on two fixed point iteration methods. J. Math. Anal. Appl. 1976;56:741–750.
55. Saejung S, Halpern's iteration in CAT(0) spaces. Fixed Point Theory Appl. 2010;2010:Article ID 471781.
56. Senter HF, Dotson WG, Approximating fixed points of nonexpansive mappings. Proc. Am. Math. Soc. 1974;44:375–380.
57. Shimizu T, Takahashi W, Fixed points of multivalued mappings in certain convex metric spaces. Topol. Meth. Nonlinear Anal. 1996;8:197–203.
58. Suantai S, Weak and strong convergence criteria of Noor iterations for asymptotically nonexpansive mappings. J. Math. Anal. Appl. 2005;311:506–517.
59. Thakur BS, Thakur D, Postolache M, Modified Picard-Mann hybrid iteration process for total asymptotically nonexpansive mappings. Fixed Point Theory Appl. 2015;2015:Article ID 140.
60. Takahashi W, A convexity in metric spaces and nonexpansive mappings. Kodai Math. Sem. Rep. 1970;22:142–149.
61. Takahashi W, Iterative methods for approximation of fixed points and their applications. J. Oper. Res. Soc. Japan 2000;43:87–108.
62. Takahashi W, Shimoji K, Convergence theorems for nonexpansive mappings and feasibility problems. Math. Comput. Model. 2000;32:1463–1471.
63. Takahashi W, Tamura T, Convergence theorems for a pair of nonexpansive mappings. J. Convex Anal. 1995;5:45–58.
64. Talman LA, Fixed points for condensing multifunctions in metric spaces with convex structure. Kodai Math. Sem. Rep. 1977;29:62–70.
65. Tan KK, Xu HK, Approximating fixed points of nonexpansive mappings by the Ishikawa iteration process. J. Math. Anal. Appl. 1993;178:301–308.
66. Xu B, Noor MA, Fixed-point iterations for asymptotically nonexpansive mappings in Banach spaces. J. Math. Anal. Appl. 2002;267:444–453.
67. Youla D, On deterministic convergence of iterations of related projection mappings. J. Vis. Commun. Image Represent. 1990;1:12–20.
68. Zhou J, Cuib Y, Fixed point theorems for mean nonexpansive mappings in CAT(0) spaces. Num. Funct. Anal. Optim. DOI:10.1080/01630563.2015.1060614.

CHAPTER 7

Metric Fixed Point Theory in Spaces with a Graph

Monther Rashed Alfuraidan
King Fahd University of Petroleum and Minerals, Department of Mathematics and Statistics, Dhahran, Saudi Arabia
Corresponding: monther@kfupm.edu.sa

Contents

http://dx.doi.org/10.1016/B978-0-12-804295-3.50007-3

Abstract

In this chapter, we discuss a new area that overlaps between metric fixed point theory and graph theory. This new area yields interesting generalizations of the Banach contraction principle in metric and modular spaces endowed with a graph. The bridge between both theories is motivated by the fact that they often arise in industrial fields such as image processing engineering, physics, computer science, economics, ladder networks, dynamic programming, control theory, stochastic filtering, statistics, telecommunications and many other applications.

7.1. Introduction

The Banach fixed point theorem, also known as the Banach contraction principle, first appeared in 1922 in Banach's thesis entitled "Sur les opérations dans les ensembles abstraits et leur application aux équations intégrales" [15]. This theorem quickly became one of the central results of analysis and possibly one of the most commonly applied theorems in mathematics. Because of its major impact, this fixed point theorem is considered as the cornerstone of metric fixed point theory.

The Banach contraction principle is notable for its simplicity: it only requires a complete metric space and a contractive condition which is simple to inspect. This explains the numerous applications in different branches of Mathematics. For example, in analysis, this theorem is essential to prove the Picard-Lindelöf theorem (Picard's existence theorem). Also, it is the main key to prove the existence of the solutions of differential and integral equations. In addition, the Banach contraction principle plays an influential role in the modeling of physical processes. As for numerical methods, the Banach contraction principle is used to prove the convergence of Newton's method. Because of its large area of influence and its significance, the Banach contraction principle is still a source of inspiration and investigations, which explains the considerable attention given to its generalizations. One of these generalizations was carried out recently in metric spaces endowed with a partial order and more generally in metric spaces endowed with a graph. This should not come as a surprise. Indeed, most of the metric spaces used in various applied areas are naturally endowed with a partial order. It is worth nothing that most of the generalizations of the Banach contraction principle focused only on the distance and the metric conditions satisfied by the mappings.

Multivalued functions arise in optimal control theory, especially differential inclusions and related subjects like game theory and economics, where the Kakutani fixed point theorem for multivalued functions has been applied to prove the existence of Nash equilibria (note: in the context of game theory and economics, a multivalued function is usually referred to as a correspondence). In physics, multivalued functions play an increasingly important role. They form the mathematical basis for Dirac's magnetic monopoles, for the theory of defects in crystals and the resulting plasticity of materials, for vortices in superfluids and superconductors, and for phase transitions in these systems, for instance melting and quark confinement. They are the origin of gauge field structures in many branches of physics. The multivalued version of the Banach contraction principle was carried out by Nadler [68]. Following this extension, many mathematicians turned their attention to the study of the fixed points of multivalued mappings.

Nonexpansive mappings are those maps which have Lipschitz constant equal to one. They are a natural extension of contractive mappings. However, the fixed point problem for nonexpansive mappings differs sharply from that of the contractive mappings. Indeed, the existence of the fixed points of nonexpansive mappings requires restrictive conditions on the domain. This explains why it took more than four decades to prove the first fixed point results for nonexpansive mappings in Banach spaces following the publication in 1965 of the works of Browder [23], Göhde [39], and Kirk [50].

Recently a new approach has been attracting some attention, dealing with the extension of the Banach contraction principle to metric spaces endowed with a partial order. This new interest followed the publication of Ran and Reurings' paper [73]. The main ideas behind the central fixed point theorem of Ref. [73] were initiated in Ref. [37]. The foremost motivation of El-Sayed and Ran [37] was the investigation of the solutions of some special matrix equations. Later on, Nieto and Rodríguez-López [70] extended the new fixed point result discovered in Ref. [73] and used this extension to solve some differential equations. Jachymski [46] gave a more general unified version of these extensions by considering graphs instead of a partial order (see also some recent papers [4–6,8–12]). The case of monotone nonexpansive mappings is being investigated extensively [7, 13, 14]. It should also be pointed out that in fact the first attempt to generalize the Banach contraction principle to partially ordered metric spaces was carried out by Turinici [79, 80].

In this chapter, the recent bridge between graph theory and metric fixed point theory is presented. In particular, the necessary foundations from both theories are established. Then I present most of the extensions of the single-valued and multivalued Banach contraction principle in Banach spaces, hyperbolic metric spaces, modular metric spaces, and modular function spaces endowed with a graph. Finally the most recent results on monotone nonexpansive mappings are considered. I will also give the two main applications that motivated this new area, i.e., matrix and differential equations.

7.2. Banach Contraction Principle

Although the basic idea behind the Banach contraction principle was known to others earlier, this fixed point theorem first appeared in explicit form in Banach's thesis [15], where it was used to establish the existence of a solution to an integral equation.

Definition 7.2.1. Let (X,d) be a metric space. A mapping $T : X \to X$ is said to be *Lipschitzian* if there exists a constant $k \geq 0$ such that for any $x, y \in X$

$$d(T(x), T(y)) \leq k\, d(x,y).$$

The constant k is also known as a Lipschitz constant of T.

Definition 7.2.2. Let (X,d) be a metric space. A Lipschitzian mapping $T : X \to X$ is said to be:

 (a) *a contraction mapping* if T has a Lipschitz constant $k < 1$;

 (b) *a nonexpansive mapping* if T has $k = 1$ as a Lipschitz constant.

Theorem 7.2.1 (Banach Contraction Principle). *Let (X,d) be a complete metric space and let $T : X \to X$ be a contraction mapping. Let $k \in [0,1)$ be a Lipschitz constant of T. Then T has a unique fixed point $\hat{x}$, and for each $x \in X$ and $n \in \mathbb{N}$, we have*

$$d\left(T^n(x), \hat{x}\right) \leq \frac{k^n}{1-k}\, d\left(T(x), x\right).$$

In particular, this implies that

$$\lim_{n \to +\infty} T^n(x) = \hat{x},$$

for each $x \in X$.

The conclusion of Theorem 7.2.1 suggests the following definition:

Definition 7.2.3. Let (X,d) be a metric space. The mapping $T : X \to X$ is said to be a *Picard operator* (in short PO) if T has a unique fixed point $\hat{x}$, and for each $x \in X$, we have $\lim\limits_{n\to+\infty} T^n(x) = \hat{x}$.

The Banach contraction principle only requires a complete metric space and a contractive mapping to have a unique fixed point. In addition, the theorem delivers an estimate which is very useful for numerical computations.

Since we will discuss the extension of the Banach contraction principle to metric spaces endowed with a partial order, we will need the following definition.

Definition 7.2.4. Let X be a set. A binary relation $\preceq$, defined on X, is said to have a *partial order* if the following conditions are satisfied: For any $x,y,z \in X$,

(a) $x \preceq x$;

(b) $x \preceq y$ and $y \preceq x \Rightarrow x = y$;

(c) $x \preceq y$ and $y \preceq z \Rightarrow x \preceq z$.

Fixed point theorems for monotone singlevalued mappings in a metric space endowed with a partial order have been widely investigated. These theorems can be seen as a combination of the two most fundamental theorems in fixed point theory: the Banach contraction principle and Tarski's fixed point theorem [77].

Theorem 7.2.2 (Tarski's Fixed Point Theorem). *Let L be a complete lattice and let $f : L \to L$ be an order-preserving mapping. Then the set of fixed points of f in L is not empty and is a complete lattice.*

As stated earlier, the study of the fixed points of singlevalued mappings in partially ordered metric spaces triggered excitement after the publication of the work of Ran and Reurings [73].

7.2.1. Ran and Reurings Fixed Point Theorem

Ran and Reurings considered the following problem:

Problem 7.2.1. Find $X \in H(n)$ (the set of all $n \times n$ Hermitian matrices) such that

$$X = Q \pm \sum_{i=1}^{m} A_1^* F(X) A_i, \qquad (7.2.1)$$

where $F : H(n) \to H(n)$ is a monotone function, i.e., $F(X_1) \leq F(X_2)$ if $X_1 \leq X_2$, which maps the set of all $n \times n$ positive definite matrices $P(n)$ into itself, $A_1,\dots,A_m$ are arbitrary $n \times n$ matrices and $Q \in P(n)$.

The study of equations like (7.2.1) is motivated by the fact that they regularly arise in the analysis of ladder networks, dynamic programming, control theory, stochastic filtering, statistics and many other applications (see Ref. [37, Chapter 7]).

Equation (7.2.1) is a fixed point problem. Indeed, consider the mappings $G_{\pm}(X)$: $H(n) \to H(n)$ defined by

$$G_{+}(X) = Q + \sum_{i=1}^{m} A_1^* F(X) A_i \text{ and } G_{-}(X) = Q - \sum_{i=1}^{m} A_1^* F(X) A_i.$$

Note that X is a solution of (7.2.1) if and only if $G_{\pm}(X) = X$, i.e., X is a fixed point of $G_{\pm}$. To investigate the existence and the uniqueness of the solutions of equation (7.2.1), we have to study the existence and the uniqueness of a fixed point of $G_{\pm}$. From a computational point of view, it will also be interesting to study the behavior of an iteration process which can be induced by the fixed point equation.

While investigating this problem, Ran and Reurings obtained the following result.

Theorem 7.2.3 [73]. *Let $(X, \preceq)$ be a partially ordered set such that every pair $x, y \in X$ has an upper and lower bound. Let d be a distance on X such that (X, d) is a complete metric space. Let $f : X \to X$ be a continuous monotone (either order preserving or order reversing) mapping. Suppose that the following conditions hold:*

(i) *There exists $k \in [0, 1)$ such that*

$$d(f(x), f(y)) \le k\, d(x, y), \quad \text{for all } x \preceq y.$$

(ii) *There exists an $x_0 \in X$ such that x_0 and $f(x_0)$ are comparable, i.e., $x_0 \preceq f(x_0)$ or $f(x_0) \preceq x_0$.*

Then f is a PO.

7.2.2. Nieto and Rodríguez-López Fixed Point Theorem

Following the publication of Theorem 7.2.3, Nieto and Rodríguez-López offered to weaken the continuity assumption of the mapping. In their investigation, they considered the following problem:

Problem 7.2.2 [70]. Find a function $u \in C_1(I, R)$, the space of all continuous functions with continuous first derivative, such that u satisfies the following equation:

$$\begin{cases} u'(t) = f(t, u(t)), & t \in I = [0, T]; \\ u(0) = u(T), \end{cases} \qquad (7.2.2)$$

where $T > 0$ and $f : I \times \mathbb{R} \to \mathbb{R}$ is a continuous function.

First, the problem (7.2.2) is equivalent to the new ODE:

$$\begin{cases} u'(t) + \lambda u(t) = f(t, u(t)) + \lambda u(t), & t \in I = [0, T]; \\ u(0) = u(T), \end{cases} \tag{7.2.3}$$

where $\lambda > 0$. The solutions to the problem (7.2.3) are exactly the solutions to the integral equation

$$u(t) = \int_0^T G(t, s) \Big[f(x, u(s)) + \lambda u(s) \Big] ds, \tag{7.2.4}$$

where $G(t, s)$ is the Green function

$$G(t, s) = \begin{cases} \frac{e^{\lambda(T+s-t)}}{e^{\lambda T} - 1}, & 0 \le s \le t \le T; \\ \frac{e^{\lambda(s-t)}}{e^{\lambda T} - 1}, & 0 \le t \le s \le T. \end{cases}$$

It is clear that the equation (7.2.4) is a fixed point equation. Indeed, define $A : C_1(I, R) \to C_1(I, R)$ by

$$[A(u)](t) = \int_0^T G(t, s) \Big[f(x, u(s)) + \lambda u(s) \Big] ds. \tag{7.2.5}$$

Clearly, we have that u is a solution to (7.2.2) if and only if u satisfies the fixed point problem $A(u) = u$. In their investigation of the existence and uniqueness of the fixed point of A, Nieto and Rodríguez-López proved the following result:

Theorem 7.2.4 [70]. *Let (X, d) be a complete metric space endowed with a partial ordering $\preceq$. Let $f : X \to X$ be an order-preserving mapping such that there exists a $k \in [0, 1)$ with*

$$d(f(x), f(y)) \le k\, d(x, y), \quad \textit{for all } x \preceq y.$$

Assume that one of the following conditions holds:

(i) *f is continuous and there exists an $x_0 \in X$ such that x_0 and $f(x_0)$ are comparable;*

(ii) *$(X, d, \preceq)$ is such that for any nondecreasing $\{x_n\}_{n \in \mathbb{N}}$, if $x_n \to x$, then $x_n \preceq x$ for $n \in \mathbb{N}$, and there exists an $x_0 \in X$ with $x_0 \preceq f(x_0)$;*

(iii) *$(X, d, \preceq)$ is such that for any nonincreasing $\{x_n\}_{n \in \mathbb{N}}$, if $x_n \to x$, then $x \preceq x_n$ for $n \in \mathbb{N}$, and there exists an $x_0 \in X$ with $f(x_0) \preceq x_0$.*

Then f has a fixed point. Moreover, if $(X, \preceq)$ is such that every pair of elements of X has an upper or a lower bound, then f is a PO.

Instead of working with partial orders, it is often natural to consider graphs which are more general. This point was used by Jachymski [46], who gave a similar result to Theorem 7.2.4 in metric spaces endowed with a graph. In the next section, the tools from graph theory necessary to state Jachymski's fixed point theorem will be established.

7.3. Basic Definitions and Properties

This section provides the basic definitions and properties of graph theory, hyperbolic metric spaces, modular function spaces and modular metric spaces which will be used throughout.

7.3.1. Graph Theory

A *graph* is an ordered pair (V, E) where V is a set and E is a binary relation on V $(E \subseteq V \times V)$. Elements of E are called *edges*. We are concerned here with directed graphs (digraphs) that have a loop at every vertex (i.e., $(a, a) \in E$ for each $a \in V$). Such digraphs are called *reflexive*. In this case $E \subseteq V \times V$ corresponds to a reflexive (and symmetric) binary relation on V. Moreover, we assume that there exists a distance function d defined on the set of vertices V. We may treat G as a weighted graph by assigning to each edge the distance between its vertices. By G^{-1} we denote the conversion of a graph G, i.e., the graph obtained from G by reversing the direction of edges. Thus we have

$$E(G^{-1}) = \{(y, x) : (x, y) \in E(G)\}.$$

A digraph G is called an *oriented graph* if whenever $(u, v) \in E(G)$, then $(v, u) \notin E(G)$. The letter $\widetilde{G}$ denotes the undirected graph obtained from G by ignoring the direction of edges. Actually, it will be more convenient for us to treat $\widetilde{G}$ as a directed graph for which the set of its edges is symmetric. Under this convention, we have

$$E(\widetilde{G}) = E(G) \cup E(G^{-1}).$$

Example 7.3.1. Let $(X, \preceq)$ be a partially ordered set. We define the *oriented graph* $G_{\preceq}$ on X as follows: the vertices of $G_{\preceq}$ are the elements of X, and two vertices $x, y \in X$ are connected by a directed edge (arc) from x to y if $x \preceq y$. Therefore $G_{\preceq}$ has no parallel arcs as $x \preceq y$ and $y \preceq x \Rightarrow x = y$.

Given a digraph $G = (V, E)$, a (di)path of G is a sequence $a_0, a_1, \ldots, a_n, \ldots$ with $(a_i, a_{i+1}) \in E(G)$ for each $i \in \mathbb{N}$. A finite path $(a_0, a_1, \ldots, a_n)$ is said to have *length* n, for $n \in \mathbb{N}$. A closed directed path of length $n > 1$ from x to y, i.e., $x = y$, is called a *directed cycle*. An *acyclic digraph* is a digraph that has no directed cycle. A digraph is *connected* if there is a finite (di)path joining any two of its vertices and it is weakly connected if $\widetilde{G}$ is connected. Given an acyclic digraph G, we can always define a *partial order* $\preceq_G$ on the set of vertices of G by $x\preceq_G y$ whenever there is a directed path from x to y.

Definition 7.3.1. A digraph G is *transitive* if

$$(x,y) \in E(G) \quad \text{and} \quad (y,z) \in E(G) \Rightarrow (x,z) \in E(G),$$

for all $x,y,z \in V(G)$.

Definition 7.3.2. A *G-interval* is any subset of the form

(a) $[a,\rightarrow) = \{x \in V(G) : (a,x) \in E(G)\}$,

(b) $(\leftarrow,a] = \{x \in V(G) : (x,a) \in E(G)\}$,

for any $a \in V(G)$.

Definition 7.3.3. Let C be a nonempty subset of X. A mapping $T : C \to C$ is said to be

(a) *G-monotone* if T is edge preserving, i.e., $(T(x),T(y)) \in E(G)$ whenever $(x,y) \in E(G)$, for any $x,y \in C$.

(b) *Banach G-contraction* or simply *G-contraction* if T is G-monotone and decreases weights of edges of G at a uniform rate, i.e., there exists a constant $k \in [0,1)$ such that for all $x,y \in C$ with $(x,y) \in E(G)$, we have $d(T(x),T(y)) \leq kd(x,y)$.

Let us finish this subsection with an example of a transitive cyclic digraph which cannot be generated by a partial order. Therefore this example enforces the idea that replacing the partial order by a graph is worthy of consideration.

Example 7.3.2. Consider the Hilbert space ℓ_2 defined by

$$\ell_2 = \left\{ (x_n) \in \mathbb{R}^{\mathbb{N}} : \sum_{n \in \mathbb{N}} |x_n|^2 < +\infty \right\}.$$

Define the digraph G on ℓ_2 by:

$$(x,y) \in E(G) \text{ if and only if } x_n \leq y_n, \ n \geq 2,$$

where $x = (x_n)$ and $y = (y_n)$ are in ℓ_2. Then G is reflexive, transitive for which G-intervals are convex and closed. Note that G contains cycles. Indeed, we have $(x,y) \in E(G)$ and $(y,x) \in E(G)$, where

$$x = (1,0,0,\cdots) \text{ and } y = (2,0,0,\cdots).$$

Therefore the graph G will not be generated by a partial order.

7.3.2. Hyperbolic Metric Spaces

Let (X,d) be a metric space and suppose that there exists a family $\mathscr{F}$ of metric segments such that any two points x,y in X are endpoints of a unique metric segment $[x,y] \in \mathscr{F}$ ($[x,y]$ is an isometric image of the real line interval $[0,d(x,y)]$). We denote by $\beta x \oplus (1-\beta)y$ the unique point z of $[x,y]$ which satisfies

$$d(x,z) = (1-\beta)d(x,y) \text{ and } d(z,y) = \beta d(x,y),$$

where $\beta \in [0,1]$. Such metric spaces with a family $\mathscr{F}$ of metric segments are usually called *convex metric spaces* [65]. Moreover, if we have

$$d\Big(\alpha p \oplus (1-\alpha)x, \alpha q \oplus (1-\alpha)y\Big) \le \alpha d(p,q) + (1-\alpha)d(x,y),$$

for all p,q,x,y in X, and $\alpha \in [0,1]$, then X is said to be a *hyperbolic metric space* (see Ref. [74]).

Obviously, normed linear spaces are hyperbolic spaces. As nonlinear examples, one can consider the Hadamard manifolds [25], the Hilbert open unit ball equipped with the hyperbolic metric [42], and the CAT(0) spaces [52, 54, 63]. We will say that a subset C of a hyperbolic metric space X is *convex* if $[x,y] \subset C$ whenever x,y are in C.

Definition 7.3.4. Let (X,d) be a hyperbolic metric space. A graph G on X is said to be *convex* if and only if for any $x,y,z,w \in X$ and $\alpha \in [0,1]$, we have

$$(x,z) \in E(G) \text{ and } (y,w) \in E(G) \Rightarrow (\alpha x \oplus (1-\alpha)y, \alpha z \oplus (1-\alpha)w) \in E(G).$$

In the sequel, we assume that (X,d) is a metric space, and G is a directed graph (digraph) with set of vertices $V(G) = X$ and the set of edges $E(G)$ contains all the loops, i.e., $(x,x) \in E(G)$, for any $x \in X$.

Definition 7.3.5. Let (X,d) be a metric space and C a nonempty subset of X.

(a) We say that a mapping $T : C \to C$ is *G-nonexpansive* if T is *G-monotone* and for any $x,y \in X$ such that $(x,y) \in E(G)$ we have

$$d(T(x),T(y)) \le d(x,y).$$

(b) We say that the multivalued mapping $T : C \to 2^C$ is *monotone increasing (respectively decreasing) G-nonexpansive* if for any $x,y \in C$ with $(x,y) \in E(G)$ and any $u \in T(x)$ there exists $v \in T(y)$ such that $(u,v) \in E(G)$ *(respectively $(v,u) \in E(G)$)* and

$$d(u,v) \le d(x,y).$$

7.3.3. Modular Function Spaces

The first attempts to generalize the classical function spaces of the Lebesgue type L^p were made in the early 1930s by Orlicz and Birnbaum in connection with orthogonal expansions. Their approach involved considering spaces of functions with some growth properties different from the power-type growth control provided by the L^p-norms. Namely, they considered the function spaces defined as follows:

$$L^{\varphi} = \left\{ f : \mathbb{R} \to \mathbb{R}; \ \exists \lambda > 0 : \rho(\lambda f) = \int_{\mathbb{R}} \varphi(\lambda|f(x)|) \, dx < +\infty \right\},$$

where $\varphi : [0,+\infty] \to [0,+\infty]$ was assumed to be a convex function increasing to infinity, i.e., the function which to some extent behaves similarly to power functions $\varphi(t) = t^p$, $p \geq 1$. The vector space L^{φ} furnishes a wonderful example of a modular function space. For more examples on modular function spaces, the reader may consult the book of Kozlowski [59].

Let Ω be a nonempty set and Σ be a nontrivial σ-algebra of subsets of Ω. Let $\mathscr{P}$ be a δ-ring of subsets of Σ, such that $E \cap A \in \mathscr{P}$ for any $E \in \mathscr{P}$ and $A \in \Sigma$. Let us assume that there exists an increasing sequence of sets $K_n \in \mathscr{P}$ such that $\Omega = \bigcup K_n$. By $\mathscr{E}$ we denote the linear space of all simple functions with supports from $\mathscr{P}$. By $\mathscr{M}_{\infty}$ we will denote the space of all extended measurable functions, i.e., all functions $f : \Omega \to [-\infty,+\infty]$ such that there exists a sequence $\{g_n\} \subset \mathscr{E}$, $|g_n| \leq |f|$ and $g_n(\omega) \to f(\omega)$ for all $\omega \in \Omega$. By $\mathbb{1}_A$ we denote the characteristic function of the set A.

Definition 7.3.6 [5]. Let $\rho : \mathscr{M}_{\infty} \to [0,+\infty]$ be a nontrivial, convex and even function. We say that ρ is *a regular function pseudomodular* if

(i) $\rho(0) = 0$;

(ii) ρ is monotone, i.e., $|f(\omega)| \leq |g(\omega)|$ for all $\omega \in \Omega$ implies $\rho(f) \leq \rho(g)$, where $f,g \in \mathscr{M}_{\infty}$;

(iii) ρ is orthogonally subadditive, i.e., $\rho(f\mathbb{1}_{A \cup B}) \leq \rho(f\mathbb{1}_A) + \rho(f\mathbb{1}_B)$ for any $A,B \in \Sigma$ such that $A \cap B \neq \emptyset$, $f \in \mathscr{M}_{\infty}$;

(iv) ρ has the Fatou property, i.e., $|f_n(\omega)| \uparrow |f(\omega)|$ for all $\omega \in \Omega$ implies $\rho(f_n) \uparrow \rho(f)$, where $f \in \mathscr{M}_{\infty}$;

(v) ρ is order continuous in $\mathscr{E}$, i.e., $g_n \in \mathscr{E}$ and $|g_n(\omega)| \downarrow 0$ implies $\rho(g_n) \downarrow 0$.

Similarly as in the case of measure spaces, we say that a set $A \in \Sigma$ is ρ-null if $\rho(g\mathbb{1}_A) = 0$ for every $g \in \mathscr{E}$. We say that a property holds ρ-*almost everywhere* if the exceptional set is ρ-null. As usual we identify any pair of measurable sets whose symmetric difference is ρ-null as well as any pair of measurable functions differing only on a ρ-null set. With this in mind we define

$$\mathcal{M}(\Omega,\Sigma,\mathscr{P},\rho) = \{f \in \mathcal{M}_\infty : |f(\omega)| < +\infty\ \rho - a.e\}, \tag{7.3.6}$$

where each $f \in \mathcal{M}(\Omega,\Sigma,\mathscr{P},\rho)$ is actually an equivalence class of functions equal to ρ-a.e. rather than an individual function. Where no confusion exists we write $\mathcal{M}$ instead of $\mathcal{M}(\Omega,\Sigma,\mathscr{P},\rho)$.

Definition 7.3.7. Let ρ be a regular convex function pseudomodular.

(a) We say that ρ is *a regular function semimodular* if $\rho(\alpha f) = 0$ for every $\alpha > 0$ implies $f = 0$ ρ-a.e.;

(b) We say that ρ is *a regular function modular* if $\rho(f) = 0$ implies $f = 0$ ρ-a.e.

The class of all nonzero regular convex function modulars defined on Ω will be denoted by $\mathfrak{R}$.

Let us denote $\rho(f,E) = \rho(f\mathbf{1}_E)$ for $f \in \mathcal{M}$ and $E \in \Sigma$. It is easy to prove that $\rho(f,E)$ is a function pseudomodular in the sense of Ref. [59, Definition 2.1.1] (more precisely, it is a function pseudomodular with the Fatou property). Therefore we can use all results of the standard theory of modular function spaces as per the framework defined by Kozlowski [59–61].

Definition 7.3.8 [59–61]. Let ρ be a convex function modular.

(a) A *modular function space* is the vector space $L_\rho(\Omega,\Sigma)$, or briefly L_ρ, defined by

$$L_\rho = \{f \in \mathcal{M} : \rho(\lambda f) \to 0 \text{ as } \lambda \to 0\}.$$

(b) The following formula defines a norm in L_ρ (frequently called the *Luxemburg norm*):

$$\|f\|_\rho = \inf\{\alpha > 0 : \rho(f/\alpha) \le 1\}.$$

In the following theorem we recall some of the properties of modular spaces.

Theorem 7.3.1 [59–61]. *Let $\rho \in \mathfrak{R}$.*

(a) $(L_\rho, \|.\|_\rho)$ *is complete and the norm $\|\cdot\|_\rho$ is monotone w. r. t. the natural order in $\mathcal{M}$.*

(b) $\|f_n\|_\rho \to 0$ *if and only if $\rho(\alpha f_n) \to 0$ for every $\alpha > 0$.*

(c) *If $\rho(\alpha f_n) \to 0$ for an $\alpha > 0$ then there exists a subsequence $\{g_n\}$ of $\{f_n\}$ such that $g_n \to 0$ ρ-a.e.*

(d) $\rho(f) \leq \liminf\limits_{n \to +\infty} \rho(f_n)$, *whenever $f_n \to f$ ρ-a.e. Note that this property will be referred to as the Fatou property.*

The following definition plays an important role in the theory of modular function spaces.

Definition 7.3.9. Let $\rho \in \Re$.

(a) We say that ρ has *the Δ_2-property* if and only if $\lim\limits_{k \to +\infty} \sup_n \rho(2f_n, D_k) = 0$ for any $\{f_n\}_{n \geq 1} \subset M$ and $D_k \in \Sigma$ such that $D_k \downarrow \emptyset$ and $\lim\limits_{k \to +\infty} \sup_n \rho(f_n, D_k) = 0$.

(b) We say that ρ has *the Δ_2-type condition* if there exists $k \in [0, +\infty)$ such that $\rho(2f) \leq k\,\rho(f)$, for any $f \in L_\rho$.

We have the following interesting result:

Theorem 7.3.2 [59]. *Let $\rho \in \Re$. The following conditions are equivalent:*

(a) *ρ has Δ_2-property,*

(b) *if $\rho(f_n) \to 0$ then $\rho(2f_n) \to 0$,*

(c) *if $\rho(\alpha f_n) \to 0$ for an $\alpha > 0$ then $\|f_n\|_\rho \to 0$, i.e., the modular convergence is equivalent to the norm convergence.*

Moreover, if ρ has the Δ_2-type condition, then ρ has Δ_2-property. In general, the converse is not true (see Ref. [61, Example 3.2.6]).

Remark 7.3.1. It is easy to check that ρ has the Δ_2-type condition if and only if, for any $\lambda > 1$, there exists $k \in [0, +\infty)$ such that

$$\rho(\lambda f) \leq k\,\rho(f), \quad \text{for any } f \in L_\rho.$$

We also use another type of convergence which is situated between norm and modular convergence.

Definition 7.3.10. Let $\rho \in \Re$.

(a) We say that $\{f_n\}$ is *ρ-convergent to f* and write $f_n \to f$ (ρ) if and only if $\rho(f_n - f) \to 0$.

(b) A sequence $\{f_n\}$ where $f_n \in L_\rho$ is called *ρ-Cauchy* if $\rho(f_n - f_m) \to 0$ as $n, m \to +\infty$.

(c) A set $B \subset L_\rho$ is called ρ-*closed* if for any sequence of $f_n \in B$, the convergence $f_n \to f$ (ρ) implies that f belongs to B.

(d) A set $B \subset L_\rho$ is called ρ-*bounded* if $\sup\{\rho(f-g); f, g \in B\} < +\infty$.

(e) A set $C \subset L_\rho$ is called ρ-*a.e. closed* if for any $\{f_n\}$ in C which ρ-a.e. converges to some f, then we must have $f \in C$.

(f) A set $C \subset L_\rho$ is called ρ-*a.e. compact* if for any $\{f_n\}$ in C, there exists a subsequence $\{f_{n_k}\}$ which ρ-a.e. converges to for some $f \in C$.

Let us note that ρ-convergence does not necessarily imply the ρ-Cauchy condition. Also, $f_n \to f$ does not imply in general $\lambda f_n \to \lambda f$, $\lambda > 1$. Using Theorem 7.3.1, it is not difficult to prove the following.

Proposition 7.3.1. *Let* $\rho \in \mathfrak{R}$.

(a) L_ρ *is* ρ-*complete.*

(b) L_ρ *is a lattice, i.e., for any* $f, g \in L_\rho$, *we have* $\max\{f, g\} \in L_\rho$ *and* $\min\{f, g\} \in L_\rho$.

(c) ρ-*balls* $B_\rho(x, r) = \{y \in L_\rho; \rho(x-y) \le r\}$ *are* ρ-*closed and* ρ-*a.e. closed.*

Using the property (c) of Theorem 7.3.1, we get the following result:

Theorem 7.3.3. *Let* $\rho \in \mathfrak{R}$ *and* $\{f_n\}$ *be a* ρ-*Cauchy sequence in* L_ρ. *Assume that* $\{f_n\}$ *is monotone increasing, i.e.,* $f_n \le f_{n+1}$ ρ-*a.e. (respectively decreasing, i.e.,* $f_{n+1} \le f_n$ ρ-*a.e.), for any* $n \ge 1$. *Then there exists* $f \in L_\rho$ *such that* $\rho(f_n - f) \to 0$ *and* $f_n \le f$ ρ-*a.e. (respectively* $f \le f_n$ ρ-*a.e.), for any* $n \ge 1$.

Definition 7.3.11 [31]. Assume that $\rho \in \mathfrak{R}$ satisfies the Δ_2-type condition. Define *the growth function* Ω by

$$\Omega(\alpha) = \sup\left\{\frac{\rho(\alpha f)}{\rho(f)} : f \in L_\rho, f \ne 0\right\}, \quad \text{for any } \alpha \ge 0.$$

The following properties were proved in Ref. [31].

Lemma 7.3.1. *Assume that* $\rho \in \mathfrak{R}$ *satisfies the* Δ_2-*type condition. Then the growth function* Ω *satisfies the following properties:*

(a) $\Omega(\alpha) < +\infty$, *for any* $\alpha > 0$;

(b) Ω *is a strictly increasing function, and* $\Omega(1) = 1$;

(c) $\Omega(\alpha\beta) \leq \Omega(\alpha)\Omega(\beta)$, *for any* $\alpha, \beta \in (0, +\infty)$;

(d) $\Omega^{-1}(\alpha)\Omega^{-1}(\beta) \leq \Omega^{-1}(\alpha\beta)$, *where* Ω^{-1} *is the function inverse of* Ω;

(e) *For any* $f \in L_\rho$, $f \neq 0$, *we have*

$$\|f\|_\rho \leq \frac{1}{\Omega^{-1}\Big(1/\rho(f)\Big)}.$$

The following technical lemma will be useful later on.

Lemma 7.3.2 [31]. *Assume that* $\rho \in \mathfrak{R}$ *satisfies the* Δ_2*-type condition. Let* $\{f_n\}$ *be a sequence in* L_ρ *such that*

$$\rho(f_{n+1} - f_n) \leq K\,\alpha^n, \quad n = 1, \cdots,$$

where K *is an arbitrary nonzero constant and* $\alpha \in (0,1)$. *Then* $\{f_n\}$ *is Cauchy for* $\|.\|_\rho$ *and* ρ*-Cauchy.*

Note that this lemma is crucial since the main assumption on $\{f_n\}$ will not be enough to imply that $\{f_n\}$ is ρ-Cauchy since ρ generally fails the triangle inequality.

7.3.4. Modular Metric Spaces

In this subsection, we introduce the concept of modular metric spaces. These spaces were introduced in Refs. [29, 30]. However, the way we approach the concept of modular metric spaces in this chapter is different. Indeed, we look at these spaces as the nonlinear version of the classical modular spaces as introduced by Nakano [69] on vector spaces and modular function spaces introduced by Musielack [67] and Orlicz [72]. Therefore most of the definitions and properties of this subsection are the nonlinear versions of the definitions and properties discussed in the previous subsection.

Let X be a nonempty set. Throughout this subsection for a function $\omega : (0, +\infty) \times X \times X \to (0, +\infty)$, we write

$$\omega_\lambda(x,y) = \omega(\lambda, x, y), \quad \text{for all } \lambda > 0 \text{ and all } x, y \in X.$$

Definition 7.3.12 [29, 30]. A function $\omega : (0, +\infty) \times X \times X \to [0, +\infty]$ is said to be a *modular* on X if it satisfies the following axioms:

(i) $x = y$ if and only if $\omega_\lambda(x,y) = 0$, for all $\lambda > 0$;

(ii) $\omega_\lambda(x,y) = \omega_\lambda(y,x)$, for all $\lambda > 0$, and $x, y \in X$;

(iii) $\omega_{\lambda+\mu}(x,y) \le \omega_\lambda(x,z) + \omega_\mu(z,y)$ for all $\lambda, \mu > 0$ and $x,y,z \in X$.

If instead of (i), we have only the condition (i'):

$$\omega_\lambda(x,x) = 0, \quad \text{for all } \lambda > 0, \ x \in X,$$

then ω is said to be a *pseudomodular* on X. A modular ω on X is said to be *regular* if the following weaker version of (i) is satisfied:

$$x = y \text{ if and only if } \omega_\lambda(x,y) = 0, \text{ for some } \lambda > 0.$$

Finally, ω is said to be *convex* if for $\lambda, \mu > 0$ and $x,y,z \in X$, it satisfies the inequality

$$\omega_{\lambda+\mu}(x,y) \le \frac{\lambda}{\lambda+\mu}\omega_\lambda(x,z) + \frac{\mu}{\lambda+\mu}\omega_\mu(z,y).$$

Note that for a pseudomodular ω on a set X, and any $x,y \in X$, the function $\lambda \to \omega_\lambda(x,y)$ is nonincreasing on $(0,+\infty)$. Indeed, if $0 < \mu < \lambda$, then

$$\omega_\lambda(x,y) \le \omega_{\lambda-\mu}(x,x) + \omega_\mu(x,y) = \omega_\mu(x,y).$$

Definition 7.3.13 [29, 30]. Let ω be a pseudomodular on X. Fix $x_0 \in X$. The two sets

$$X_\omega = X_\omega(x_0) = \{x \in X : \omega_\lambda(x,x_0) \to 0 \text{ as } \lambda \to +\infty\}$$

and

$$X_\omega^* = X_\omega^*(x_0) = \{x \in X : \exists \lambda = \lambda(x) > 0 \text{ such that } \omega_\lambda(x,x_0) < +\infty\}$$

are said to be *modular spaces* (around x_0).

We obviously have $X_\omega \subset X_\omega^*$. In general this inclusion may be proper. It follows from Refs. [29, 30] that if ω is a modular on X, then the modular space X_ω can be equipped with a (nontrivial) distance generated by ω and given by

$$d_\omega(x,y) = \inf\{\lambda > 0 : \omega_\lambda(x,y) \le \lambda\}, \quad \text{for any } x,y \in X_\omega.$$

If ω is a convex modular on X, according to Refs. [29, 30] the two modular spaces coincide, i.e., $X_\omega^* = X_\omega$, and this common set can be endowed with the distance d_ω^* given by

$$d_\omega^*(x,y) = \inf\{\lambda > 0 : \omega_\lambda(x,y) \le 1\}, \quad \text{for any } x,y \in X_\omega.$$

These distances will be called *Luxemburg distances*.

Example 7.3.3. Consider the modular function space L_ρ, where ρ is a convex function modular. Define the functional $\omega : L_\rho \times L_\rho \times (0,+\infty) \to [0,+\infty]$ by

$$\omega_\lambda(f,g) = \rho\left(\frac{f-g}{\lambda}\right), \quad \text{for all } \lambda > 0 \text{ and } f,g \in L_\rho.$$

Then ω is a modular on L_ρ. Moreover, the distance d_ω^* is exactly the distance generated by the Luxemburg norm on L_ρ.

For more examples on modular metric spaces, the reader may consult Refs. [1, 2, 29, 30].

Definition 7.3.14. Let X_ω be a modular metric space.

(a) The sequence $\{x_n\}_{n \in \mathbb{N}}$ in X_ω is said to be ω-*convergent* to $x \in X_\omega$ if and only if $\omega_1(x_n, x) \to 0$, as $n \to +\infty$. x will be called the ω-*limit of* $\{x_n\}$.

(b) The sequence $\{x_n\}_{n \in \mathbb{N}}$ in X_ω is said to be ω-*Cauchy* if $\omega_1(x_m, x_n) \to 0$ as $m, n \to +\infty$.

(c) A subset C of X_ω is said to be ω-*closed* if the ω-limit of a ω-convergent sequence of C always belong to C.

(d) A subset C of X_ω is said to be ω-*complete* if any ω-Cauchy sequence in C is a ω-convergent sequence and its ω-limit is in C.

(e) A subset C of X_ω is said to be ω-*bounded* if we have

$$\delta_\omega(C) = \sup\{\omega_1(x,y); x, y \in C\} < +\infty.$$

In general, if $\lim\limits_{n \to +\infty} \omega_\lambda(x_n, x) = 0$ for some $\lambda > 0$, then we may not have $\lim\limits_{n \to +\infty} \omega_\lambda$ $(x_n, x) = 0$, for all $\lambda > 0$. Therefore, as is done in modular function spaces, we say that ω satisfies the Δ_2-*condition* whenever

$$\lim\limits_{n \to +\infty} \omega_\lambda(x_n, x) = 0 \text{ for some } \lambda > 0 \text{ implies } \lim\limits_{n \to +\infty} \omega_\lambda(x_n, x) = 0 \text{ for all } \lambda > 0.$$

In Refs. [1, 2, 29, 30], one can find a discussion about the connection between ω-convergence and metric convergence with respect to the Luxemburg distances. In particular, we have

$$\lim\limits_{n \to +\infty} d_\omega(x_n, x) = 0 \text{ if and only if } \lim\limits_{n \to +\infty} \omega_\lambda(x_n, x) = 0, \text{ for all } \lambda > 0,$$

for any $\{x_n\} \in X_\omega$ and $x \in X_\omega$. In particular, we have that ω-convergence and d_ω-convergence are equivalent if and only if the modular ω satisfies the Δ_2-condition. Moreover, if the modular ω is convex, then d_ω^* and d_ω are equivalent, which implies

$$\lim_{n \to +\infty} d_\omega^*(x_n, x) = 0 \text{ if and only if } \lim_{n \to +\infty} \omega_\lambda(x_n, x) = 0, \text{ for all } \lambda > 0,$$

for any $\{x_n\} \in X_\omega$ and $x \in X_\omega$ [1, 2, 29, 30].

Definition 7.3.15 [2]. Let (X, ω) be a modular metric space. We will say that ω satisfies the Δ_2-type condition if, for any $\alpha > 0$, there exists $C > 0$ such that

$$\omega_{\lambda/\alpha}(x, y) \leq C\omega_\lambda(x, y), \quad \text{for any } \lambda > 0 \text{ and } x, y \in X_\omega.$$

Note that if ω satisfies the Δ_2-type condition, then ω satisfies the Δ_2-condition. The above definition will allow us to introduce the growth function in the modular metric spaces as was done in the linear case.

Definition 7.3.16 [2]. Let (X, ω) be a modular metric space. Define *the growth function* Ω by

$$\Omega(\alpha) = \sup \left\{ \frac{\omega_{\lambda/\alpha}(x, y)}{\omega_\lambda(x, y)} : \lambda > 0, \ x, y \in X_\omega, \ x \neq y \right\}, \quad \text{for any } \alpha > 0.$$

The following properties were proved in Ref. [2].

Lemma 7.3.3 [2, Lemma 2.1]. *Let (X, ω) be a modular metric space. Assume that ω is a convex regular modular which satisfies the Δ_2-type condition. Then*

(a) $\Omega(\alpha) < +\infty$, *for any $\alpha > 0$;*

(b) Ω *is a strictly increasing function, and $\Omega(1) = 1$;*

(c) $\Omega(\alpha\beta) \leq \Omega(\alpha)\, \Omega(\beta)$, *for any $\alpha, \beta \in (0, +\infty)$;*

(d) $\Omega^{-1}(\alpha)\, \Omega^{-1}(\beta) \leq \Omega^{-1}(\alpha\beta)$, *where Ω^{-1} is the function inverse of Ω;*

(e) *for any $x, y \in X_\omega$, $x \neq y$, we have*

$$d_\omega^*(x, y) \leq \frac{1}{\Omega^{-1}\left(1/\omega_1(x, y)\right)}.$$

The following technical lemma will be useful later on.

Lemma 7.3.4 [2]. *Let (X, ω) be a modular metric space. Assume that ω is a convex regular modular metric which satisfies the Δ_2-type condition. Let $\{x_n\}$ be a sequence in X_ω such that*

$$\omega_1(x_{n+1}, x_n) \le K \alpha^n, \quad n = 1, \cdots, \tag{7.3.7}$$

where K is an arbitrary nonzero constant and $\alpha \in (0,1)$. Then $\{x_n\}$ is Cauchy for both ω and d_ω^*.

Note that this lemma is crucial since the main assumption (7.3.7) on $\{x_n\}$ will not be enough to imply that $\{x_n\}$ is ω-Cauchy since ω generally fails the triangle inequality.

7.4. Banach Contraction Principle in Metric Spaces with a Graph

Jachymski is credited as being the first one to extend the main fixed point results of Refs. [70, 73] from a metric space endowed with a partial order to the case of a metric space endowed with a graph [46, 64]. Jachymski obtained the following result:

Theorem 7.4.1. *Let (X,d) be a complete metric space and let the triplet (X,d,G) have the following property:*

(P) *For any sequence $\{x_n\}_{n \in \mathbb{N}}$ in X, if $x_n \to x$ as $n \to +\infty$ and $(x_n, x_{n+1}) \in E(G)$, then $(x_n, x) \in E(G)$, for all $n \in \mathbb{N}$.*

Let $f : X \to X$ be a G-contraction. Set $X_f = \{x \in X; (x, f(x)) \in E(G)\}$. Then the following statements hold:

(a) *$Fix(f) \ne \emptyset$ if and only if $X_f \ne \emptyset$;*

(b) *if $X_f \ne \emptyset$ and G is weakly connected, then f is a PO;*

(c) *for any $x \in X_f$, $f \mid_{[x]_{\widetilde{G}}}$ is a PO.*

7.5. Caristi's Fixed Point Theorem

Caristi's fixed point theorem is perhaps one of the most elegant extensions of the Banach contraction principle. In this section, we discuss Caristi's fixed point theorem for mappings defined in metric spaces endowed with a graph. This work should be seen as a generalization of the classical Caristi fixed point theorem. It extends some recent works on the extension of the Banach contraction principle to metric spaces with a graph.

First, we state Caristi's fixed point theorem.

Theorem 7.5.1 [77]. *Let (M,d) be a metric space. Any map $T : M \to M$ has a fixed point provided that M is complete and there exists a lower semicontinuous function*

$\phi : M \to [0,+\infty)$ *such that*

$$d(x,Tx) \leq \phi(x) - \phi(Tx), \quad \textit{for every } x \in M. \tag{7.5.8}$$

This general fixed point theorem has found many applications in nonlinear analysis. It is shown, for example, that this theorem yields essentially all the known inwardness results [26, 27, 43]. Recall that inwardness conditions are the ones which assert that, in some sense, points from the domain are mapped toward the domain. Possibly the weakest of the inwardness conditions, the Leray-Schauder boundary condition is the assumption that a map points x of the boundary ∂M anywhere except to the outward part of the ray originating at some interior point of C and passing through x.

The proofs given for Caristi's fixed point theorem vary and use different techniques (see Refs. [21, 24, 26, 56]). It is worth mentioning that because of Caristi's result being equivalent with Ekeland's variational principle [36], many authors refer to it as the Caristi-Ekeland fixed point theorem. For more on Ekeland's variational principle and the equivalence between the Caristi-Ekeland fixed point theorem and the completeness of metric spaces, the reader is advised to read Ref. [76].

7.5.1. Caristi's Theorem in Partially Ordered Metric Spaces

Perhaps one of the most interesting examples of the use of metric fixed point theorems is the proof of the existence of solutions to differential equations. The general approach is to convert such equations to integral equations which describe exactly a fixed point of a mapping. The metric spaces in which such mappings act are usually a function space. Putting a norm (in the case of a vector space) or a distance gives us a metric structure rich enough to use the Banach contraction principle or other known fixed point theorems. But one structure naturally enjoyed by such function spaces is rarely used. Indeed, we have an order on the functions inherited from the order of $\mathbb{R}$. In the classical use of the Banach contraction principle, the focus is on the metric behavior of the mapping. The connection with the natural order is usually ignored. In Refs. [70, 73], the authors gave interesting examples where the order is used in combination with the metric conditions.

Clearly from Theorem 7.2.4, one may see that the contractive nature of the mapping T is restricted to the comparable elements of $(X, \preceq)$ not the entire set X. The detailed investigation of the problem in Section 7.2.2 shows that some mappings may exist which are not contractive on the entire set X. Therefore the Banach contraction principle will not work in this situation. The analog to Caristi's fixed point theorem is the following result:

Theorem 7.5.2 [12]. *Let $(X, \preceq)$ be a partially ordered set and suppose that there exists a distance d in X such that (X,d) is a complete metric space. Let $T : X \to$*

X be a continuous and monotone increasing mapping. Assume there exists a lower semicontinuous function $\phi : X \to [0, +\infty)$ *such that*

$$d(x, T(x)) \leq \phi(x) - \phi(T(x)), \quad \text{whenever } T(x) \preceq x.$$

Then T has a fixed point if and only if there exists $x_0 \in X$ *with* $T(x_0) \preceq x_0$.

Proof. Clearly, if x_0 is a fixed point of T, i.e., $T(x_0) = x_0$, then we have $T(x_0) \preceq x_0$. Assume that there exists $x_0 \in X$ such that $T(x_0) \preceq x_0$. Since T is monotone increasing, we have $T^{n+1}(x_0) \preceq T^n(x_0)$, for any $n \geq 1$. Hence

$$d(T^n(x_0), T^{n+1}(x_0)) \leq \phi(T^n(x_0)) - \phi(T^{n+1}(x_0)), \quad n = 1, 2, \cdots.$$

It is clear that the sequence of positive numbers $\{\phi(T^n(x_0))\}$ is decreasing. Let $\phi_0 = \lim_{n \to +\infty} \phi(T^n(x_0))$. For any $n, h \geq 1$, we have

$$d(T^n(x_0), T^{n+h}(x_0)) \leq \sum_{k=0}^{h-1} d(T^{n+k}(x_0), T^{n+k+1}(x_0)) \leq \phi(T^n(x_0)) - \phi(T^{n+h}(x_0)).$$

Therefore $\{T^n(x_0)\}$ is a Cauchy sequence in X. Since X is complete, there exists $\bar{x} \in X$ such that $\lim_{n \to +\infty} T^n(x_0) = \bar{x}$. Since T is continuous, we conclude that $T(\bar{x}) = \bar{x}$, i.e., $\bar{x}$ is a fixed point of T.   □

The continuity assumption of T may be relaxed if we assume that X satisfies the property (OSC).

Definition 7.5.1. Let $(X, \preceq)$ be a partially ordered set. Let d be a distance defined on X. We say that X *satisfies the property (OSC)* if and only if for any convergent decreasing sequence $\{x_n\}$ in X, i.e., $x_{n+1} \preceq x_n$ for any $n \geq 1$, we have $\lim_{m \to +\infty} x_m = \inf\{x_n, n \geq 1\}$.

We have the following improvement to Theorem 7.5.2.

Theorem 7.5.3 [12]. *Let* $(X, \preceq)$ *be a partially ordered set and suppose that there exists a distance d in X such that* (X, d) *is a complete metric space. Assume that X satisfies the property (OSC). Let* $T : X \to X$ *be a monotone increasing mapping. Assume there exists a lower semicontinuous function* $\phi : X \to [0, +\infty)$ *such that*

$$d(x, T(x)) \leq \phi(x) - \phi(T(x)), \quad \text{whenever } T(x) \preceq x.$$

Assume there exists $x_0 \in X$ *such that* $T(x_0) \preceq x_0$. *Then T has a fixed point.*

Proof. Let $x_0 \in X$ such that $T(x_0) \preceq x_0$. Write $x_n = T^n(x_0)$, $n \geq 1$. Then we have $\{x_n\}$ is decreasing and $\lim_{n \to +\infty} x_n = x_\omega$ exists in X. Since we did not assume T continuous, then x_ω may not be a fixed point of T. The idea is to build a transfinite orbit to eventually catch the fixed point. Note that since X satisfies (OSC), then we have $x_\omega = \inf\{x_n, n \geq 1\}$. Since T is monotone increasing, then we will have $T(x_\omega) \preceq x_\omega$. These basic facts so far will help us seek the transfinite orbit $\{x_\alpha\}_{\alpha \Gamma}$, where Γ is the set of all ordinals. This transfinite orbit must satisfy the following properties:

(a) $T(x_\alpha) = x_{\alpha+1}$, for any $\alpha \in \Gamma$;

(b) $x_\alpha = \inf\{x_\beta, \beta < \alpha\}$, if α is a limit ordinal;

(c) $x_\alpha \preceq x_\beta$, whenever $\beta < \alpha$;

(d) $d(x_\alpha, x_\beta) \leq \phi(x_\beta) - \phi(x_\alpha)$, whenever $\beta < \alpha$.

Clearly, the above properties are satisfied for any $\alpha \in \{0, 1, \cdots, \omega\}$. Let α be an ordinal number. Assume that the properties (a)–(d) are satisfied by $\{x_\beta\}_{\beta < \alpha}$. We have two cases:

CASE (I) If $\alpha = \beta + 1$, then set $x_\alpha = T(x_\beta)$.

CASE (II) Assume α is a limit ordinal. Set $\phi_0 = \inf\{\phi(x_\beta), \beta < \alpha\}$. Then one can easily find an increasing sequence of ordinals $\{\beta_n\}$, with $\beta_n < \alpha$, such that $\lim_{n \to +\infty} \phi(x_{\beta_n}) = \phi_0$. Property (d) will force $\{x_{\beta_n}\}$ to be Cauchy. Since X is complete, then $\lim_{n \to +\infty} x_{\beta_n} = \bar{x}$ exists in X. The property (OSC) will then imply $\bar{x} = \inf\{x_{\beta_n}, n \geq 1\}$. Let us show that $\bar{x} = \inf\{x_\beta : \beta < \alpha\}$. Let $\beta < \alpha$. If $\beta_n < \beta$, for all $n \geq 1$, then we have

$$d(x_\beta, x_{\beta_n}) \leq \phi(x_{\beta_n}) - \phi(x_\beta), \quad n = 1, 2, \cdots.$$

But $\phi(x_\beta) \succeq \phi_0 = \lim_{n \to +\infty} \phi(x_{\beta_n}) \succeq \phi(x_\beta)$. Hence $\phi(x_\beta) = \phi_0$ which implies that $\lim_{n \to +\infty} x_{\beta_n} = x_\beta$. Hence $x_\beta = \bar{x}$. Assume otherwise that there exists $n_0 \geq 1$ such that $\beta < \beta_{n_0}$. Hence $x_{\beta_{n_0}} \preceq x_\beta$ which implies $\bar{x} \preceq x_\beta$. In any case, we have $\bar{x} \preceq x_\beta$, for any $\beta < \alpha$. Therefore we have

$$\bar{x} = \inf\{x_{\beta_n} : n \geq 1\} \preceq \inf\{x_\beta : \beta < \alpha\} \preceq \inf\{x_{\beta_n} : n \geq 1\}.$$

Hence $\bar{x} = \inf\{x_\beta : \beta < \alpha\}$. Set $x_\alpha = \bar{x}$. Let us prove that $\{x_\beta : \beta \leq \alpha\}$ satisfies all properties (a)–(d). Clearly, (a) and (b) are satisfied. Let us focus on (c) and (d). Let $\beta < \alpha$. We need to show that $x_\alpha \preceq x_\beta$. If α is a limit ordinal, this is obvious. Assume that $\alpha - 1$ exists. We have two cases: if $\alpha - 2$ exists, then we have $x_{\alpha-1} \preceq x_{\alpha-2}$. Since T is monotone increasing, then $T(x_{\alpha-1}) \preceq T(x_{\alpha-2})$, i.e., $x_\alpha \preceq x_{\alpha-1}$. Otherwise if $\alpha - 2$ is an ordinal limit, then $x_{\alpha-2} = \inf\{x_\gamma, \ \gamma < \alpha - 2\}$. Since T is monotone increasing, then we have

$$x_{\alpha-1} = T(x_{\alpha-2}) \preceq x_{\gamma+1}, \quad \text{for any } \gamma < \alpha - 2,$$

which implies $x_\alpha \preceq x_{\gamma+2}$, for any $\gamma < \alpha - 2$. Therefore we have $x_\alpha \preceq x_\beta$, which completes the proof of (c). Let us prove (d). Let $\beta < \alpha$. First assume that $\alpha - 1$ exists. Then in the proof of (c), we saw that $x_\alpha = T(x_{\alpha-1}) \preceq x_{\alpha-1}$. Our assumption on T will then imply

$$d(x_{\alpha-1}, x_\alpha) \leq \phi(x_{\alpha-1}) - \phi(x_\alpha).$$

If $\beta = \alpha - 1$ we are done. Otherwise if $\beta < \alpha - 1$, then we use the induction assumption to get

$$d(x_\beta, x_{\alpha-1}) \leq \phi(x_\beta) - \phi(x_{\alpha-1}).$$

The triangle inequality will then imply

$$d(x_\beta, x_\alpha) \leq \phi(x_\beta) - \phi(x_\alpha).$$

Next we assume that α is a limit ordinal. Then there exists an increasing sequence of ordinals $\{\beta_n\}$, with $\beta_n < \alpha$, such that $x_\alpha = \lim\limits_{n \to +\infty} x_{\beta_n}$. Given $\beta < \alpha$, assume we have $\beta_n < \beta$, for all $n \geq 1$. In this case, we have seen that $x_\alpha = x_\beta$. Otherwise let us assume there exists $n_0 \geq 1$ such that $\beta < \beta_{n_0}$. In this case, from our induction assumption and the triangle inequality, we get

$$d(x_\beta, x_{\beta_n}) \leq \phi(x_\beta) - \phi(x_{\beta_n}), \quad n \geq n_0.$$

Using the lower semicontinuity of ϕ, we conclude that

$$d(x_\beta, x_\alpha) \leq \phi(x_\beta) - \phi(x_\alpha),$$

which completes the proof of (d). By the transfinite induction we conclude that the transfinite orbit $\{x_\alpha\}$ exists, which satisfies the properties (a)–(d). Since $\mathbb{R}$ has a finite cardinality, there exists $\alpha \in \Gamma$ such that $\phi(x_\alpha) = \phi(x_{\alpha+1})$. Property (d) will then force $x_{\alpha+1} = x_\alpha$, i.e., $T(x_\alpha) = x_\alpha$. Therefore T has a fixed point. $\square$

One may wonder if Theorem 7.5.3 is a true extension of the main results of Refs. [46, 70, 73]. The following example shows that this is the case.

Example 7.5.1 [12]. Let $X = L^1([0,1], dx)$ with the natural pointwise order generated by $\mathbb{R}$. Let $C = \{f \in X, \ f(t) \geq 0 \text{ a.e. } t \in [0,1]\}$ be the positive cone of X. Define

$T : C \to C$ by

$$T(f)(t) = \begin{cases} f(t), & \text{if } f(t) > 1/2, \\ 0, & \text{if } f(t) \leq 1/2. \end{cases}$$

First note that C is a closed subset of X. Hence C is complete for the norm-1 distance. Also, it is easy to check that the property (OSC) holds in this case. Note that for any $f \in C$, we have $0 \leq T(f) \leq f$. Also, we have $T^2(f) = T(T(f)) = T(f)$, i.e., $T(f)$ is a fixed point of T for any $f \in C$. Note that for any $f \in C$, we have

$$d(f, T(f)) = \int_0^1 |f(t) - T(f)(t)| dt = \int_0^1 f(t) dt - \int_0^1 T(f)(t) dt = \|f\| - \|T(f)\|.$$

Therefore all the assumptions of Theorem 7.5.3 are satisfied. But T fails to satisfy the assumptions of Refs. [46, 70, 73]. Indeed, if we take

$$f(t) = \frac{1}{2} \text{ and } f_n(t) = \frac{1}{2} + \frac{1}{n}, \quad n \geq 1,$$

then we have $T(f) = 0$ and $T(f_n) = f_n$, for any $n \geq 1$. Therefore T is not continuous since $\{f_n\}$ converges uniformly (and in norm-1 as well) to f. Note also that $f \leq f_n$, for $n \geq 1$. So no Lipschitz condition will be satisfied by T with respect to the partial order of C.

7.5.2. Caristi's Theorem in Metric Spaces with a Graph

In this subsection, we give the graph versions of the two main results of the previous subsection.

Definition 7.5.2. A mapping $T : X \to X$ is said to be a *Caristi G-mapping* if there exists a lower semicontinuous function $\phi : X \to [0, +\infty)$ such that

$$d(x, Tx) \leq \phi(x) - \phi(Tx), \quad \text{whenever} \quad (T(x), x) \in E(G).$$

Theorem 7.5.4 [12]. *Let G be an oriented graph on the set X with $E(G)$ containing all loops and suppose that there exists a distance d in X such that (X, d) is a complete metric space. Let $T : X \to X$ be continuous, G-monotone, and a G-Caristi mapping. Then T has a fixed point if and only if there exists $x_0 \in X$, with $(T(x_0), x_0) \in E(G)$.*

Proof. G has all the loops. In particular, if x_0 is a fixed point of T, i.e., $T(x_0) = x_0$, then we have $(T(x_0), x_0) \in E(G)$. Assume that there exists $x_0 \in X$ such that $(T(x_0), x_0) \in E(G)$. Since T is G-monotone, we have $(T^{n+1}(x_0), T^n(x_0)) \in E(G)$, for any $n \geq 1$. Hence

$$d(T^n(x_0), T^{n+1}(x_0)) \leq \phi(T^n(x_0)) - \phi(T^{n+1}(x_0)), \quad n = 1, 2, \cdots.$$

It is clear that the sequence of positive numbers $\{\phi(T^n(x_0))\}$ is decreasing. Let $\phi_0 = \lim_{n \to +\infty} \phi(T^n(x_0))$. For any $n, h \geq 1$, we have

$$d\left(T^n(x_0), T^{n+h}(x_0)\right) \leq \sum_{k=0}^{h-1} d\left(T^{n+k}(x_0), T^{n+k+1}(x_0)\right) \leq \phi\left(T^n(x_0)\right) - \phi\left(T^{n+h}(x_0)\right).$$

Therefore $\{T^n(x_0)\}$ is a Cauchy sequence in X. Since X is complete, there exists $\bar{x} \in X$ such that $\lim_{n \to +\infty} T^n(x_0) = \bar{x}$. Since T is continuous, we conclude that $T(\bar{x}) = \bar{x}$, i.e., $\bar{x}$ is a fixed point of T. $\qquad\square$

The following definition is needed to prove the analog to Theorem 7.5.3.

Definition 7.5.3. Let G be an acyclic oriented graph on the set X with $E(G)$ containing all loops. We say that G *satisfies the property (OSC)* if and only if $(X, \preceq_G)$ satisfies (OSC).

The analog to Theorem 7.5.3 may be stated as:

Theorem 7.5.5 [12]. *Let G be an oriented graph on the set X with $E(G)$ containing all loops and suppose that there exists a metric d in X such that (X, d) is a complete metric space. Assume that G satisfies the property (OSC). Let $T : X \to X$ be a G-monotone and a Caristi G-mapping. Then T has a fixed point if and only if there exists $x_0 \in X$, with $(T(x_0), x_0) \in E(G)$.*

Proof. The proof of Theorem 7.5.5 is similar to the proof of Theorem 7.5.3. $\qquad\square$

7.5.3. Minimal Points and Fixed Point Property in a Graph

Let A be an abstract set partially ordered by $\preceq$. We will say that $a \in A$ is a minimal element of A if and only if $b \preceq a$ implies $b = a$. The concept of a minimal element is crucial in the proof given to the Caristi's fixed point theorem.

Theorem 7.5.6 [47]. *Let $(A, \preceq)$ be a partially ordered set. Then the following statements are equivalent:*

(a) *A contains a minimal element.*

(b) *Any multivalued map T defined on A such that, for any $x \in A$, there exists $y \in Tx$ with $y \preceq x$, has a fixed point, i.e., there exists a in A such that $a \in T(a)$.*

Remark 7.5.1. Recall that Taskovic [78] showed that Zorn's lemma is equivalent to:

(TT)   Let $\mathscr{F}$ be a family of self-mappings defined on a partially ordered set A such that

$$x \preceq f(x) \quad (\text{respectively } f(x) \preceq x), \quad \text{for all } x \in A \text{ and all } f \in \mathscr{F}.$$

If each chain in A has an upper bound (respectively lower bound), then the family $\mathscr{F}$ has a common fixed point.

Therefore Theorem 7.5.6 is different from the remark as the theorem considers the existence of minimal elements, which in general does not imply that any chain has a lower bound. In the next result, we discuss a common fixed point theorem. Let $\phi : X \to [0, +\infty)$ be a map and define the order $\preceq_\phi$ (see Refs. [20–22]) on X by:

$$x \preceq_\phi y \quad \text{if and only if} \quad d(x,y) \le \phi(y) - \phi(x), \quad \text{for any } x, y \in X.$$

It is straightforward to see that $(X, \preceq_\phi)$ is a partially ordered set. However, it is not clear what are the minimal assumptions on X and ϕ which ensure the existence of a minimal element. In particular, if X is complete and ϕ is lower semicontinuous, then any decreasing chain in $(X, \preceq_\phi)$ has a lower bound. Indeed, let $\{x_\alpha\}_{\alpha \in \Gamma}$, where Γ is a totally ordered set, be a decreasing chain. Denote by $\preceq$ the order of Γ. Then $\{\phi(x_\alpha)\}_{\alpha \in \Gamma}$ is a decreasing net of positive numbers. Indeed, let α and β in Γ such that $\alpha \preceq \beta$. Since $\{x_\alpha\}$ is decreasing, we have $x_\beta \preceq_\phi x_\alpha$. Therefore we must have

$$d(x_\alpha, x_\beta) \le \phi(x_\alpha) - \phi(x_\beta),$$

which implies $\phi(x_\beta) \le \phi(x_\alpha)$. Set $\phi_0 = \inf\{\phi(x_\alpha); \ \alpha \in \Gamma\}$. Let $\{\alpha_n\}_{n \ge 1}$ be an increasing sequence of elements from Γ such that

$$\lim_{n \to +\infty} \phi(x_{\alpha_n}) = \phi_0.$$

Using the definition of $\preceq_\phi$, one can easily show that $\{x_{\alpha_n}\}$ is a Cauchy sequence and therefore converges to $x \in X$. Finally, it is straightforward to see that $x \preceq_\phi x_{\alpha_n}$ for all $n \ge 1$. Indeed, we have $x_{\alpha_m} \preceq_\phi x_{\alpha_n}$, for $n \le m$. Hence

$$d(x_{\alpha_m}, x_{\alpha_n}) \le \phi(x_{\alpha_n}) - \phi(x_{\alpha_m})$$

holds, which implies

$$\phi(x_{\alpha_m}) \le \phi(x_{\alpha_n}) - d(x_{\alpha_m}, x_{\alpha_n}).$$

If we let $m \to +\infty$, using the lower semicontinuity of ϕ we get

$$\phi(x) \le \liminf_{m \to +\infty} \phi(x_{\alpha_m}) \le \phi(x_{\alpha_n}) - d(x, x_{\alpha_n}), \quad \text{for any } n \ge 1.$$

Clearly, this implies $x \preceq_\phi x_{\alpha_n}$, for any $n \ge 1$. We claim that $x \preceq_\phi x_\alpha$, for any $\alpha \in \Gamma$. Let $\alpha \in \Gamma$. Assume that there exists $n \ge 1$ such that $\alpha \preceq \alpha_n$. Then, we must have

$x_{\alpha_n} \preceq_\phi x_\alpha$. Since $x \preceq_\phi x_{\alpha_n}$, we conclude that $x \preceq_\phi x_\alpha$. Otherwise, we assume that $\alpha_n \preceq \alpha$, for any $n \geq 1$. Then we have $x_\alpha \preceq_\phi x_{\alpha_n}$ which implies

$$d(x_\alpha, x_{\alpha_n}) \leq \phi(x_{\alpha_n}) - \phi(x_\alpha), \quad \text{for any } n \geq 1.$$

If we let $n \to +\infty$, and using the fact $\phi_0 \leq \phi(x_\alpha)$, we get

$$d(x_\alpha, x) \leq \phi_0 - \phi(x_\alpha) \leq 0.$$

Clearly, this forces $x_\alpha = x$. Therefore we have proved that $x \preceq_\phi x_\alpha$, for all $\alpha \in \Gamma$, which means that x is a lower bound for $\{x_\alpha\}_{\alpha \in \Gamma}$. Zorn's lemma will therefore imply that $(X, \preceq_\phi)$ has minimal elements.

Corollary 7.5.1 [47]. *Let (X, d) be a metric space and $\phi : X \to [0, +\infty)$ be a map. Consider the partially ordered set $(X, \preceq_\phi)$. Assume that $a \in X$ is a minimal element. Then any map $T : X \to X$ such that for all $x \in X$*

$$d(x, Tx) \leq \phi(x) - \phi(Tx)$$

(i.e., $Tx \preceq_\phi x$) fixes a, i.e., $Ta = a$.

One can now give the graph version of the above corollary as follows:

Corollary 7.5.2 [4]. *Let G be an acyclic oriented graph with $E(G)$ containing all loops. Suppose that there exists a distance d in $V(G)$ such that $(V(G), d)$ is a metric space and $\phi : V(G) \to [0, +\infty)$ is a map. Consider the partially ordered set $(V(G), (\preceq_G)_\phi)$. Assume that $a \in V(G)$ is a minimal element. Then any map $T : V(G) \to V(G)$ such that for all $x \in V(G)$ with $(x, Tx) \in E(G)$*

$$d(x, Tx) \leq \phi(x) - \phi(Tx)$$

(i.e., $Tx(\preceq_G)_\phi x$) fixes a, i.e., $Ta = a$.

This corollary can be seen as a generalization of Caristi's result endowed with a graph (see Theorem 7.5.2). Indeed, the regular assumptions made in Caristi's theorem imply that any chain (for $\preceq_\phi$) has a lower bound, which is stronger than having a minimal element. In fact, Corollary 7.5.2 contains implicitly a conclusion of the existence of a common fixed point. For a similar conclusion in the partial order version, see Refs. [21, 47].

7.5.4. Kirk's Problem via Graphs

In an attempt to generalize Caristi's fixed point theorem, Kirk [27] has raised the problem of whether a map $T : X \to X$ such that for all $x \in X$

$$\eta(d(x,Tx)) \le \phi(x) - \phi(Tx),$$

for some positive function η, has a fixed point. In fact Kirk's original question was stated when $\eta(t) = t^p$, for some $p > 1$. Khamsi [47] gave a good example which answers Kirk's problem in the negative where the order is used implicitly. A similar example is presented here, but in terms of a graph.

Example 7.5.2 [4]. Let G be a graph with vertex set $V(G) = \{x_n; n \ge 1\} \subset [0,+\infty)$ defined by

$$x_n = 1 + \frac{1}{2} + \frac{1}{3} + \cdots + \frac{1}{n}, \quad \text{for all } n \ge 1,$$

and $E(G) = \{(x_i,x_i),(x_i,x_{i+1}); \ i = 1,2,3,\dots\}$. Then $V(G)$ is a closed subset of $[0,+\infty)$ and therefore is complete. Let $V(G)$ be endowed with metric $d : V(G) \times V(G) \to \mathbb{R}^+$ defined by the standard distance in $\mathbb{R}$ and $T : V(G) \to V(G)$ be defined by $Tx_n = x_{n+1}$ for all $n \ge 1$. The graph of G is shown in the figure below:

$$x_1 \qquad x_2 \qquad x_3 \qquad \qquad x_n \qquad x_{n+1}$$

Then

$$d(x_n,Tx_n)^p = \frac{1}{(n+1)^p} = \phi(x_n) - \phi(Tx_n),$$

where $\phi(x_n) = \sum\limits_{i=n+1}^{+\infty} \frac{1}{i^p}$, for all $n \ge 1$. An easy computation shows that ϕ is lower semicontinuous on $V(G)$. Furthermore, one can show that T is nonexpansive, i.e., $d(Tx,Ty) \le d(x,y)$, for all $x,y \in V(G)$. And it is obvious that T has no fixed point.

Though the above example gives a negative answer to Kirk's problem, some positive partial answers may also be found. Note that the order approach to Caristi's traditional result is no longer possible. Indeed, if we define on the metric space $V(G)$ the relation $\preceq_{\eta,\phi}$ (defined below), then $\preceq_{\eta,\phi}$ is reflexive and antisymmetric. But it is not in general transitive. Of course if η is subadditive, i.e., $\eta(a+b) \le \eta(a) + \eta(b)$ for any $a,b \in [0,+\infty)$, then $\preceq_{\eta,\phi}$ is transitive. So one may wonder how to approach this general case when $\preceq_{\eta,\phi}$ is not transitive and therefore $(V(G),\preceq_{\eta,\phi})$ is not a partially ordered set.

Definition 7.5.4 [4]. We say that a map $\eta : [0,+\infty) \to [0,+\infty)$ is a *well-behaved map* if it is nondecreasing, continuous, and such that there exist $c > 0$ and $\delta_0 > 0$ such that for any $t \in [0,\delta_0]$ we have $\eta(t) \geq ct$.

Notice that as η is continuous, there exists an $\varepsilon_0 > 0$ such that $\eta^{-1}([0,\varepsilon_0]) \subset [0,\delta_0]$.

Definition 7.5.5 [4]. Let G be an acyclic oriented graph with $E(G)$ containing all loops. Suppose that there exists a metric d in $V(G)$ such that $(V(G),d)$ is a metric space. Let $\phi : V(G) \to [0,+\infty)$ and $\eta : [0,+\infty) \to [0,+\infty)$ be two functions. Define the relation $\preceq_{\phi,\eta}$ by

$$x \preceq_{\phi,\eta} y \quad \Leftrightarrow \quad \eta(d(x,y)) \leq \phi(y) - \phi(x).$$

One says that G has a *minimal vertex (root)* $x_* \in V(G)$ if x_* is a minimal element in $(V(G), \preceq_{\phi,\eta})$, i.e., if $x \preceq_{\phi,\eta} x_*$ then we must have $x = x_*$.

Theorem 7.5.7 [4]. *Let G be an acyclic oriented graph with $E(G)$ containing all loops. Suppose that there exists a metric d in $V(G)$ such that $(V(G),d)$ is a complete metric space. Let η be a well-behaved map and $\phi : V(G) \to [0,+\infty)$ be a lower semicontinuous map, then G has a root x_*.*

Proof. Set $\phi_0 = \inf\{\phi(x) : x \in V(G)\}$. For any $\varepsilon > 0$, set

$$V(G)_\varepsilon = \{x \in V(G) : \phi(x) \leq \phi_0 + \varepsilon\}.$$

Since ϕ is lower semicontinuous, then $V(G)_\varepsilon$ is a closed nonempty subset of $V(G)$. Also note that if $x,y \in V(G)_\varepsilon$ and $x \preceq_{\phi,\eta} y$, then $\eta(d(x,y)) \leq \phi(y) - \phi(x)$, which implies

$$\phi_0 \leq \phi(x) \leq \phi(y) \leq \phi_0 + \varepsilon.$$

Hence $\eta(d(x,y)) \leq \varepsilon$. Using c, ε_0, and δ_0 associated with the well-behaved map η, we get

$$c\, d(x,y) \leq \eta(d(x,y)) \leq \phi(y) - \phi(x),$$

for any $x,y \in V(G)_{\varepsilon_0}$ with $x \preceq_{\phi,\eta} y$. On $V(G)_{\varepsilon_0}$, we define the new relation $\preceq_*$ by

$$x \preceq_* y \quad \Leftrightarrow \quad d(x,y) \leq \frac{1}{c}\phi(y) - \frac{1}{c}\phi(x).$$

Clearly, $(V(G)_{\varepsilon_0}, \preceq_*)$ is a partially ordered set with all necessary assumptions to secure the existence of a minimal element x_* for $\preceq_*$. We claim that x_* is also a minimal element for the relation $\preceq_{\phi,\eta}$ in $V(G)$ and hence a root of the graph G. Indeed, let $x \in V(G)$ be such that $x \preceq_{\phi,\eta} x_*$. Then we have $\eta(d(x,x_*)) \leq \phi(x_*) - \phi(x)$. In particular, we have $\phi(x) \leq \phi(x_*)$, which implies $\phi(x) \leq \phi_0 + \varepsilon_0$; i.e., $x \in V(G)_{\varepsilon_0}$. As before, we have $\eta(d(x,x_*) \leq \varepsilon_0$, which implies

$$c\, d(x,x_*) \leq \eta(d(x,x_*)) \leq \phi(x_*) - \phi(x).$$

Hence we have $x \preceq_* x_*$. Since x_* is a minimal element in $(V(G)_{\varepsilon_0}, \preceq_*)$, we get $x = x_*$.

$\square$

Definition 7.5.6 [4]. One says that a mapping $T : V(G) \to V(G)$ is a *Caristi-Kirk G-mapping* if there exist a lower semicontinuous function $\phi : V(G) \to [0,+\infty)$ and a well-behaved map η such that

$$\eta(d(x,Tx)) \leq \phi(x) - \phi(Tx),$$

whenever $(T(x),x) \in E(G)$.

The next result is a positive partial answer to Kirk's problem via a graph and should be seen as an analog to Corollary 7.5.2.

Theorem 7.5.8 [4]. *Let G be an acyclic oriented graph with $E(G)$ containing all loops. Suppose that there exists a distance d in $V(G)$ such that $(V(G),d)$ is a complete metric space. Let $T : V(G) \to V(G)$ be G-monotone and a Caristi-Kirk G-mapping. Assume that $(T(x),x) \in E(G)$ for all $x \in V(G)$. Then T has a fixed point.*

Proof. Since T is a G-monotone and a Caristi-Kirk G-mapping, and as $(T(x),x) \in E(G)$ for all $x \in V(G)$, there exist a lower semicontinuous function $\phi : V(G) \to [0,+\infty)$ and a well-behaved map η such that

$$\eta(d(x,Tx)) \leq \phi(x) - \phi(Tx), \quad \text{for all } x \in V(G).$$

Consider the binary relation $\preceq_{\phi,\eta}$. Theorem 7.5.7 implies the existence of a root x_* of G. Since $(T(x_*),x_*) \in E(G)$, then we have $T(x_*) \preceq_{\phi,\eta} x_*$. Since x_* is a root, then we must have $T(x_*) = x_*$, i.e., x_* is a fixed point of T.

$\square$

This is an amazing result because the relation $\preceq_{\phi,\eta}$ is not a partial order. Moreover, the minimal point is fixed by any map T that is G-monotone and a Caristi-Kirk G-mapping such that $(T(x),x) \in E(G)$ for all $x \in V(G)$. So the fixed point is independent of the map T and only depends on the functions ϕ and η.

7.5.5. Applications of Metric Spaces with a Graph

Graph theory is considered to be one of the most important branches of mathematics. For example, it plays a crucial role in structural models. The structural arrangements of various objects or technologies lead to new inventions and modifications. This section gives an idea of the implementation of our main results in computer science applications that use graph theoretical concepts.

A graph model for fault-tolerant computing systems:

This section is based on graph theory, where it is used to model the fault-tolerant system. Here, the computer is represented as S and the algorithm to be executed by S is known as A. Both S and A are represented by means of graphs whose vertices represent computing facilities. Algorithm A is executable by S if A is isomorphic to a subgraph of S. Hayes [45] has presented a graph model and algorithms for computing systems for fault-tolerant systems. These graphs show the computing facility of a particular computation and the interconnection among them. This model is applied directly to the minimum configuration or structure required to achieve fault tolerance to a specified degree. The model is represented in the form of a facility graph. A facility graph is a graph G whose vertices represent system facilities and whose edges represent access links between facilities [45]. A facility here is said to be a hardware or software component of any system that can fail independently. Hardware facilities include control units, arithmetic processors, storage units and input/output equipment. Software facilities include compilers, application programs, library routines, operating systems, etc. Since each facility can access some other facilities, the real time systems are represented as a facility graph. The following is a labeled directed facility graph. Facility types are indicated by numbers in parentheses. The graph indicates the types of facilities accessed by other facilities. The vertex x_1 access the vertices x_2 and x_3. Similarly, the vertex x_5 with facility type t_2 accesses the facility types t_1, t_3 and t_2 of vertices x_2, x_4 and x_6 respectively.

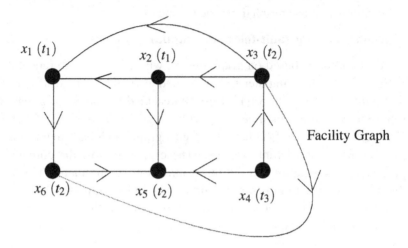

Facility Graph

Graphical representation of an algorithm

An algorithm will be defined in the form of a facility graph whose vertices represent the facilities required to execute the algorithm and whose edges represent the links required among these facilities. An algorithm A is executable by a computing system S if A is isomorphic to a subgraph of S. This means that there is a one to one mapping from the vertices of A into the vertices of S that preserves vertex labels and adjacencies between vertices. This implies that S contains all the facilities and connections between facilities required by A. So, A can be embedded in S.

A *k-fault* F in a system S is the removal of any k vertices $\{x_1, x_2, x_3, x_4 \ldots, x_k\}$ from S. All edges connected to these vertices are also removed. The resultant graph will be denoted by SF. The basic concepts relating to this fault tolerance system are given in Ref. [45].

(a) A system S is fault tolerant with respect to algorithm A and fault F if A is executable by SF.

(b) S is fault tolerant with respect to a set of algorithms $\{A_1, A_2,, A_3, \ldots, A_p\}$ and a set of faults $\{F_1, F_2, \ldots, F_q\}$ if A_i is executable by SF_j for all i and j where $1 \leq i \leq p$.

If S is k-fault tolerant with respect to A, then S is j-fault tolerant with respect to A, for all j where $0 \leq j \leq k$.

Now let G be an acyclic oriented facility graph with $E(G)$ containing all loops. Suppose that there exists a metric d in $V(G)$ such that $(V(G), d)$ is a complete metric space. Let η be a well-behaved map and $\phi : V(G) \to [0, +\infty)$ be a lower semicontinuous map, then by Theorem 7.5.7, G has a root x_*, i.e., all system faults will be accessed from the root vertex (mother keyboard of the computer) and $SF = \{x^*\}$.

7.6. The Contraction Principle in Modular Metric Spaces with a Graph

The aim of this section is to discuss the existence of fixed points for Lipschitzian mappings defined on some subsets of modular metric spaces endowed with a graph. Abdou and Khamsi [1] have defined and investigated the fixed point property in the framework of modular metric spaces and introduced the analog of the Banach contraction principle in modular metric spaces.

Let us start this section with the necessary graph theory terminology. Let (X, ω) be a modular metric space and M be a nonempty subset of X_ω. Let Δ denote the diagonal of the Cartesian product $M \times M$. Consider a directed graph G_ω such that the set $V(G_\omega)$ of its vertices coincides with M, and the set $E(G_\omega)$ of its edges contains all loops, i.e., $E(G_\omega) \supseteq \Delta$. We assume G_ω has no parallel edges (arcs), so we can identify G_ω with the pair $(V(G_\omega), E(G_\omega))$.

Definition 7.6.1. Let (X, ω) be a modular metric space and M be a nonempty subset of X_ω. A mapping $T : M \to M$ is called:

(a) G_ω-*contraction* if T is G_ω-monotone, i.e.,

$$\forall x, y \in M; \ ((x, y) \in E(G_\omega) \Rightarrow (T(x), T(y)) \in E(G_\omega)),$$

and if there exists a constant $\alpha \in [0, 1)$ such that

$$\omega_1(T(x), T(y)) \leq \alpha \, \omega_1(x, y), \ \text{for any} \ (x, y) \in E(G_\omega).$$

(b) $(\varepsilon, \alpha) - G_\omega$-*uniformly local contraction* if T is G_ω-monotone and there exists a constant $\alpha \in [0, 1)$ such that for any $(x, y) \in E(G_\omega)$

$$\omega_1(T(x), T(y)) \leq \alpha \, \omega_1(x, y), \quad \text{whenever} \ \omega_1(x, y) < \varepsilon.$$

A point $x \in M$ is called a *fixed point* of T whenever $T(x) = x$. The set of fixed points of T will be denoted by $Fix(T)$.

Throughout this section, we assume that (X, ω) is a modular metric space. Let M be a nonempty subset of X_ω and G_ω be a directed graph such that $V(G_\omega) = M$ and $E(G_\omega) \supseteq \Delta$. Our first result can be seen as an extension of Jachymski's fixed point theorem (Theorem 7.4.1) to modular metric spaces. As Jachymski [46] did, we introduce the following property:

Property 7.6.1. We say that the triple $(M, d_\omega^*, G_\omega)$ has Property 7.6.1 if for any sequence $\{x_n\}_{n\in\mathbb{N}}$ in M, if $x_n \to x$ as $n \to +\infty$ and $(x_n, x_{n+1}) \in E(G_\omega)$, then $(x_n, x) \in E(G_\omega)$, for all $n \in \mathbb{N}$.

Note that Property 7.6.1 is precisely the Nieto et al. [71] hypothesis relaxing continuity assumption as in Theorem 7.2.4 (ii, iii) rephrased in terms of edges.

Theorem 7.6.1 [6]. *Let (X, ω) be a modular metric space with a graph G_ω. Suppose that ω is a convex regular modular metric which satisfies the Δ_2-type condition. Assume that $M = V(G_\omega)$ is a nonempty ω-bounded, ω-complete subset of X_ω and the triple $(M, d_\omega^*, G_\omega)$ has Property 7.6.1. Let $T : M \to M$ be a G_ω-contraction map such that $M_T := \{x \in M; (x, T(x)) \in E(G_\omega)\}$ is nonempty. Then the following statements hold:*

(a) *For any $x \in M_T$, $T|_{[x]_{\widetilde{G}_\omega}}$ has a fixed point.*

(b) *If G_ω is weakly connected, then T has a fixed point in M.*

(c) *If $M' := \bigcup\{[x]_{\widetilde{G}_\omega} : x \in M_T\}$, then $T|_{M'}$ has a fixed point in M.*

Proof. (a) Let $x_0 \in M_T$, i.e., $(x_0, T(x_0)) \in E(G_\omega)$. Since T is a G_ω-contraction, there exists a constant $\alpha \in [0, 1)$ such that $(T(x_0), T(T(x_0))) \in E(G_\omega)$ and

$$\omega_1(T(x_0), T(T(x_0))) \leq \alpha\, \omega_1(x_0, T(x_0)).$$

By induction, we construct a sequence $\{x_n\}$ such that $x_{n+1} := T(x_n)$, $(x_n, x_{n+1}) \in E(G_\omega)$ and

$$\omega_1(x_{n+1}, x_n) \leq \alpha\, \omega_1(x_n, x_{n-1}) \leq \alpha^n\, \omega_1(x_0, x_1), \quad \text{for any } n \geq 1.$$

Since M is ω-bounded, we have

$$\omega_1(x_{n+1}, x_n) \leq \delta_\omega(M)\alpha^n, \quad \text{for any } n \geq 1.$$

The technical Lemma 7.3.4 implies that $\{x_n\}$ is ω-Cauchy. Since M is ω-complete, therefore $\{x_n\}$ ω-converges to some point $x \in M$. By Property 7.6.1, $(x_n, x) \in E(G_\omega)$ for all n, and hence

$$\omega_1(x_{n+1}, T(x)) \leq \alpha\, \omega_1(x_n, x).$$

We conclude that $\lim_{n\to+\infty} \omega_1(x_{n+1}, T(x)) = 0$. Using the properties of ω, we have

$$\omega_2(x, T(x)) \leq \omega_1(x, x_{n+1}) + \omega_1(x_{n+1}, T(x)), \quad \text{for all } n \geq 1.$$

This implies $\omega_2(x, T(x)) = 0$. Therefore $x = T(x)$, i.e., x is a fixed point of T. As $(x_0, x) \in E(G_\omega)$, we have $x \in [x_0]_{\widetilde{G}_\omega}$.

(b) Since $M_T \neq \emptyset$, there exists an $x_0 \in M_T$, and since G_ω is weakly connected, then $[x_0]_{\widetilde{G}_\omega} = M$ and by Property 7.6.1, T has a fixed point.

(c) This follows easily from (a) and (b).   □

Edelstein [35] has extended the classical fixed point theorem for contractions to the case when X is a complete ε-chainable metric space and the mapping $T : X \rightarrow X$ is an (ε, k)-uniformly local contraction. Here we investigate Edelstein's result in modular metric spaces endowed with a graph. First let us introduce the ε-chainable concept in modular metric spaces with a graph. The definition given here is slightly different from the one used for classical metric spaces since the modulars generally fail the triangle inequality (see also Ref. [17]).

Definition 7.6.2 [6].   Let (X, ω) be a modular metric space and $M = V(G_\omega)$ be a nonempty subset of X_ω. M is said to be *finitely ε-chainable* (where $\varepsilon > 0$ is fixed) if and only if there exists $N \geq 1$ such that for any $a, b \in M$ with $(a, b) \in E(G_\omega)$ there is an N, ε-chain from a to b (that is, a finite set of vertices $x_0, x_1, \ldots, x_N \in V(G_\omega) = M$ such that $x_0 = a$, $x_N = b$, $(x_i, x_{i+1}) \in E(G_\omega)$ and $\omega_1(x_i, x_{i+1}) < \varepsilon$, for all $i = 0, 1, 2, \ldots, N - 1$).

We have the following result.

Theorem 7.6.2 [6]. *Let (X, ω) be a modular metric space. Suppose that ω is a convex regular modular metric which satisfies the Δ_2-type condition. Assume that $M = V(G_\omega)$ is a nonempty ω-complete and ω-bounded subset of X_ω which is finitely ε-chainable, for some fixed $\varepsilon > 0$. Suppose that the triple $(M, d_\omega^*, G_\omega)$ has Property 7.6.1. Let $T : M \rightarrow M$ be a $(\varepsilon, \alpha) - G_\omega$-uniformly local contraction map. Then T has a fixed point in M, the vertex set of the graph.*

Proof. Since M is finitely ε-chainable, there exists $N \geq 1$ such that for any $a, b \in M$ with $(a, b) \in E(G_\omega)$ there is a finite set of vertices $x_0, x_1, \ldots, x_N \in M$ such that $x_0 = a$, $x_N = b$, $(x_i, x_{i+1}) \in E(G_\omega)$ and $\omega_1(x_i, x_{i+1}) < \varepsilon$, for all $i = 0, 1, 2, \ldots, N - 1$. For any $x, y \in M$ define

$$\omega^*(x, y) = \inf \left\{ \sum_{i=0}^{i=N-1} \omega_1(x_i, x_{i+1}) \right\},$$

where the infimum is taken over all N, ε-chains $x_0, x_1, \ldots, x_N$ from x to y. Since M is finitely ε-chainable it follows that $\omega^*(x, y) < +\infty$, for any $x, y \in M$. Using the basic properties of ω, we get

$$\omega_N(x, y) \leq \omega^*(x, y),$$

for any $x, y \in M$ with $(x, y) \in E(G_\omega)$. Moreover, if $\omega_1(x, y) < \varepsilon$, then we have $\omega^*(x, y) \leq \omega_1(x, y)$, for any $x, y \in M$ with $(x, y) \in E(G_\omega)$. Fix $x \in M$. Set $z_0 = x$ and $z_1 = T(z_0)$ with $(z_0, z_1) \in E(G_\omega)$. Let $x_0, \cdots, x_N$ be an N, ε-chain from z_0 to z_1. Such an N, ε-chain exists since M is finitely ε-chainable. Since T is a $(\varepsilon, \alpha) - G_\omega$-uniformly local contraction map, there exists a constant $\alpha \in [0, 1)$ such that

$$\omega_1(T(x_i), T(x_{i+1})) \leq \alpha\, \omega_1(x_i, x_{i+1}) < \alpha\, \varepsilon < \varepsilon, \quad \text{for every } i.$$

Clearly, this implies that $T(x_0), T(x_1), \ldots, T(x_N)$ is an N, ε-chain from $T(z_0)$ to $T(z_1)$ and

$$\omega^*(z_1, z_2) \leq \alpha\, \omega^*(z_0, z_1),$$

where $z_2 = T(z_1)$. By induction, we construct the sequence $\{z_n\} \in M$ with $(z_n, z_{n+1}) \in E(G_\omega)$ such that

$$\omega^*(z_n, z_{n+1}) \leq \alpha\, \omega^*(z_{n-1}, z_n), \quad \text{for any } n \geq 1,$$

where $z_{n+1} = T(z_n)$. Obviously, we have

$$\omega^*(z_n, z_{n+1}) \leq \alpha^n\, \omega^*(z_0, z_1), \quad \text{for any } n \geq 1.$$

Since ω satisfies the Δ_2-type condition, there exists $C > 0$ such that

$$\omega_1(z_n, z_{n+1}) \leq C\, \omega_N(z_n, z_{n+1}) \leq C\, \omega^*(z_n, z_{n+1}) \leq C\, \alpha^n\, \omega^*(z_0, z_1), \quad \text{for any } n \geq 1.$$

Lemma 7.3.4 implies that $\{z_n\}$ is ω-Cauchy. Since M is ω-complete, then $\{z_n\}$ ω-converges to some $z \in M$. We claim that z is a fixed point of T. By Property 7.6.1, $(z_n, z) \in E(G_\omega)$ for any $n \geq 1$. Using the ideas developed above, we have

$$\omega^*(z_{n+1}, T(z)) \leq \alpha\, \omega^*(z_n, z), \quad \text{for any } n \geq 1.$$

Since

$$\omega_{N+1}(T(z), z) \leq \omega_1(z_{n+1}, z) + \omega_N(z_{n+1}, T(z))$$

and

$$\omega_N(z_{n+1}, T(z)) \leq \omega^*(z_{n+1}, T(z)) \leq \alpha\, \omega^*(z_n, z),$$

for any $n \geq 1$, this will imply that $\omega_{N+1}(T(z), z) = 0$. Indeed, there exists n_0 such that for any $n \geq n_0$ we have $\omega_1(z_n, z) < \varepsilon$. From the definition of ω^*, we have $\omega^*(z_n, z) < \omega_1(z_n, z)$, for $n \geq n_0$. Hence $\lim_{n \to +\infty} \omega^*(z_n, z) = 0$. Therefore $\omega_{N+1}(T(z), z) = 0$ holds. Since ω is regular, we get $z = T(z)$, i.e., z is a fixed point of T as claimed. $\square$

7.7. Monotone Pointwise Contractions in Banach Spaces with a Graph

The notion of asymptotic pointwise mappings was introduced in Refs. [53, 55, 57]. The use of the ultrapower technique was useful in proving some related fixed point results. Kirk and Xu [57] gave simple and elementary proofs for the existence of fixed point theorems for asymptotic pointwise mappings without the use of ultrapowers. In Ref. [44] most of these results were extended to metric spaces.

In this section, we investigate the fixed point theory of pointwise G-monotone contraction mappings. In particular, we will extend the main result of Ref. [57] to the case of G-monotone mappings.

Definition 7.7.1. Let $(X, \|.\|)$ be a Banach space. x is called a *weak-cluster point* of a sequence $\{x_n\}_{n\in\mathbb{N}}$ in X, if there exists a subsequence $\{x_{\varphi(n)}\}_{n\in\mathbb{N}}$ such that $\{x_{\varphi(n)}\}_{n\in\mathbb{N}}$ converges weakly to x.

Let $(X, \|.\|)$ be a Banach space and G a reflexive digraph defined on X. As Jachymski [46] did, we introduce the following property:

Property 7.7.1. We say that $E(G)$ has Property 7.7.1 if for any sequence $\{x_n\}_{n\in\mathbb{N}}$ in X such that $(x_n, x_{n+1}) \in E(G)$ for $n \in \mathbb{N}$ with x a weak-cluster point of $\{x_n\}_{n\in\mathbb{N}}$, then there exists a subsequence $\{x_{\varphi(n)}\}_{n\in\mathbb{N}}$ which converges weakly to x and $(x_{\varphi(n)}, x) \in E(G)$, for every $n \in \mathbb{N}$.

Note that if G is a reflexive transitive digraph defined on X, then Property 7.7.1 implies the following property:

Property 7.7.2. For any sequence $\{x_n\}_{n\geq 1}$ in X such that $(x_n, x_{n+1}) \in E(G)$ for $n \geq 1$ with x a weak-cluster point of $\{x_n\}_{n\geq 1}$, we have $(x_n, x) \in E(G)$, for every $n \in \mathbb{N}$.

Next we give the definition of G-monotone pointwise Lipschitzian mappings.

Definition 7.7.2. Let (X, d) be a metric space and G be a reflexive digraph defined on X. Let C be a nonempty subset of X. A mapping $T : C \to C$ is said to be *G-monotone pointwise Lipschitzian* if T is G-monotone and, for any $x \in C$, there exists $k(x) \in [0, +\infty)$ such that

$$d(T(x), T(y)) \leq k(x)\, d(x,y), \quad \text{for any } y \in C,$$

such that $(x,y) \in E(\widetilde{G})$. If $k(x) \in [0,1)$, for any $x \in C$, then T is said to be a *G-monotone pointwise contraction mapping*. If $k(x) \leq 1$, for any $x \in C$, then T is said to be a *G-monotone nonexpansive mapping*, i.e., for any $x, y \in C$, we have

$$d(T(x), T(y)) \leq d(x, y).$$

It is clear that the pointwise contractive concept was introduced to extend the contractive behavior in the Banach contraction principle.

Example 7.7.1. After Kirk [51], we consider K a bounded closed convex subset of the Hilbert space ℓ_2. Let $F : K \to K$ be such that F is continuously Fréchet differentiable on a convex open set containing K. Then F is a pointwise contraction on K if and only if $\|F'_x\| < 1$, for each $x \in K$, where F'_x denotes the Fréchet derivative of F at x. Next consider the metric space

$$M = \{0, 1\} \times K = \{(0, x), (1, x); x \in K\}.$$

The distance d on M is defined by

$$d((\varepsilon_1, x_1), (\varepsilon_2, x_2)) = |\varepsilon_1 - \varepsilon_2| + \|x_1 - x_2\|.$$

Let G be a graph with M as its vertex set and its edge set $E(G)$ defined by the following two conditions:

(a) $(0, x)$ and $(1, y)$ are not connected for any $x, y \in K$;

(b) (ε, x) and (ε, y) are connected if and only if $x \leq y$ (using the natural pointwise order in ℓ_2), for any $\varepsilon \in \{0, 1\}$ and $x, y \in K$.

Define the mapping $T : M \to M$ by

$$T((\varepsilon, x)) = (1 - \varepsilon, F(x)).$$

We choose F to be monotone on K, i.e., $x \leq y$ implies that $F(x) \leq F(y)$ for any $x, y \in K$. Then T is a G-monotone pointwise contraction on M. Indeed, any two vertices of G are connected if and only if they have the same first component. Next we notice

$$\begin{aligned} d(T((\varepsilon, x)), T((\varepsilon, y))) &= d((1 - \varepsilon, F(x)), (1 - \varepsilon, F(y))) \\ &= \|F(x) - F(y)\| \\ &\leq \alpha(x) \|x - y\| \\ &= \alpha(x) d((\varepsilon, x), (\varepsilon, y)), \end{aligned}$$

for any $\varepsilon \in \{0, 1\}$ and $x, y \in K$, such that x and y are comparable, i.e., $x \leq y$ or $y \leq x$, where $\alpha(x) \in [0, 1)$. Clearly we used the fact that F is a pointwise contraction on K. But T is not a pointwise contraction on M since

$$d(T((0, x)), T((1, x))) = d((0, x), (1, x)) = 1, \quad \text{for any } x \in K.$$

The fundamental fixed point result for pointwise contraction mappings is the following theorem.

Theorem 7.7.1 [53, 57]. *Let C be a weakly compact convex subset of a Banach space X. Let $T : C \to C$ be a pointwise contraction. Then T has a unique fixed point $z \in C$. Moreover, the orbit $\{T^n(x)\}_{n \in \mathbb{N}}$ converges to z, for each $x \in C$.*

Note that if T is a G-monotone pointwise Lipschitzian mapping, then it is not necessarily continuous by contrast to the case of pointwise Lipschitzian mappings. Since the main focus of this section is about the existence of the fixed points, we have the following result:

Theorem 7.7.2 [3, 10]. *Let (X, d) be a metric space and G be a reflexive digraph defined on X. Let C be a nonempty subset of X. Let $T : C \to C$ be a G-monotone pointwise contraction. If $a \in Fix(T)$, then for any $x \in X$ such that $(a, x) \in E(G)$, we have that $\{T^n(x)\}_{n \geq 1}$ converges to a. In particular, if a and b are two fixed points of T and $(a, b) \in E(G)$, then we must have $a = b$.*

Proof. Let $a \in Fix(T)$ and $(a, x) \in E(G)$. Since T is G-monotone, we have $(a, T^n(x)) \in E(G)$, for any $n \geq 1$. Using the definition of a pointwise contraction we get

$$d(T^n(x), a) \leq k(a) \, d(T^{n-1}(x), a) \leq k(a)^n \, d(x, a), \quad \text{for any } n \geq 1.$$

Since $k(a) < 1$, we conclude that $\{T^n(x)\}_{n \in \mathbb{N}}$ converges to a. Obviously if a and b are two fixed points of T and $(a, b) \in E(G)$, then we have $\{T^n(b)\}_{n \in \mathbb{N}} = \{b\}_{n \in \mathbb{N}}$ converges to a, which implies $a = b$. □

Remark 7.7.1. In both Banach and metric spaces [44, 53], the pointwise contraction mappings have at most one fixed point. But in the case of G-monotone pointwise contraction mappings, we may have more than one fixed point. Indeed, Jachymski [46] proved that G-contractions have a fixed point in each component. Since we do not assume the weak connectivity of the digraph G, we may have more than one component, which implies the possibility of having more than one fixed point.

The crucial part in dealing with pointwise contractions is the existence of the fixed point. Usually it takes more assumptions than the Banach contraction principle. Since Theorem 7.7.1 is for the linear case, we will assume that $(X, \|.\|)$ is a Banach space and G is a reflexive digraph defined on X. Moreover, the linear convexity of X is assumed to be compatible with the graph structure in the following sense:

Property 7.7.3. If $(x,y) \in E(G)$ and $(w,z) \in E(G)$, then

$$(\alpha x + (1-\alpha) w, \alpha y + (1-\alpha) z) \in E(G), \quad \text{for all } x,y,w,z \in X \text{ and } \alpha \in [0,1].$$

Note that the classical proof of Theorem 7.7.1 will not work in the setting of G-monotone mappings. The main difficulty encountered in this setting has to do with the fact that the mappings do not show good behavior on the entire sets. They exhibit good behavior only on connected points. For this reason, the investigation given here is based on a constructive approach initiated by Krasnoselskii [41].

Lemma 7.7.1 [3, 10]. *Let $(X, \|.\|)$ be a Banach space and G a reflexive digraph defined on X. Assume that $E(G)$ has Property 7.7.3. Let C be a nonempty convex subset of X. Let $T : C \to C$ be a G-monotone mapping. Fix $\lambda \in (0,1)$ and $x_1 \in C$. Consider the Krasnoselskii iteration sequence $\{x_n\}_{n \geq 1} \subset C$ defined by*

$$x_{n+1} = (1-\lambda)x_n + \lambda T(x_n), \quad n \geq 1. \tag{7.7.9}$$

(a) *If $(x_1, T(x_1)) \in E(G)$, then we have $(x_n, x_{n+1}) \in E(G)$, for any $n \geq 1$.*

(b) *If $(T(x_1), x_1) \in E(G)$, then we have $(x_{n+1}, x_n) \in E(G)$, for any $n \geq 1$.*

Proof. We prove only (a). The proof of (b) is similar to that of (a) and therefore it is omitted. As $(x_1, T(x_1)) \in E(G)$ and $(x_1, x_1) \in E(G)$, we have by Property 7.7.3

$$\left((1-\lambda)x_1 + \lambda x_1, (1-\lambda)x_1 + \lambda T(x_1)\right) \in E(G),$$

i.e., $(x_1, x_2) \in E(G)$. Now assume that $(x_{n-1}, x_n) \in E(G)$ for $n > 1$. Since T is G-monotone, we have $(T(x_{n-1}), T(x_n)) \in E(G)$. Using Property 7.7.3 again, we get

$$\left((1-\lambda)x_{n-1} + \lambda T(x_{n-1}), (1-\lambda)x_n + \lambda T(x_n)\right) \in E(G),$$

i.e., $(x_n, x_{n+1}) \in E(G)$. Hence, by induction, we have $(x_n, x_{n+1}) \in E(G)$ for all $n \geq 1$. $\qquad \square$

In order to show that the main property is satisfied by the sequence defined by (7.7.9), we need the following result, which may be found in Refs. [40, 41]. For the sake of competence, we will give its (graph version) proof here.

Lemma 7.7.2. *Let $(X, \|.\|)$ be a Banach space and G be a reflexive digraph defined on X. Assume that $E(G)$ has Property 7.7.3. Let C be a nonempty convex subset of X. Let $T : C \to C$ be a G-monotone nonexpansive mapping. Assume there exists $x_1 \in C$ such that $(x_1, T(x_1)) \in E(\widetilde{G})$. Consider the sequence $\{x_n\}_{n \geq 1}$ defined by (7.7.9). Then we have*

$$(1+n\lambda)\|T(x_i)-x_i\| \leq \|T(x_{i+n})-x_i\|+$$
$$(1-\lambda)^{-n}\Big(\|T(x_i)-x_i\|-\|T(x_{i+n})-x_{i+n}\|\Big), \quad (7.7.10)$$

for any $i,n \in \mathbb{N}$.

Proof. Without loss of any generality, we may assume $(x_1,T(x_1)) \in E(G)$. We will prove this inequality by induction on $n \in \mathbb{N}$. The inequality is obvious when $n = 0$. Fix $n \geq 1$ and assume the inequality holds for any $i \in \mathbb{N}$. In particular, we have

$$(1+n\lambda)\|T(x_{i+1})-x_{i+1}\| \leq \|T(x_{i+1+n})-x_{i+1}\|+(1-\lambda)^{-n}\|T(x_{i+1})-x_{i+1}\|$$
$$-(1-\lambda)^{-n}\|T(x_{i+1+n})-x_{i+1+n}\|.$$

Lemma 7.7.1 implies that $(x_m,x_{m+1}) \in E(G)$ for any $m \geq 1$. Since T is a G-monotone nonexpansive mapping, we get

$$\|T(x_{m+1})-T(x_m)\| \leq \|x_{m+1}-x_m\|$$

and $(T(x_m),T(x_{m+1})) \in E(G)$, for any $m \geq 1$. Since

$$\|T(x_{i+1+n})-x_{i+1}\| \leq (1-\lambda)\|T(x_{i+n+1})-x_i\|+\lambda\|T(x_{i+n+1})-T(x_i)\|$$
$$\leq (1-\lambda)\|T(x_{i+n+1})-x_i\|+\lambda\sum_{k=0}^{n}\|T(x_{i+k+1})-T(x_{i+k})\|$$
$$\leq (1-\lambda)\|T(x_{i+n+1})-x_i\|+\lambda\sum_{k=0}^{n}\|x_{i+k+1}-x_{i+k}\|,$$

we get

$$(1+n\lambda)\|T(x_{i+1})-x_{i+1}\| \leq (1-\lambda)\|T(x_{i+n+1})-x_i\|+\lambda\sum_{k=0}^{n}\|x_{i+k+1}-x_{i+k}\|+$$
$$(1-\lambda)^{-n}\Big(\|T(x_{i+1})-x_{i+1}\|-\|T(x_{i+1+n})-x_{i+1+n}\|\Big),$$

which implies

$$\|T(x_{i+n+1})-x_i\| \geq \frac{(1+n\lambda)}{(1-\lambda)}\|T(x_{i+1})-x_{i+1}\|-\frac{\lambda}{(1-\lambda)}\sum_{k=0}^{n}\|x_{i+k+1}-x_{i+k}\|-$$
$$(1-\lambda)^{-n-1}\Big(\|T(x_{i+1})-x_{i+1}\|-\|T(x_{i+1+n})-x_{i+1+n}\|\Big).$$

Note that $(\|T(x_m)-x_m\|)_{m\geq 1}$ is a decreasing sequence. Indeed, we have

$$\|x_{m+1}-x_m\| = \lambda\|T(x_m)-x_m\|, \ m \geq 1.$$

So to show that $(\|T(x_m)-x_m\|)_{m\geq 1}$ is decreasing, we only need to prove that $(\|x_{m+1}-x_m\|)_{m\geq 1}$ is decreasing, which is true since

$$\|x_{m+2} - x_{m+1}\| \leq (1-\lambda)\|x_{m+1} - x_m\| + \lambda\|T(x_{m+1}) - T(x_m)\|$$
$$\leq (1-\lambda)\|x_{m+1} - x_m\| + \lambda\|x_{m+1} - x_m\| = \|x_{m+1} - x_m\|,$$

for any $m \geq 1$. Using this fact and $1 + n\lambda \leq (1-\lambda)^{-n}$, we get

$$\|T(x_{i+n+1}) - x_i\| \geq (1-\lambda)^{-n-1}\left[\|T(x_{i+n+1}) - x_{i+n+1}\| - \|T(x_{i+1}) - x_{i+1}\|\right]$$
$$+ \frac{(1+n\lambda)}{(1-\lambda)}\|T(x_{i+1}) - x_{i+1}\| - \frac{\lambda^2(n+1)}{(1-\lambda)}\|T(x_i) - x_i\|$$
$$= (1-\lambda)^{-n-1}\left[\|T(x_{i+n+1}) - x_{i+n+1}\| - \|T(x_i) - x_i\|\right]$$
$$+ \left(\frac{(1+n\lambda)}{(1-\lambda)} - (1-\lambda)^{-n-1}\right)\|T(x_{i+1}) - x_{i+1}\|$$
$$+ \left((1-\lambda)^{-n-1} - \frac{\lambda^2(n+1)}{(1-\lambda)}\right)\|T(x_i) - x_i\|$$
$$\geq (1-\lambda)^{-n-1}\left[\|T(x_{i+n+1}) - x_{i+n+1}\| - \|T(x_i) - x_i\|\right]$$
$$+ \left(\frac{(1+n\lambda)}{(1-\lambda)} - (1-\lambda)^{-n-1}\right)\|T(x_i) - x_i\|$$
$$+ \left((1-\lambda)^{-n-1} - \frac{\lambda^2(n+1)}{(1-\lambda)}\right)\|T(x_i) - x_i\|$$
$$= (1-\lambda)^{-n-1}\left[\|T(x_{i+n+1}) - x_{i+n+1}\| - \|T(x_i) - x_i\|\right]$$
$$+ \left(1 + (n+1)\lambda\right)\|T(x_i) - x_i\|.$$

This is our inequality when we take $n+1$ instead of n. Therefore by induction the inequality 7.7.10 is true for any $i, n \in \mathbb{N}$. □

As a direct consequence of Lemma 7.7.2, we get the following result:

Theorem 7.7.3 [3, 10]. *Let $(X, \|.\|)$ be a Banach space and G a reflexive digraph defined on X. Assume that $E(G)$ has Property 7.7.3. Let C be a bounded nonempty convex subset of X. Let $T : C \to C$ be a G-monotone nonexpansive mapping. Assume there exists $x_1 \in C$ such that $(x_1, T(x_1)) \in E(\widetilde{G})$. Consider the sequence $\{x_n\}_{n\geq 1}$ defined by (7.7.9). Then $\lim_{n\to+\infty} \|x_n - T(x_n)\| = 0$.*

Proof. Using Lemma 7.7.2, we know that the inequality

$$(1+n\lambda)\|T(x_i) - x_i\| \leq \|T(x_{i+n}) - x_i\| +$$
$$(1-\lambda)^{-n}(\|T(x_i) - x_i\| - \|T(x_{i+n}) - x_{i+n}\|)$$

holds for any $i, n \in \mathbb{N}$. Since $(\|x_n - T(x_n)\|)_{n\geq 1}$ is decreasing, we set $\lim_{n\to+\infty} \|x_n - T(x_n)\| = R$. Then if we let $i \to +\infty$ in the above inequality, we obtain

$$(1+n\lambda)R \leq \delta(C),$$

for any $n \in \mathbb{N}$, where $\delta(C) = \sup\{\|x - y\|; x, y \in C\} < +\infty$. Obviously this will imply a contradiction if we assume $R \neq 0$. Therefore we must have

$$\lim_{n \to +\infty} \|x_n - T(x_n)\| = 0.$$

$\square$

A sequence such as $\{x_n\}_{n \geq 1}$ is known as an approximate fixed point sequence of T. Assume that there exists $x_1 \in C$ such that $(x_1, T(x_1)) \in E(G)$. Let x be a weak-cluster point of $\{x_n\}$. Since $E(G)$ has Property 7.7.1 and $(x_n, x_{n+1}) \in E(G)$, for any $n \geq 1$, there exists a subsequence $\{x_{\varphi(n)}\}$ of $\{x_n\}$ such that $\{x_{\varphi(n)}\}$ weakly converges to x and $(x_{\varphi(n)}, x) \in E(G)$, for any $n \geq 1$. If we assume that T is a G-monotone pointwise contraction, then T is a G-monotone nonexpansive mapping. Moreover, we have

$$\|T(x_{\varphi(n)}) - T(x)\| \leq k(x) \|x_{\varphi(n)} - x\|, \quad n \geq 1. \tag{7.7.11}$$

Assume that X satisfies the large Opial property, which states that for any sequence $\{y_n\} \subset X$ which weakly converges to y, we have

$$\liminf_{n \to +\infty} \|y_n - y\| \leq \liminf_{n \to +\infty} \|y_n - z\|, \quad \text{for any } z \in X.$$

Using the inequality (7.7.11), we get

$$\liminf_{n \to +\infty} \|T(x_{\varphi(n)}) - T(x)\| \leq k(x) \liminf_{n \to +\infty} \|x_{\varphi(n)} - x\|.$$

Using Theorem 7.7.3, we conclude that

$$\liminf_{n \to +\infty} \|x_{\varphi(n)} - T(x)\| \leq k(x) \liminf_{n \to +\infty} \|x_{\varphi(n)} - x\|.$$

The large Opial property implies that

$$\liminf_{n \to +\infty} \|x_{\varphi(n)} - x\| \leq \liminf_{n \to +\infty} \|x_{\varphi(n)} - T(x)\| \leq k(x) \liminf_{n \to +\infty} \|x_{\varphi(n)} - x\|.$$

Since $k(x) < 1$, we obtain $\liminf_{n \to +\infty} \|x_{\varphi(n)} - x\| = 0$. Combined with the conclusion of Theorem 7.7.3, we get $T(x) = x$, i.e., x is a fixed point of T. In other words, we proved the following result:

Theorem 7.7.4 [3, 10]. *Let $(X, \|.\|)$ be a Banach space and G be a reflexive digraph defined on X. Assume that $E(G)$ has Properties 7.7.1 and 7.7.3. Assume X satisfies the large Opial property. Let C be a weakly compact nonempty convex subset of X. Let $T : C \to C$ be a G-monotone pointwise contraction. Assume there exists $x_1 \in C$ such that $(x_1, T(x_1)) \in E(G)$. Then T has a fixed point.*

We do not know whether the conclusion of Theorem 7.7.4 holds if X does not satisfy the large Opial condition. However, if the digraph G is transitive, we may get the conclusion of Theorem 7.7.4 without this condition. Indeed, instead of assuming Property 7.7.1, we will assume that G-intervals are closed. Using the convexity properties of G, we conclude that G-intervals are convex and closed. Therefore they are also weakly closed. Under these assumptions, we have the following result, which is the analog to Theorem 7.7.1.

Theorem 7.7.5 [3, 10]. *Let $(X, \|.\|)$ be a Banach space and G be a reflexive transitive digraph defined on X. Assume that $E(G)$ has Property 7.7.3 and G-intervals are closed and convex. Let C be a weakly compact nonempty convex subset of X. Let $T : C \to C$ be a G-monotone pointwise contraction. Assume there exists $x_1 \in C$ such that $(x_1, T(x_1)) \in E(\widetilde{G})$. Then T has a fixed point.*

Proof. Without loss of any generality, we may assume that $(x_1, T(x_1)) \in E(G)$. Let $\{x_n\}$ be the sequence generated by x_1 and defined by (7.7.9). Since C is weakly compact, then $\{x_n\}$ has a weak-cluster point $z \in C$. Since G is transitive, then we have $(x_n, z) \in E(G)$, for any $n \geq 1$. Since T is G-monotone, we get $(T(x_n), T(z)) \in E(G)$, for any $n \geq 1$. Using the conclusion of Theorem 7.7.3, we know that z is also a weak-cluster point of $\{T(x_n)\}$. Since the G-intervals are weakly closed, we conclude that $(z, T(z)) \in E(G)$. Consider the set

$$C_z = [z, \to) \cap C = \{x \in C : (z, x) \in E(G)\}.$$

Then C_z is a nonempty closed convex subset of C. Hence C_z is weakly compact. Let $x \in C_z$, then we have $(T(z), T(x)) \in E(G)$ since T is G-monotone. Using the transitivity of G, we get $(z, T(x)) \in E(G)$, i.e., $T(x) \in C_z$. Next, we consider the type function $\tau : C_z \to [0, +\infty)$ defined by

$$\tau(x) = \limsup_{n \to +\infty} \|x_n - x\|.$$

It is obvious that τ is convex and continuous. Since C_z is weakly compact and convex, we conclude that there exists $z \in C_z$ such that $\tau(z) = \inf\{\tau(x); x \in C_z\}$. Since $(x_n, z) \in E(G)$, by the transitivity of G, we get $(x_n, z) \in E(G)$, for any $n \geq 1$. Since T is a G-monotone pointwise contraction, we get

$$\|T(x_n) - T(z)\| \leq k(z) \|x_n - z\|, \quad n = 1, 2, \ldots.$$

Hence

$$\limsup_{n \to +\infty} \|T(x_n) - T(z)\| \leq k(z) \limsup_{n \to +\infty} \|x_n - z\|.$$

Using the conclusion of Theorem 7.7.3, we get

$$\limsup_{n\to+\infty}\|x_n - T(z)\| = \limsup_{n\to+\infty}\|T(x_n) - T(z)\| \leq k(z)\limsup_{n\to+\infty}\|x_n - z\|,$$

i.e., $\tau(T(z)) \leq k(z)\tau(z)$. Since $\tau(z) \leq \tau(T(z))$ and $k(z) < 1$, we get $\tau(z) = 0$, i.e., $\limsup_{n\to+\infty}\|x_n - z\| = 0$. So $\{x_n\}$ converges to z. Since $\tau(T(z)) \leq k(z)\tau(z)$, we get $\tau(T(z)) = 0$, which also implies that $\{x_n\}$ converges to $T(z)$. Therefore we must have $T(z) = z$, i.e., z is a fixed point of T. $\qquad\square$

7.8. Monotone Ćirić Quasi-Contraction Mappings

In this section, we obtain sufficient conditions for the existence of fixed points for monotone quasi-contraction mappings in metric and modular metric spaces endowed with a graph [9]. This is the extension of Ran and Reurings [73] and Jachymski [46] fixed point theorems for monotone contraction mappings in partially ordered metric spaces and in metric spaces endowed with a graph to the case of quasi-contraction mappings introduced by Ćirić.

7.8.1. Monotone Quasi-Contraction Mappings in Metric Spaces with a Graph

As a generalization of the Banach contraction principle, Ćirić [28] introduced the concept of quasi-contraction mappings. In this subsection, we investigate monotone mappings which are quasi-contraction mappings. Throughout this subsection we assume that (X,d) is a metric space and G is a reflexive transitive digraph defined on X. Moreover, we assume that G-intervals are closed and $E(G)$ has Property 7.8.1.

Property 7.8.1. We say that $E(G)$ (or the triple (X,d,G)) has Property 7.8.1 if for any $\{x_n\}_{n\geq 1}$ in X, if $x_n \to x$ and $(x_n, x_{n+1}) \in E(G)$, for $n \in \mathbb{N}$, then there is a subsequence $\{x_{k_n}\}_{n\in\mathbb{N}}$ with $(x_{k_n}, x) \in E(G)$, for every $n \in \mathbb{N}$.

Note that if G is a reflexive transitive digraph defined on X, then Property 7.8.1 implies the following property:

Property 7.8.2. For any $\{x_n\}_{n\geq 1}$ in X, if $x_n \to x$ and $(x_n, x_{n+1}) \in E(G)$, for $n \in \mathbb{N}$, then $(x_n, x) \in E(G)$, for every $n \in \mathbb{N}$.

Definition 7.8.1 [9]. Let C be a nonempty subset of X. A mapping $T : C \to C$ is called a *G-monotone quasi-contraction* if T is G-monotone and there exists $k < 1$ such that for any $x,y \in C$ with $(x,y) \in E(G)$, we have

$$d(T(x), T(y)) \leq k \max \Big(d(x,y); d(x, T(x)); d(y, T(y)); d(x, T(y)); d(y, T(x)) \Big).$$

In the sequel, we prove the existence of a fixed point theorem for such mappings. First, let T and C be as in Definition 7.8.1. For any $x \in C$, define the orbit $\mathcal{O}(x) = \{x, T(x), T^2(x), \cdots\}$, and its diameter by

$$\delta(x) = \sup \{ d(T^n(x), T^m(x)) : n, m \in \mathbb{N} \}.$$

The following technical lemma is crucial to prove the main result of this subsection.

Lemma 7.8.1 [9]. *Let (X, d) and G be as above. Let C be a nonempty subset of X and $T : C \to C$ be a G-monotone quasi-contraction mapping. Let $x \in C$ be such that $(x, T(x)) \in E(G)$ and $\delta(x) < +\infty$. Then*

$$\delta(T^n(x)) \leq k^n \delta(x), \quad \text{for any } n \geq 1,$$

where $k < 1$ is the constant associated with the G-monotone quasi-contraction definition of T. Therefore

$$d(T^n(x), T^{n+m}(x)) \leq k^n \delta(x), \quad \text{for any } n, m \in \mathbb{N}.$$

Proof. Since T is G-monotone, then $(T^n(x), T^{n+1}(x)) \in E(G)$, for any $n \in \mathbb{N}$. By transitivity of the graph G, we have $(T^n(x), T^m(x)) \in E(G)$, for any $n, m \in \mathbb{N}$. Hence

$$\begin{aligned}
d(T^n(x), T^m(x)) \leq \ & k \max \Big(d(T^{n-1}(x), T^{m-1}(x)); d(T^{n-1}(x), T^n(x)); \\
& d(T^{m-1}(x), T^m(x)); d(T^{n-1}(x), T^m(x)); d(T^n(x), T^{m-1}(x)) \Big)
\end{aligned}$$

for any $n, m \geq 1$. This obviously implies that

$$\delta(T^n(x)) \leq k \, \delta(T^{n-1}(x)), \quad n \geq 1.$$

Hence

$$\delta(T^n(x)) \leq k^n \, \delta(x), \quad n \geq 1.$$

This implies that

$$d(T^n(x), T^{n+m}(x)) \leq \delta(T^n(x)) \leq k^n \, \delta(x), \quad \text{for any } n, m \in \mathbb{N}.$$

$\square$

Using Lemma 7.8.1, we have the following result.

Theorem 7.8.1 [9]. *Let (X, d) and G be as above. Assume that (X, d) is complete. Let C be a closed nonempty subset of X and $T : C \to C$ be a G-monotone quasi-contraction mapping. Let $x \in C$ be such that $(x, T(x)) \in E(G)$ and $\delta(x) < +\infty$. Then*

(a) $\{T^n(x)\}$ *converges to* $z \in C$, *which is a fixed point of* T, *and* $(x,z) \in E(G)$. *Moreover, we have*

$$d(T^n(x),z) \le k^n \, \delta(x), \quad n \ge 1.$$

(b) *If* w *is a fixed point of* T *such that* $(x,w) \in E(G)$, *then* $w = z$.

Proof. Let us prove (a). Lemma 7.8.1 implies that $\{T^n(x)\}$ is Cauchy. Since X is complete and C is closed, then there exists $z \in C$ such that $\{T^n(x)\}$ converges to z. Since

$$d(T^n(x),T^{n+m}(x)) \le k^n \, \delta(x), \quad n,m \in \mathbb{N},$$

we let $m \to +\infty$ to get

$$d(T^n(x),z) \le k^n \, \delta(x), \quad n \ge 1.$$

Since T is G-monotone, we get $(T^n(x),T^{n+1}(x)) \in E(G)$, for any $n \ge 1$. By Property 7.8.2, we conclude that $(T^n(x),z) \in E(G)$, for any $n \ge 1$. In particular, we have $(x,z) \in E(G)$. In order to show that z is a fixed point of T, note that we have

$$d(T^n(x),T(z)) \le k \max \Big(d(T^{n-1}(x),z); d(T^{n-1}(x),T^n(x));$$
$$d(z,T(z)); d(T^{n-1}(x),T(z)); d(T^n(x),z) \Big),$$

for any $n \ge 1$. If we let $n \to +\infty$, we get $d(z,T(z)) \le kd(z,T(z))$, which forces $d(z,T(z)) = 0$ since $k < 1$. Therefore we have $T(z) = z$.

Next, we show (b). Let $z^* \in C$ be a fixed point of T such that $(x,z^*) \in E(G)$. Then we have

$$d(T^n(x),z^*) \le k \max \Big(d(T^{n-1}(x),z^*); d(T^{n-1}(x),T^n(x)); d(T^n(x),z^*) \Big),$$

for any $n \ge 2$. If we let $n \to +\infty$, we get

$$d(z,z^*) = \limsup_{n \to +\infty} d(T^n(x),z^*) \le k \limsup_{n \to +\infty} d(T^n(x),z^*) = k \, d(z,z^*).$$

Since $k < 1$, we get $d(z,z^*) = 0$, i.e., $z = z^*$. $\square$

In the next subsection, we discuss the validity of Theorem 7.8.1 in modular metric spaces. This is a very important class of spaces since they are similar to metric spaces in their structure but without the triangle inequality and offer a wide range of applications.

7.8.2. Monotone Quasi-Contraction Mappings in Modular Metric Spaces with a Graph

Let us start this subsection with the necessary background. For detailed information, see Section 7.3.4.

Definition 7.8.2. Let X_ω be a modular metric space. ω is said to satisfy *the Fatou property* if and only if, for any sequence $\{x_n\}_{n \in \mathbb{N}}$ in X_ω ω-convergent to x, we have

$$\omega_1(x,y) \leq \liminf_{n \to +\infty} \omega_1(x_n, y), \quad \text{for any } y \in X_\omega.$$

Definition 7.8.3 [9]. Let (X, ω) be a modular metric space and G be a reflexive digraph defined on X. Let C be a nonempty subset of X. The mapping $T : C \to C$ is said to be *G-monotone ω-quasi-contraction* if T is G-monotone and there exists $k < 1$ such that for any $x, y \in C$ with $(x,y) \in E(G)$, we have

$$\omega_1(T(x), T(y)) \leq k \, \max\Big(\omega_1(x,y); \omega_1(x, T(x)); \omega_1(y, T(y)); \omega_1(x, T(y)); \omega_1(y, T(x)) \Big).$$

Let (X, ω) be a modular metric space and G be a reflexive digraph defined on X. The modular version of Property 7.8.1 is the following property:

Property 7.8.3. We say that $E(G)$ has Property 7.8.3 if for any $\{x_n\}_{n \in \mathbb{N}}$ in X, if x_n ω-converges to x and $(x_n, x_{n+1}) \in E(G)$, for $n \in \mathbb{N}$, then there is a subsequence $\{x_{k_n}\}_{n \in \mathbb{N}}$ with $(x_{k_n}, x) \in E(G)$, for $n \in \mathbb{N}$.

Note that if G is a reflexive transitive digraph defined on X, then Property 7.8.3 implies the following property:

Property 7.8.4. For any $\{x_n\}_{n \in \mathbb{N}}$ in X, if x_n ω-converges to x and $(x_n, x_{n+1}) \in E(G)$, for $n \in \mathbb{N}$, then $(x_n, x) \in E(G)$, for every $n \in \mathbb{N}$.

Throughout this subsection, we assume that (X, ω) is a modular metric space, G is a reflexive transitive digraph defined on X and $E(G)$ has Property 7.8.3. We also assume that ω is regular and satisfies the Fatou property. In the sequel, we prove an analog to Theorem 7.8.1 in modular metric spaces. For any $x \in C$, define the orbit $\mathscr{O}(x) = \{x, T(x), T^2(x), \ldots\}$, and its diameter by

$$\delta_\omega(x) = \sup\{\omega_1(T^n(x), T^m(x)) : n, m \in \mathbb{N}\}.$$

The following technical lemma is crucial to prove the main result of this subsection. It is the modular version of Lemma 7.8.1. Its proof is omitted.

Lemma 7.8.2 [9]. *Let (X, ω) and G be as above. Let C be a nonempty subset of X and $T : C \to C$ be a G-monotone ω-quasi-contraction mapping. Let $x \in C$ be such that $(x, T(x)) \in E(G)$ and $\delta_\omega(x) < +\infty$. Then*

$$\delta_\omega(T^n(x)) \le k^n \delta_\omega(x), \quad \text{for any } n \ge 1,$$

where $k < 1$ is the constant associated with the G-monotone ω-quasi-contraction definition of T. Therefore

$$\omega_1(T^n(x), T^{n+m}(x)) \le k^n \, \delta_\omega(x), \quad \text{for any } n, m \in \mathbb{N}.$$

Using Lemma 7.8.2, we prove the main result of this subsection.

Theorem 7.8.2 [9]. *Let (X, ω) and G be as above. Let C be a nonempty subset of X which is ω-complete. Let $T : C \to C$ be a G-monotone ω-quasi-contraction mapping. Let $x \in C$ be such that $(x, T(x)) \in E(G)$ and $\delta_\omega(x) < +\infty$. Then*

(a) *$\{T^n(x)\}$ ω-converges to $z \in C$ which is a fixed point of T and $(x, z) \in E(G)$, provided $\omega_1(z, T(z)) < +\infty$ and $\omega_1(x, T(z)) < +\infty$. Moreover, we have*

$$\omega_1(T^n(x), z) \le k^n \, \delta_\omega(x), \quad \text{for any } n \in \mathbb{N}.$$

(b) *If w is a fixed point of T such that $(x, w) \in E(G)$ and $\omega_1(T^n(x), w) < +\infty$, for any $n \in \mathbb{N}$, then $z = w$.*

Proof. Let us prove (a). Lemma 7.8.2 implies that $\{T^n(x)\}$ is ω-Cauchy. Since C is ω-complete, then there exists $z \in C$ such that $\{T^n(x)\}$ ω-converges to z. Since

$$\omega_1(T^n(x), T^{n+m}(x)) \le k^n \, \delta_\omega(x), \quad \text{for any } n, m \in \mathbb{N},$$

the Fatou property (once we let $m \to +\infty$) will imply

$$\omega_1(T^n(x), z) \le k^n \, \delta_\omega(x), \quad n \in \mathbb{N}.$$

Since T is G-monotone, we have $(T^n(x), T^{n+1}(x)) \in E(G)$, for any $n \in \mathbb{N}$. By Property 7.8.4, we get $(T^n(x), z) \in E(G)$, for any $n \ge \mathbb{N}$. In particular, we have $(x, z) \in E(G)$. Next, we assume $\omega_1(z, T(z)) < +\infty$ and $\omega_1(x, T(z)) < +\infty$. Let us prove that z is a fixed point of T. By induction, we have $\omega_1(T^n(x), T(z)) < +\infty$, and

$$(\diamond) \quad \omega_1(T^n(x), T(z)) \le k \max \Big(\omega_1(T^{n-1}(x), z); \omega_1(T^{n-1}(x), T^n(x)); $$
$$\omega_1(T(z), z); \omega_1(T^{n-1}(x), T(z)); \omega_1(T^n(x), z) \Big)$$

for any $n \geq 1$. Consider $r(y) = \limsup_{n \to +\infty} \omega_1(T^n(x), y)$, for $y \in C$. From $(\diamond)$, we get

$$
\begin{aligned}
\omega_1(T^n(x), T(z)) &\leq k \max\left(k^{n-1} \delta_\omega(x); \omega_1(T(z), z); \omega_1(T^{n-1}(x), T(z)); k^n \delta_\omega(x)\right) \\
&= k \max\left(k^{n-1} \delta_\omega(x); \omega_1(T(z), z); \omega_1(T^{n-1}(x), T(z))\right) \\
&\leq k^n \delta_\omega(x) + k \omega_1(T(z), z) + k \omega_1(T^{n-1}(x), T(z)) \\
&\leq \delta_\omega(x) + \omega_1(T(z), z) + k \omega_1(T^{n-1}(x), T(z))
\end{aligned}
$$

for any $n \in \mathbb{N}$. By induction, we obtain

$$
\omega_1(T^n(x), T(z)) \leq \frac{1}{1-k}\left(\delta_\omega(x) + \omega_1(T(z), z)\right) + k^n \omega_1(x, T(z)),
$$

for any $n \in \mathbb{N}$, which implies

$$
r(T(z)) \leq \frac{1}{1-k}\left(\delta_\omega(x) + \omega_1(T(z), z)\right) < +\infty.
$$

So if we let $n \to +\infty$ in the inequality

$$
\omega_1(T^n(x), T(z)) \leq k \max\left(k^{n-1} \delta_\omega(x); \omega_1(T(z), z); \omega_1(T^{n-1}(x), T(z))\right),
$$

we get

$$
r(T(z)) \leq k \max\left(\omega_1(z, T(z)), r(T(z))\right).
$$

Since ω satisfies the Fatou property, we get $\omega_1(z, T(z)) \leq r(T(z))$, which implies

$$
r(T(z)) \leq k \max\left(\omega_1(z, T(z)), r(T(z))\right) = k\, r(T(z)).
$$

Since $k < 1$, we conclude $r(T(z)) = 0$, which implies $\omega_1(z, T(z)) = 0$. Since ω is regular, we get $T(z) = z$.

Next, we show (b). Let $w \in C$ be a fixed point of T such that $(x, w) \in E(G)$ and $\omega_1(T^n(x), w) < +\infty$, for any $n \in \mathbb{N}$. Then, by induction, we get

$$
\omega_1(T^n(x), w) \leq k \max\left(\omega_1(T^{n-1}(x), w); \omega_1(T^{n-1}(x), T^n(x)); \omega_1(T^n(x), w)\right),
$$

for any $n \geq 2$. Note that if

$$
\max\left(\omega_1(T^{n-1}(x), w); \omega_1(T^{n-1}(x), T^n(x)); \omega_1(T^n(x), w)\right) = \omega_1(T^n(x), w),
$$

for some $n \in \mathbb{N}$, we get $\omega_1(T^n(x), w) \leq k \omega_1(T^n(x), w)$. Since $k < 1$, we get $\omega_1(T^n(x), w) = 0$. So $T^n(x) = w$, which implies $T^{n+m}(x) = w$, for any $m \geq 0$, since w is a fixed point of T. This clearly will force $z = w$. Assume otherwise that

$$
\max\left(\omega_1(T^{n-1}(x), w); \omega_1(T^{n-1}(x), T^n(x)); \omega_1(T^n(x), w)\right) \neq \omega_1(T^n(x), w),
$$

for any $n \geq 2$. In this case, we get

$$\omega_1(T^n(x),w) \leq k \max\left(\omega_1(T^{n-1}(x),w);k^{n-1}\delta_\omega(x)\right), \quad \text{for any } n \geq 2.$$

Hence

$$\omega_1(T^n(x),w) \leq k\,\omega_1(T^{n-1}(x),w)+k^n\,\delta_\omega(x) \leq k\,\omega_1(T^{n-1}(x),w)+\delta_\omega(x),$$

which implies by induction

$$\omega_1(T^n(x),w) \leq k^n\,\omega_1(x,w)+\frac{1}{1-k}\,\delta_\omega(x), \quad \text{for any } n \in \mathbb{N}.$$

In particular, we have $\limsup\limits_{n\to+\infty}\omega_1(T^n(x),w) < +\infty$. Using the inequality

$$\omega_1(T^n(x),w) \leq k \max\left(\omega_1(T^{n-1}(x),w);k^{n-1}\delta_\omega(x)\right), \quad \text{for any } n \geq 2,$$

we obtain

$$\limsup\limits_{n\to+\infty}\omega_1(T^n(x),w) \leq k \limsup\limits_{n\to+\infty}\omega_1(T^n(x),w).$$

Since $k < 1$, we get $\limsup\limits_{n\to+\infty}\omega_1(T^n(x),w) = 0$, i.e., $\{T^n(x)\}$ converges to w. The uniqueness of the ω-limit implies that $z = w$. Indeed, we have

$$\omega_2(z,w) \leq \omega_1(T^n(x),z)+\omega_1(T^n(x),w), \quad n \in \mathbb{N}.$$

If we let $n \to +\infty$, we get $\omega_2(z,w) = 0$. Since ω is regular, we get $z = w$. □

Note that under the assumptions of Theorem 7.8.2, if w is another fixed point of T such that $(w,z) \in E(G)$ and $\omega_1(z,w) < +\infty$, then we have

$$\omega_1(z,w) = \omega_1(T(z),T(w)) \leq k\,\omega_1(z,w),$$

which implies $z = w$, since $k < 1$.

7.9. Monotone Nonexpansive Mappings in Banach Spaces with a Graph

In this section, we study the case of nonexpansive mappings defined in Banach spaces endowed with a graph. The fixed point theory for such mappings is rich and varied. It finds many applications in nonlinear functional analysis [23].

Throughout, we assume that $(X,\|.\|)$ is a Banach space and τ is a Hausdorff vector space topology on X which is weaker than the norm topology. Let C be a nonempty, convex and bounded subset of X not reduced to one point. Let G be a directed graph such that $V(G) = C$ and $E(G) \supseteq \Delta$. Assume that G-intervals are convex. Recall that a G-interval is any of the subsets $[a,\to) = \{x \in C; (a,x) \in E(G)\}$ and $(\leftarrow,b] = \{x \in C; (x,b) \in E(G)\}$, for any $a,b \in C$.

Definition 7.9.1 [7]. Let C be a nonempty subset of X. A mapping $T : C \to C$ is called *G-monotone nonexpansive* if T is G-monotone and

$$\|T(x) - T(y)\| \le \|x - y\|, \quad \text{whenever } (x, y) \in E(G),$$

for any $x, y \in C$.

Remark 7.9.1. For examples of metric spaces endowed with a graph and G-monotone mappings which are Lipschitzian with respect to the graph, we refer the reader to the examples found in Ref. [19].

Let $T : C \to C$ be a G-monotone nonexpansive mapping. Fix $\lambda \in (0, 1)$. Let $x_0 \in C$ be such that $(x_0, T(x_0)) \in E(G)$. Consider the Krasnoselskii iteration sequence $\{x_n\}_{n \in \mathbb{N}} \subset C$ defined by (7.7.9), i.e.,

$$x_{n+1} = \lambda x_n + (1 - \lambda) T(x_n), \quad \text{for any } n \in \mathbb{N}.$$

By induction, we can easily see that the following hold for any $n \in \mathbb{N}$:

(a) (x_n, x_{n+1}), $(x_n, T(x_n))$ and $(T(x_n), T(x_{n+1}))$ are in $E(G)$,

(b) $\|T(x_{n+1}) - T(x_n)\| \le \|x_{n+1} - x_n\|$.

From Lemma 7.7.2, we know that

$$(1 + n\lambda)\|T(x_i) - x_i\| \le \|T(x_{i+n}) - x_i\| + (1 - \lambda)^{-n}\Big(\|T(x_i) - x_i\| - \|T(x_{i+n}) - x_{i+n}\|\Big),$$

for any $i, n \in \mathbb{N}$. Theorem 7.7.3 will imply that

$$\lim_{n \to +\infty} \|x_n - T(x_n)\| = 0,$$

i.e., $\{x_n\}$ is an approximate fixed point sequence of T.

Remark 7.9.2. We may let λ change with $n \in \mathbb{N}$. In this case, the sequence $\{x_n\}$ is defined by

$$x_{n+1} = \lambda_n x_n + (1 - \lambda_n) T(x_n), \quad n = 0, 1, \dots.$$

Under suitable assumptions on the sequence $\{\lambda_n\}$, we will have the same conclusions on $\{x_n\}$.

Before we state the main result of this section, let us recall the definition of the τ-Opial condition.

Definition 7.9.2. X is said to satisfy *the τ-Opial condition* if whenever any sequence $\{y_n\}$ in X τ-converges to y, we have

$$\liminf_{n\to+\infty}\|y_n - y\| < \liminf_{n\to+\infty}\|y_n - z\|,$$

for any $z \in X$ such that $z \neq y$.

The analog to Property 7.8.1 stated in terms of the topology τ is the following property.

Property 7.9.1. The triple $(X, \|.\|, G)$ has Property 7.9.1 if and only if for any sequence $\{x_n\}_{n\in\mathbb{N}}$ in C such that $(x_n, x_{n+1}) \in E(G)$, for any $n \in \mathbb{N}$, and if a subsequence $\{x_{k_n}\}$ τ-converges to x, then $(x_{k_n}, x) \in E(G)$, for all $n \in \mathbb{N}$.

We have the following result:

Theorem 7.9.1 [7]. *Let X be a Banach space which satisfies the τ-Opial condition. Let C be a bounded convex τ-compact nonempty subset of X not reduced to one point. Assume that $(X, \|.\|, G)$ has Property 7.9.1 and the G-intervals are convex. Let $T : C \to C$ be a G-monotone nonexpansive mapping. Assume there exists $x_0 \in C$ such that $(x_0, T(x_0)) \in E(G)$. Then T has a fixed point.*

Proof. Consider the Krasnoselskii sequence $\{x_n\}$ generated by x_0 and $\lambda \in (0,1)$. Since C is τ-compact, then $\{x_n\}$ will have a subsequence $\{x_{k_n}\}$ which τ-converges to some point $w \in C$. Using the Property 7.9.1 and the property (a), we get $(x_{k_n}, w) \in E(G)$, for any $n \in \mathbb{N}$. Since $\lim_{n\to+\infty}\|x_n - T(x_n)\| = 0$, we get

$$\lim_{n\to+\infty}\left| \|x_{k_n} - x\| - \|T(x_{k_n}) - x\| \right| \leq \lim_{n\to+\infty}\|x_{k_n} - T(x_{k_n})\| = 0.$$

Hence

$$\liminf_{n\to+\infty}\|T(x_{k_n}) - x\| = \liminf_{n\to+\infty}\|x_{k_n} - x\|, \quad \text{for any } x \in C.$$

In particular, we have

$$\liminf_{n\to+\infty}\|T(x_{k_n}) - T(w)\| \leq \liminf_{n\to+\infty}\|x_{k_n} - w\|,$$

since $(x_{k_n}, w) \in E(G)$. Finally, if X satisfies the τ-Opial condition, then we must have $T(w) = w$, i.e., w is a fixed point of T. $\qquad\square$

Remark 7.9.3. The existence of $x_0 \in C$ such that $(x_0, T(x_0)) \in E(G)$ was crucial. Indeed, the Krasnoselskii sequence $\{x_n\}$ will satisfy $(x_n, x_{n+1}) \in E(G)$, for every $n \in \mathbb{N}$.

However, if $(T(x_0), x_0) \in E(G)$, then we will have $(x_{n+1}, x_n) \in E(G)$, for every $n \in \mathbb{N}$. In this case, we need to revise Property 7.9.1 into the new property defined by:

Property (7.9.1*). The triple $(X, \|.\|, G)$ has Property (7.9.1*) if and only if for any sequence $\{x_n\}_{n \in \mathbb{N}}$ in X such that $(x_{n+1}, x_n) \in E(G)$, for any $n \in \mathbb{N}$, and if a subsequence $\{x_{k_n}\}$ τ-converges to x, then $(x, x_{k_n}) \in E(G)$, for all $n \in \mathbb{N}$.

The following results are direct consequences of Theorem 7.9.1.

Corollary 7.9.1 [7]. *Let C be a bounded closed convex nonempty subset of ℓ_p, $1 < p < +\infty$. Let τ be the weak topology. Let G be the digraph defined on ℓ_p by $(\{\alpha_n\}, \{\beta_n\}) \in E(G)$ if and only if $\alpha_n \leq \beta_n$, for any $n \geq 1$. Then any G-monotone nonexpansive mapping $T : C \to C$ has a fixed point provided there exists a point $x_0 \in C$ such that $(x_0, T(x_0)) \in E(G)$ or $(T(x_0), x_0) \in E(G)$.*

Recall the definition of ℓ_p, $1 \leq p < +\infty$, spaces:

$$\ell_p = \left\{ (x_n) \in \mathbb{R}^{\mathbb{N}} : \sum_{n \in \mathbb{N}} |x_n|^p < +\infty \right\}.$$

Remark 7.9.4. The case of $p = 1$ is not interesting for the weak topology since ℓ_1 is a Schur Banach space. But if we consider the weak* topology $\sigma(\ell_1, c_0)$ on ℓ_1 or the pointwise convergence topology, then ℓ_1 satisfies the weak* Opial condition. In this case, we have a similar conclusion to Corollary 3.4.1 for ℓ_1.

Corollary 7.9.2 [7]. *Let C be a bounded closed convex nonempty subset of L_p, $1 \leq p < +\infty$. Let τ be the almost everywhere convergence topology. Let G be the digraph defined on L_p by $(f, g) \in E(G)$ if and only if $f(t) \leq g(t)$ almost everywhere. Assume C is compact almost everywhere. Then any G-monotone nonexpansive mapping $T : C \to C$ has a fixed point provided there exists a point $f_0 \in C$ such that $(f_0, T(f_0)) \in E(G)$ or $(T(f_0), f_0) \in E(G)$.*

7.10. Monotone Nonexpansive Mappings in Hyperbolic Metric Spaces with a Graph

In this section, we discuss some existence results for nonexpansive G-monotone mappings in hyperbolic metric spaces.

Theorem 7.10.1 [13]. *Let (X, d) be a complete hyperbolic metric space and suppose that the triple (X, d, G) has Property 7.8.1. Assume G is convex. Let C be a nonempty, closed, convex and bounded subset of X. Let $T : C \to C$ be a G-nonexpansive mapping.*

Assume $C_T := \{x \in C : (x, T(x)) \in E(G)\} \neq \emptyset$. Then

$$\inf\{d(x, T(x)) : x \in C\} = 0.$$

In particular, there exists an approximate fixed point sequence $\{x_n\}$ in C of T, i.e.,

$$\lim_{n \to +\infty} d(x_n, T(x_n)) = 0.$$

Proof. Fix $a \in C$. Let $\lambda \in (0,1)$ and define $T_\lambda : C \to C$ by

$$T_\lambda(x) = \lambda a \oplus (1 - \lambda) T(x).$$

If $(x, y) \in E(G)$, then we have $(T(x), T(y)) \in E(G)$, since T is G-monotone. Moreover, since G is convex and $(a, a) \in E(G)$, we get

$$(T_\lambda(x), T_\lambda(y)) = (\lambda a \oplus (1 - \lambda) T(x), \lambda a \oplus (1 - \lambda) T(y)) \in E(G),$$

i.e., T_λ is G-monotone, and

$$d(\lambda a \oplus (1 - \lambda) T(x), \lambda a \oplus (1 - \lambda) T(y)) \leq (1 - \lambda) d(T(x), T(y)) \leq (1 - \lambda) d(x, y),$$

i.e., $d(T_\lambda(x), T_\lambda(y)) \leq (1 - \lambda) d(x, y)$. In other words, T_λ is a G-contraction. It is easy to see that $C_T \subset C_{T_\lambda}$. Hence C_{T_λ} is not empty. Theorem 7.4.1 implies the existence of a fixed point w_λ of T_λ in C. So, we have

$$w_\lambda = \lambda a \oplus (1 - \lambda) T(w_\lambda),$$

which implies

$$d(w_\lambda, T(w_\lambda)) \leq \lambda d(a, T(w_\lambda)) \leq \lambda \, \delta(C),$$

where $\delta(C) = \sup\{d(x, y) : x, y \in C\}$ is the diameter of C. Set $x_n = w_{1/n}$, for $n \geq 1$. Then we have $d(x_n, T(x_n)) \leq \delta(C)/n$, for $n \geq 1$. In particular, we have

$$\inf\{d(x, T(x)) : x \in X\} \leq \lim_{n \to +\infty} d(x_n, T(x_n)) = 0.$$

The proof of Theorem 7.10.1 is therefore complete. $\qquad\square$

In order to obtain a fixed point existence result for G-nonexpansive mappings, we need some extra assumptions.

Definition 7.10.1 [13]. We will say that a nonempty subset C of X is G-*compact* if and only if for any $\{x_n\}_{n \geq 1}$ in C, if $(x_n, x_{n+1}) \in E(G)$, for $n \geq 1$, then there exists a subsequence $\{x_{k_n}\}$ of $\{x_n\}$ which is convergent to a point in C.

Remark 7.10.1. Note that G-compactness does not necessarily imply compactness. Indeed, consider the metric set X, a subset of $\mathbb{R}^3$, built on a cone routed at the origin.

All rays are bounded and compact. But X is unbounded. Define the graph G on X by $(x,y) \in E(G)$ if and only if x and y are on the same ray. Then any sequence $\{x_n\} \in X$ such that $(x_n, x_{n+1}) \in E(G)$, for $n \geq 1$, will belong to a ray. Hence $\{x_n\}$ has a convergent subsequence. This shows that X is G-compact but is not compact.

Theorem 7.10.2 [13]. *Let (X,d) be a complete hyperbolic metric space and suppose that the triple (X,d,G) has Property 7.8.1. Assume G is convex and transitive. Let C be a nonempty, G-compact and convex subset of X. Let $T : C \to C$ be a G-nonexpansive mapping. Assume $C_T := \{x \in C : (x, T(x)) \in E(G)\} \neq \emptyset$. Then T has a fixed point.*

Proof. Since C_T is not empty, choose $x_0 \in C_T$. Let (λ_n) be a sequence of numbers in $(0,1)$ such that $\lim\limits_{n \to +\infty} \lambda_n = 0$. As we did in the proof of Theorem 7.10.1, define the mapping $T_1 : C \to C$ by

$$T_1(x) = \lambda_1 x_0 \oplus (1 - \lambda_1) T(x).$$

Since $(x_0, T(x_0)) \in E(G)$, we get $(x_0, T_1(x_0)) \in E(G)$. Since T_1 is G-monotone, we obtain $(T_1^n(x_0), T_1^{n+1}(x_0)) \in E(G)$ and

$$d(T_1^n(x_0), T_1^{n+1}(x_0)) \leq \lambda_1^n d(x_0, T_1(x_0)), \quad \text{for } n \geq 1.$$

Hence $\{T_1^n(x_0)\}$ is a Cauchy sequence. Since C is G-compact, we conclude that $\{T_1^n(x_0)\}$ is convergent. Set $\lim\limits_{n \to +\infty} T_1^n(x_0) = x_1$. Since G is transitive and has Property 7.8.1, then G has Property 7.8.2, which implies that $(x_0, x_1) \in E(G)$. By induction, we construct a sequence $\{x_n\}$ such that x_{n+1} is a fixed point of $T_{n+1} : C \to C$ defined by

$$T_{n+1}(x) = \lambda_{n+1} x_n \oplus (1 - \lambda_{n+1}) T(x),$$

obtained as the limit of $\{T_{n+1}^k(x_n)\}_{k \geq 1}$. In particular, we have $(x_n, x_{n+1}) \in E(G)$, for any $n \geq 1$. Since C is G-compact, there exists a subsequence $\{x_{k_n}\}$ which converges to $w \in C$. Property 7.8.2 implies that $(x_{k_n}, w) \in E(G)$. Using the G-nonexpansiveness of T, we conclude that

$$d(T(x_{k_n}), T(w)) \leq d(x_{k_n}, w), \text{ for } n \geq 1.$$

Hence $\{T(x_{k_n})\}$ converges to $T(w)$. And since x_{n+1} is a fixed point of T_{n+1}, we get

$$x_{n+1} = \lambda_{n+1} x_n \oplus (1 - \lambda_{n+1}) T(x_{n+1}),$$

which implies

$$d(x_{n+1}, T(x_{n+1})) \leq \lambda_{n+1} d(x_n, T(x_{n+1})) \leq \lambda_{n+1} \delta(C), \quad \text{for } n \geq 1,$$

which implies

$$\lim_{n \to +\infty} d(x_n, T(x_n)) = 0.$$

Hence $\{T(x_{k_n})\}$ converges to w as well. Therefore we must have $T(w) = w$, i.e., T has a fixed point.   □

7.11. The Contraction Principle for Monotone Multivalued Mappings

In this section, we study the concept of monotone multivalued contractions in metric spaces, modular function spaces and modular metric spaces with a graph.

7.11.1. Monotone Multivalued Mappings in Metric Spaces with a Graph

Let (X,d) be a metric space and $J : X \to 2^X$ be a multivalued mapping. In this subsection, we discuss the definition of G-contraction mappings introduced by Beg et al. [16] and show that it is restrictive and fails to give the main result of Ref. [16]. Moreover, a new definition of G-contraction [11] is given and sufficient conditions obtained for the existence of fixed points for such mappings.

Generalizing the Banach contraction principle for multivalued mappings, Nadler obtained the following result.

Theorem 7.11.1 [68]. *Let (X,d) be a complete metric space. Denote by $CB(X)$ the set of all nonempty closed bounded subsets of X. Let $F : X \to CB(X)$ be a multivalued mapping. If there exists $k \in [0,1)$ such that*

$$H(F(x), F(y)) \le k\, d(x,y), \quad \text{for all } x,y \in X,$$

where H is the Pompeiu-Hausdorff metric on $CB(X)$, then F has a fixed point in X, i.e., there exists $x \in X$ such that $x \in F(x)$.

A number of extensions and generalizations of Nadler's Theorem have been obtained by different authors (see, for instance, Refs. [34, 38, 58] and references cited therein).

The aim of this subsection is twofold: first, give a correct definition of monotone multivalued mappings; second, extend the conclusion of Theorem 7.2.3 to the case of monotone multivalued mappings in metric spaces endowed with a graph.

Definition 7.11.1 [18]. Let (X,d) be a metric space and $CB(X)$ be the class of all nonempty closed and bounded subsets of X. *The Pompeiu-Hausdorff distance* [18] on

$CB(X)$ induced by d is defined by

$$H(U,W) := \max \left\{ \sup_{w \in W} d(w,U), \sup_{u \in U} d(u,W) \right\}, \quad \text{for } U,W \in CB(X),$$

where $d(u,W) := \inf_{w \in W} d(u,w)$.

Definition 7.11.2 [68]. Let (X,d) be a metric space and $CB(X)$ be the class of all nonempty closed and bounded subsets of X. A multivalued map $J : X \to CB(X)$ is called a *contraction* if there exists $k \in [0,1)$ such that

$$H(J(x),J(y)) \leq k\,d(x,y), \quad \text{for all } x,y \in X.$$

A point $x \in X$ is said to be a fixed point of J if $x \in J(x)$. The set of fixed points of J will be denoted by $Fix(J)$.

Example 7.11.1. Let $I = [0,1]$ denote the unit interval of real numbers (with the usual metric) and let $f : I \to I$ be given by

$$f(x) = \begin{cases} \frac{1}{2}x + \frac{1}{2} & 0 \leq x \leq \frac{1}{2}, \\ -\frac{1}{2}x + \frac{1}{2} & \frac{1}{2} \leq x \leq 1. \end{cases} \tag{7.11.12}$$

Define $F : I \to 2^I$ by $F(x) = \{0\} \cup \{f(x)\}$ for each $x \in I$. It is easy to verify that F is a multivalued contraction mapping with set of fixed points $\{0, \frac{2}{3}\}$.

Example 7.11.2. Let $I^2 = \{(x,y) : 0 \leq x \leq 1 \text{ and } 0 \leq y \leq 1\}$ and let $F : I^2 \to CB(I^2)$ be defined by $F(x,y)$ as the line segment in I^2 from the point $(\frac{1}{2}x,0)$ to the point $(\frac{1}{2}x,1)$ for each $(x,y) \in I^2$. It is easy to see that F is a multivalued contraction mapping with set of fixed points $\{(0,y) : 0 \leq y \leq 1\}$.

Next we introduce the concept of monotone multivalued mappings. Beg et al. [16] offered the following definition:

Definition 7.11.3 [16, Definition 2.6]. Let (X,d) be a metric space. Let $F : X \to CB(X)$ be a multivalued mapping with nonempty closed and bounded values. The mapping F is said to be *a G-contraction* if there exists $k \in [0,1)$ such that

$$H(F(x),F(y)) \leq k\,d(x,y), \text{ for all } (x,y) \in E(G),$$

and if $u \in F(x)$ and $v \in F(y)$ are such that

$$d(u,v) \leq k\,d(x,y) + \alpha, \text{ for each } \alpha > 0,$$

then $(u,v) \in E(G)$.

In particular, this definition implies that if $u \in F(x)$ and $v \in F(y)$ are such that

$$d(u,v) \leq k\, d(x,y),$$

then $(u,v) \in E(G)$, which is very restrictive. In fact, in the proof of Theorem 3.1 in Ref. [16], there is absolutely no reason for $(x_1, x_2) \in E(G)$. The definition (7.11.4 below) of G-contraction multivalued mappings, inspired by the definition of contraction multivalued mappings in Refs. [48, 75], is more appropriate. In the sequel, we assume that (X, d) is a metric space and G is a reflexive directed graph (digraph) with set of vertices $V(G) = X$ and set of edges $E(G)$.

Definition 7.11.4 [11]. Let (X, d) be a metric space. A multivalued mapping $T : X \to 2^X$ is said to be a *monotone increasing G-contraction* if there exists $k \in [0, 1)$ such that for any $u, v \in X$ with $(u, v) \in E(G)$ and any $U \in T(u)$ there exists $V \in T(v)$ such that $(U, V) \in E(G)$ and

$$d(U,V) \leq k\, d(u,v).$$

As we have seen before, we need the graph to have appropriate behavior for monotone sequences. Recall Property 7.8.2:

> *For any sequence $\{x_n\}_{n \in \mathbb{N}}$ in X, if $x_n \to x$ and $(x_n, x_{n+1}) \in E(G)$ for $n \in \mathbb{N}$, then $(x_n, x) \in E(G)$.*

First, we begin with the following theorem that gives the existence of a fixed point for monotone multivalued mappings in metric spaces endowed with a graph.

Theorem 7.11.2 [11]. *Let (X, d) be a complete metric space and suppose that the triple (X, d, G) has Property 7.8.2. Let $T : X \to CB(X)$ be a monotone increasing G-contraction mapping. If $X_T := \{x \in X : (x, u) \in E(G)$ for some $u \in T(x)\}$ is not empty, then the following statements hold:*

(a) *For any $x \in X_T$, $T|_{[x]_{\widetilde{G}}}$ has a fixed point.*

(b) *If G is weakly connected, then T has a fixed point in G.*

(c) *If $X' := \bigcup\{[x]_{\widetilde{G}} : x \in X_T\}$, then $T|_{X'}$ has a fixed point in X.*

Proof. (a) Let $x_0 \in X_T$, then there exists $x_1 \in T(x_0)$ such that $(x_0, x_1) \in E(G)$. Since T is a monotone increasing G-contraction, there exists $x_2 \in T(x_1)$ such that $(x_1, x_2) \in E(G)$ and

$$d(x_1, x_2) \leq k\, d(x_0, x_1),$$

where $k < 1$ is associated with the definition of T being a monotone increasing G-contraction. Without loss of generality, we may assume $k > 0$. By induction, we construct a sequence $\{x_n\}$ such that $x_{n+1} \in T(x_n)$, $(x_n, x_{n+1}) \in E(G)$ and

$$d(x_n, x_{n+1}) \le k\, d(x_n, x_{n-1}) \le k^n\, d(x_0, x_1), \quad \text{for any } n \ge 1.$$

Since $\sum\limits_{n=0}^{+\infty} d(x_n, x_{n+1}) \le d(x_0, x_1) \sum\limits_{n=0}^{+\infty} k^n < +\infty$, we conclude that $\{x_n\}$ is a Cauchy sequence, and hence converges to some $x \in X$ since X is a complete metric space. We claim that $x \in T(x)$, i.e., x is a fixed point of T. Indeed, using Property 7.8.2 we know that $(x_n, x) \in E(G)$ for any $n \in \mathbb{N}$. Then using the definition of G-contraction, there exists $y_n \in T(x)$ such that $(x_{n+1}, y_n) \in E(G)$ and

$$d(x_{n+1}, y_n) \le k\, d(x_n, x), \quad \text{for any } n \in \mathbb{N}.$$

Hence

$$d(y_n, x) \le d(y_n, x_{n+1}) + d(x_{n+1}, x) \le k\, d(x_n, x) + d(x_{n+1}, x), \quad \text{for any } n \in \mathbb{N}.$$

This implies that $\{y_n\}$ converges to x. Since $T(x)$ is closed, we get $x \in T(x)$ as claimed. As $(x_n, x) \in E(G)$, for every $n \ge 0$, we conclude that $(x_0, x_1, \ldots, x_n, x)$ is a path in G and so $x \in [x_0]_{\widetilde{G}}$.

(b) Since $X_T \ne \emptyset$, there exists an $x_0 \in X_T$, and since G is weakly connected, then $[x_0]_{\widetilde{G}} = X$ and by (a), mapping T has a fixed point.

(c) This follows easily from (a) and (b). □

Remark 7.11.1 [11]. The missing information in Theorem 7.11.2 is the uniqueness of the fixed point. According to Theorem 7.4.1, we do have a partial positive answer to this question when T is single valued. But in the multivalued case, this conclusion may fail. Indeed, take $A \in CB(X)$ with $u, v \in A$ such that $(u, v) \in E(G)$ and $u \ne v$. Then u and v are fixed points of the map $T : X \to CB(X)$ defined by $T(x) = A$, for any $x \in X$.

If we assume G is such that $E(G) := X \times X$ then clearly G is connected and our Theorem 7.11.2 gives Nadler's theorem (Theorem 7.11.1).

The following is a direct consequence of Theorem 7.11.2.

Corollary 7.11.1 [11]. *Let (X, d) be a complete metric space. Assume that the triple (X, d, G) has Property 7.8.2. If G is weakly connected then every G-contraction $T : X \to CB(X)$ such that $(x_0, x_1) \in E(G)$, for some $x_0 \in X$ and $x_1 \in T(x_0)$, has a fixed point.*

Example 7.11.3. Let $X = \{0,1,2,3,4\} = V(G)$ and

$$E(G) = \{(0,0),(1,1),(2,2),(3,3),(0,1),(0,2),(0,3),(1,2),(1,3),(2,3)\}.$$

Let $V(G)$ be endowed with metric $d : X \times X \to \mathbb{R}^+$ defined by

$$d(0,0) = d(1,1) = d(2,2) = d(3,3) = 0,$$

$$d(0,1) = d(1,0) = \frac{1}{4},$$

$$d(0,2) = d(2,0) = d(1,2) = d(2,1) = d(1,3) = \cdots = d(3,2) = \frac{4}{5}.$$

The graph of G is shown in the figure below:

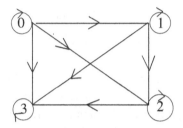

The Pompeiu-Hausdorff weight assigned to $U, W \in CB(X)$ is:

$$H(U,W) = \begin{cases} \frac{1}{4}, & \text{if } U, W \subseteq \{0,1\} \text{ with } U \neq W, \\ \frac{4}{5}, & \text{if } U \text{ or } W \text{ (or both) } \nsubseteq \{0,1\} \text{ with } U \neq W, \\ 0, & \text{if } U = W. \end{cases} \quad (7.11.13)$$

Define $T : X \to CB(X)$ as follows:

$$T(x) = \begin{cases} \{0\}, & \text{if } x \in \{0,1\}, \\ \{1\}, & \text{if } x \in \{2,3\}. \end{cases} \quad (7.11.14)$$

Note that, for all $x, y \in X$ with edge between x and y, there is an edge between $T(x)$ and $T(y)$. If there is a path between x and y, then there is a path between $T(x)$ and $T(y)$. Moreover, T is a G-contraction with all other assumptions of Theorem 7.11.2 satisfied and T has 0 as a fixed point.

7.11.2. The Contraction Principle for Multivalued Mappings in Modular Function Spaces with a Graph

In this subsection we discuss the existence of fixed points of G-contraction mappings in modular function spaces. These results are the modular version of the Jachymski fixed point result (Theorem 7.4.1) for mappings defined in a metric space endowed with a graph. Recall that the fixed point theory in modular function spaces was initiated by Khamsi et al. [49].

In the sequel, we assume that $\rho \in \mathfrak{R}$ is a convex σ-finite modular function, C is a nonempty subset of the modular function space L_ρ and G is a reflexive directed graph (digraph) with a set of vertices of G being the elements of C. We also assume that G has no parallel edges (arcs) and so we can identify G with the pair $(V(G), E(G))$. We denote by $\mathscr{C}_\rho(C)$ the collection of all nonempty ρ-closed subsets of C and by $\mathscr{K}_\rho(C)$ the collection of all nonempty ρ-compact subsets of C.

Definition 7.11.5 [5]. Let $\rho \in \mathfrak{R}$ and $C \subset L_\rho$ be nonempty, ρ-closed and ρ-bounded. A multivalued mapping $T : C \to 2^C$ is said to be a *monotone increasing G-contraction* if there exists $\alpha \in [0, 1)$ such that for any $f, h \in C$ with $(f, h) \in E(G)$ and any $F \in T(f)$ there exists $H \in T(h)$ such that $(F, H) \in E(G)$ and

$$\rho(F - H) \leq \alpha \, \rho(f - h).$$

We begin with the following theorem that gives the existence of a fixed point for monotone multivalued mappings in modular spaces endowed with a graph. The key feature in this theorem is that the Lipschitzian condition on the nonlinear map is only assumed to hold for elements that are comparable in the natural partial order of L_ρ.

Definition 7.11.6. Let $\mathscr{C}_\rho(C)$ be the set of nonempty ρ-closed subsets of a set C. A multivalued mapping $T : C \to \mathscr{C}_\rho(C)$ is a *monotone increasing ρ-contraction* if there exists $\alpha \in [0, 1)$ such that for all $f, g \in C$ with $f \leq g$ ρ-a.e. and any $F \in T(f)$, there exists $G \in T(g)$ such that $F \leq G$ ρ-a.e. and

$$\rho(F - G) \leq \alpha \, \rho(f - g).$$

Theorem 7.11.3 [5]. *Let $\rho \in \mathfrak{R}$. Let $C \subset L_\rho$ be a nonempty and ρ-closed subset. Assume that ρ satisfies the Δ_2-type condition. Let $T : C \to \mathscr{C}_\rho(C)$ be a monotone increasing ρ-contraction mapping and $C_T := \{f \in C : f \leq g$ ρ-a.e. for some $g \in T(f)\}$. If $C_T \neq \emptyset$, then T has a fixed point in C.*

Proof. Fix $f_0 \in C_T$. Then there exists $f_1 \in T(f_0)$ such that $f_0 \leq f_1$ ρ-a.e. Since T is monotone increasing ρ-contraction, there exists $f_2 \in T(f_1)$, $f_1 \leq f_2$ ρ-a.e., such that

$$\rho(f_2 - f_1) \leq \alpha \, \rho(f_1 - f_0),$$

where $\alpha < 1$ is associated with the definition of T being a monotone increasing ρ-contraction. Without loss of generality, we may assume $\alpha > 0$. By induction, we construct a sequence $\{f_n\}$ such that $f_{n+1} \in T(f_n)$, $f_n \leq f_{n+1}$ ρ-a.e. and

$$\rho(f_{n+1} - f_n) \leq \alpha\, \rho(f_n - f_{n-1}) \leq \alpha^n \rho(f_1 - f_0), \quad \text{for any } n \geq 1.$$

Lemma 7.3.4 implies that $\{f_n\}$ is ρ-Cauchy and converges to some $f \in C$ since L_ρ is ρ-complete. We claim that $f \in T(f)$, i.e., f is a fixed point of T. Indeed, Theorem 7.3.3 implies that $f_n \leq f$ ρ-a.e., for any $n \geq 0$. Since T is a monotone increasing ρ-contraction, there exists $g_n \in T(f)$ such that $f_{n+1} \leq g_n$ and

$$\rho(f_{n+1} - g_n) \leq \alpha\, \rho(f_n - f).$$

Hence $\{f_{n+1} - g_n\}$ ρ-converges to 0. Since ρ satisfies the Δ_2-type condition, we conclude that $\{\|f_{n+1} - g_n\|_\rho\}$ converges to 0. Hence $\{g_n\}$ also ρ-converges to f. Since $T(f)$ is ρ-closed, we conclude that $f \in T(f)$. □

Note that the fixed point may not be unique. Indeed, if we take A to be any nonempty ρ-closed subset of C, then the multivalued map $T : C \to \mathscr{C}_\rho(C)$ defined by $T(f) = A$, for any $f \in C$, is a monotone increasing ρ-contraction mapping. The set of fixed points of T is exactly the set A.

Next, we give the graph version of the results found above. As we have seen before, we need the graph to have a nice behavior for monotone sequences. Therefore the following property will be needed:

Property 7.11.1. We say that (L_ρ, G) has Property 7.11.1 if and only if for any sequence $\{f_n\}_{n\in\mathbb{N}}$ in L_ρ, if f_n ρ-converges to f and $(f_n, f_{n+1}) \in E(G)$ for $n \in \mathbb{N}$, then $(f_n, f) \in E(G)$.

Theorem 7.11.4 [5]. *Let $\rho \in \mathfrak{R}$. Assume that (L_ρ, G) has Property 7.11.1 and ρ satisfies the Δ_2-type condition. Let C be a nonempty, ρ-closed and ρ-bounded subset of L_ρ. Let $T : C = V(G) \to \mathscr{C}_\rho(C)$ be a monotone increasing G-contraction mapping and $C_T := \{f \in C : (f, g) \in E(G) \text{ for some } g \in T(f)\}$. If $C_T \neq \emptyset$, then the following statements hold:*

(a) *For any $f \in C_T$, $T|_{[f]_{\widetilde{G}}}$ has a fixed point.*

(b) *If $f \in C$ with $(\bar{f}, f) \in E(G)$, where $\bar{f}$ is a fixed point of T, then there exists a sequence $\{f_n\}$ such that $f_{n+1} \in T(f_n)$, for every $n \geq 0$, and $\{f_n\}$ ρ-converges to $\bar{f}$.*

(c) *If G is weakly connected, then T has a fixed point in G.*

(d) *If $C' := \bigcup\{[f]_{\widetilde{G}_\rho} : f \in C_T\}$, then $T|_{C'}$ has a fixed point in C.*

Proof. (a) Let $f_0 \in C_T$, then there exists an $f_1 \in T(f_0)$ with $(f_0, f_1) \in E(G)$. Since T is a monotone increasing G-contraction, there exists $f_2 \in T(f_1)$ such that $(f_1, f_2) \in E(G)$ and

$$\rho(f_1, f_2) \leq \alpha\rho(f_0, f_1),$$

where $\alpha < 1$ is the constant associated with the contraction definition of T. Similarly, there exists $f_3 \in T(f_2)$ such that $(f_2, f_3) \in E(G)$ and

$$\rho(f_2, f_3) \leq \alpha\,\rho(f_1, f_2).$$

By induction we build $\{f_n\}$ in C with $f_{n+1} \in T(f_n)$ and $(f_n, f_{n+1}) \in E(G)$ such that

$$\rho(f_{n+1}, f_n) \leq \alpha\rho(f_n, f_{n-1}), \quad \text{for every } n \geq 1.$$

Hence

$$\rho(f_{n+1}, f_n) \leq \alpha^n \rho(f_1, f_0), \quad \text{for every } n \geq 0.$$

Lemma 7.3.4 implies that $\{f_n\}$ is ρ-Cauchy. Since L_ρ is ρ-complete and C is ρ-closed, therefore $\{f_n\}$ ρ-converges to some point $g \in C$. Since $(f_n, f_{n+1}) \in E(G)$, for every $n \geq 1$, then $(f_n, g) \in E(G)$ by Property 7.11.1. Since T is a monotone increasing G-contraction, there exists $g_n \in T(g)$ such that

$$\rho(f_{n+1}, g_n) \leq \alpha\,\rho(f_n, g), \quad \text{for every } n \geq 1.$$

Hence

$$\rho\left(\frac{g_n - g}{2}\right) \leq \rho(g_n - f_{n+1}) + \rho(f_{n+1} - g)$$
$$\leq \alpha\,\rho(f_n - g) + \rho(f_{n+1} - g), \quad \text{for every } n \geq 1.$$

Since $\{f_n\}$ ρ-converges to g, we conclude that $\lim\limits_{n \to +\infty} \rho((g_n - g)/2) = 0$. The Δ_2- type condition satisfied by ρ implies that $\lim\limits_{n \to +\infty} \rho(g_n - g) = 0$, i.e., $\{g_n\}$ ρ-converges to g. Since $T(g)$ is ρ-closed, we conclude that $g \in T(g)$, i.e., g is a fixed point of T.

(b) Let $f \in C$ such that $(\bar{f}, f) \in E(G)$. Since T is a monotone increasing G-contraction, then there exists $f_1 \in T(f)$ such that $(\bar{f}, f_1) \in E(G)$ and

$$\rho(\bar{f} - f_1) \leq \alpha\rho(\bar{f} - f).$$

By induction, we construct a sequence $\{f_n\}$ such that $f_{n+1} \in T(f_n)$, $(\bar{f}, f_n) \in E(G)$, and

$$\rho(\bar{f} - f_{n+1}) \leq \alpha\rho(\bar{f} - f_n), \quad \text{for every } n \geq 0.$$

Hence, we have

$$\rho(\bar{f} - f_n) \leq \alpha^n \, \rho(\bar{f} - f), \quad \text{for every } n \geq 0.$$

Since $\alpha < 1$, we conclude that $\{f_n\}$ ρ-converges to $\bar{f}$.

(c) Since $C_T \neq \emptyset$, there exists an $f_0 \in C_T$, and since G is weakly connected, then $[f_0]_{\tilde{G}_\rho} = C$ and, by (a), mapping T has a fixed point.

(d) This follows easily from (a) and (c).   □

Remark 7.11.2 [5]. If we assume G is such that $E(G) := C \times C$ then clearly G is connected and Theorem 7.11.4 gives Nadler's theorem (Theorem 7.11.1) in modular function spaces [33]. Moreover, if T is single valued, then we get the Banach contraction principle and if T is multivalued then we get the corrected version of the analog of the main result of Ref. [16] in modular function spaces.

The following is a direct consequence of Theorem 7.11.4.

Corollary 7.11.2 [5]. *Let $\rho \in \mathfrak{R}$. Let C be a nonempty and ρ-closed subset of L_ρ. Assume that ρ satisfies the Δ_2-type condition and C is ρ-bounded. Assume that G is weakly connected and (L_ρ, G) has Property 7.11.1. Let $T : C = V(G) \to \mathscr{C}_\rho(C)$ be a monotone increasing G-contraction multivalued mapping such that there exists $f_0, f_1 \in T(f_0)$ with $(f_0, f_1) \in E(G)$. Then T has a fixed point.*

Let us give an example which will illustrate the role of the above-defined notions.

Example 7.11.4 [5]. Consider $L^\infty[0,1]$ (the space of all bounded measurable functions on $[0,1]$ or rather all measurable functions which are bounded except possibly on a subset of measure zero). Notice that we identify functions which are equivalent, i.e., we should say that the elements of $L^\infty[0,1]$ are not discrete functions but rather equivalent classes of functions. Then $L^\infty[0,1]$ is a normed linear space with

$$\| f \| = \| f \|_\infty = \inf\{M : | f(t) | \leq M a.e\} = \inf\{M : m\{t : f(t) > M\} = 0\}.$$

Let $C = \{f_0, f_1, f_2\}$, where

(i) f_0 is the step function on $[0,1]$ with partition $0 = x_0 < x_1 < x_2 < \cdots < x_n = 1$ such that $f_0 = c_i = \frac{1}{i+1}$ on (x_{i-1}, x_i) and $f_0(x_i) = d_i$ is an arbitrary real number. Note that $\|f_0\|_\infty = \max | c_i | = \frac{1}{2}$.

(ii) $f_1 = e^x$ on $[0,1]$. So $\| f_1 \|_\infty = e$.

(iii) $f_2 = X^2$ on $[0,1]$. So $\| f_2 \|_\infty = 1$.

Let $\rho := \| . \|_\infty$ and $T : C \to \mathscr{C}_\rho(C)$ be defined as follows:

$$T(f) = \begin{cases} f_0, & \text{if } f = f_0, \\ f_2, & \text{if } f = f_1, f_2. \end{cases}$$

Therefore $E(G) = \{(f_0, f_0), (f_1, f_1), (f_2, f_2), (f_0, f_1), (f_2, f_1)\}$. The digraph of G is shown in the figure below:

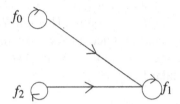

Now, for all $(f, g) \in E(G)$, T is a G-contraction. Also, all other assumptions of Theorem 7.11.4 are satisfied and T has a fixed point.

7.11.3. The Contraction Principle for Multivalued Mappings in Modular Metric Spaces with a Graph

In this subsection we study the existence of fixed points for contraction multivalued mappings in modular metric spaces endowed with a graph. The notion of a modular metric on an arbitrary set and the corresponding modular spaces, generalizing classical modulars over linear spaces like Orlicz spaces, were recently introduced. The results of this subsection can be seen as a generalization of Nadler's and Edelstein's fixed point theorems to modular metric spaces endowed with a graph. In particular, we obtain a multivalued version of the result of Theorem 4.1 in Ref. [2] to modular metric spaces endowed with a graph. We also extend the results of Nadler [68], Mizoguchi and Takahashi [66] to modular metric spaces with a graph. The necessary background of this subsection has already been introduced in Section 7.3.4.

The following result, where the directed graph G_ω is defined on a subset M of the modular metric space (X, ω), can be seen as a generalization of Nadler's fixed point result [68] for modular metric spaces endowed with a graph.

Let (X, ω) be a modular metric space and G_ω be a reflexive digraph defined on a nonempty subset M of X. Recall that $E(G_\omega)$ has Property 7.8.4 if:

for any $\{x_n\}_{n \geq 1}$ in M, if x_n ω-converges to x and $(x_n, x_{n+1}) \in E(G_\omega)$, for $n \geq 1$, then $(x_n, x) \in E(G_\omega)$, for every $n \geq 1$.

Theorem 7.11.5 [8]. *Let (X, ω) be a modular metric space. Suppose that ω is a convex regular modular which satisfies the Δ_2-type condition. Assume that $M =$*

$V(G_\omega)$ is a nonempty ω-complete subset of X_ω and $E(G_\omega)$ has Property 7.8.4. Let $T : M \to CB(M)$ be a G_ω-contraction map and $M_T := \{x \in M : (x,y) \in E(G_\omega)$ for some $y \in T(x)\}$. If $M_T \neq \emptyset$, then T has a fixed point.

Proof. Let $x_0 \in M_T$, then there exists an $x_1 \in T(x_0)$ with $(x_0,x_1) \in E(G_\omega)$. Since T is a G_ω-contraction, there exists $x_2 \in T(x_1)$ such that $(x_1,x_2) \in E(G_\omega)$ and

$$\omega_1(x_1,x_2) \leq k\, \omega_1(x_0,x_1).$$

Similarly, there exists $x_3 \in T(x_2)$ such that $(x_2,x_3) \in E(G_\omega)$ and

$$\omega_1(x_2,x_3) \leq k\, \omega_1(x_1,x_2).$$

By induction, we build $\{x_n\}$ in M with $x_{n+1} \in T(x_n)$ and $(x_n,x_{n+1}) \in E(G_\omega)$ such that

$$\omega_1(x_{n+1},x_n) \leq k\, \omega_1(x_n,x_{n-1}), \quad \text{for every } n \geq 1.$$

Hence

$$\omega_1(x_{n+1},x_n) \leq k^n \omega_1(x_1,x_0), \quad \text{for every } n \geq 0.$$

The technical Lemma 7.3.4 implies that $\{x_n\}$ is ω-Cauchy. Since M is ω-complete, therefore $\{x_n\}$ ω-converges to some point $z \in M$. Since $(x_n,x_{n+1}) \in E(G_\omega)$, for every $n \geq 1$, then $(x_n,z) \in E(G_\omega)$ by Property 7.8.4. Since T is a G_ω-contraction, there exists $z_n \in T(z)$ such that

$$\omega_1(x_{n+1},z_n) \leq k\, \omega_1(x_n,z), \quad \text{for every } n \geq 1.$$

Hence

$$\omega_2(z_n,z) \leq \omega_1(z_n,x_{n+1}) + \omega_1(x_{n+1},z) \leq k\, \omega_1(x_n,z) + \omega_1(x_{n+1},z), \quad \text{for every } n \geq 1.$$

Since $\{x_n\}$ ω-converges to z, we conclude that $\lim_{n\to+\infty} \omega_2(z_n,z) = 0$. The Δ_2-type condition satisfied by ω implies that $\lim_{n\to+\infty} \omega_1(z_n,z) = 0$, i.e., $\{z_n\}$ ω-converges to z. Since $T(z)$ is ω-closed, we conclude that $z \in T(z)$, i.e., z is a fixed point of T. This completes the proof of Theorem 7.11.5. $\qquad\qquad\qquad\square$

Edelstein [35] has extended the classical fixed point theorem for contractions to the case when X is a complete ε-chainable metric space and the mapping $T : X \to X$ is an (ε,k)-uniformly local contraction. This result was extended by Nadler [68] to multivalued mappings. Here we investigate Nadler's result in modular metric spaces endowed with a graph. First, let us recall the ε-chainable concept in modular metric spaces with a graph. The definition given here is slightly different from the one used for classical metric spaces since the modular functions may not obey the triangle inequality.

Let (X, ω) be a modular metric space and $M = V(G_\omega)$ be a nonempty subset of X_ω. M is said to be finitely ε-chainable (where $\varepsilon > 0$ is fixed) if and only if there exists $N \geq 1$ such that for any $a, b \in M$ with $(a, b) \in E(G_\omega)$ there is an N, ε-chain from a to b (that is, a finite set of vertices $x_0, x_1, \ldots, x_N \in V(G_\omega) = M$ such that $x_0 = a$, $x_N = b$, $(x_i, x_{i+1}) \in E(G_\omega)$ and $\omega_1(x_i, x_{i+1}) < \varepsilon$, for all $i = 0, 1, 2, \ldots, N-1$).

We have the following result.

Theorem 7.11.6 [8]. *Let (X, ω) be a modular metric space. Suppose that ω is a convex regular modular which satisfies the Δ_2-type condition. Assume that $M = V(G_\omega)$ is a nonempty, ω-complete, ω-bounded subset of X_ω which is finitely ε-chainable, for some fixed $\varepsilon > 0$, and $E(G_\omega)$ has Property 7.8.4. Let $T : M \to CB(M)$ be an $(\varepsilon, k) - G_\omega$-uniformly local contraction map. If $M_T := \{x \in M : (x, y) \in E(G_\omega)$ for some $y \in T(x)\} \neq \emptyset$, then T has a fixed point.*

Proof. Since M is finitely ε-chainable, there exists $N \geq 1$ such that for any $a, b \in M$ with $(a, b) \in E(G_\omega)$ there is a finite set of vertices $x_0, x_1, \ldots, x_N \in M$ such that $x_0 = a$, $x_N = b$, $(x_i, x_{i+1}) \in E(G_\omega)$ and $\omega_1(x_i, x_{i+1}) < \varepsilon$, for all $i = 0, 1, 2, \ldots, N-1$. For any $x, y \in M$ with $(x, y) \in E(G_\omega)$, define

$$\omega^*(x, y) = \inf \left\{ \sum_{i=0}^{i=N-1} \omega_1(x_i, x_{i+1}) \right\},$$

where the infimum is taken over all N, ε-chains $x_0, x_1, \ldots, x_N$ from x to y. Since M is N, ε-chainable, then $\omega^*(x, y) < +\infty$, for any $x, y \in M$ with $(x, y) \in E(G_\omega)$. Using the basic properties of ω, we get

$$\omega_N(x, y) \leq \omega^*(x, y),$$

for any $x, y \in M$ with $(x, y) \in E(G_\omega)$. Moreover, if $\omega_1(x, y) < \varepsilon$, then we have $\omega^*(x, y) \leq \omega_1(x, y)$, for any $x, y \in M$ with $(x, y) \in E(G_\omega)$. Fix $x \in M_T$. Set $z_0 = x$. Choose $z_1 \in T(z_0)$ with $(z_0, z_1) \in E(G_\omega)$. Let $x_0, x_1, \ldots, x_N$ be an N, ε-chain from z_0 to z_1. Since T is an $(\varepsilon, k) - G_\omega$-uniformly local contraction map, there exist $y_0, y_1, \ldots, y_N$ in M such that:

(a) $y_i \in T(x_i)$, for any $i = 1, \ldots, N$.

(b) $(y_i, y_{i+1}) \in E(G_\omega)$, for any $i = 0, \ldots, N-1$.

(c) $\omega_1(y_i, y_{i+1}) \leq k \, \omega_1(x_i, x_{i+1})$, for any $i = 0, \ldots, N-1$.

It is easy to check that $y_0 = z_0, y_1, \ldots, y_N$ is an N, ε-chain from z_0 to $y_N \in T(z_1)$. Set $y_N = z_2$. Using the fact that T is an $(\varepsilon, k) - G_\omega$-uniformly local contraction map, we get

$$\omega^*(z_1, z_2) \leq k \, \omega^*(z_0, z_1).$$

By induction, we construct a sequence $\{z_n\} \in M$ with $(z_n, z_{n+1}) \in E(G_\omega)$ such that

$$\omega^*(z_n, z_{n+1}) \leq k\,\omega^*(z_{n-1}, z_n),$$

and $z_{n+1} \in T(z_n)$, for any $n \geq 1$. Obviously, we have $\omega^*(z_n, z_{n+1}) \leq k^n \omega^*(z_0, z_1)$, for any $n \geq 1$. Since ω satisfies the Δ_2-type condition, there exists $C > 0$ such that

$$\omega_1(z_n, z_{n+1}) \leq C\omega_N(z_n, z_{n+1}) \leq C\omega^*(z_n, z_{n+1}) \leq Ck^n\omega^*(z_0, z_1),$$

for any $n \geq 1$. Lemma 7.3.4 implies that $\{z_n\}$ is ω-Cauchy. Since M is ω-complete, then $\{z_n\}$ ω-converges to some $z \in M$. We claim that z is a fixed point of T. Indeed, we first note that $(z_n, z) \in E(G_\omega)$ for any $n \geq 1$ by Property 7.8.4. Using the ideas developed above, there exists $v_n \in T(z)$ such that

$$\omega^*(z_{n+1}, v_n) \leq k\,\omega^*(z_n, z), \quad \text{for any } n \geq 1.$$

By ω properties, we have

$$\omega_{N+1}(v_n, z) \leq \omega_1(z_{n+1}, z) + \omega_N(z_{n+1}, v_n) \leq \omega_1(z_{n+1}, z) + k\,\omega^*(z_n, z),$$

for any $n \geq 1$. Since $\{z_n\}$ ω-converges to z, there exists $n_0 \geq 1$ such that, for any $n \geq n_0$, we have $\omega_1(z_n, z) < \varepsilon$. Hence $\omega^*(z_n, z) \leq \omega_1(z_n, z)$, for any $n \geq n_0$, which implies

$$\omega_{N+1}(v_n, z) \leq \omega_1(z_{n+1}, z) + k\,\omega_1(z_n, z), \quad \text{for any } n \geq n_0.$$

Therefore we have $\lim\limits_{n \to +\infty} \omega_{N+1}(v_n, z) = 0$. The Δ_2-type condition satisfied by ω implies that $\lim\limits_{n \to +\infty} \omega_1(v_n, z) = 0$, i.e., $\{v_n\}$ ω-converges to z. Since $v_n \in T(z)$ and $T(z)$ is ω-closed, we conclude that $z \in T(z)$, i.e., z is a fixed point of T. This completes the proof of Theorem 7.11.6. $\square$

7.12. Monotone Nonexpansive Multivalued Mappings

In this section we study the concept of monotone nonexpansive multivalued mappings in hyperbolic metric spaces and modular function spaces with a graph.

7.12.1. Monotone Nonexpansive Multivalued Mappings in Hyperbolic Metric Spaces with a Graph

In this subsection we discuss some existence results for monotone nonexpansive multivalued mappings in hyperbolic metric spaces endowed with a graph. First, we give an analog to Theorem 7.10.1 for these mappings.

Theorem 7.12.1 [13]. *Let (X, d) be a complete hyperbolic metric space and suppose that the triple (X, d, G) has Property 7.8.1. Assume G is convex. Let C be a nonempty, closed, convex and bounded subset of X. Set $\mathscr{C}(C)$ to be the set of all nonempty* *closed subsets of C. Let $T : C \to \mathscr{C}(C)$ be a monotone increasing G-nonexpansive mapping. If $C_T := \{x \in C : (x,y) \in E(G)$ for some $y \in T(x)\}$ is not empty, then T has an approximate fixed point sequence $\{x_n\} \in C$, that is, for any $n \geq 1$, there exists $y_n \in T(x_n)$ such that $\lim_{n \to +\infty} d(x_n, y_n) = 0$. In particular, we have*

$$\lim_{n \to +\infty} dist(x_n, T(x_n)) = \lim_{n \to +\infty} \inf\{d(x_n, y) : y \in T(x_n)\} = 0.$$

Proof. Fix $\lambda \in (0,1)$ and $x_0 \in C$. Define the multivalued map T_λ on C by

$$T_\lambda(x) = \lambda \, x_0 \oplus (1-\lambda) \, T(x) = \{\lambda \, x_0 \oplus (1-\lambda)y : y \in T(x)\}.$$

Note that $T_\lambda(x)$ is a nonempty and closed subset of C, for any $x \in C$. Let us show that T_λ is a monotone increasing G-contraction. Let $x, y \in C$ such that $(x,y) \in E(G)$. Since T is a monotone increasing G-nonexpansive mapping, for any $x^* \in T(x)$ there exists $y^* \in T(y)$ such that $(x^*, y^*) \in E(G)$ and $d(x^*, y^*) \leq d(x,y)$. Hence

$$d\left(\lambda \, x_0 \oplus (1-\lambda) \, x^*, \lambda \, x_0 \oplus (1-\lambda) \, y^*\right) \leq (1-\lambda)d(x^*, y^*) \leq (1-\lambda) \, d(x,y),$$

which proves our claim. Since G is convex, we get $(\lambda \, x_0 \oplus (1-\lambda)x^*, \lambda \, x_0 \oplus (1-\lambda) \, y^*) \in E(G)$. This clearly shows that T_λ is a monotone increasing G-contraction as claimed. Note that we have $C_T \subset C_{T_\lambda}$, which implies that C_{T_λ} is nonempty. Using Theorem 7.11.2 we conclude that T_λ has a fixed point $x_\lambda \in C$, i.e., $x_\lambda \in T_\lambda(x_\lambda)$. Thus there exists $y_\lambda \in T(x_\lambda)$ such that

$$x_\lambda = \lambda \, x_0 \oplus (1-\lambda) \, y_\lambda.$$

In particular, we have

$$d(x_\lambda, y_\lambda) \leq \lambda \, d(x_0, y_\lambda)) \leq \lambda \, \delta(C),$$

which implies $dist(x_\lambda, T(x_\lambda)) \leq \lambda \, \delta(C)$. If we choose $\lambda = \frac{1}{n}$, for $n \geq 1$, there exists $x_n \in C$ and $y_n \in T(x_n)$ such that $d(x_n, y_n) \leq \delta(C)/n$, which implies

$$dist(x_n, T(x_n)) \leq \frac{1}{n} \, \delta(C).$$

The proof of Theorem 7.12.1 is therefore complete. $\square$

The multivalued version of Theorem 7.10.2 may be stated as follows.

Theorem 7.12.2 [13]. *Let (X,d) be a complete hyperbolic metric space and suppose that the triple (X,d,G) has Property 7.8.1. Assume G is convex and transitive. Let C be a nonempty, G-compact and convex subset of X. Then any $T : C \to \mathscr{C}(C)$ monotone increasing G-nonexpansive mapping has a fixed point provided $C_T := \{x \in C : (x,y) \in E(G)$ for some $y \in T(x)\}$ is not empty.*

Proof. Since C_T is not empty, choose $x_0 \in C_T$. Let (λ_n) be a sequence of numbers in $(0,1)$ such that $\lim_{n \to +\infty} \lambda_n = 0$. As we did in the proof of Theorem 7.12.1, define the mapping $T_1 : C \to C$ by

$$T_1(x) = \lambda_1 x_0 \oplus (1 - \lambda_1)T(x).$$

Since $C_T \subset C_{T_1}$, there exists $y_0 \in T_1(x_0)$ such that $(x_0, y_0) \in E(G)$. Using the properties of T_1, there exists $y_2 \in T_1(y_1)$ such that $(y_1, y_2) \in E(G)$ and

$$d(y_1, y_2) \leq (1 - \lambda_1)d(x_0, y_1).$$

By induction, we build a sequence $\{y_n\}$, with $y_0 = x_0$, such that $y_{n+1} \in T_1(y_n)$, $(y_n, y_{n+1}) \in E(G)$ and

$$d(y_n, y_{n+1}) \leq (1 - \lambda_1)\, d(y_{n-1}, y_n) \leq (1 - \lambda_1)^n\, d(x_0, y_1) \leq (1 - \lambda_1)^n \delta(C),$$

for $n \geq 1$. So $\{y_n\}$ is Cauchy. Set $\lim_{n \to +\infty} y_n = x_1 \in C$. Since G is transitive and has Property 7.8.1, then G has Property 7.8.2, which implies that $(y_n, x_1) \in E(G)$, for any n. In particular, we have $(x_0, x_1) \in E(G)$. Using the properties of T_1, for any n there exists $z_n \in T(x_1)$ such that

$$d(y_{n+1}, z_n) \leq (1 - \lambda_1)\, d(y_n, x_1).$$

Clearly this implies that $\{z_n\}$ converges to x_1 as well. Since $T(x_1)$ is closed, we conclude that $x_1 \in T(x_1)$, i.e., x_1 is a fixed point of T_1. By induction, we construct a sequence $\{x_n\}$ in C such that x_{n+1} is a fixed point of T_{n+1} defined by

$$T_{n+1}(x) = \lambda_{n+1} x_n \oplus (1 - \lambda_{n+1})T(x),$$

and $(x_n, x_{n+1}) \in E(G)$. Since C is G-compact, there exists a subsequence $\{x_{k_n}\}$ which converges to $w \in C$. Property 7.8.2 implies that $(x_n, w) \in E(G)$. Since x_n is a fixed point of T_n, there exists $z_n \in T(x_n)$ such that

$$x_n = \lambda_n\, x_{n-1} \oplus (1 - \lambda_n)\, z_n, \quad \text{for any } n \geq 1.$$

Note that $d(x_n, z_n) \leq \lambda_n d(x_{n_1}, z_n) \leq \lambda_n \delta(C)$, for any $n \geq 1$. In particular, we have $\lim_{n \to +\infty} d(x_n, z_n) = 0$. Hence $\{z_{k_n}\}$ also converges to w. Using the G-nonexpansiveness of T, since $(x_{k_n}, w) \in E(G)$, there exists $w_n \in T(w)$ such that $d(z_{k_n}, w_n) \leq d(x_{k_n}, w)$, for any n. Therefore we have that (w_n) converges to w. Since $T(w)$ is closed, we conclude that $w \in T(w)$, i.e., w is a fixed point of T. $\square$

7.12.2. Monotone Nonexpansive Multivalued Mappings in Modular Function Spaces with a Graph

In this subsection we discuss the existence of fixed points of monotone nonexpansive multivalued mappings in modular function spaces endowed with a graph. These results are the modular version of Jachymski's fixed point result (Theorem 7.4.1) for mappings defined in metric spaces endowed with a graph.

In the sequel, we assume that $\rho \in \Re$ is a convex σ-finite modular function, C is a nonempty subset of the modular function space L_ρ and G is a reflexive directed graph (digraph) with a set of vertices of G being the elements of C. We also assume that G has no parallel edges (arcs) and so we can identify G with the pair $(V(G), E(G))$. We denote by $\mathscr{C}_\rho(C)$ the collection of all nonempty ρ-closed subsets of C and by $\mathscr{K}_\rho(C)$ the collection of all nonempty ρ-compact subsets of C.

Definition 7.12.1 [5]. Let $\rho \in \Re$ and $C \subset L_\rho$ be nonempty ρ-closed and ρ-bounded. A multivalued mapping $T : C \to 2^C$ is *monotone increasing G-nonexpansive* if for any $f, h \in C$ with $(f, h) \in E(G)$ and any $F \in T(f)$ there exists $H \in T(h)$ such that $(F, H) \in E(G)$ and

$$\rho(F - H) \leq \rho(f - h).$$

An easy consequence of Theorem 7.11.3 is the following result.

Proposition 7.12.1 [5]. *Let $\rho \in \Re$. Let C be a nonempty and ρ-closed convex subset of L_ρ. Assume that ρ satisfies the Δ_2-type condition and C is ρ-bounded. Let $T : C \to \mathscr{C}_\rho(C)$ be a monotone increasing ρ-nonexpansive mapping and $C_T := \{f \in C : f \leq g \rho\text{-a.e. for some } g \in T(f)\}$. If $C_T \neq \emptyset$, then T has an approximate fixed point sequence $\{f_n\} \in C$, that is, for any $n \geq 1$, there exists $g_n \in T(f_n)$ such that*

$$\lim_{n \to +\infty} \rho(f_n - g_n) = 0.$$

In particular, we have $\lim_{n \to +\infty} dist_\rho(f_n, T(f_n)) = 0$, where

$$dist_\rho(f_n, T(f_n)) = \inf \{\rho(f_n - g) : g \in T(f_n)\}.$$

Proof. Fix $\lambda \in (0, 1)$ and $h_0 \in C$. Define the multivalued map T_λ on C by

$$T_\lambda(f) = \lambda h_0 + (1 - \lambda) T(f) = \{\lambda h_0 + (1 - \lambda)g : g \in T(f)\}.$$

Note that $T_\lambda(f)$ is a nonempty and ρ-closed subset of C. Let us show that T_λ is a monotone increasing ρ-contraction. Let $f, g \in C$ such that $f \leq g \rho$-a.e. Since T is a monotone increasing ρ-nonexpansive mapping, for any $F \in T(f)$ there exists $G \in T(g)$ such that $F \leq G \rho$-a.e. and $\rho(F - G) \leq \rho(f - g)$. Since

$$\rho\Big((\lambda h_0 + (1 - \lambda) F) - (\lambda h_0 + (1 - \lambda) G)\Big) = \rho\Big((1 - \lambda)(F - G)\Big)$$

and ρ is convex, we get

$$\rho\Big((\lambda\,h_0+(1-\lambda)F)-(\lambda h_0+(1-\lambda)G)\Big) \le (1-\lambda)\rho(F-G).$$

Using the basic properties of the partial order of L_ρ, we get $\lambda\,h_0+(1-\lambda)F \le \lambda\,h_0+(1-\lambda)\,G\,\rho$-a.e. This clearly shows that T_λ is a monotone increasing ρ-contraction as claimed. Note that if C_T is not empty, then C_{T_λ} is also nonempty. Using Theorem 7.11.3 we conclude that T_λ has a fixed point $f_\lambda \in C$, i.e., $f_\lambda \in T_\lambda(f_\lambda)$. Thus there exists $F_\lambda \in T(f_\lambda)$ such that

$$f_\lambda = \lambda h_0+(1-\lambda)F_\lambda.$$

In particular, we have

$$\rho(f_\lambda - F_\lambda) = \rho(\lambda(h_0 - F_\lambda)) \le \lambda\,\delta_\rho(C),$$

which implies $dist_\rho(f_\lambda, T(f_\lambda)) \le \lambda\,\delta_\rho(C)$. If we choose $\lambda = \frac{1}{n}$, for $n \ge 1$, there exists $f_n \in C$ and $F_n \in T(f_n)$ such that $\rho(f_n - F_n) \le \frac{1}{n}\,\delta_\rho(C)$, which implies

$$dist_\rho(f_n, T(f_n)) \le \frac{1}{n}\,\delta_\rho(C).$$

The proof of Proposition 7.12.1 is therefore complete.   □

Remark 7.12.1 [5]. We can modify slightly the above proof to show that the approximate fixed point sequence $\{f_n\}$ and its associated sequence $\{F_n\}$ satisfy $f_n \le f_{n+1} \le F_{n+1}\,\rho$-a.e. Indeed, set $\{\lambda_n\} = \{1/(n+1)\}_{n\ge1}$. Let $f_0 \in C_T$. Then from the above proof, there exists a fixed point $f_1 = \lambda_1 f_0+(1-\lambda_1)F_1$, with $f_0 \le f_1\,\rho$-a.e. It is easy to check that $f_0 \le f_1 \le F_1\,\rho$-a.e. Clearly $f_1 \in C_T$. By induction, we build the sequences $\{f_n\}$ and $\{F_n\}$, with

$$F_n \in T(f_n), \quad f_{n+1} = \lambda_{n+1}f_n+(1-\lambda_{n+1})F_{n+1},$$

and

$$f_n \le f_{n+1} \le F_{n+1}, \quad \text{for every } n \ge 1.$$

Since $\lambda_n \to 0$ as $n \to +\infty$, we conclude that $\{f_n\}$ is an approximate fixed point sequence of T.

Using the above results, we are now ready to prove the main fixed point theorem for ρ-nonexpansive monotone multivalued mappings. This theorem may be seen as the monotone version of Theorem 2.4 of Ref. [62]. Note that Kutbi and Latif [62] must assume that ρ is additive to be able to reach their conclusion. To avoid the additivity of ρ, we need the following property.

Definition 7.12.2. Let $\rho \in \mathfrak{R}$. We will say that L_ρ satisfies *the ρ-a.e.-Opial property* if for every (f_n) in L_ρ ρ-a.e. convergent to 0, such that there exists $\beta > 1$ for which $\sup_n \rho(\beta \, f_n) < +\infty$, then we have

$$\liminf_{n \to +\infty} \rho(f_n) < \liminf_{n \to +\infty} \rho(f_n + f)$$

for every $f \in L_\rho$ not equal to 0.

Theorem 7.12.3 [5]. *Let $\rho \in \mathfrak{R}$. Let $C \subset L_\rho$ be a nonempty, ρ-closed and ρ-bounded convex subset. Assume that ρ satisfies the Δ_2-type condition, L_ρ satisfies the ρ-a.e.-Opial property and C is ρ-a.e. compact. Then each monotone increasing ρ-nonexpansive map $T : C \to \mathcal{K}_\rho(C)$ has a fixed point.*

Proof. Proposition 7.12.1 and Remark 7.12.1 ensure the existence of a sequence $\{f_n\}$ in C and a sequence $\{F_n\}$ such that $F_n \in T(f_n)$, $f_n \le f_{n+1} \le F_{n+1}$ ρ-a.e., for every $n \ge 1$, and $\lim_{n \to +\infty} \rho(f_n - F_n) = 0$. Without loss of generality, we may assume that f_n ρ-a.e. converges to $f \in C$ and F_n ρ-a.e. converges to $F \in C$. The Fatou property implies

$$\rho(f - F) \le \liminf_{n \to +\infty} \rho(f_n - F_n) = 0.$$

Hence $f = F$. Clearly, we have $f_n \le f$ ρ-a.e., for any $n \ge 1$. Since T is a monotone increasing ρ-nonexpansive map, then there exists a sequence $\{H_n\}$ in $T(f)$ such that $F_n \le H_n$ ρ-a.e. and

$$\rho(F_n - H_n) \le \rho(f_n - f), \quad \text{for all } n \ge 1.$$

Since $T(f)$ is ρ-compact, we may assume that $\{H_n\}$ is ρ-convergent to some $h \in T(f)$. Since ρ satisfies the Δ_2-condition, Lemma 4.2 in Ref. [32] implies

$$\liminf_{n \to +\infty} \rho(f_n - h) = \liminf_{n \to +\infty} \rho(f_n - F_n + F_n - H_n + H_n - h) = \liminf_{n \to +\infty} \rho(F_n - H_n).$$

Since $\rho(F_n - H_n) \le \rho(f_n - f)$, we get

$$\liminf_{n \to +\infty} \rho(f_n - h) \le \liminf_{n \to +\infty} \rho(f_n - f).$$

Since C is ρ-bounded and ρ satisfies the Δ_2-condition, the ρ-a.e.-Opial property implies that $f = h$, i.e., $f \in T(f)$. Hence f is a fixed point of T. $\square$

As an application of Theorem 7.11.4, we have the following result whose proof is similar to Proposition 7.12.1.

Proposition 7.12.2 [5]. *Let $\rho \in \mathfrak{R}$. Let $C \subset L_\rho$ be a nonempty, ρ-bounded and ρ-closed convex subset. Assume that ρ satisfies the Δ_2-type condition and (L_ρ, G) has*

Property 7.11.1. Let $T : C \to \mathscr{C}_\rho(C)$ be a monotone increasing G-nonexpansive map. Then there exists an approximate fixed points sequence $\{f_n\}$ in C, i.e., for any $n \geq 1$ there exists $F_n \in T(f_n)$ such that

$$\lim_{n \to +\infty} \rho(f_n - F_n) = 0.$$

ACKNOWLEDGMENTS

The author acknowledges King Fahd University of Petroleum and Minerals for supporting all of his research that is related to this chapter.

REFERENCES

1. Abdou AN, Khamsi MA, Fixed points results of pointwise contractions in modular metric spaces. Fixed Point Theory Appl. 2013;163 doi:10.1186/1687-1812-2013-163.
2. Abdou AN, Khamsi MA, Fixed points of multivalued contraction mappings in modular metric spaces. Fixed Point Theory Appl. 2014;249 doi:10.1186/1687-1812-2014-249.
3. Abdou AN, Khamsi MA, On monotone pointwise contractions in Banach and metric spaces. Fixed Point Theory Appl. 2015;135 doi:10.1186/s13663-015-0381-7.
4. Alfuraidan MR, Remarks on Caristi's fixed point theorem in metric spaces with a graph. Fixed Point Theory Appl. 2014;240 doi:10.1186/1687-1812-2014-240.
5. Alfuraidan MR, Fixed points of multivalued mappings in modular function spaces with a graph. Fixed Point Theory Appl. 2015;42 doi:10.1186/s13663-015-0292-7.
6. Alfuraidan MR, The contraction principle for mappings on a modular metric space with a graph. Fixed Point Theory Appl. 2015;46 doi:10.1186/s13663-015-0296-3.
7. Alfuraidan MR, Fixed points of monotone nonexpansive mappings with a graph. Fixed Point Theory Appl. 2015;49 doi:10.1186/s13663-015-0299-0.
8. Alfuraidan MR, The contraction principle for multivalued mappings on a modular metric space with a graph. Can. Math. Bull. 2015;29 doi:10.4153/CMB-2015-029-x.
9. Alfuraidan MR, On monotone Ćirić quasi-contraction mappings with a graph. Fixed Point Theory Appl. 2015;93 doi:10.1186/s13663-015-0341-2.
10. Alfuraidan MR, On monotone pointwise contractions in Banach spaces with a graph. Fixed Point Theory Appl. 2015;139 doi:10.1186/s13663-015-0390-6.
11. Alfuraidan MR, Remarks on monotone multivalued mappings on a metric space with a graph. J. Inequal. Appl. 2015;202 doi:10.1186/s13660-015-0712-6.
12. Alfuraidan MR, Khamsi MA, Caristi fixed point theorem in metric spaces with a graph. Abstr. Appl. Anal. 2014;303484 doi:10.1155/2014/303484.
13. Alfuraidan MR, Khamsi MA, Fixed points of monotone nonexpansive mappings on a hyperbolic metric space with a graph. Fixed Point Theory Appl. 2015;44 doi:10.1186/s13663-015-0294-5.
14. Bachar M, Khamsi MA, Fixed points of monotone nonexpansive mappings. Preprint.
15. Banach S, Sur les opérations dans les ensembles abstraits et leur application aux équations intégrales. Fund. Math. 1922;3:133–181.
16. Beg I, Butt AR, Radojević S, The contraction principle for set valued mappings on a metric space with a graph. Comput. Math. Appl. 2010;60:1214–1219.
17. Ben-El-Mechaiekh H, The Ran-Reurings fixed point theorem without partial order: a simple proof. J. Fixed Point Theory Appl. 2014;16(1):373–383.
18. Berinde V, Păcurar M, The role of the Pompeiu-Hausdorff metric in fixed point theory. Creative Math. Inform. 2013;22(2):143–150.
19. Bojor F, Fixed point theorems for Reich type contractions on metric spaces with a graph. Nonlinear Anal. 2012;75(9):3895–3901.
20. Brondsted A, On a lemma of Bishop and Phelps. Pacific J. Math. 1974;55(2):335–341.

21. Brondsted A, Fixed point and partial orders. Proc. Am. Math. Soc. 1976;60:365–366.
22. Brondsted A, Common fixed points and partial orders. Proc. Am. Math. Soc. 1979;77:365–368.
23. Browder FE, Nonexpansive nonlinear operators in a Banach space. Proc. Natl. Acad. Sci. U.S.A. 1965:1041–1044.
24. Browder FE, On a theorem of Caristi and Kirk. In: Fixed Point Theory and Its Applications (Proc. Sem., Dalhousie University, 1975), pp. 23–27. San Diego: Academic Press;1976.
25. Busemann H, Spaces with non-positive curvature. Acta Math. 1948;80:259–310.
26. Caristi J, Fixed point theorems for mappings satisfying inwardness conditions. Trans. Am. Math. Soc. 1976;215:241–251.
27. Caristi J, Fixed point theory and inwardness conditions. In: Applied Nonlinear Analysis 1979:479–483.
28. Ćirić LB, A generalization of Banach's contraction principle. Proc. Am. Math. Soc. 1974;45:267–273.
29. Chistyakov VV, Modular metric spaces, I. Basic concepts. Nonlinear Anal. 2010;72(1):1–14.
30. Chistyakov VV, Modular metric spaces, II. Application to superposition operators. Nonlinear Anal. 2010;72(1):15–30.
31. Dominguez-Benavides T, Khamsi MA, Samadi S, Uniformly Lipschitzian mappings in modular function spaces. Nonlinear Anal., Theory Methods Appl. 2001;46(2):267–278.
32. Dominguez-Benavides T, Khamsi MA, Samadi S, Asymptotically regular mappings in modular function spaces. Sci. Math. Japon. 2001;53(2):295–304.
33. Dhompongsa S, Dominguez Benavides T, Kaewcharoen A, Panyanak B, Fixed point theorems for multivalued mappings in modular function spaces. Sci. Math. Japon. 2006;63(2):161–169.
34. Echenique F, A short and constructive proof of Tarskis fixed point theorem. Int. J. Game Theory 2005;33(2):215–218.
35. Edelstein M, An extension of Banach's contraction principle. Proc. Am. Math. Soc. 1961;12:7–10.
36. Ekeland I, Sur les problemes variationnels. Comptes Rendus Acad. Sci. Paris 1972;275:1057–1059.
37. El-Sayed SM, Ran ACM, On an iteration method for solving a class of nonlinear matrix equations. SIAM J. Martix Anal Appl. 2002;23(3):632–645.
38. Feng Y, Liu S, Fixed point theorems for multivalued contractive mappings and multivalued Caristi type mappings. J. Math. Anal. Appl. 2006;317:103–112.
39. Göhde D, Zum Prinzip der kontraktiven Abbildung. Math. Nachr. 1965;30:251–258.
40. Goebel K, Kirk WA, Iteration processes for nonexpansive mappings. Contemp. Math. 1983;21:115–123.
41. Goebel K, Kirk WA, Topics in Metric Fixed Point Theory. Cambridge Stud. Adv. Math. vol. 28. Cambridge: Cambridge University Press;1990.
42. Goebel K, Reich S, Uniform Convexity, Hyperbolic Geometry, and Nonexpansive Mappings. Series of Monographs and Textbooks in Pure and Applied Mathematics. vol. 83. New York:Dekker;1984.
43. Halpern B, Bergman G, A fixed point theorem for inward and outward maps. Trans. Am. Math. Soc. 1968;130:353–358.
44. Hussain N, Khamsi MA, On asymptotic pointwise contractions in metric spaces. Nonlinear Anal., Theory Methods Appl. 2009;71:4423–4429.
45. Hayes JP, A graph model for fault tolerant computing systems. IEEE Trans. Comput. 1976;25:875–884.
46. Jachymski J, The contraction principle for mappings on a metric space with a graph. Proc. Am. Math. Soc. 2008;1(136):1359–1373.
47. Khamsi MA, Remarks on Caristi's fixed point theorem. Nonlinear Anal. 2009;71(1–2):227–231.
48. Khamsi MA, Misane D, Disjunctive signed logic programs. Fundam. Inform. 1996;32:349–357.
49. Khamsi MA, Kozlowski WM, Reich S, Fixed point theory in modular function spaces. Nonlinear Anal., Theory Meth. Appl. 1990;14:935–953.
50. Kirk WA, A fixed point theorem for mappings which do not increase distances. Am. Math. Monthly. 1965;72:1004–1006.
51. Kirk WA, Mappings of generalized contractive type. J. Math. Anal. Appl. 1970;32:567–572.
52. Kirk WA, Fixed point theory for nonexpansive mappings, Vol. I and II. Lecture Notes in Mathematics, vol. 886, pp. 485–505. Berlin: Springer; 1981.

53. Kirk WA, Fixed points of asymptotic contractions. J. Math. Anal. Appl. 2003;277:645–650.
54. Kirk WA, A fixed point theorem in CAT(0) spaces and R-trees. Fixed Point Theory Appl. 2004;(4):309–316.
55. Kirk WA, Asymptotic pointwise contractions. In: Plenary Lecture: The 8th International Conference on Fixed Point Theory and Its Applications, Chiang Mai University, 2007:16–22.
56. Kirk WA, Caristi J, Mapping theorems in metric and Banach spaces. Bull. L'Acad. Polon. Sci. 1975;25:891–894.
57. Kirk WA, Xu H-K, Asymptotic pointwise contractions. Nonlinear Anal. 2008;69:4706–4712.
58. Klim D, Wardowski D, Fixed point theorems for set-valued contractions in complete metric spaces. J. Math. Anal. Appl. 2007;334:132–139.
59. Kozlowski WM, Modular Function Spaces. Series of Monographs and Textbooks in Pure and Applied Mathematics, vol. 122. New York: Dekker;1988.
60. Kozlowski WM, Notes on modular function spaces I. Comment. Math. 1988;28:91–104.
61. Kozlowski WM, Notes on modular function spaces II. Comment. Math. 1988;28:105–120.
62. Kutbi MA, Latif A, Fixed points of multivalued maps in modular function spaces. Fixed Point Theory Appl. 2009;786357 doi:10.1155/2009/786357.
63. Leustean L, A quadratic rate of asymptotic regularity for CAT(0)-spaces. J. Math. Anal. Appl. 2007;325:386–399.
64. Lukawska GG, Jachymski J, IFS on a metric space with a graph structure and extension of the Kelisky-Rivlin theorem. J. Math. Anal. Appl. 2009;356:453–463.
65. Menger K, Untersuchungen ber allgemeine Metrik. Math. Ann. 1928;100:75–163.
66. Mizoguchi N, Takahashi W, Fixed point theorems for multivalued mappings on complete metric spaces. J. Math. Anal. Appl. 1989;141 doi:10.1016/0022-247X(89)90214-X.
67. Musielak J, Orlicz Spaces and Modular Spaces. Lecture Notes in Math., vol. 1034. Berlin: Springer;1983.
68. Nadler SB, Multivalued contraction mappings. Pacific J. Math. 1969;30:475–488.
69. Nakano H, Modulared Semi-Ordered Linear Spaces. Tokyo: Maruzen;1950.
70. Nieto JJ, Rodríguez-López R, Contractive mapping theorems in partially ordered sets and applications to ordinary differential equations. Order 2005;22(3):223–239.
71. Nieto JJ, Pouso RL, Rodríguez-López R, Fixed point theorems in ordered abstract spaces. Proc. Am. Math. Soc. 2007;135:2505–2517.
72. Orlicz W, Collected Papers. Part I, II. PWN: Warsaw;1988.
73. Ran ACM, Reurings MCB, A fixed point theorem in partially ordered sets and some applications to matrix equations. Proc. Am. Math. Soc. 2004;132:1435–1443.
74. Reich S, Shafrir I, Nonexpansive iterations in hyperbolic spaces. Nonlinear Anal. 1990;15:537–558.
75. Straccia U, Ojeda-Aciegoy M, Damásioz CV, On fixed-points of multi-valued functions on complete lattices and their application to generalized logic programs. SIAM J. Comput. 2009;38(5):1881–1911.
76. Sullivan F, A characterization of complete metric spaces. Proc. Am. Math. Sot. 1981;85:345–346.
77. Tarski A, A lattice theoretical fixed point and its application. Pacific J. Math. 1955;5:285–309.
78. Taskovic MR, On an equivalent of the axiom of choice and its applications. Math. Japon. 1986;31(6):979–991.
79. Turinici M, Fixed points for monotone iteratively local contractions. Dem. Math. 1986;19:171–180.
80. Turinici M, Ran and Reurings theorems in ordered metric spaces. J. Indian Math. Soc. 2011;78:207–2014.

CHAPTER 8

The Use of Retractions in the Fixed Point Theory for Ordered Sets

Bernd S. W. Schröder

The University of Southern Mississippi, Department of Mathematics, 118 College Avenue, #5045, Hattiesburg, MS 39406, USA,
Corresponding: bernd.schroeder@usm.edu

Contents

Abstract

This chapter gives an overview how retractions are used to prove fixed point results in ordered sets. The primary focus is on comparative retractions, that is, retractions r so that, for each x, the image $r(x)$ is comparable to the preimage x. We start with infinite ordered sets and the classical Abian-Brown Theorem, which establishes that, for every order-preserving self map f of a chain-complete ordered set, if there is an $x \leq f(x)$, then f has a fixed point. Subsequently, using comparative retractions, we prove that the unit ball in L^p has the (order-theoretical) fixed point property, that is, every order-preserving self map has a fixed point. On a finite ordered set, a comparative retraction is the composition of comparative retractions that each remove a single point. Such a point is called irreducible and the fixed point property is not affected by the presence or absence of irreducible points. An ordered set that can be reduced, by successive removal of irreducible points, to a singleton is called dismantlable by irreducibles. We exhibit the relation between ordered sets that are dismantlable by irreducibles and the application of constraint propagation methods to find fixed point free

http://dx.doi.org/10.1016/B978-0-12-804295-3.50008-5

order-preserving self maps. Closely related to irreducible points are points that are removed by a, not necessarily comparative, retraction that removes a single point. These points are called retractable points. There is a fixed point theorem for retractable points that generalizes the one for irreducible points. However, connectedly collapsible ordered sets, a natural class of ordered sets that is defined based on this theorem, are computationally more challenging than ordered sets that are dismantlable by irreducibles. Whereas the definition of dismantlability can be directly verified in polynomial time, direct verification of the definition of connected collapsibility is worst-case exponential. For graphs, the natural analogue of the fixed point property for ordered sets is the fixed clique property. Although there is no analogue of the Abian-Brown Theorem for the fixed clique property, there are analogues of the fixed point theorems for irreducible and for retractable vertices. For simplicial complexes, the natural analogue of the fixed point property for ordered sets is the fixed simplex property. Although there is an analogue of the fixed point theorem for irreducible vertices, there is no full analogue of the corresponding theorem for retractable vertices. We conclude with the connection between the fixed point property for ordered sets and the iteration of the clique graph operator on the comparability graph. Specifically, it is shown that, for ordered sets that are dismantlable by irreducibles, iteration of the clique graph operator on the comparability graph leads to a graph with one vertex. It is also shown that, if iteration of the clique graph operator on the comparability graph leads to a graph with one vertex, then the ordered set has the fixed point property.

8.1. Introduction

An *ordered set*, or, *partially ordered set*, consists of a set P and a reflexive, antisymmetric and transitive relation $\leq$ on P, which is called the *order relation*. Unless there is the possibility of confusing several order relations, we will refer to the underlying set P as the ordered set. Familiar examples of ordered sets include the number systems $\mathbb{N}$, $\mathbb{Z}$, $\mathbb{Q}$ and $\mathbb{R}$ with their usual orders, as well as spaces of real valued functions, ordered by the pointwise order.

The morphisms between ordered sets P and Q are the *order-preserving functions*, that is, functions $f : P \rightarrow Q$ so that $x \leq y$ (in P) implies $f(x) \leq f(y)$ (in Q). As most of this volume is concerned with analytical/topological results, we start with a connection to analysis.[1] Section 8.2 shows that, in L^p-spaces, the unit ball is chain-complete. Chain-completeness is crucial for the widely used Abian-Brown Theorem (see Theorem 8.3.1) which we will investigate in Section 8.3. The Abian-Brown Theorem states that, in a chain-complete ordered set P, for an order-preserving self map $f : P \rightarrow P$, existence of a point $x \leq f(x)$ implies existence of a fixed point. Together with its proof, it is the foundation for many applications of order-theoretical methods in analysis to date. Retractions are now a way to obtain more sophisticated results.

---

[1]Readers whose primary interest is in finite structures and who are familiar with ordered sets can move directly from the introduction to Section 8.5.

We say that an ordered set P has the *fixed point property* if and only if every order-preserving self map $f : P \to P$ has a *fixed point* $x = f(x)$. Retractions are a standard tool when investigating fixed point properties for morphisms of any kind. Note that, in an ordered set P, any subset $Q \subseteq P$ is an ordered set with the induced order $\leq_Q := \leq |_{Q \times Q}$ obtained by restricting $\leq$ to Q.

Definition 8.1.1. Let P be an ordered set. Then an order-preserving function $r : P \to P$ is called a *retraction* if and only if $r^2 = r$. The ordered subset $r[P]$ is called a *retract* of P.

Proposition 8.1.1. *Let P be an ordered set with the fixed point property and let $r : P \to P$ be a retraction. Then $r[P]$ has the fixed point property.*

Proof. (The following standard argument applies to any fixed point property for morphisms and to retractions that are morphisms of the same type.) Let P have the fixed point property and let $f : r[P] \to r[P]$ be order-preserving. Then $f \circ r : P \to P$ is order-preserving too, and hence it has a fixed point $x = f(r(x))$. Now $x \in f[r[P]] \subseteq r[P]$ implies $x = r(x)$. Therefore $x = f(r(x)) = f(x)$ is a fixed point of f. (Note that the argument only used that the function r is idempotent and a morphism of the same type as f.) $\qquad\qquad\qquad\qquad\qquad\qquad\qquad\qquad\qquad\qquad\qquad\qquad\qquad\quad$ $\square$

Proposition 8.1.1 is quite natural: If you have a structure with a good property, then appropriate substructures should inherit the good property. One way to motivate this chapter is the desire for theorems that address the converse. That is, we would like to have theorems in which a good property of substructures implies a good property for the larger structure, which is a much rarer situation.

Comparative retractions (see Definition 8.4.2) are a class of retractions for which such a converse is possible. This converse is investigated in Section 8.4, specifically in Theorem 8.4.1. A natural consequence is that, in any L^p-space, the unit ball has the order-theoretical fixed point property (see Theorem 8.4.2).

With Theorem 8.4.2 being the culmination of our investigation of infinite ordered sets in Sections 8.2, 8.3 and 8.4, the chapter switches attention to finite structures starting in Section 8.5. In finite ordered sets, comparative retractions are compositions of retractions that remove an irreducible point (see Definition 8.5.2). Because of Theorem 8.5.1, any ordered set that can be dismantled by successive removal of irreducible points to a singleton (see Definition 8.5.4) has the fixed point property.

Computational complexity is a key concern in the finite setting. Theorem 8.6.1 in Section 8.6 shows that, for ordered sets that are dismantlable by irreducibles, there is a performance guarantee for a polynomial algorithm that establishes the existence or nonexistence of a fixed point free order-preserving self map. Section 8.7 focuses on retractable points (see Definition 8.7.1) for which there is a similar fixed point theorem as for irreducible points (see Theorem 8.7.1). At the end of Section 8.7, we ask if computational performance guarantees similar to Theorem 8.6.1 hold for retractable points. Example 8.8.1 in Section 8.8 shows that proofs of deeper results should be quite nontrivial.

Switching to graphs in Section 8.9, the natural analogue of the fixed point property is the fixed clique property (see Definition 8.9.3). Theorem 8.9.1 shows that the fixed clique property for the comparability graph implies the fixed point property for the ordered set, but not vice versa. Theorem 8.9.2 then provides an analogue of Theorem 8.7.1 for the fixed clique property. Note that this is possible even though, by Example 8.9.1, there is no analogue of the Abian-Brown Theorem for the fixed clique property.

More general than graphs and the fixed clique property, in Section 8.10 we consider simplicial complexes and the fixed simplex property (see Definition 8.10.5). Theorem 8.10.1 establishes a connection between the fixed simplex property and the fixed point property other than the one through the chain complex of an ordered set. Theorem 8.10.2 is an analogue of Theorem 8.5.1 for the fixed simplex property and Example 8.10.1 shows that there is no full analogue of Theorem 8.7.1 for the fixed simplex property. The connection to algebraic topology and topology could fill multiple separate chapters. However, in Section 8.11, we at least show in Theorem 8.11.1 that the topological fixed point property for the topological realization of an ordered set implies the order-theoretical fixed point property, but not vice versa. For more on this connection, Ref. [41] would be a good start.

We conclude with an investigation of the connection between the fixed point property and the iteration of the clique graph operator on the comparability graph in Sections 8.12 and 8.13. Theorem 8.12.1 shows that certain analogue of comparative retractions on a graph induce similar retractions on the clique graph, which leads to an analogue of Theorem 8.5.1 for the clique graph operator in Theorem 8.12.2. Finally, Theorem 8.13.1 shows that, if iteration of the clique graph operator on the comparability graph leads to a graph with one vertex, then the ordered set has the fixed point property.

Theorem 8.9.2 as well as Examples 8.8.1, 8.9.1, 8.10.1, and Example 8.12.6 (due to M. Pizaña) are heretofore unpublished. The multitude of open questions throughout this chapter shows that this area provides many opportunities for further contributions.

8.2. Chain-Complete Ordered Sets

In a finite ordered set, it is easily proved that, if f is an order-preserving self map, then existence of a point $x \leq f(x)$ implies existence of a fixed point: Simply iterate the map until the iterations become stationary. The Abian-Brown Theorem (see Theorem

8.3.1) shows that, in chain-complete ordered sets, the same result holds true even when the ordered set is infinite. The idea for the proof is essentially the same too: We iterate the function, and we use chain-completeness to get past limit ordinals in the iteration. Let us first recall the definition of a chain.

Definition 8.2.1. Let C be an ordered set. Then C is called a *chain* or *linearly ordered* if and only if, for all $x, y \in C$, we have that $x \le y$ or $y \le x$.

Chains are the most well-known ordered sets: The familiar number systems $\mathbb{N}$, $\mathbb{Z}$, $\mathbb{Q}$ and $\mathbb{R}$ as well as ordinal numbers are linearly ordered. For the definition of chain-completeness, we also need the idea of suprema and infima.

Definition 8.2.2. Let P be an ordered set and let $A \subseteq P$. Then $u \in P$ is called an *upper bound* of A if and only if, for all $a \in A$, we have that $a \le u$. Moreover, $s \in P$ is called the *supremum* or *lowest upper bound* of A if and only if s is an upper bound of A and, for all upper bounds u of A, we have $s \le u$. The supremum of A is also denoted $\bigvee A$.

Dually, $\ell \in P$ is called a *lower bound* of A if and only if, for all $a \in A$, we have that $a \ge \ell$. Moreover, $i \in P$ is called the *infimum* or *greatest lower bound* of A if and only if i is a lower bound of A and, for all lower bounds ℓ of A, we have $i \ge \ell$. The infimum of A is also denoted $\bigwedge A$.

Suprema and infima are familiar from the real numbers, because, in $\mathbb{R}$, every subset with an upper bound has a supremum. We speak of the supremum and the infimum, because it can be proved (and it is a good exercise to do so) that a subset has at most one supremum. Note that suprema and infima are "dual" notions. We can obtain the definition of the infimum from the definition of the supremum by reversing all inequalities, which is what duality is all about: Essentially, many order-theoretical ideas work the same way whether we use the original ordered set or whether we turn the ordered set upside down. Hence many theorems are stated with what could be considered an "upward bias" (or a "downward bias") and it is understood that we can obtain the dual result by reversing all inequalities. For example, the Abian-Brown Theorem (see Theorem 8.3.1) is stated with an "upward bias" and the dual result can easily be obtained (another good exercise).

We can now state the definition of chain-completeness.

Definition 8.2.3. Let P be an ordered set. Then P is called *chain-complete* if and only if each nonempty subchain $C \subseteq P$ has a supremum and an infimum.

Let us consider an example of a chain-complete ordered set in analysis. Every L^p-

space is ordered by the pointwise almost everywhere order. Because L^p-spaces cannot be chain-complete (consider the chain consisting of the functions $f_n := n\mathbf{1}_F$, where $\mathbf{1}_F$ is the indicator function of a fixed set F of finite nonzero measure and $n \in \mathbb{N}$ is arbitrary), we first need to establish a condition that is as close to chain-completeness as we can get.

Lemma 8.2.1. *Let (Ω, Σ, μ) be a measure space, let $p \in [1, \infty)$, let $u > 0$ and let $C \subseteq L^p(\Omega, \Sigma, \mu)$ be a chain in $L^p(\Omega, \Sigma, \mu)$ so that, for all $c \in C$, we have $\|c\|_p \leq u$. Then C has a supremum d in $L^p(\Omega, \Sigma, \mu)$ and $\|d\|_p \leq u$.*

Proof. Because we can pick an element $c_0 \in C$ and consider the chain $\{c - c_0 : c \in C, c \geq c_0\}$ instead of C, we can assume, without loss of generality, that $c \geq 0$ for all $c \in C$. There is nothing to prove if C has a largest element, so we can assume that C does not have a largest element.

Because $\|c\|_p \leq u$ for all $c \in C$, we have that $s := \sup_{c \in C} \|c\|_p$ exists. For each $n \in \mathbb{N}$, choose a $c_n \in C$ so that $\|c_n\|_p \geq s - \frac{1}{n}$. Without loss of generality, we can assume that $c_1 \leq c_2 \leq c_3 \leq \cdots$. Then, for each $c \in C$, there is an $n_c \in \mathbb{N}$ so that $\|c\|_p \leq \|c_{n_c}\|_p$ and hence $c \leq c_{n_c}$.

Let d be the pointwise almost everywhere supremum of the c_n. Then $d \geq c$ for all $c \in C$. By the Monotone Convergence Theorem, we have that $\|d\|_p = \lim_{n \to \infty} \|c_n\|_p = s \leq u < \infty$, so $d \in L^p(\Omega, \Sigma, \mu)$. Now let v be any upper bound of C. Then $v \geq c$ pointwise almost everywhere for each $c \in C$ and hence $v \geq d$ pointwise almost everywhere. Thus d is the supremum of C in $L^p(\Omega, \Sigma, \mu)$. $\square$

It is an easy consequence of the above that closed and bounded subsets of L^p are chain-complete.

Proposition 8.2.1. *Let (Ω, Σ, μ) be a measure space and let $p \in [1, \infty)$. Then the unit ball in $L^p(\Omega, \Sigma, \mu)$ is chain-complete.*

Note that, in spaces of continuous functions with the uniform norm, the unit ball will not be chain-complete: For example, on the interval $[-1, 1]$, consider the sequence of continuous functions that are equal to 0 on $[-1, 0]$, equal to 1 on $\left(\frac{1}{n}, 1\right]$ and equal to nx on $\left(0, \frac{1}{n}\right]$. For this sequence, there is no continuous function that could act as the supremum.

8.3. The Abian-Brown Theorem

We can now prove that, in chain-complete ordered sets, the existence of a point that is comparable to its image implies the existence of a fixed point. Recall that an ordered set is *well ordered* if and only if every nonempty subset has a smallest element. In a well ordered set, the element c^- is the *immediate predecessor* of c if and only if $c^- < c$ and there is no element z so that $c^- < z < c$.

Theorem 8.3.1 (The Abian–Brown Theorem) [2]. *Let P be a chain-complete ordered set and let $f : P \to P$ be order-preserving. If there is a $p \in P$ with $p \le f(p)$, then f has a fixed point above p.*

Proof. (This proof follows the argument in Ref. [1].) A well ordered subset C of P is called an *f-chain starting at p* if and only if the following hold:

1. The point p is the smallest element of C.

2. Using ordinal number terminology for elements of C, we have that, for every $c \in C \setminus \{p\}$,

$$c = \begin{cases} f(c^-); & \text{if } c \text{ has an immediate predecessor } c^-, \\ \bigvee_P \{x \in C : x < c\}; & \text{if } c \text{ does not have an immediate predecessor.} \end{cases}$$

A chain C_1 is called an *initial segment* of a chain C_2 if and only if $C_1 \subseteq C_2$ and there is no $b \in C_2 \setminus C_1$ and $u \in C_1$ so that $b < u$. We will now prove that, if C, K are two f-chains starting at p, then either C is an initial segment of K or K is an initial segment of C. Let C, K be f-chains starting at p and let $\mathscr{I}$ be the set of all chains $I \subseteq C \cap K$ so that I is an initial segment of C and of K. Then $\mathscr{I}$ is not empty, because $\{p\}$ is in $\mathscr{I}$. Consider $H := \bigcup \mathscr{I}$. Clearly $H \subseteq C \cap K$. Let $c \in C \setminus H = \bigcap_{I \in \mathscr{I}} C \setminus I$. Then c is an upper bound of all $I \in \mathscr{I}$ and thus it is an upper bound of H. Hence H is an initial segment of C. Similarly, H is an initial segment of K. Now assume $K \setminus H \ne \emptyset$ and $C \setminus H \ne \emptyset$. If H has a largest element h, then $f(h)$ is the immediate successor of h in C and in K. Hence $H \cup \{f(h)\}$ is an initial segment of C and of K, which is a contradiction. If H does not have a largest element, then $\bigvee H$ exists in P and it is the supremum of H in C and in K. Hence, in this case, $H \cup \{\bigvee H\}$ is an initial segment of C and of K, which is a contradiction. Thus $H = C$ or $H = K$ and one of C and K is an initial segment of the other.

Because, for any two f-chains starting at p, one must be an initial segment of the other, the union U of the set of all f-chains starting at p is again an f-chain starting at p. This set U is the unique *maximal f-chain* starting at p. (Note that we did not need Zorn's Lemma to prove the existence of this maximal f chain starting at p.) Because P is chain-complete, U must have a supremum s and because U is the maximal f-chain starting at p, we must have $s \in U$ and $f(s) = s$. $\qquad \square$

It can actually be proved that the fixed point constructed in the proof of the Abian-Brown Theorem is the *smallest* fixed point of f above p. Because we will not need this fact here, we leave it as a good exercise for the reader.

A typical application of the Abian-Browm Theorem in analysis is to show that an order-preserving operator T for which there is a function u so that $Tu \geq u$ has a fixed point above u. This is done by explicitly demanding that the iterates in a T-chain are bounded above or by imposing conditions that assure boundedness and convergence.

Corollary 8.3.1. (Compare with Theorem 5.3 in Ref. [20], where additional conditions are imposed to assure continuity of the solution.) *Let (Ω, Σ, μ) be a measure space and consider the Hammerstein integral equation*

$$u(t) = v(t) + \int_\Omega k(t,s) f(s,u) \, d\mu(s), \qquad t \in \Omega,$$

where all functions assume values in $\mathbb{R}$. Assume that the following hold:

(i) *$k : \Omega \times \Omega \to [0, \infty)$ only assumes nonnegative values and is measurable and bounded;*

(ii) *$f : \Omega \times L^p(\Omega, \Sigma, \mu) \to L^p(\Omega, \Sigma, \mu)$ is such that $f(s, \cdot)$ is order-preserving for almost every $s \in \Omega$;*

(iii) *There exist $a, b \in L^p(\Omega, \Sigma, \mu)$ so that $a \leq b$, $a(t) \leq v(t) + \int_\Omega k(t,s) f(s,a) \, d\mu(s)$, and $v(t) + \int_\Omega k(t,s) f(s,b) \, d\mu(s) \leq b(t)$.*

Then the above integral equation has a solution u so that $a \leq u \leq b$.

Proof. Consider the set $[a,b] := \{u \in L^p(\Omega, \Sigma, \mu) : a \leq u \leq b\}$. Because every element is bounded above by b and below by a, we conclude via Lemma 8.2.1 and its dual that $[a,b]$ is chain-complete. By assumptions (i) and (ii), the operator

$$Tu := v(t) + \int_\Omega k(t,s) f(s,u) \, d\mu(s)$$

is order-preserving. By assumption (iii), T maps $[a,b]$ to itself and, in particular, $Ta \geq a$. (Technically, we would also need to combine all three assumptions to show T is well defined). Hence, by the Abian-Brown Theorem, T has a fixed point in $[a,b]$ and this fixed point is a solution to the indicated equation.   □

Note that Corollary 8.3.1 vitally depends on conditions that assure that the operator in question maps a chain-complete set with a smallest and a largest element to itself. From the point of view of the Abian-Brown Theorem, these conditions assure that there is a point so that $a \leq Ta$ and that the T-chain starting at a has a supremum. This need for a starting point for a T-chain and for a way to guarantee that the T-chain does not grow beyond all bounds in a function space is attached to many results that use order-theoretical methods in analysis. On one hand, this means that the main step needed to apply various results is to find a situation in which a point $a \leq Ta$ exists. On the other hand, there are not too many ways in which the iterates $T^n u$ of an order-preserving map on an element u can behave.

(a) It could be that there are $n, k \in \mathbb{N}$ so that $T^{n+k}u$ is comparable to $T^n u$, say, without loss of generality, $T^{n+k}u \geq T^n u$. In this case, we could assume, without loss of generality, that $T^{n+k}u \geq T^n u$ for all $n \in \mathbb{N}$. In a function space like $L^p(\Omega, \Sigma, \mu)$, we would then obtain a point $x \leq Tx$ as follows: Let x be the pointwise a.e. supremum of $\{Tu, T^2 u, \ldots, T^k u\}$. Then Tx is an upper bound of $\{T^2 u, T^3 u, \ldots, T^{k+1}u\}$ and, because $T^{1+k}u \geq Tu$, it is an upper bound of $\{Tu, T^2 u, \ldots, T^k u\}$. Because x is the supremum of $\{Tu, T^2 u, \ldots, T^k u\}$, we infer $Tx \geq x$.

(b) The only other possibility is that no two $T^m u$ and $T^n u$ with $m \neq n$ are comparable. In this case, if none of the sets $A_m := \{T^n u : n \geq m\}$ has upper or lower bounds, then there is no fixed point of T that is comparable to any of the $T^n u$. It would be interesting to determine if there are useful results in the case where one of the A_m has an upper or a lower bound.

Sometimes, the starting point $u \leq Tu$ for the Abian-Brown Theorem can be constructed from other conditions. In Ref. [21], which treats a slightly different Hammerstein integral equation, a certain upper bound on f assures the existence of a function $u \leq Tu$. The proof uses the Banach contraction principle to obtain an upper bound for a T-chain. Recently, the Banach contraction principle has been used as an explicit condition that only applies to functions that are comparable, that is, $a \leq b$ implies $\|Ta - Tb\|_p < c\|a - b\|_p$. For an overview of such conditions that connect order- and graph-theoretical ideas to the Banach contraction principle, consider Ref. [24], which summarizes a wide variety of conditions in a small number of succinct theorems, as well as Chapter 7 in this volume.

8.4. Comparative Retractions

As we return to retractions, recall that, by Proposition 8.1.1, retracts of ordered sets with the fixed point property have the fixed point property too. Rival [35] demonstrated a situation in which the fixed point property for the retract implies the fixed point property for the surrounding ordered set. Theorem 8.4.1 below is the natural generalization of Rival's seminal result to chain-complete ordered sets.

Definition 8.4.1. Let P be an ordered set and let $p, q \in P$. We write $p \sim q$ if and only if $p \leq q$ or $p \geq q$.

Definition 8.4.2. Let P be an ordered set. A retraction $r : P \to P$ is called a *comparative retraction* if and only if, for all $p \in P$, we have $r(p) \sim p$.

Theorem 8.4.1. (Compare with Proposition 1 in Ref. [35].) *Let P be a chain-complete ordered set and let $r : P \to P$ be a comparative retraction. Then P has the fixed point property if and only if $r[P]$ has the fixed point property.*

Proof. The direction "$\Rightarrow$" follows from Proposition 8.1.1.

For the converse "$\Leftarrow$," let $r[P]$ have the fixed point property and let $f : P \to P$ be order-preserving. Then $r \circ f|_{r[P]}$ is an order-preserving self map of $r[P]$. Because $r[P]$ has the fixed point property, there is an $x \in r[P]$ so that $r \circ f|_{r[P]}(x) = x$. If $f(x) = x$, then x is a fixed point of f. In the case where $f(x) \neq x$, note that, because r is a comparative retraction, we have that $x = r \circ f|_{r[P]}(x)$ is comparable to $f|_{r[P]}(x) = f(x)$. By the Abian-Brown Theorem (Theorem 8.3.1) or its dual, f must have a fixed point.   □

Theorem 8.4.1 can be applied repeatedly until a certain subset is reached or until there are no more comparative retractions. This is the idea of dismantlability.

Definition 8.4.3. Let P be a chain-complete ordered set. Then we say that P is *dismantlable* to $Q \subseteq P$ if and only if there are subsets $P = P_0 \supseteq P_1 \supseteq \cdots \supseteq P_n = Q$ and comparative retractions $r_i : P_{i-1} \to P_i$ so that $r[P_{i-1}] = P_i$. If Q is a singleton, we will simply call P *dismantlable*.

Clearly, if a chain-complete ordered set is dismantlable to an ordered set with the fixed point property, then the original ordered set has the fixed point property too.

Corollary 8.4.1. *Let P be a chain-complete ordered set that is dismantlable to an ordered set Q. Then P has the fixed point property if and only if Q has the fixed point property. In particular, dismantlable chain-complete ordered sets have the fixed point property.*

Proof. This result follows from Theorem 8.4.1 if we can establish that retracts of chain-complete ordered sets are chain-complete.

So let P be a chain-complete ordered set, let $r : P \to r[P]$ be a retraction and let $C \subseteq r[P]$ be a nonempty chain. Then C has a supremum s in P. Because r is order-preserving, $r(s)$ is an upper bound of $C = r[C]$. Let $u \in r[P]$ be an upper bound of C. Because s is the supremum of C in P, we infer $u \geq s$. Because r is order-preserving, $u = r(u) \geq r(s)$ follows. Hence $r(s)$ is the supremum of C in $r[P]$.

Existence of the infimum is established dually. Hence $r[P]$ is chain-complete.   □

There may be more than one comparative retraction on a given ordered set, which leads to different ways to dismantle an ordered set. Hence a given ordered set P can usually be dismantled to different ordered subsets Q_i. When considering the fixed point property, the dismantling process is typically run until there is no nontrivial comparative retraction on the remaining ordered set. It can be shown that, if we use comparative retractions on chain-complete ordered sets and the process terminates after finitely many steps, this resulting ordered set is unique up to isomorphism (see Ref. [40]).

Theorem 8.4.2 below is a different spin on the idea of applying order-theoretical ideas in analysis. It does not have an explicit hypothesis regarding the existence of a function u so that $Tu \geq u$ to start the process. Instead, the appropriate comparative retractions assure that fixed points will exist. The proof in Ref. [9], which looks quite different from the argument given here and does not mention retractions, constructs the fixed point by, essentially, "tracing back a sequence of dismantling retractions." That is, it first constructs a fixed point of $r_1 \circ T|_{P_1}$ and, from that, it constructs a fixed point of T. (This process can be extended to dismantling sequences of arbitrary length.)

Theorem 8.4.2. (Also see Ref. [9], Theorem 2.44.) *Let (Ω, Σ, μ) be a measure space and let $p \in [1, \infty)$. Every order-preserving self map of the unit ball in $L^p(\Omega, \Sigma, \mu)$ has a fixed point.*

Proof. By Proposition 8.2.1, the unit ball $B_1(0)$ in $L^p(\Omega, \Sigma, \mu)$ is chain-complete. The function $r_1 : B_1(0) \to \{f \in B_1(0) : f \geq 0\}$ defined by $r_1(f) := \max\{f, 0\}$, where the maximum is pointwise a.e., is a comparative retraction. Moreover, the function $r_2 : \{f \in B_1(0) : f \geq 0\} \to \{0\}$ defined by $r_2(f) := 0$ is a comparative retraction too. Hence $B_1(0)$ is dismantlable to a singleton, which means that it has the (order-theoretical) fixed point property. Therefore every order-preserving self map of $B_1(0)$ has a fixed point. $\qquad\square$

Note that Theorem 8.4.2 also holds for $p = \infty$ with the same proof, because the pointwise a.e. supremum of functions with bounded uniform norm is a bounded function too. This insight does not contradict the remark after Proposition 8.2.1, because we do not demand continuity of the functions. More examples of spaces in which Theorem 8.4.2 holds, as well as applications of these results, are given in Ref. [9]. To see that Theorem 8.4.2 provides the existence of fixed points for functions for which fixed points are not obvious to find (unless we wish to mimic the proof of Theorem 8.4.2), consider the following example.

Example 8.4.1. Let ℓ^1 be the space of sequences $\{a_n\}_{n=1}^{\infty}$ of real numbers so that

$\sum_{n=1}^{\infty} |a_n| < \infty$, equipped with the usual ℓ^1-norm $\|\{a_n\}_{n=1}^{\infty}\|_1 := \sum_{n=1}^{\infty} |a_n|$. Let $\text{Hom}(\mathbb{R}, (-1,1))$ be the set of order-preserving functions from the real numbers $\mathbb{R}$ to the interval $(-1,1)$ and let $\Phi : \ell^1 \to \text{Hom}(\mathbb{R}, (-1,1))$ be an order-preserving function.

As is customary, let $\hat{a}$ denote the absence of the element in the sequence. Define $g : \ell^1 \to \ell^1$ by $\pi_m \circ g(\{a_n\}_{n=1}^{\infty}) := \frac{1}{2^m} \Phi(a_1, \ldots, a_{m-1}, \widehat{a_m}, a_{m+1}, \ldots)[a_m]$, that is, the mth component of $g(\{a_n\}_{n=1}^{\infty})$ is obtained by applying to a_m the function that Φ returns for the sequence obtained from $\{a_n\}_{n=1}^{\infty}$ by deleting the mth entry and then multiplying by $\frac{1}{2^m}$.

Clearly, each function g as defined above is order-preserving. Because each codomain is contained in the unit ball of ℓ^1, each function g maps the unit ball of ℓ^1 to itself. Thus, by Theorem 8.4.2, each such function g has a fixed point. Aside from applying Theorem 8.4.2 or mimicking its proof, the author is unaware how to prove this result.

For applications of the Abian-Brown Theorem and results like Theorem 8.4.2 in analysis and game theory, Ref. [9] is a quite comprehensive resource. It would be interesting to know if there are applications of dismantlability to analysis that go beyond the Abian-Brown Theorem and results like Theorem 8.4.2.

We conclude this section by noting that comparative retractions are also involved in an interesting structural result due to Li and Milner [30]: Every chain-complete ordered set that does not contain an infinite subset of mutually incomparable points can be dismantled to a finite ordered set!

8.5. Irreducible Points

Rival's original version of Theorem 8.4.1 was for ordered sets without infinite chains. In such ordered sets, the existence of a comparative retraction is equivalent to the existence of an irreducible point. We start with the idea of a lower/upper cover.

Definition 8.5.1. Let P be an ordered set and let $\ell, u \in P$. Then u is called an *upper cover* of ℓ and ℓ is called a *lower cover* of u if and only if $\ell < u$ and there is no $z \in P$ so that $\ell < z < u$. In this case, we also write $\ell \prec u$.

The relation $\prec$ is used to represent finite ordered sets with a *Hasse diagram* as in Figure 8.3. For each element of the ordered set, a point is placed in the plane. For points $\ell \prec u$, the y-coordinate of the point representing u must be greater than the y-coordinate of the point representing ℓ and the two points are joined by a straight line or sometimes a curve. So, in Figure 8.3, for example, c is an upper cover of a_1 and a_1 is a lower cover of c, and there are no covering relations between any of a_1, b, a_2 and

f, g, h. Elements p and q satisfy the relation $p \leq q$ if and only if there is a path from p to q that may go through other elements of the set, but for which all segments are traversed in the upward direction. So, in Figure 8.3, for example, we have $a_1 \leq f$.

Definition 8.5.2. Let P be an ordered set without infinite chains and let $x \in P$. Then x is called *irreducible* if and only if x has exactly one upper cover or x has exactly one lower cover.

Irreducible points are easy to find in a Hasse diagram. The ordered set in Figure 8.3 has no irreducible points. The leftmost ordered set in Figure 8.1 has exactly one irreducible point, which happens to have a unique lower cover.

Readers who started reading with this section may skip Proposition 8.5.1 and continue reading below it. Also, for a finite ordered set, the Abian-Brown Theorem simply states that, if $x \leq f(x)$, then the iteration $x \leq f(x) \leq f^2(x) \leq \cdots$ must ultimately end in $f^n(x) = f^{n+1}(x)$ for some $n \in \mathbb{N}$, that is, it must end with a fixed point.

Definition 8.5.3. Let P be an ordered set. Then $m \in P$ is a *minimal element* if and only if, for all $x \in P$, $x \leq m$ implies $x = m$. The dual notion of a minimal element is a *maximal element*.

Proposition 8.5.1. *Let P be an ordered set without infinite chains. Then there is a nontrivial comparative retraction $r : P \to P$ if and only if P has an irreducible point.*

Proof. For the direction "$\Rightarrow$," first note that, if r is a comparative retraction, then $r_u(x) := \max\{r(x), x\}$ is a retraction so that, for all $x \in P$, we have $r(x) \geq x$. Dually, $r_l(x) := \min\{r(x), x\}$ is a retraction so that, for all $x \in P$, we have $r(x) \leq x$. Hence we can assume, without loss of generality, that $r(x) \leq x$ for all $x \in P$ and $r(p) < p$ for some $p \in P$.

Because P does not have infinite chains, every subset of P has a minimal element. Let m be a minimal element of the subset $\{p \in P : r(p) < p\}$. Then, for all $x < m$, we have $x = r(x) \leq r(m) < m$. Therefore $r(m)$ is the unique lower cover of m.

For the direction "$\Leftarrow$," note that, if $x \in P$ has, without loss of generality, a unique lower cover ℓ, then the function r that maps x to ℓ and fixes all other elements of P is a comparative retraction from P to $P \setminus \{x\}$. $\qquad \square$

Rival's theorem can easily be derived from Theorem 8.4.1 and the "$\Leftarrow$" direction of the proof of Proposition 8.5.1. We prove it directly here.

Theorem 8.5.1 [35]. *Let P be a finite ordered set and let $x \in P$ be irreducible. Then P has the fixed point property if and only if $P \setminus \{x\}$ has the fixed point property.*

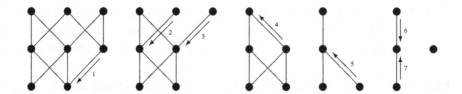

Figure 8.1 An eight-point ordered set that is dismantlable by irreducibles. Arrows indicate re-tractions to unique upper or lower covers. Going from left to right, each ordered set is obtained from the previous one by removing the points at the base of each arrow.

Proof. Without loss of generality, assume that x has a unique lower cover ℓ. First note that the function $r : P \to P \setminus \{x\}$ that maps x to its unique lower cover ℓ and that keeps all other points fixed is a retraction. In particular, this means that the direction "$\Rightarrow$" follows from Proposition 8.1.1.

For the converse "$\Leftarrow$," let $P \setminus \{x\}$ have the fixed point property and let $f : P \to P$ be order-preserving. Then $r \circ f|_{P \setminus \{x\}}$ is an order-preserving self map of $P \setminus \{x\}$. Because $P \setminus \{x\}$ has the fixed point property, there is a $p \in P \setminus \{x\}$ so that $r \circ f|_{P \setminus \{x\}}(p) = p$. If $f(p) = p$, then p is a fixed point of f. In the case where $f(p) \neq p$, we must have $p = x$ and $f(p) = \ell$. Hence $x \geq \ell = f(x)$ and, by the dual of the Abian-Brown Theorem (Theorem 8.3.1), f must have a fixed point.   □

Clearly, the removal of irreducible points can be iterated (see Figure 8.1).

Definition 8.5.4. Let P be a finite ordered set. Then we say that P is *dismantlable* or *dismantlable by irreducibles* to $Q \subseteq P$ if and only if there are subsets $P = P_0 \supseteq P_1 \supseteq \cdots \supseteq P_n = Q$ so that, for $i = 1, \ldots, n$, the ordered set P_i is obtained from P_{i-1} by removing an irreducible point. If Q is a singleton, we will simply call P *dismantlable* or *dismantlable by irreducibles*.

Corollary 8.5.1. *Let P be a finite ordered set that is dismantlable to an ordered set Q. Then P has the fixed point property if and only if Q has the fixed point property. In particular, dismantlable finite ordered sets have the fixed point property.*

Proposition 8.5.1/Theorem 8.5.1 is the reason why, for finite ordered sets, disman-tlability is also called dismantlability by irreducibles. Similar to infinite ordered sets, if we dismantle until there are no irreducible points left, the resulting set is unique up to isomorphism (see Refs. [11, 14, 40]). Dismantlability by irreducibles is a nice structural tool for the fixed point property: Ordered sets of height 1 or of width 2 have the fixed point property if and only if they are dismantlable by irreducibles (see

Refs. [15, 35]). Moreover, an ordered set with up to eight elements has the fixed point property if and only if it is dismantlable (see Ref. [37]).

8.6. Constraint Propagation

Dismantlability by irreducibles also has nice algorithmic consequences. In Ref. [48], it was shown, albeit in different terminology, that dismantlability by irreducibles means that enforcing $(1,1)$-consistency on the corresponding expanded constraint network for deciding whether P has a fixed point free order-preserving self map produces an empty network. Although this could be seen as "merely" providing another polynomial algorithm for dismantlable ordered sets, it may point the way towards deeper results and useful algorithms (see the remarks after Theorem 8.7.2). Because the number of definitions needed can be a bit overwhelming, the brief overview presented here will focus solely on $(1,1)$-consistency and on networks for the fixed point property. For a more comprehensive introduction to constraint propagation for constraint satisfaction problems in general, see Refs. [6, 36, 46]. For the connection to the fixed point property, see Ref. [41].

Definition 8.6.1. Let P be a finite ordered set. The *expanded constraint network* (terminology consistent with Ref. [31]) for deciding whether P has a fixed point free order-preserving self map is a graph $N(P)$ with the following vertices and edges. (Also see Figure 8.2, even if Figure 8.2 confirms that expanded constraint networks are not necessarily visual tools.)

(a) The vertices are the pairs (x,y) of points $x,y \in P$ so that $x \not\sim y$.

(b) The set $\{(x,y),(u,v)\}$ is an edge of the expanded constraint network if and only if $x \neq u$ and one of the following holds: $x \not\sim u$, or $x < u$ and $y \leq v$, or $x > u$ and $y \geq v$.

So, the vertices are rather trivial fixed point free order-preserving partial self maps: They are functions with domain size 1. Similarly, edges are fixed point free order-preserving partial self maps with domain size 2. A *clique* in a graph is a set C of vertices, so that, for any two distinct $a,b \in C$, we have that $\{a,b\}$ is an edge. The ordered set P has a fixed point free order-preserving self map if and only if the expanded constraint network $N(P)$ contains a clique of size $|P|$: Clearly, a fixed point free order-preserving self map f is a set of pairs $f = \{(x,y) : x \in P\}$ so that any single pair (x,y) satisfies (a) above and so that any doubleton $\{(x,y),(u,v)\} \subseteq f$ satisfies (b) above. Conversely, a clique C of size $|P|$ will contain, for each $x \in P$, one (x,y) so that (a) above holds and, for any two $x,u \in P$ and $\{(x,y),(u,v)\} \subseteq C$ we have, by (b) above, that $x \leq u$ implies $y \leq v$, and $u \leq x$ implies $v \leq y$. Hence C is, in fact, a fixed point free order-preserving self map of P.

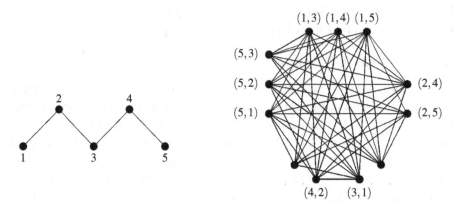

Figure 8.2 The expanded constraint network $N(P)$ (right) for a 5-fence (left). Note that, for example, the vertex $(3,1)$ is not $(1,1)$-consistent (see Definition 8.6.2).

The idea of constraint propagation is now to use steps that can be performed in polynomial time to remove vertices and edges that cannot be part of a clique with $|P|$ vertices. If the removal procedure ultimately produces a graph with no edges, then there is no fixed point free order-preserving self map. If the removal procedure terminates with a graph that has edges, at least the search space has been somewhat reduced. However, exhaustive search would still need to be used to determine if there is a clique of size $|P|$ or not. This is not surprising, as the problem whether a given finite ordered set has a fixed point free order-preserving self map is NP-complete (see Ref. [12]). Nonetheless, constraint propagation often provides significant speedup in computer searches and, in the cases in which it terminates with a graph without edges, it provides the solution in polynomial time. It would therefore be interesting to have results that guarantee that, in certain cases, constraint propagation produces an empty network (which the author calls "terminating correctly"). We will consider the enforcement of $(1,1)$-consistency here and will allude to more sophisticated techniques in the open questions at the end of Section 8.7.

Definition 8.6.2. Let P be a finite ordered set. In the expanded constraint network $N(P)$, a vertex (x,y) is called *(1,1)-consistent* if and only if, for all $u \neq x$, there is a v so that $\{(x,y),(u,v)\}$ is an edge.

Obviously, vertices that are *not* $(1,1)$-consistent cannot be part of a clique of size $|P|$. Hence these vertices can be removed without affecting the cliques of size $|P|$ in the network. Enforcing $(1,1)$-consistency on a network means to successively remove vertices that are not $(1,1)$-consistent until, in the remaining network, all vertices are $(1,1)$-consistent. It can be shown that the final network that remains does not depend on the order of removal for the vertices that are not $(1,1)$-consistent. It is also a good exercise to show that $(1,1)$-consistency can be enforced on a network in polynomial time.

It turns out that the presence or absence of irreducible points in P is irrelevant to an algorithm that enforces $(1,1)$-consistency.

Definition 8.6.3. Let P be an ordered set and let $x \in P$. We define the *up-set* of x to be $\uparrow x := \{p \in P : p \geq x\}$. Dually, we define the *down-set* of x to be $\downarrow x := \{p \in P : p \leq x\}$. Moreover, we define the *neighborhood* of x to be $\updownarrow x := \{p \in P : p \sim x\}$.

Theorem 8.6.1 [48]. *Let P be a finite ordered set with an irreducible point a. If enforcing $(1,1)$-consistency on $N(P \setminus \{a\})$ produces an empty network, then enforcing $(1,1)$-consistency on $N(P)$ produces an empty network too.*

Proof. Without loss of generality, assume that a has a unique lower cover ℓ. First consider vertices (x,a) of $N(P)$ so that $x > \ell$ and $x \not\geq a$. Suppose for a contradiction that there is a vertex (ℓ,y) so that $\{(x,a),(\ell,y)\}$ is an edge. Then we would have that $y \leq a$ and therefore $y \sim \ell$, which is not possible. Hence none of the vertices (x,a) with $x > \ell$ and $x \not\geq a$ is $(1,1)$-consistent in $N(P)$. Let H be the network obtained from $N(P)$ by removing all vertices (x,a) with $x > \ell$ and $x \not\geq a$. Note that every doubleton set $\{(x,y),(u,v)\}$ so that $a \notin \{x,y,u,v\}$ is an edge of H if and only if it is an edge of $N(P \setminus \{a\})$.

Let (z,a) be a vertex of H. Then, because vertices (z,a) with $z > \ell$ and $z \not\geq a$ have been removed, $z \not> \ell$, and, because $\ell < a$ and by definition of $N(P)$, $z \not\leq \ell$. Therefore if (z,a) is a vertex of H, then (z,ℓ) is a vertex of H. Conversely, if (z,ℓ) is a vertex of H, then $z \not> \ell$ and $z \not\sim a$ and hence (z,a) is a vertex of H. Hence (z,a) is a vertex of H if and only if (z,ℓ) is a vertex of H.

Let $r : P \to P \setminus \{a\}$ be the retraction that maps a to ℓ and that fixes all other points, and let $\{(z,a),(u,v)\}$ be an edge of H. Then $z \leq u$ implies $a \leq v$, and $u \leq z$ implies $v \leq a$. Therefore $z \leq u$ implies $\ell \leq r(v)$, and $u \leq z$ implies $r(v) \leq \ell$. We conclude that, if $\{(z,a),(u,v)\}$ is an edge of H, then $\{(z,\ell),(u,r(v))\}$ is an edge of H too.

By hypothesis, there is an order $c_1, \ldots, c_m$ in which the vertices of $N(P \setminus \{a\})$ can be listed so that, for $k = 1, \ldots, m$, we have that c_k is not $(1,1)$-consistent in $N(P \setminus \{a\}) - c_1, \ldots, c_{k-1}$. For $k = 1, \ldots, m$, if $c_k = (z,\ell)$, let $d_k := (z,a)$ and, if the second component of c_k is not ℓ, let $d_k = c_k$. We claim that, for $k = 1, \ldots, m$, the vertices c_k and d_k are not $(1,1)$-consistent in $H - c_1, d_1, \ldots, c_{k-1}, d_{k-1}$. Inductively, let $k \geq 1$ and assume that, for $j = 1, \ldots, k - 1$, we have already established that the vertices c_j and d_j are not $(1,1)$-consistent in $H - c_1, d_1, \ldots, c_{j-1}, d_{j-1}$. Note that, because we remove vertices and because each d_j is either equal to c_k or has second component a, every doubleton set $\{(x,y),(u,v)\}$ so that $a \notin \{x,y,u,v\}$ is an edge of $H - c_1, d_1, \ldots, c_{k-1}, d_{k-1}$ if and only if it is an edge of $N(P \setminus \{a\}) - c_1, \ldots, c_{k-1}$. Now consider the vertices c_k and d_k.

Because $c_k =: (z,y)$ is not $(1,1)$-consistent in $N(P \setminus \{a\}) - c_1, \ldots, c_{k-1}$, there is a $u \in P \setminus \{z,a\}$ so that there is no edge of the form $\{(z,y),(u,v)\}$ in $N(P \setminus \{a\}) - c_1, \ldots, c_{k-1}$. Suppose for a contradiction that there is an edge of the form $\{(z,y),(u,v)\}$ in $H - c_1, d_1, \ldots, c_{k-1}, d_{k-1}$. Then we must have $v = a$. However, then (u,ℓ) is a vertex of H too. Hence (u,ℓ) is a vertex of $N(P \setminus \{a\})$ and $\{(z,y),(u,\ell)\} = \{(z,r(y)),(u,\ell)\}$ is an edge of H and thus of $N(P \setminus \{a\})$. Hence, because there is no edge of the form $\{(z,y),(u,v)\}$ in $N(P \setminus \{a\}) - c_1, \ldots, c_{k-1}$, there is a $j < k$ so that $(u,\ell) = c_j$ and then $(u,v) = (u,a) = d_j$, a contradiction to $\{(z,y),(u,v)\}$ being an edge in $H - c_1, d_1, \ldots, c_{k-1}, d_{k-1}$. Thus c_k is not $(1,1)$-consistent in $H - c_1, d_1, \ldots, c_{k-1}, d_{k-1}$.

If $d_k = c_k$, then, trivially, d_k is not $(1,1)$-consistent in $H - c_1, d_1, \ldots, c_{k-1}, d_{k-1}$ either. If $d_k \neq c_k$, then $c_k = (z,\ell)$ and $d_k = (z,a)$. Suppose, for a contradiction that there is an edge of the form $\{(z,a),(u,v)\}$ in $H - c_1, d_1, \ldots, c_{k-1}, d_{k-1}$. Then $\{(z,\ell),(u,r(v))\}$ is an edge in H. Because $H - c_1, d_1, \ldots, c_{k-1}, d_{k-1}$ has no edges of the form $\{(z,\ell),(u,r(v))\}$, we conclude that there is a $j < k$ so that $(u,r(v)) = c_j$. Because $\{(z,a),(u,v)\}$ is an edge in $H - c_1, d_1, \ldots, c_{k-1}, d_{k-1}$, we must have $v \neq r(v)$ and hence $v = a$. Now, because $(u,\ell) = (u,r(v)) = c_j$, we have $(u,a) = d_j$, a contradiction to $\{(z,a),(u,a)\}$ being an edge in $H - c_1, d_1, \ldots, c_{k-1}, d_{k-1}$. Thus d_k is not $(1,1)$-consistent in $H - c_1, d_1, \ldots, c_{k-1}, d_{k-1}$.

We have proved that, for $k = 1, \ldots, m$, the vertices c_k and d_k are not $(1,1)$-consistent in $H - c_1, d_1, \ldots, c_{k-1}, d_{k-1}$. In particular, the vertex d_k is not $(1,1)$-consistent in $H - c_1, d_1, \ldots, c_{k-1}, d_{k-1}, c_k$. Similarly, if $(x_1,a), \ldots, (x_M,a)$ is any enumeration of the vertices (x,a) with $x > \ell$ and $x \not\geq a$, then, for $j = 1, \ldots, M$, the vertex (x_j,a) is not $(1,1)$-consistent in $N(P) - (x_1,a), \ldots, (x_{j-1},a)$. Hence, one way to enforce $(1,1)$-consistency on $N(P)$ is to start by removing the vertices (x,a) with $x > \ell$ and $x \not\geq a$ in any order and to continue by removing the vertices $c_1, d_1, \ldots, c_m, d_m$ in this order.

The above process terminates with the network $H - c_1, d_1, \ldots, c_m, d_m$. In this network, only vertices with first component a remain, and they are all isolated and hence not $(1,1)$-consistent. We conclude that enforcing $(1,1)$-consistency on $N(P)$ produces an empty network. $\qquad\square$

The converse of Theorem 8.6.1 is true too (see Ref. [41] or [48]). However, for our purposes, we will not need it. Theorem 8.6.1 provides a performance guarantee for enforcing $(1,1)$-consistency on networks $N(P)$.

Corollary 8.6.1. *Let P be a finite ordered set that is dismantlable by irreducibles. Then enforcing $(1,1)$-consistency on the network $N(P)$ produces an empty graph.*

It turns out (see Refs. [48]) that dismantlable ordered sets are exactly the ordered sets P for which enforcing $(1,1)$-consistency on $N(P)$ produces an empty graph. This means that, for further investigation of the fixed point property with constraint propagation algorithms, other notions of consistency are needed. We will consider the idea of $(2,1)$-consistency of edges in Definition 8.7.4 below.

Corollary 8.6.2. *Let P be a finite ordered set of width 2 or height 1. Then enforcing $(1,1)$-consistency on the network $N(P)$ and checking whether the remaining network has vertices or not provides a polynomial time algorithm to determine whether P has a fixed point free order-preserving self map or not.*

Corollary 8.6.2 shows that the very general technique of enforcing $(1,1)$-consistency provides a polynomial time solution to some natural special cases of an NP-complete problem. This is nice, but not surprising. Ordered sets of width 2 or height 1 have the fixed point property if and only if they are dismantlable by irreducibles and dismantlability by irreducibles can be checked in polynomial time: Checking if a given point in an ordered set P is irreducible takes at most $|P|$ steps. Thus it takes at most $|P|^2$ steps to determine if there is an irreducible point. Hence, in at most $|P|^3$ steps, we can determine if a finite ordered set is dismantlable by irreducibles.

Consequently, the primary attraction of Theorem 8.6.1 and Corollary 8.6.2 may well be that they show that a general technique performs well under special circumstances. The author only knows of such results in the context of the fixed point property for ordered sets (see Refs. [10, 41, 48]) and in one specific context in constraint satisfaction (see Refs. [47]). It would be interesting to see if such results could be expanded to more general contexts, including ordered sets for which the fixed point property cannot be determined as easily as for dismantlable ordered sets. In the next section, we will discuss the current state of these efforts.

8.7. Retractable Points

Retractable points are a natural generalization of irreducible points.

Definition 8.7.1 [39]. Let P be an ordered set and let $a, b \in P$. If

$$\uparrow a \setminus \{a\} \subseteq \uparrow b \quad \text{and} \quad \downarrow a \setminus \{a\} \subseteq \downarrow b,$$

then a is called *retractable* to b.

Note that a retractable point a is irreducible if and only if a is comparable to $b = r(a)$ (good exercise), but that there are retractable points that are not irreducible

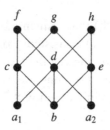

Figure 8.3 The only (up to isomorphism and duality) nondismantlable ordered set with ≤ 9 elements and the fixed point property (see Ref. [37]). Note that there are no irreducible points, but a_1 and a_2 are retractable to b.

(see Figure 8.3). The analogue of Theorem 8.5.1 for retractable points requires two conditions rather than one. As long as the ordered sets in question are small enough, such as the ordered set in Figure 8.3 or the ordered set P_0 in Figure 8.4, it is still an effective tool for checking the fixed point property.

Theorem 8.7.1 [39]. *Let P be a finite ordered set and $a \in P$ be retractable to $b \in P$. Then P has the fixed point property if and only if the ordered sets $P \setminus \{a\}$ and $\updownarrow a \setminus \{a\}$ both have the fixed point property.*

Proof. First note that the function $r : P \to P \setminus \{a\}$ that maps a to b and that keeps all other points fixed is a retraction. In particular, for the direction "$\Rightarrow$," this means that the fixed point property for $P \setminus \{a\}$ follows from Proposition 8.1.1. Regarding the fixed point property for $\updownarrow a \setminus \{a\}$, note that the function that maps all points that are not comparable to a to the point b and that fixes all points that are comparable to a is a retraction of P onto $\updownarrow a \cup \{b\}$. If $\updownarrow a \setminus \{a\}$ had a fixed point free order-preserving self map, then so would $\updownarrow a \cup \{b\}$: Use the fixed point free order-preserving self map on $\updownarrow a \setminus \{a\}$, which, by the Abian-Brown Theorem, cannot map upper bounds of a to lower bounds of a or vice versa, and extend it by mapping a to b and b to a. This is a contradiction to P having the fixed point property.

For the converse "$\Leftarrow$," let $P \setminus \{a\}$ and $\updownarrow a \setminus \{a\}$ both have the fixed point property and let $f : P \to P$ be order-preserving. Then $r \circ f|_{P \setminus \{a\}}$ is an order-preserving self map of $P \setminus \{a\}$. Because $P \setminus \{a\}$ has the fixed point property, there is a $p \in P \setminus \{a\}$ so that $r \circ f|_{P \setminus \{a\}}(p) = p$. If $f(p) = p$, then p is a fixed point of f. In the case where $f(p) \neq p$, we must have $p = b$ and $f(p) = a$, that is, $f(b) = a$. By the Abian-Brown Theorem and its dual, f has a fixed point if an upper or lower bound of a is mapped to a. If this is not the case, then f maps $\updownarrow a \setminus \{a\}$ to itself and this ordered set has the fixed point property. Hence f has a fixed point in every case. $\square$

Theorem 8.7.1 provides some more insight for small ordered sets. The unique (up to isomorphism and duality) nondismantlable ordered set with at most 9 points and the fixed point property (see Figure 8.3) has two retractable points. Indeed, rather than looking for nondismantlable ordered sets with the fixed point property, we can focus on ordered sets without retractable points. Computer searches reveal that the proverbial "combinatorial explosion" may be delayed by using Theorem 8.7.1, but it is by no means prevented: All counts for the rest of this paragraph are up to isomorphism and duality. There are 4 ordered sets with 10 points that have the fixed point property and no retractable point (this can also be verified by checking the sets in Ref. [37]). There are 8 such ordered sets with 11 points (this is verified in Ref. [39]). There are 107 such sets with 12 points. In the early 1990s, after writing Refs. [39], the author attempted a proof of this fact in the style of Refs. [37, 39] but gave up. Years later, using lists of all ordered sets of a certain size and the program in Ref. [44], the numbers could be computed without writing a tedious proof. The numbers for the next sizes show that, without further theorems to reduce the number of ordered sets to be considered, an explicit classification of small ordered sets with the fixed point property seems to be out of reach and, in light of the fact that computer verification is possible, futile: There are 964 ordered sets with 13 points, the fixed point property and no retractable point, and there are 24,546 such sets with 14 points. (The numbers of ordered sets with the fixed point property and without irreducible points are, of course, even larger.)

Nonetheless, ordered sets for which repeated application of Theorem 8.7.1 can be used to check the fixed point property are an interesting class to consider (also see Theorem 8.7.2 below). Because two subsets, rather than one, must be checked to determine if an ordered set with a retractable point has the fixed point property, the natural generalization of dismantlability has potentially exponential character.

Definition 8.7.2. A finite ordered set P is called *collapsible* if and only if $|P| \in \{0, 1\}$ or there is a point x such that the following hold:

(i) $P \setminus \{x\}$ is a retract of P.

(ii) $P \setminus \{x\}$ is collapsible.

(iii) $\updownarrow x \setminus \{x\}$ is collapsible.

Definition 8.7.3. A finite ordered set P is called *connectedly collapsible* if and only if $|P| = 1$ or there is a point x such that the following hold:

(i) $P \setminus \{x\}$ is a retract of P.

(ii) $P \setminus \{x\}$ is connectedly collapsible.

(iii) $\updownarrow x \setminus \{x\}$ is connectedly collapsible.

Clearly, dismantlable finite ordered sets are connectedly collapsible: When x has, say, a unique lower cover ℓ, the element ℓ is comparable to all elements of the set $\updownarrow x \setminus \{x\}$, which means that $\updownarrow x \setminus \{x\}$ is dismantlable. (The precise proof by induction is another nice exercise.) Aside from dismantlable ordered sets, the natural classes of ordered sets of linear or interval dimension 2 provide examples of ordered sets that are, by default, collapsible. Because it would require a departure from the main thrust of this chapter, we forego a proof of the next theorem. For linear dimension 2, the proof is in the proof of Theorem 2.6 in Ref. [16]. For interval dimension 2, see the proof of Theorem 8.4.5 in the first edition of Ref. [41] or Theorem 11.22 in the second edition of Ref. [41].

Theorem 8.7.2. *Let P be an ordered set of interval dimension 2. Then P contains a point of rank 1 with a unique lower cover or P contains a minimal element a that is retractable to another minimal element b. Hence ordered sets of interval dimension 2 are collapsible.*

In Ref. [43], the author has proved that there is a polynomial time algorithm to determine whether an ordered set P of interval dimension 2 has a fixed point free order-preserving self map. Note that dimension 2 (see Ref. [45]) as well as interval dimension 2 (see Ref. [25]) can be verified in polynomial time. Hence, on one hand, existence of a polynomial algorithm to check the fixed point property for these ordered sets may be considered natural. On the other hand, the result is interesting because the algorithm verifies (connected) collapsibility for a subclass of the class of all (connectedly) collapsible ordered sets in polynomial time. Note that Example 8.8.1 below shows that, unlike for dismantlable ordered sets, the result is not as simple as Corollary 8.6.2.

For all connectedly collapsible ordered sets that the author has checked, enforcing $(2,1)$-consistency (see Definition 8.7.4 below) on the expanded constraint network provided the correct answer (that is, it terminated with an empty network). Moreover, to the author's knowledge, the smallest example of *any* ordered set for which enforcing $(2,1)$-consistency does not provide the correct answer is an ordered set with over 400 elements. Hence, enforcing $(2,1)$-consistency appears to be a strong tool indeed for the fixed point property.

Definition 8.7.4. Let P be a finite ordered set. In the expanded constraint network $N(P)$, an edge $\{(x,y),(z,w)\}$ is called *(2,1)-consistent* if and only if, for every $u \notin \{x,z\}$, there is a v so that each of $\{(x,y),(u,v)\}$ and $\{(z,w),(u,v)\}$ is an edge.

As with $(1,1)$-consistency for vertices, an edge that is *not* $(2,1)$-consistent cannot be part of a $|P|$-clique. Enforcing $(2,1)$-consistency means to remove edges that are not $(2,1)$-consistent until this is no longer possible. As for $(1,1)$-consistency, the order in which edges that are not $(2,1)$-consistent are removed does not affect the final network and $(2,1)$-consistency can be enforced in polynomial time. Enforcing $(2,1)$-consistency would now provide a simple polynomial algorithm to check if a collapsible ordered set has the fixed point property if only the following were true.

Question 8.7.1. Let P be a finite ordered set and let $a \in P$ be retractable. Is it true that, if enforcing $(2,1)$-consistency on $N(P \setminus \{a\})$ and on $N(\updownarrow a \setminus \{a\})$ produces an empty network, then enforcing $(2,1)$-consistency on $N(P)$ produces an empty network?

In the course of writing Ref. [43], the author has attempted to answer Question 8.7.1 multiple times (and an early version of Ref. [43] used expanded constraint networks to prove the result), but neither a positive answer nor a counterexample materialized. The author conjectures that the answer is "yes," but would be enthusiastic about an answer either way.

Of course, enforcing $(2,1)$-consistency would be but one attempt at a polynomial decision algorithm for the fixed point property for collapsible ordered sets. More generally, we could ask the following.

Question 8.7.2. What is the complexity of determining if a given collapsible ordered set has the fixed point property?

8.8. Verifying Connected Collapsibility Directly

A more general continuation of the line of investigation related to Questions 8.7.1 and 8.7.2 is the following.

Question 8.8.1. What is the complexity of determining if a given finite ordered set is (connectedly) collapsible? Currently, it is not even known whether the problem is NP. The author would not be surprised if the determination is NP-hard.

We have seen that dismantlability by irreducibles can be verified directly with the definition in polynomial time. Surely, direct verification with the definition is always a natural starting point for algorithms. The direct search for a retractable point a takes $|P|^2$ steps. However, if we simply attempt to directly verify the definition, then, after a retractable point a has been found, both $P \setminus \{a\}$ and $\updownarrow a \setminus \{a\}$ must be checked if they are (connectedly) collapsible. Therefore it is easy to find examples in which *poor* choices of retractable points a lead to exponential verification times in this fashion. Below, we present a sequence of connectedly collapsible ordered sets for which *any* way to choose the retractable points in the various stages leads to an exponential effort

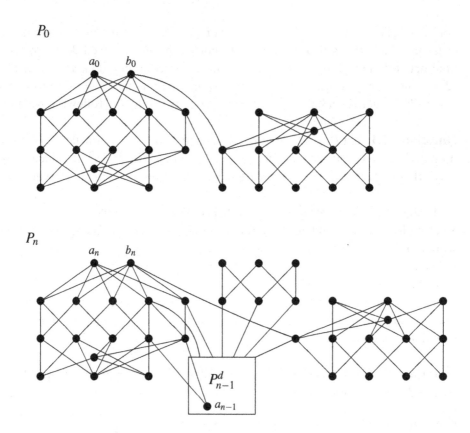

Figure 8.4 Constructing the ordered sets P_n. The ordered set P_{n-1}^d is the dual of the ordered set P_{n-1}, that is, it is the ordered set P_{n-1} with all comparabilities reversed. The connections that are attached to the box with P_{n-1}^d indicate that the point is an upper bound of P_{n-1}^d. The connections that attached to a_{n-1} indicate that the point in question is an upper cover of a_{n-1} and of no other points of P_{n-1}^d. (Note that a_{n-1} is minimal in P_{n-1}^d.)

in the *direct* verification of connected collapsibility. Hence there is no obvious answer to Question 8.8.1. For any ordered set P, the *dual* ordered set P^d is obtained from P be reversing all comparabilities, that is, $x \leq_{pd} y$ if and only if $x \geq_P y$.

Example 8.8.1. Consider the sequence of ordered sets P_n defined recursively as indicated in Figure 8.4. We start with the ordered set P_0 shown at the top of Figure 8.4, and, at every subsequent stage, we surround the dual ordered set P_{n-1}^d of P_{n-1} with the structure indicated at the bottom of Figure 8.4 to obtain P_n. Clearly, because we start with 30 points and add 35 points at each stage, we have $|P_n| = 35n + 30$.

Now note that, in each set P_n, only the point a_n is retractable, and it is retractable

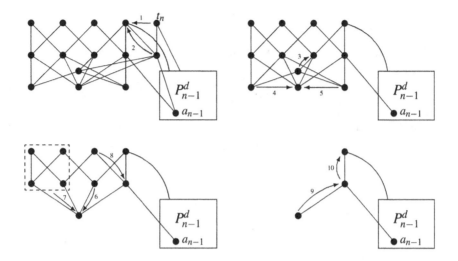

Figure 8.5 The set $\downarrow a_n \setminus \{a_n\}$ of strict lower bounds of a_n. Arrows indicate retractions of retractable points. Each ordered set is obtained from the previous one by removing the points at the base of each arrow.

to b_n. For $n = 0$, this can be verified "by inspection," though it does take a minute. Similarly, this fact can be verified "by inspection" for $n > 0$. The similarity with P_0 will help here and it reveals that the only point other than a_n that could be retractable is a_{n-1} and it could only be retracted to b_{n-1}. However, the extra upper bounds for a_{n-1}, which is minimal in P_{n-1}^d, prevent this from being the case.

To discuss the amount of effort that direct verification of the definition would take, let e_n be the minimum effort needed to determine *via direct verification of the definition* if P_n or P_n^d is collapsible. We will now give a lower bound for e_n and the discussion will also verify that P_n is connectedly collapsible. For $n = 0$, checking connected collapsibility is another nice, if a bit tedious, exercise. To check connected collapsibility of P_n, we can therefore assume inductively that P_{n-1}, and hence P_{n-1}^d, is connectedly collapsible. Because we check via direct verification, we must retract a_n in the first step and then check collapsibility of the set $\downarrow a_n \setminus \{a_n\}$ (see Figure 8.5) and of $P_n \setminus \{a_n\}$ (see Figure 8.6).

First consider $\downarrow a_n \setminus \{a_n\}$ (see Figure 8.5). Retracting the only retractable point t_n in $\downarrow a_n \setminus \{a_n\}$ in the first step requires checking if $\downarrow t_n \setminus \{t_n\}$, which is equal to P_{n-1}^d with an additional irreducible upper cover for a_{n-1} that must be removed first, is collapsible. By definition, this will require an effort of at least e_{n-1}. Although there

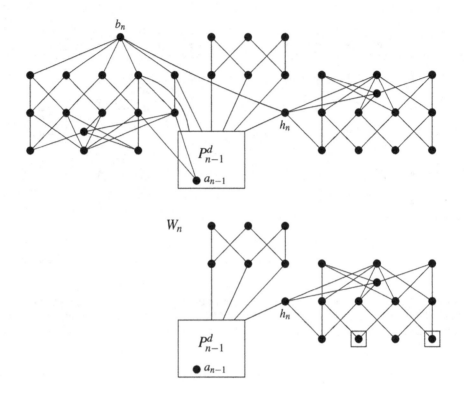

Figure 8.6 The set $P_n \setminus \{a_n\}$.

is further effort required, this lower bound on the effort will suffice for this part of the verification. To prove that $\downarrow a_n \setminus \{a_n\}$ is connectedly collapsible, after removing t_n, we remove two irreducible elements as indicated by arrows 2 and 3 in Figure 8.5, and then the resulting retractable minimal elements as indicated by arrows 4 and 5 in Figure 8.5. The sets of strict upper bounds of these elements are fences and hence dismantlable. Subsequently, we remove further elements with unique lower covers as indicated by arrows 6, 7 and 8 in Figure 8.5. Finally, we remove elements with unique upper covers as indicated by arrows 9 and 10 in Figure 8.5, which leaves us with a set that has a largest element, and which is therefore dismantlable. Hence $\downarrow a_n \setminus \{a_n\}$ is connectedly collapsible.

Now consider the set $P_n \setminus \{a_n\}$ (see top of Figure 8.6). All lower covers of b_n, except for h_n, have a unique upper cover. We continue removing irreducible lower covers of b_n. This allows us to remove all lower bounds of b_n except for those in $\downarrow h_n$. Moreover, independent of the order of removal, until all extra upper covers of a_{n-1}

have been retracted to upper bounds of P_{n-1}^d, at each stage, all retractable elements are in $\downarrow b_n \setminus (\downarrow h_n \cup \{b_n\})$. Consider the case where we remove retractable points until $\downarrow b_n \setminus (\downarrow h_n \cup \{b_n\})$ has been removed. Then h_n is the unique lower cover of b_n and we can retract b_n to h_n. Now (see the ordered set W_n at the bottom of Figure 8.6), the only retractable points will be in P_{n-1}^d. Moreover, the collapsing cannot continue until P_{n-1}^d is collapsed into a singleton. So we must verify collapsibility of the embedded copy of P_{n-1}^d, which will require effort e_{n-1}. Again, this estimate is all we need to establish the desired lower bound. Note that collapsing this copy of P_{n-1}^d to establish connected collapsibility is unproblematic: Let U be the ordered subset of W_n that consists of the 6-crown and of $\uparrow h_n$. The center-deleted neighborhoods $\updownarrow_{W_n} x \setminus \{x\}$ of points $x \in P_{n-1}^d$ in the ordered set W_n consist of the center-deleted P_{n-1}^d-neighborhood of x with the ordered set U attached on top. Because no point of the 6-crown is retractable in U, ordered sets of this form are connectedly collapsible if and only if the center-deleted P_{n-1}^d-neighborhood is connectedly collapsible.

If, in the process of removing points, we had decided to start removing some retractable elements of P_{n-1}^d after all upper covers of a_{n-1} that were not in P_{n-1}^d had been removed, but before all elements of $\downarrow b_n \setminus \downarrow h_n$ were removed, the ordered set U would have been a little larger and may have changed as we alternate between removing points in P_{n-1}^d and points in $\downarrow b_n \setminus \downarrow h_n$, but there would not have been any way to circumvent the need for an effort of at least e_{n-1} needed to directly verify connected collapsibility of P_{n-1}^d.

Once P_{n-1}^d is collapsed to a singleton, verifying connected collapsibility of the remaining 20-element set is a good, if by now routine, exercise.

In summary, we have established inductively that P_n is connectedly collapsible and that the minimum effort for checking collapsibility *by direct verification of the definition* satisfies $e_n \geq 2e_{n-1}$. Hence $e_n \geq 2^n$ and the *direct* verification of collapsibility for the sets P_n is *best case* exponential. Because the outlined procedure is in fact exponential, we have that the effort of direct verification is exactly exponential.

It can be argued that P_{n-1}^d is being collapsed twice in the suggested procedure. A "smarter" algorithm might detect this fact if we simply first check for embedded copies of P_{n-1}^d so that every element that is not in P_{n-1}^d either is an upper bound, a lower bound or not comparable to any element of P_{n-1}^d. (In this situation, we say that P_{n-1}^d is *order-autonomous*.) Because we are dealing with the same set and P_{n-1}^d is investigated in the checking of connected collapsibility of $\downarrow a_n \setminus \{a_n\}$, this might even be possible without using an isomorphism checker. However, this possible speedup can be destroyed by attaching h_n not as an upper bound of P_{n-1}^d, but as an upper bound of a subset of P_{n-1}^d that, in the process of collapsing P_{n-1}^d, will be dismantled to a singleton. For example, instead of attaching h_n as an upper bound of P_{n-1}^d, attaching

h_n as an upper bound of only one or two maximal elements of P_{n-1}^d (such as the boxed elements in Figure 8.6; recall that, in the dual ordered set, minimal elements become maximal) would assure that, in the process of collapsing $P_n \setminus \{a_n\}$, the only subsets of set P_{n-1}^d that are order-autonomous are dismantlable subsets with fewer than 15 elements, which means the "smarter" algorithm would not provide significant, if any, speedup. (In this case, b_n would not be irreducible until P_{n-1}^d had been reduced to a subset of $\downarrow h_n$, but this is a negligible change.)

Unlike for order-autonomous subsets, there are no convenient theorems to handle situations such as those described above. Moreover, with $p(\cdot)$ being a fixed polynomial, we could alter the construction of P_n to constructing sets Q_n by allowing, at each step, the addition of $p(n)$ additional points subject to the following rules: First of all, if any additional point is comparable to an element in P_n, then it is not below any element of P_n. In this fashion, the set of strict lower bounds of a_n in Q_n is the same as in Figure 8.5, but with Q_{n-1}^d instead of P_{n-1}^d. Second, the points are attached in such a way that the resulting set remains connectedly collapsible and so that the effort required for direct verification does not decrease. (This effort would decrease drastically, for example, if we were to attach a largest element. On the other hand, it would not decrease if we attached a dismantlable ordered set D in such a way that the only element of D that is an upper cover of an element not in D is an element c to which D can be dismantled, and a_{n-1} is its unique lower bound in Q_{n-1}^d. Similar to the 6-crown in the construction of the P_n, we could also add ordered sets that are not connectedly collapsible and which can only be removed after all or part of Q_{n-1}^d has been removed.) This leads to sequences of sets in which the size increases polynomially in n, but the effort for direct verification of connected collapsibility remains exponential in n.

Therefore it stands to reason that even "smarter" algorithms that rely on *direct* verification of collapsibility will be worst case exponential using the sequence of examples given here or modifications thereof.

It is shown in Ref. [43] that there is no example similar to Example 8.8.1 that satisfies the notion of collapsibility suggested by Theorem 8.7.2.

8.9. Graphs

For an ordered set, there are sufficient conditions for the fixed point property that refer to the comparability graph (this section), to the chain complex (see Section 8.10), or to the topological realization of the chain complex (see Section 8.11). In this chapter, we will address some combinatorial aspects of these conditions that are related to retractions. The indicated translation to simplicial complexes and to topological spaces is more often used to involve the tools of algebraic topology, such as acyclicity, as was shown in Baclawski and Björner's seminal paper [4], or discrete Morse functions and Whitehead collapsibility, as was recently shown in Ref. [3]. The tools required for these observations are substantial and, for more details, the reader is referred to Ref. [41] or to texts on algebraic topology and discrete Morse Theory. Note that our combinatorial results complement the results from these deeper investigations and cannot be completely replaced by using tools from other fields.

Definition 8.9.1. The *comparability graph* $C(P)$ of an ordered set P has as its vertices the elements of P and there is an edge between two distinct vertices x and y if and only if x is comparable to y.

There are at least two types of homomorphisms for graphs. Graph homomorphisms typically are edge-preserving maps that are not allowed to map two vertices $x \sim y$ to the same image. Although these graph homomorphisms appear to be more widely used in graph theory (examples include Hamiltonian cycles as well as colorings; Ref. [23] is a very good introduction) and although they can be related to the fixed point property for ordered sets too (see Ref. [42]), for our purposes the more natural relative of order-preserving functions is the following.

Definition 8.9.2. Let $G = (V,E)$ be a graph. A function $f : V \to V$ will be called a *simplicial endomorphism* of G if and only if, for all $x \sim y$ in G, we have $f(x) \sim f(y)$ or $f(x) = f(y)$.

The author chose this terminology because these functions are exactly the simplicial maps (see Definition 8.10.4 below) on the clique complex (see Definition 8.10.3 below). Any graph with at least one edge $\{x,y\}$ has a simplicial endomorphism without fixed points (map vertices $\neq x$ to x and map x to y). Thus the natural analogue of the fixed point property in this setting is the following.

Definition 8.9.3. A *clique* in a graph is a set C of vertices so that any two distinct elements of C are adjacent. A graph G has the *fixed clique property* if and only if, for every simplicial endomorphism of G, there is a clique C so that $f[C] = C$.

Theorem 8.9.1. *Let P be a finite ordered set. If $C(P)$ has the fixed clique property, then P has the fixed point property. The converse is not true.*

Proof. Let $C(P)$ have the fixed clique property and let $f : P \to P$ be order-preserving. Then $x \sim y$ in $C(P)$ implies $f(x) = f(y)$ or $f(x) \sim f(y)$ in $C(P)$, that is, f is a simplicial endomorphism of $C(P)$. Hence there is a clique H in $C(P)$ that is fixed by f. However,

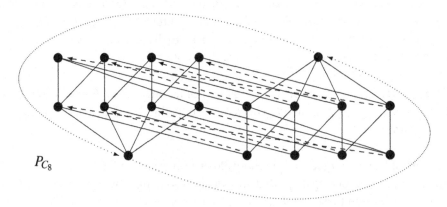

P_{C_8}

Figure 8.7 An ordered set with the fixed point property so that the comparability graph has a fixed clique free simplicial endomorphism (partly indicated by the arrows).

that means that H is a chain in P that is mapped to itself by f. Hence for every $x \in H$ we have that $f(x)$ is comparable to x. By the Abian-Brown Theorem, f must have a fixed point.

Regarding the converse, consider the ordered set in Figure 8.7. It can be verified with a computer search, for example using Ref. [44], that this ordered set has the fixed point property. (Any written proof the author knows is surprisingly complex. Corollary 6.6 in Ref. [38] applies to generalizations of the set and can be used to provide such a proof.) Now recall that a simplicial endomorphism only needs to preserve adjacencies. Hence the function indicated by the arrows in Figure 8.7, extended so that the top left 8 elements, which form an *8-crown*, are mapped to the bottom right 8 elements in analogous and adjacency preserving fashion, is a simplicial endomorphism that does not fix any cliques. (Another such map is to first rotate the ordered set by 180 degrees around the center and then to rotate the 8-crowns.)   □

Example 8.9.1. For the fixed clique property, the Abian-Brown Theorem fails, and quite spectacularly: The graph G in Figure 8.8 does not have the fixed clique property. Mapping each vertex of each of the two cycles on vertices 1–8 and on vertices 9–12 to the next one on the respective cycle provides a simplicial endomorphism that does not fix any cliques in G. However, for every simplicial endomorphism of G, there is a vertex x so that $x \sim f(x)$ or $x = f(x)$. Verification of this fact through enforcing $(2, 1)$-consistency (see Definition 8.7.4) on the corresponding expanded constraint network (defined similar to Definition 8.6.1) with the program in Ref. [44] takes less than a sec-

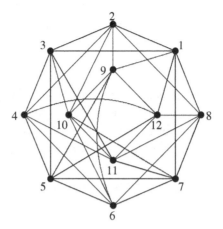

Figure 8.8 A graph for which every simplicial endomorphism f has a vertex with $x \sim f(x)$ or $x = f(x)$ and which does not have the fixed clique property.

ond: The algorithm in Ref. [44] iterates the natural idea of sequentially checking each edge for $(2,1)$-consistency and removing those that are not. This produces an empty network in six iterations, thus showing that there is no simplicial endomorphism f of G so that, for all vertices x, we have $x \nsim f(x)$ and $x \neq f(x)$. The author expects a direct proof of this fact to be quite tedious, and not just because he has tried and failed: Direct verification of the fixed point property for the ordered sets in Ref. [37] is quite a task, as is shown in Section 2 of this reference [37]. For the ordered sets in Ref. [39], direct verifications were not included because they were even longer than for the sets in Ref. [37]. Now note that, for all these sets, enforcing $(2,1)$-consistency on the corresponding expanded constraint network, which, in many ways, mirrors a direct proof by systematically ruling out combinations of assignments, takes three iterations of checking the network's edges and removing those that are not $(2,1)$-consistent. As noted, for the network for the graph in Figure 8.8, it takes six iterations. Moreover, the first two iterations remove comparatively few of the network's edges, namely 3.27% and 5.94% respectively. This means that a direct verification would have very few "good" starting points and would need to consider a significant number of combinations beyond the first few choices. Another indication of the potential difficulties for a direct verification is that the graph in Figure 8.8 has 28,065,828 simplicial endomorphisms, compared to each of the ordered sets in Refs. [37] and [39] having fewer than 220,000 order-preserving self maps.

Lemma 8.9.1 below shows that weaker analogue of the Abian-Brown Theorem exist if we allow stronger hypotheses.

Definition 8.9.4. Let $G = (V, E)$ be a finite graph and let $a \in V$. We define the *neighborhood* of a to be $N(a) := \{v \in V : a \sim v\}$ and the *center-included neighborhood* of a to be $N[a] := N(a) \cup \{a\}$.

Although the above appears to be standard graph-theoretical notation, the author always needs to be very careful to read the parentheses/brackets correctly. Also note that, in graph theory, the vertex itself is not considered to be part of the neighborhood, whereas in ordered sets, the point is considered to be part of its neighborhood.

Lemma 8.9.1. *(A lemma for the fixed clique property that is similar to the Abian-Brown Theorem.) Let $G = (V, E)$ be a graph, let $a \in V$ and let $f : V \to V$ be a simplicial endomorphism so that $f[N[a]] \subseteq N[a]$. Then f has a fixed clique.*

Proof. Because $f[N[a]] \subseteq N[a]$, inductively, for all $n \in \mathbb{N}$, we have that $f^n[N[a]] \subseteq N[a]$. In particular, this means that a is adjacent to all $f^n(a)$. It follows that, for all $n \in \mathbb{N}$, $f(a)$ is adjacent to all $f^{1+n}(a)$. Again, inductively, we conclude that, for all $m, n \in \mathbb{N}$, $f^m(a)$ is adjacent to all $f^{m+n}(a)$. Because V is finite, there are $k, \ell \in \mathbb{N}$ so that $f^k(a) = f^{k+\ell}(a)$. Hence $\{f^k(a), \ldots, f^{k+\ell-1}(a)\}$ is a fixed clique of f. $\qquad\square$

The graph-theoretical analogue of irreducible points are dominated vertices and the analogue of retractable points are retractable vertices.

Definition 8.9.5. Let $G = (V, E)$ be a graph and let $a, b \in V$ be two distinct vertices. Then a is *dominated* by b if and only if $a \sim b$ and, for all $v \in V$, we have that $a \sim v$ implies $b \sim v$.

Definition 8.9.6. Let $G = (V, E)$ be a graph and let $a, b \in V$ be two distinct vertices. Then a is *retractable* to b if and only if, for all $v \in V$, we have that $v \sim a$ implies $v \sim b$.

Definition 8.9.7. Let $G = (V, E)$ be a graph. A simplicial endomorphism $r : V \to V$ so that $r^2 = r$ is called a (simplicial) *retraction*.

A dominated vertex is, in particular, retractable to any vertex that dominates it. If a is retractable to b, then a is dominated by b if and only if $a \sim b$. Moreover, if a is retractable to b, then the function $r : V \to V$ that maps a to b and fixes all other vertices is a (simplicial) retraction. We will now prove the analogue of Theorem 8.7.1 for the fixed clique property.

Lemma 8.9.2. (Compare with Ref. [37], Lemmas 5 and 8). *Let $G = (V, E)$ be a graph. Let the pairwise disjoint subsets $P_1, \ldots, P_n \subseteq V$ be such that $p_i \in P_i$, $p_j \in P_j$ and $i \neq j$ implies $p_i \sim p_j$. Then there is a retraction*

$$r : V \to P_1 \cup \cdots \cup P_n \cup \{v \in V : v \sim p \text{ for all } p \in P_1 \cup \cdots \cup P_n\}.$$

Proof. For each $i \in \{1, \ldots, n\}$, choose a point $t_i \in P_i$. For each $x \in V \setminus (P_1 \cup \cdots \cup P_n)$ for which there is an $i \in \{1, \ldots, n\}$ such that $x \not\sim p_i$ for all $p_i \in P_i$ choose a $j_x \in \{1, \ldots, n\}$ such that $x \not\sim p_{j_x}$ for all $p_{j_x} \in P_{j_x}$. Define

$$r(x) := \begin{cases} x; & \text{if } x \in P_1 \cup \cdots \cup P_n \cup \{v \in V : v \sim p \text{ for all } p \in P_1 \cup \cdots \cup P_n\}, \\ t_{j_x}; & \text{if } x \notin P_1 \cup \cdots \cup P_n \cup \{v \in V : v \sim p \text{ for all } p \in P_1 \cup \cdots \cup P_n\}. \end{cases}$$

Then $r^2 = r$. To see that r is a simplicial endomorphism, let $x \sim y$ and $x \neq y$. If $r(x) = x$, then we are trivially done if $r(y) = y$. Consider the case that $r(x) = x$ and $r(y) \neq y$, which means $r(y) = t_{j_y}$. If $r(x) = x \notin P_1 \cup \cdots \cup P_n$ and $r(y) = t_{j_y}$ we are done, because x must be adjacent to all points in $P_1 \cup \cdots \cup P_n$ and $t_{j_y} \in P_1 \cup \cdots \cup P_n$. If $r(x) = x \in P_1 \cup \cdots \cup P_n$ and $r(y) = t_{j_y}$, then $x \notin P_{j_y}$ and hence in this case we have $x \sim t_{j_y} = r(y)$ too. Finally, if $r(x) \neq x$ and $r(y) \neq y$, then $r(x) = t_{j_x}$ and $r(y) = t_{j_y}$ and, if $j_x \neq j_y$, we have $r(x) \sim r(y)$, while, if $j_x = j_y$, we have $r(x) = t_{j_x} = t_{j_y} = r(y)$. Thus r is a retraction. □

Recall that, for a graph $G = (V, E)$ and a set $W \subseteq V$, $G[W]$ denotes the *induced subgraph* of the graph $G = (V, E)$ on the vertices in the set $W \subseteq V$: The vertices of $G[W]$ are given by W and the edge set of $G[W]$ is $\{\{x, y\} \in E : x, y \in W\}$.

Theorem 8.9.2. *Let $G = (V, E)$ be a finite graph and let $a \in V$ be retractable to $b \in V$. Then G has the fixed clique property if and only if the induced subgraphs $G[V \setminus \{a\}]$ and $G[N(a)]$ both have the fixed clique property.*

Proof. The proof of the direction "$\Rightarrow$" is completely analogous to the proof of the direction "$\Rightarrow$" in Theorem 8.7.1, so we leave it as an exercise.

For the converse "$\Leftarrow$," first note that the case $b \sim a$ will follow from Theorem 8.10.2 below. Because the proof of Theorem 8.10.2 is independent from the arguments here, we will assume here that $b \not\sim a$.

Let $G[V \setminus \{a\}]$ and $G[N(a)]$ both have the fixed clique property and let $f : V \to V$ be a simplicial endomorphism. Let $r : V \to V$ be the retraction that maps a to b and fixes all other vertices of G. Consider the simplicial endomorphism $r \circ f|_{V \setminus \{a\}}$ of $G[V \setminus \{a\}]$. Because $G[V \setminus \{a\}]$ has the fixed clique property, there is a fixed clique $C \subseteq V \setminus \{a\}$ of $r \circ f|_{V \setminus \{a\}}$. In the case where $f[C] = C$, we are done. If $f[C] \neq C$, we must have that $b \in C$ and that there is a $c \in C$ so that $f(c) = a$. If $f(a) = a$, there is nothing to prove, so, for the following, we can assume that $f(a) \neq a$.

For the case $f(b) = a$, we infer that $f[N(a)] \subseteq N[a]$, that is, the image of the center-deleted neighborhood of a is contained in the center-included neighborhood of a. Hence, if $f(a) \in N[a]$, we have $f[N[a]] \subseteq N[a]$ and, by Lemma 8.9.1, f has a fixed clique. If $f(a) \notin N[a]$, then every element of $f[N(a)]$ is adjacent to $f(a)$ and to $a = f(b)$, which is not adjacent to $f(a)$. Hence $f[N(a)] \subseteq N(a)$ and, because $G[N(a)]$ has the fixed clique property, f has a fixed clique in $N(a)$. This concludes the case $f(b) = a$. For the following, we can therefore assume that $f(b) \neq a$.

If $|C| = 1$, we infer $f(b) = a$, which was already considered. For the following, we can therefore assume that $|C| > 1$.

For all $x \in C \setminus \{c\} \neq \emptyset$, we have $x \sim c$ and hence $f(x) \sim f(c) = a$. Because $r \circ f|_{V \setminus \{a\}}[C] = C$, we can assume without loss of generality that $C = \{b, f(b), \ldots, f^k(b) = c\}$. Hence, for $j = 1, \ldots, k$, each element $f^j(b)$ is adjacent to $a = f^{k+1}(b)$. Consequently, $f^2(b), \ldots, f^k(b) = c, f^{k+1}(b) = f(c) = a$ are adjacent to $f^{k+2}(b) = f(a) \neq a$. In particular, we have $a \sim f(a)$, which implies $b \sim f(a) = f^{k+2}(b)$.

Inductively, let $m \geq k+2$, consider the sequence $b, f(b), f^2(b), \ldots, f^m(b)$, and assume that we have proved that the only pairs in this sequence that could be nonadjacent are of the form $\{f^j(b), f^{j+\ell(k+1)}(b)\}$. Then $f^{m+1}(b)$ is adjacent to all $f^{m+1-j}(b)$ with $j \in \{1, \ldots, m\} \setminus \{\ell(k+1) : \ell \in \mathbb{N}\}$. If a is among these elements, then $f^{m+1}(b)$ is also adjacent to b. If $a = f^{k+1}(b)$ is not among these elements, then $f^k(b)$ is not adjacent to $f^m(b)$. Hence $m - k = \ell(k+1)$ and $m+1 = (\ell+1)(k+1)$. Therefore, in either case, the only pairs in the sequence $b, f(b), f^2(b), \ldots, f^{m+1}(b)$ that could be nonadjacent are of the form $\{f^j(b), f^{j+\ell(k+1)}(b)\}$. Note, though, that $f^{m+1}(b)$ *could* be adjacent to $f^{m+1-j(k+1)}(b)$ for some values j.

For the sequence $\{f^n(b)\}_{n=0}^{\infty}$ there are $n_0, p \in \mathbb{N}$, both chosen as small as possible, so that, for all $n \geq n_0$, we have $f^n(b) = f^{n+p}(b)$. Let $n \geq n_0$. Clearly, the set $\{f^n(b), \ldots, f^{n+p-1}(b)\}$ is fixed by f. If $\{f^n(b), \ldots, f^{n+p-1}(b)\}$ is a clique, then f has a fixed clique. For the following, we can assume that $\{f^n(b), \ldots, f^{n+p-1}(b)\}$ is not a clique.

Suppose, for a contradiction, that p is not a multiple of $(k+1)$. If $p < k+1$, then $\{f^n(b), \ldots, f^{n+p-1}(b)\}$ would be a clique, which was excluded. Hence $p > k+1$. Then, for all $j \in \mathbb{N}$ so that $j(k+1) < p$, we have that $f^n(b)$ is adjacent to $f^{n+p-j(k+1)}(b)$, which implies that the vertex $f^{n+j(k+1)}(b)$ is adjacent to $f^n(b) = f^{j(k+1)}\left(f^{n+p-j(k+1)}(b)\right)$. Then, for all $i, j \in \mathbb{N}$ so that $i + j(k+1) < p$, we have that $f^{n+i}(b)$ is adjacent to $f^{n+i+j(k+1)}(b)$ and $\{f^n(b), \ldots, f^{n+p-1}(b)\}$ is a clique, which was excluded. Hence $p = \ell(k+1)$ for some ℓ and $G[\{f^n(b), \ldots, f^{n+p-1}(b)\}]$ consists of $k+1$ copies $G[P_0], \ldots, G[P_k]$ of a graph H with ℓ vertices so that H is not complete and so that, if $i \neq j$, $v \in P_i$, $w \in P_j$, then $v \sim w$.

Let $A := \{v \in V : v \sim f^i(b) \text{ for } i = n, \ldots, n+p-1\}$. Because f maps the set of vertices $\{f^n(b), \ldots, f^{n+p-1}(b)\}$ onto itself, we have that $f[A] \subseteq A$. If $a \notin A$ and $a \neq f^j(b)$

for any $j \in \{n, \ldots, n+p-1\}$, then, by Lemma 8.9.2, $G[\{f^n(b), \ldots, f^{n+p-1}(b)\} \cup A]$ is a retract of $G[V \setminus \{a\}]$ that is mapped to itself by f. Because $G[V \setminus \{a\}]$ has the fixed clique property, f must fix a clique in $G[\{f^n(b), \ldots, f^{n+p-1}(b)\} \cup A]$.

If $a \in A$, then $f^n(b), \ldots, f^{n+p-1}(b)$ are adjacent to a and hence to b. This implies that $\{f^n(b), \ldots, f^{n+p-1}(b)\}$ is a clique, which was already discussed. This leaves the case that $a = f^j(b)$ for some $j \in \{n, \ldots, n+p-1\}$.

If $a = f^j(b)$ for some $j \in \{n, \ldots, n+p-1\}$, assume without loss of generality that $f^n(b) = a$. Let $B := \{f^j(b) : n < j \le n+p-1, j \ne n+\ell(k+1)\}$. Then, again by Lemma 8.9.2, $G[B \cup A]$ is a retract of $G[N(a)]$, which means it has the fixed clique property. Because B has a simplicial endomorphism without a fixed clique (map each $f^j(b)$ to $f^{j+1}(b)$ if $f^{j+1}(b) \in B$ and map it to $f^{j+2}(b)$ otherwise), $G[A]$ must have the fixed clique property and f must have a fixed clique in A.   □

Aside from replacing "ordered set" with "graph" and "irreducible point" with "dominated vertex," the definitions of dismantlable, collapsible and connectedly collapsible graphs are the same as the definitions of their counterparts in ordered sets (see Definitions 8.5.4, 8.7.2 and 8.7.3). By Theorem 8.9.2, connectedly collapsible graphs have the fixed clique property. This result is also a consequence of deeper results from algebraic topology, because connectedly collapsible graphs are acyclic in the sense of algebraic topology. However, these results do not override Theorem 8.9.2. The combinatorial nature of Theorem 8.9.2 means that it can be applied to graphs that are not acyclic in the sense of algebraic topology (and to which therefore the results from algebraic topology do not apply), as long as they have the fixed clique property.

8.10. Simplicial Complexes

Simplicial complexes are central to homology and to discrete Morse Theory. This section can only serve as a gentle introduction to these structures. Interested readers are encouraged to pursue the above-mentioned deeper theories. A starting point can be found in Ref. [41].

Definition 8.10.1. A *simplicial complex* on a finite set V is a set $\Sigma \subseteq \mathscr{P}(V) \setminus \{\emptyset\}$ of nonempty subsets of V so that, for all $\sigma \in \Sigma$, every nonempty subset of σ is in Σ too. The elements of V are called the *vertices* of Σ and the elements of Σ are the *simplices* of Σ.

Note that some authors include the empty set as part of the simplicial complex, which leads to the same results reading slightly differently in different publications. There is a natural way to associate a simplicial complex with an ordered set or with a graph. In each case, the verification that the structure is indeed a simplicial complex is routine.

Definition 8.10.2. Let P be a finite ordered set. The *chain complex* $\Delta(P)$ of P is the simplicial complex Σ that consists of all nonempty chains of P.

Definition 8.10.3. Let G be a finite graph. The *clique complex* of G is the simplicial complex Σ that consists of the nonempty cliques of G.

Note that the chain complex of an ordered set P is also the clique complex of the ordered set's comparability graph $C(P)$. Moreover, the clique complexes of graphs are exactly those simplicial complexes Σ for which, for all vertices $v_1, \ldots, v_n$ so that $\{v_i, v_j\} \in \Sigma$ for all i, j, we have that $\{v_1, \ldots, v_n\} \in \Sigma$. The natural relatives of order-preserving maps and simplicial endomorphisms are simplicial maps.

Definition 8.10.4. Let Σ be a simplicial complex on the vertex set V. A function $f : V \to V$ is called a *simplicial map* if and only if, for all $\sigma \in \Sigma$, we have that $f[\sigma] \in \Sigma$.

When $f : P \to P$ is an order-preserving self map of the ordered set P, then it maps chains to chains, which means it is also a simplicial map of the chain complex $\Delta(P)$. There are simplicial maps on the chain complex that are not order-preserving (see Figure 8.7). Similarly, when $f : V \to V$ is a simplicial endomorphism of the graph G, then f maps cliques to cliques, which means that f is a simplicial map on the clique complex. For graphs, however, simplicial endomorphisms of the graph and simplicial maps on the clique complex are one and the same. The natural generalization of the fixed point and fixed clique properties is the fixed simplex property.

Definition 8.10.5. Let Σ be a simplicial complex on the set V. We say that Σ has the *fixed simplex property* if and only if, for every simplicial map $f : V \to V$, there is a simplex $\sigma \in \Sigma$ with $f[\sigma] = \sigma$.

Because, for graphs and their clique complexes, simplicial endomorphisms and simplicial maps are the same, we have that a graph has the fixed clique property if and only if the clique complex has the fixed simplex property. This also means that the connection between the fixed point property for an ordered set and the fixed simplex property for the chain complex has already been investigated in Theorem 8.9.1.

Corollary 8.10.1. *Let P be a finite ordered set. If the chain complex of P has the fixed simplex property, then P has the fixed point property, but not conversely.*

Because it is a set of sets, a simplicial complex is also an ordered set ordered by inclusion. This fact presents another connection between the fixed point property and the fixed simplex property.

Theorem 8.10.1. *A finite simplicial complex Σ has the fixed simplex property if and only if Σ, viewed as an ordered set ordered by inclusion, has the fixed point property.*

Proof. For the direction "$\Rightarrow$," let Σ have the fixed simplex property and let $f : \Sigma \to \Sigma$ be a containment order-preserving function. For each $x \in V$, find a $g(x) \in V$ so that $g(x) \in f(x)$. Then g is a simplicial map: Indeed, let $\sigma \in \Sigma$. Then $g[\sigma] \subseteq f(\sigma)$, which means that $g[\sigma]$ is a simplex. Because Σ has the fixed simplex property, there is a $\sigma \in \Sigma$ so that $g[\sigma] = \sigma$. Hence $\sigma = g[\sigma] \subseteq f(\sigma)$ and, by the Abian-Brown Theorem, f must have a fixed point.

For the direction "$\Leftarrow$," let Σ, viewed as an ordered set with the containment order, have the fixed point property and let $f : V \to V$ be a simplicial map. Define $F : \Sigma \to \Sigma$ by $F(\sigma) := f[\sigma]$. Clearly, F is order-preserving and hence there is a $\sigma \in \Sigma$ so that $f[\sigma] = F(\sigma) = \sigma$. Hence Σ, as a simplicial complex, has the fixed simplex property.

$\square$

Theorem 8.10.1 may at first seem as if we have come full circle back to the fixed point property. Note, however, that sets Σ are very special ordered sets indeed. We can now turn to the analogue of irreducible points/dominated vertices for simplicial complexes.

Definition 8.10.6. Let Σ be a simplicial complex on the vertex set V. Then $a \in V$ is called *irreducible* if and only if there is a $b \in V \setminus \{a\}$ such that, for all simplices $\sigma \in \Sigma$ that contain a, the set $\sigma \cup \{b\}$ is a simplex too.

Definition 8.10.7. Let Σ be a simplicial complex on the vertex set V and let $W \subseteq V$. Then $\Sigma|W$ is the simplicial complex on the vertex set W that consists of all the simplices of Σ that are contained in W.

Theorem 8.10.2. (Also see Ref. [5], Théorème 3.1, which proves this result for clique complexes of graphs.) *Let Σ be a simplicial complex on the finite set V and let $a \in V$ be irreducible. Then Σ has the fixed simplex property if and only if $\Sigma|V \setminus \{a\}$ has the fixed simplex property.*

Proof. On Σ, viewed as an ordered set via set containment, the function r_1 that maps every simplex σ that contains a to $\sigma \cup \{b\}$ and fixes all other simplices is a retraction from Σ to $\Sigma \setminus \{\sigma \in \Sigma : a \in \sigma, b \notin \sigma\}$ so that, for all $\sigma \in \Sigma$, we have $\sigma \subseteq r_1(\sigma)$. Moreover, on $\Sigma \setminus \{\sigma \in \Sigma : a \in \sigma, b \notin \sigma\}$, the function r_2 that maps every simplex σ that contains a and b to $\sigma \setminus \{a\}$ and fixes all other simplices is a retraction from $\Sigma \setminus \{\sigma \in \Sigma : a \in \sigma, b \notin \sigma\}$ to $\Sigma|V \setminus \{a\}$ so that, for all $\sigma \in \Sigma \setminus \{\sigma \in \Sigma : a \in \sigma, b \notin \sigma\}$, we have $\sigma \supseteq r_2(\sigma)$.

Because both r_1 and r_2 are comparative retractions, by Theorem 8.4.1 (or by Theorem 8.5.1 after verifying that, on a finite ordered set, every comparative retraction is a composition of retractions that remove irreducible points), we conclude that Σ, as an ordered set, has the fixed point property if and only if $\Sigma|V \setminus \{a\}$, as an ordered set, has the fixed point property. By Theorem 8.10.1, we can now conclude that Σ has the fixed simplex property if and only if $\Sigma|V \setminus \{a\}$ has the fixed simplex property. $\square$

Theorem 8.10.2 means that dismantlability can be defined for simplicial complexes just like for graphs and for ordered sets and that dismantlable simplicial complexes have the fixed simplex property. It is now natural to ask if there is an analogue of Theorems 8.7.1 and 8.9.2 for the fixed simplex property too. Retractable vertices are easily defined. A *simplicial retraction* is simply a simplicial map so that $r^2 = r$.

Definition 8.10.8. Let Σ be a simplicial complex on V. Then a vertex a is called *retractable* to $b \in V \setminus \{a\}$ if and only if the function $r : V \to V$ that maps a to b and fixes all other vertices is a simplicial retraction of Σ onto $\Sigma|V \setminus \{a\}$.

Note that every irreducible vertex in a simplicial complex is retractable, but not vice versa. Example 8.10.1 below contains a retractable vertex that is not irreducible. Example 8.10.1 also shows that a *complete* analogue of Theorems 8.7.1 and 8.9.2 is not possible for simplicial complexes: One of the two conditions on the smaller complexes is not necessary. Hence, although a graph has the fixed clique property if and only if its clique complex has the fixed simplex property, there are combinatorial differences between the behavior of clique complexes and the behavior of general simplicial complexes. For the analogue of the (more interesting) sufficiency part of Theorems 8.7.1 and 8.9.2, the author is still looking for a proof (see Question 8.10.1 below).

Example 8.10.1. Consider the simplicial complex Σ on $V = \{a, b, c, d, e, f\}$ that consists of $\{a, c, f\}$, $\{a, c, d\}$, $\{a, d, e\}$, $\{a, e, f\}$, $\{d, e, f\}$, $\{b, c, f\}$, $\{b, c, d\}$, $\{b, d, e\}$, $\{b, e, f\}$, and all their nonempty subsets (also see Figure 8.9). In Σ, the vertex a is retractable to the vertex b: Consider the function $r : V \to V$ that maps a to b and fixes all other vertices and let $\sigma \in \Sigma$. If $a \notin \sigma$, then $r(\sigma) = \sigma$. If $a \in \sigma$, then σ is a subset of $\{a, c, f\}$, $\{a, c, d\}$, $\{a, d, e\}$, or $\{a, e, f\}$, and $r(\sigma)$ is a subset of $\{b, c, f\}$, $\{b, c, d\}$,

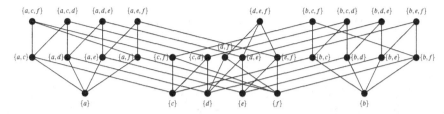

Figure 8.9 The simplices of a simplicial complex with the fixed simplex property and a re-tractable vertex a.

$\{b,d,e\}$, or $\{b,e,f\}$, which means that $r(\sigma)$ is a simplex of Σ and r is a simplicial map. Because r is idempotent, it is a retraction.

We will now prove that Σ has the fixed simplex property, because, as an ordered set with the containment order, Σ has the fixed point property: This fact can easily be verified with the program in Ref. [44], so we will be brief. Let $g : \Sigma \to \Sigma$ be a containment order-preserving function. Without loss of generality, we can assume that g maps singletons to singletons. First note that $\Sigma \setminus \{\{a\}\}$ (not a typo, we remove the singleton set $\{a\}$, which is a single element of the set Σ that is ordered by inclusion) is dismantlable. Therefore we can assume that $\{a\} \in g[\Sigma]$, because otherwise $g|_{\Sigma \setminus \{\{a\}\}}$ has a fixed point. Similarly, we can assume that $\{b\}, \{c\}, \{d\}, \{f\} \in g[\Sigma]$. Moreover, because $\{d,f\}$ is irreducible, we can assume that $\{d,f\} \notin g[\Sigma]$.

In an ordered set, the *distance* from a point x to a point y is the smallest n so that there is a so-called *fence* $x = x_0 < x_1 > x_2 < \cdots x_n = y$ or $x = x_0 > x_1 < x_2 > \cdots x_n = y$ from x to y. Because g is order-preserving, the distance from $g(x)$ to $g(y)$ is at most the distance from x to y. Moreover, if the distance of the images is the same as the distance of the preimages, then the strict comparabilities in a shortest fence joining x and y are preserved in a shortest fence joining $g(x)$ and $g(y)$. Note that the distance between any two elements of Σ is at most 4. Moreover, the two elements $\{a\}$ and $\{b\}$ and the two elements $\{c\}$ and $\{e\}$ are the only pairs of elements of Σ so that the distance is 4 and so that any fence joining them is of the form $x_0 < x_1 > x_2 < x_3 > x_4$. Thus, because $\{a\}, \{b\} \in g[\Sigma]$, we have $g[\{\{a\}, \{b\}\}] = \{\{a\}, \{b\}\}$ or $g[\{\{c\}, \{e\}\}] = \{\{a\}, \{b\}\}$.

First consider the case $g[\{\{a\}, \{b\}\}] = \{\{a\}, \{b\}\}$. We can assume that $g(\{a\}) = \{b\}$ and $g(\{b\}) = \{a\}$, because otherwise g has a fixed point. Similar distance considerations show that $g[\uparrow \{a\} \setminus \{\{a\}\}] \subseteq \uparrow \{b\} \setminus \{\{b\}\}$ and $g[\uparrow \{b\} \setminus \{\{b\}\}] \subseteq \uparrow \{a\} \setminus \{\{a\}\}$. Hence g maps the set $\{\{c\}, \{d\}, \{e\}, \{f\}, \{c,f\}, \{c,d\}, \{d,e\}, \{e,f\}\}$ to itself. If g is surjective on this set, then g must fix $\{d,e,f\}$. If not, then g maps a dismantlable subset (in fact, a fence) to itself, which means it has a fixed point. Thus, for the remainder of the argument, we can assume $g[\{\{a\}, \{b\}\}] \neq \{\{a\}, \{b\}\}$ and $g[\{\{c\}, \{e\}\}] = \{\{a\}, \{b\}\}$.

Because every shortest fence from $\{c\}$ to $\{e\}$ must contain one of $\{d\}$ and $\{f\}$, we infer that $g[\{\{d\},\{f\}\}] \subseteq \{\{c\},\{d\},\{e\},\{f\}\}$. Note that $g[\{\{d\},\{f\}\}] \subseteq \{\{d\},\{f\}\}$ leads to a fixed point. Hence we can assume that $g[\{\{d\},\{f\}\}] \not\subseteq \{\{d\},\{f\}\}$.

Consider the case that $\{e\} \in g(\Sigma)$. Then we must have $g[\{\{a\},\{b\}\}] = \{\{c\},\{e\}\}$. Because both $\{d\}$ and $\{f\}$ can be on a shortest fence from $\{a\}$ to $\{b\}$ and because every shortest fence from $\{c\}$ to $\{e\}$ must contain one of them, we infer that $g[\{\{d\}, \{f\}\}] \subseteq \{\{d\},\{f\}\}$, which was already excluded. Hence we can assume $\{e\} \notin g(\Sigma)$.

Thus $g[\{\{d\},\{f\}\}] \subseteq \{\{c\},\{d\},\{f\}\}$ and $g[\{\{d\},\{f\}\}] \not\subseteq \{\{d\},\{f\}\}$, which implies, without loss of generality, $g(\{d\}) = \{c\}$. Because $g[\{\{c\},\{e\}\}] = \{\{a\},\{b\}\}$ and because every maximal element is above at least one of c,d,e, this means that no maximal element (and hence no element at all) is mapped to $\{d,e,f\}$. Thus $g[\Sigma] \subseteq \Sigma \setminus \{\{e\},\{d,e,f\},\{d,f\}\}$ and the latter set is dismantlable.

Thus, in every possible case, g has a fixed point. Hence, considered as an ordered set with the containment order, Σ has the fixed point property, which means that, as a simplicial complex, Σ has the fixed simplex property.

The verification that the fixed simplex property is preserved by simplicial retractions is a modification of the proof of Proposition 8.1.1. Hence, regarding an analogue of Theorems 8.7.1 and 8.9.2, if a is retractable and Σ has the fixed simplex property, then $\Sigma|V \setminus \{a\}$ has the fixed simplex property. However, as we will see, the analogue of the other condition on the center-deleted neighborhood of a in Theorems 8.7.1 and 8.9.2 need not be satisfied when Σ has the fixed simplex property.

A first concern regarding this condition would be that it is not clear what the natural analogue of the neighborhood of a point would be for a simplicial complex. This example will address this concern, as it works for the two most natural choices.

For simplicial complexes, typically the analogue of the center-deleted neighborhood of a point in an ordered set (or of a vertex in a graph) is the *link* of the vertex a, defined as $Lk_\Sigma(a) := \{\tau \in \Sigma | \tau \cup \{a\} \in \Sigma, a \notin \tau\}$. In a graph, these are all the cliques in the center-deleted neighborhood of a. The difference in a simplicial complex is that there may be simplices σ that are unions of simplices $\tau \in Lk_\Sigma(a)$, but so that $\sigma \cup \{a\} \notin \Sigma$. (For example, consider $\{d,e,f\}$ in Figure 8.9.) In a clique complex, this is not possible. In the complex Σ in this example, $Lk_\Sigma(a)$ does not have the fixed simplex property, because, as an ordered set ordered by inclusion, it is an 8-crown.

Another natural analogue of the center-deleted neighborhood of a would be to take $\Sigma|N$, where $N = \{v \in V \setminus \{a\} : \{a,v\} \in \Sigma\}$. In our example, this would be $\Sigma|\{c,d,e,f\}$, which does not have the fixed simplex property, because, as an ordered set, it does not have the fixed point property: It can be dismantled to the 6-crown consisting of $\{c\}$, $\{d\}$, $\{f\}$, $\{c,f\}$, $\{c,d\}$, $\{d,e,f\}$.

Hence, either way, the analogue of the second condition in Theorems 8.7.1 and 8.9.2 is not satisfied, even though Σ has the fixed simplex property and a is retractable.

As noted earlier, the other direction from Theorems 8.7.1 and 8.9.2 is an open question.

Question 8.10.1. Let Σ be a simplicial complex on the vertex set V, let $a \in V$ be retractable and let $\Sigma|V \setminus \{a\}$ as well as $Lk(a)$ have the fixed simplex property. Does this imply that Σ has the fixed simplex property? The author has a proof that, if there is a counterexample for simplicial complexes, then there is a counterexample with an automorphism that maps the n-dimensional octahedral graph $\mathcal{O}_n$ from Example 8.12.5 below to itself. The proof is similar to the proof of Theorem 8.9.2, except that the last part cannot easily be translated to simplicial complexes.

8.11. Topological Realizations

The final step towards topology is to turn a simplicial complex into a topological space on which we can work with the topological fixed point property. Indicating this connection is a fitting conclusion to this overview on connecting the fixed point property for ordered sets to similar properties for related structures.

Definition 8.11.1. Let Σ be a simplicial complex. A *topological realization* $\mathcal{R}(\Sigma)$ of Σ is a set of simplices in $\mathbb{R}^d$ so that the vertices $v_1, \ldots, v_n$ of Σ are identified with points $P(v_1), \ldots, P(v_n)$ in $\mathbb{R}^d$ so that all convex hulls of images (under P) of simplices in Σ are nontrivial, so that any two of these convex hulls satisfy a containment relation if and only if their corresponding simplices of Σ satisfy the same containment relation, and so that the simplices in $\mathcal{R}(\Sigma)$ are exactly the convex hulls of the images of the simplices in Σ.

Proposition 8.11.1. *Let Σ be a simplicial complex on the vertex set V and let $f : V \to V$ be a simplicial map. Then f can be extended to an affine map $F : \mathcal{R}(\Sigma) \to \mathcal{R}(\Sigma)$.*

Proof. Every point in $\mathcal{R}(\Sigma)$ is of the form $Q = \sum_{j=1}^{k} \lambda_j(Q) P(v_{i_j})$, with $\lambda_j(Q) > 0$, $\sum_{j=1}^{k} \lambda_j(Q) = 1$ and $\{v_{i_1}, \ldots, v_{i_k}\} \in \Sigma$. The affine map corresponding to f is $F(Q) := \sum_{j=1}^{k} \lambda_j(Q) P(f(v_{i_j}))$. □

Theorem 8.11.1. *If the topological realization of the simplicial complex Σ on the vertex set V has the topological fixed point property (that is, every continuous self map has a fixed point), then Σ has the fixed simplex property. The converse is not true.*

Proof. Let $\mathcal{R}(\Sigma)$ have the topological fixed point property and let $f : V \to V$ be a simplicial map. Consider the corresponding affine map F from Proposition 8.11.1. This

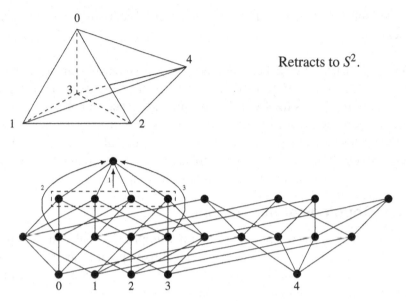

Retracts to S^2.

Figure 8.10 A simplicial complex with the fixed simplex property and a topological realization that does not have the topological fixed point property.

function F is continuous, so it has a fixed point $Q = \sum_{j=1}^{k} \lambda_j(Q)P(v_{i_j})$. Without loss of generality let us assume that the $\lambda_j(Q)$ are strictly greater than 0. Because F maps the set $\{P(v) : v \in V\}$ to itself, F must map $\{P(v_{i_j}) : j = 1, \ldots, k\}$ to itself. Because the $\lambda_j(Q)$ are strictly greater than 0, f must fix the simplex $\{v_{i_j} : j = 1, \ldots, k\} \in \Sigma$.

Regarding the converse being not true, consider Figure 8.10. The simplicial complex Σ on $V = \{0, 1, 2, 3, 4\}$ at the bottom of the figure has the fixed simplex property, because, as an ordered set with the containment order, Σ dismantles as indicated by the arrows to the ordered set in Figure 8.7, which has the fixed point property. Its topological realization, indicated above the simplicial complex in Figure 8.10, consists of the solid tetrahedron with vertices $0, 1, 2, 3$ as indicated, with the four triangles with vertices $0, 1, 4$; $0, 3, 4$; $1, 2, 4$; and $2, 3, 4$ attached. This solid contains a hole in the middle and it can indeed be retracted to the two-dimensional unit sphere $S^2 \subset \mathbb{R}^3$. Because topological retractions preserve the topological fixed point property and S^2 does not have the topological fixed point property, the topological realization of Σ does not have the topological fixed point property.   □

8.12. Iterated Clique Graphs

For the final two sections of this chapter, in this section we turn to the relationship between iteration of the clique graph operator and the retractions we have investigated so far, and, in Section 8.13, to the relationship between iteration of the clique graph operator and the fixed point property for ordered sets. The study of iterated clique graphs started in Ref. [19] and has since grown into a substantial body of literature, as a quick MathSCI search reveals.

Definition 8.12.1. Let $G = (V, E)$ be a graph. We define the *clique graph* $k(G)$ of G to be the following graph:

(i) The vertices of $k(G)$ are the maximal (with respect to inclusion) cliques of G.

(ii) Two vertices v, w of $k(G)$ (which are maximal cliques of G) are adjacent in $k(G)$ if and only if $v \cap w \neq \emptyset$.

Clearly, the clique graph operator can be iterated. That is, we can compute the clique graph of the clique graph $k(k(G))$ and the clique graph of that graph and so forth. It is natural to ask how this sequence of graphs will behave.

Definition 8.12.2. We define $k^n(G)$ recursively by $k^0(G) := G$ and $k^{n+1}(G) = k(k^n(G))$.

(a) A graph $G = (V, E)$ is called *K-invariant* if and only if $k(G)$ is isomorphic to G.

(b) A graph $G = (V, E)$ is called *K-periodic* with period p if and only if $k^p(G)$ is isomorphic to G.

(c) A graph $G = (V, E)$ is called *K-bounded* if and only if there is an $N \in \mathbb{N}$ so that, for every $n \in \mathbb{N}$, the graph $k^n(G)$ has at most N vertices.

(d) A graph $G = (V, E)$ is called *K-null* if and only if there is an $n \in \mathbb{N}$ so that the graph $k^n(G)$ has one vertex.

(e) A graph $G = (V, E)$ is called *K-divergent* if and only if the number of vertices of the graphs $k^n(G)$ is unbounded.

To familiarize ourselves with these ideas, let us consider some examples that show that all the above cases can occur.

Example 8.12.1. Let $m \in \mathbb{N}$ and let $2 \leq n < \frac{m}{3}$. Define the graph $C_m^n = (V, E)$ as a graph with vertex set $V = \{v_1, \ldots, v_m\}$ so that, for all i, we have $v_i \sim v_{i+j \pmod{m}}$ if and only if $j \in \{1, \ldots, n\}$ (also see Figure 8.11, top left). Then C_m^n is K-invariant.

Proof. By definition, v_i and v_j are adjacent in C_m^n if and only if, on the cycle $v_1 \sim v_2 \sim \cdots \sim v_m \sim v_1$, the distance between v_i and v_j is $\leq n$. Because $n < \frac{m}{3}$, with index arithmetic being modulo m, every clique in C_m^n is contained in a, not necessarily unique,

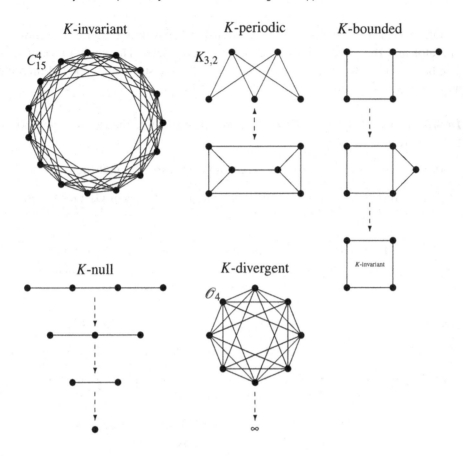

Figure 8.11 Picturing the different possibilities for iterated clique graphs.

maximal clique $\{v_i,\ldots,v_{i+n}\}$ of size $n+1$. That is, the maximal cliques of C_m^n are exactly the cliques $\{v_i,\ldots,v_{i+n}\}$ for $i=1,\ldots,m$. Moreover, two cliques $\{v_i,\ldots,v_{i+n}\}$ and $\{v_j,\ldots,v_{j+n}\}$ intersect if and only if, on the cycle $v_1 \sim v_2 \sim \cdots \sim v_m \sim v_1$, the distance between v_i and v_j is $\leq n$. Hence $v_i \mapsto \{v_i,\ldots,v_{i+n}\}$ is an isomorphism from C_m^n to $k(C_m^n)$.   □

It is shown in Ref. [7] that the graphs C_m^n are the only K-invariant Helly circular arc graphs. There are many other examples of K-invariant graphs (see, for example, Ref. [8]).

Example 8.12.2. Let $K_{n,m}$ be a complete bipartite graph with $n \geq 2$ and $m \geq 3$ (also see Figure 8.11, top center). Then $k(K_{n,m})$ is the line graph of $K_{n,m}$, consisting of n cliques with m vertices, m cliques with n vertices and every n-clique intersects every

m-clique. In particular, $k(K_{n,m})$ is not isomorphic to $K_{n,m}$. Moreover, the maximal cliques in $k(K_{n,m})$ are in bijective correspondence with the vertices of $K_{n,m}$ and two maximal cliques in $k(K_{n,m})$ intersect if and only if their corresponding vertices are adjacent. Hence $k^2(K_{n,m})$ is isomorphic to $K_{n,m}$. We conclude that $K_{n,m}$ is K-periodic with period 2, but not K-invariant.

The paper [13] presents, for every $p \geq 2$, examples of K-periodic graphs with period p.

Example 8.12.3. Let $G = (V,E)$ be the graph with vertex set $V = \{c_1, c_2, c_3, c_4, p\}$ and the adjacencies $c_1 \sim c_2 \sim c_3 \sim c_4 \sim c_1$ and $p \sim c_1$ (see Figure 8.11, top right). The clique graph of G is the line graph of G, which consists of five vertices $\{c_2, c_3\}$, $\{c_3, c_4\}$, $\{c_4, c_1\}$, $\{c_1, c_2\}$, $\{c_1, p\}$. The first four of these vertices of $k(G)$ form a 4-cycle, the last three of them form a 3-cycle, and these are all the adjacencies in $k(G)$. The maximal cliques in $k(G)$ are the 3-cycle and the three edges of the 4-cycle that are not part of the 3-cycle. The second iterated clique graph $k^2(G)$ of G is thus a 4-cycle and all further iterated clique graphs of G are 4-cycles too. We conclude that G is K-bounded, but not K-periodic.

Example 8.12.4. Let P_n be a path $v_0 \sim v_1 \sim \cdots \sim v_n$ of length n (see Figure 8.11, bottom left). Then $k(P_n)$ is a path of length $n - 1$. Hence $k^n(P_n)$ is a singleton and P_n is K-null.

Example 8.12.5. The n-dimensional octahedral graph $\mathscr{O}_n$ is the complementary graph of n pairwise disjoint copies of an edge $v_1 \sim v_2$ (also see Figure 8.11, bottom center). That is, if the vertices of $\mathscr{O}_n$ are $v_1, v_2, \ldots, v_{2n-1}, v_{2n}$, then all vertices are adjacent to each other *except* that $v_{2i-1} \nsim v_{2i}$ for $i = 1, \ldots, n$.

The maximal cliques of $\mathscr{O}_n$ are exactly the subsets C of $\{1, \ldots, 2n\}$ so that, for every $i \in \{1, \ldots, n\}$, we have that $|C \cap \{v_{2i-1}, v_{2i}\}| = 1$. For every such C, there is exactly one C', defined by, for every $i \in \{1, \ldots, n\}$, $C' \cap \{v_{2i-1}, v_{2i}\} := \{v_{2i-1}, v_{2i}\} \setminus C$, so that $C \cap C' = \emptyset$. Hence $k(\mathscr{O}_n)$ is isomorphic to $\mathscr{O}_{2^{n-1}}$.

In particular, this means that the number of vertices in the iterated clique graphs for $\mathscr{O}_n$ with $n \geq 3$ grows rather rapidly: Note that $\mathscr{O}_n$ has $m = 2n$ vertices. The graph $k(\mathscr{O}_n)$ is isomorphic to $\mathscr{O}_{2^{n-1}}$ and has $2^n = \left(2^{\frac{1}{2}}\right)^{2n} = \sqrt{2}^m$ vertices. Hence $k^2(\mathscr{O}_n) = k(k(\mathscr{O}_n))$ has $\sqrt{2}^{\sqrt{2}^m}$ vertices, $k^3(\mathscr{O}_n) = k(k(k(\mathscr{O}_n)))$ has $\sqrt{2}^{\sqrt{2}^{\sqrt{2}^m}}$ vertices and so on.

Therefore $\mathscr{O}_n$ with $n \geq 3$ is K-divergent.

Example 8.12.5 appears to be a standard example of a K-divergent family of graphs. Note that the graphs $\mathscr{O}_n$ are also the comparability graphs of what, in ordered sets, is called a 4-crown-tower: In a 4-crown-tower with points $p_1, \ldots, p_{2n}$, the comparabilities are $\{p_1, p_2\} \leq \{p_3, p_4\} \leq \cdots \leq \{p_{2n-1}, p_{2n}\}$ plus those comparabilities forced by transitivity.

Clearly, the growth of $k^m(\mathcal{O}_n)$ is quite fast. Interestingly enough, there are also graphs whose iterated clique graphs are growing much more slowly, namely linearly (see Ref. [26]).

Returning to retractions, we first show that clique graphs of retracts are isomorphic to retracts of clique graphs.

Proposition 8.12.1 [32, Proposition 3]. *Let $G = (V, E)$ be a graph and let $r : V \to V$ be a retraction. Let $W := r[V]$. Then $k(G)$ has a retract that is isomorphic to $k(G[W])$.*

Proof. For any vertex w of $k(G[W])$, that is, for every maximal clique in $G[W]$, choose $f(w)$ to be a maximal clique in G that contains w. (For some vertices/maximal cliques w, there may be multiple choices for $f(w)$, in which case we simply choose one of them.) For every vertex w of $k(G[W])$, we have that $r[f(w)]$ is a clique in $G[W]$ that contains w. Hence, by maximality of w, we have $r[f(w)] = w$. In particular, this implies that the function f is injective.

Now let $w_1 \neq w_2$ be distinct vertices of $k(G[W])$. If $w_1 \sim_{k(G[W])} w_2$, then $w_1 \cap w_2 \neq \emptyset$, which clearly implies $f(w_1) \cap f(w_2) \neq \emptyset$, that is, $f(w_1) \sim_{k(G)} f(w_2)$. Conversely, if $f(w_1) \sim_{k(G)} f(w_2)$, then $f(w_1) \cap f(w_2) \neq \emptyset$, which implies $w_1 \cap w_2 = r[f(w_1)] \cap r[f(w_2)] \supseteq r[f(w_1) \cap f(w_2)] \neq \emptyset$, that is, $w_1 \sim_{k(G[W])} w_2$. Hence f is an embedding of $k(G[W])$ into $k(G)$. Call the image graph $k(G)_W$.

Similar to the definition of f, for every, not necessarily maximal, clique c in $G[W]$, choose a maximal clique $g(c)$ in G that contains c and that is of the form $g(c) = f(w)$ for some maximal clique w of $G[W]$. Define the function R on the vertices v of $k(G)$, that is, on the maximal cliques v of G, by $R(v) := g(r[v])$. Then R is a well-defined function from $k(G)$ to $k(G)_W$. Recall that, for every maximal clique w of $G[W]$, we have that $r[f(w)] = w$, which means that (by injectivity of f) $R(f(w)) = g(r[f(w)]) = g(w) = f(w)$. Therefore R is surjective onto the vertices of $k(G)_W$ and idempotent. Moreover, if $v_1 \sim_{k(G)} v_2$, then $v_1 \cap v_2 \neq \emptyset$, which implies $g(r[v_1]) \cap g(r[v_2]) \supseteq r[v_1] \cap r[v_2] \supseteq r[v_1 \cap v_2] \neq \emptyset$, and hence $R(v_1) \sim_{k(G)_W} R(v_2)$. We have proved that R is a retraction of $k(G)$ onto $k(G)_W$, which is isomorphic to $k(G[W])$. $\square$

We conclude this section with a proof that dismantlable graphs are K-null. To do this, we need to introduce another type of retraction that generalizes the retractions that remove a single dominated vertex.

Definition 8.12.3. Let $G = (V, E)$ be a finite graph and let $r : V \to V$ be a retraction.

Then r is called a *#-retraction* if and only if every vertex of $V \setminus r[V]$ is dominated by some vertex of $r[V]$.

Every #-retraction is a composition of retractions on G that remove one dominated vertex of G. Conversely, if $r_1, \ldots, r_n$ are retractions that remove the dominated vertices $a_1, \ldots, a_n$ and every a_i is dominated by a vertex in $V \setminus \{a_1, \ldots, a_n\}$, then the composition of the r_i is a #-retraction. Consequently, the #-retracts of G are exactly the induced subgraphs $G[W]$ so that every vertex of G is dominated by a vertex in W. If we consider the natural analogue for ordered sets, the class of #-retractions on an ordered set properly contains the class of retractions that remove irreducible points, and it is itself properly contained in the class of comparative retractions. The proof of Theorem 8.12.2 will show why #-retractions are exactly the right notion of retraction to be used in this setting.

Theorem 8.12.1 [17, Theorem 3]. *Let $G = (V, E)$ be a finite graph and let $r : V \to V$ be a #-retraction onto the induced subgraph $G[W]$ of G. Then $k(G)$ can be #-retracted onto an induced subgraph of $k(G)$ that is isomorphic to $k(G[W])$.*

Proof. The function f from the proof of Proposition 8.12.1 is an embedding of $k(G[W])$ into $k(G)$ as the retract $k(G)_W$. Because we do not know if the retraction R from the proof of Proposition 8.12.1 is a #-retraction, we prove that every vertex of $k(G)$ is dominated by a vertex of $k(G)_W$.

Let Q be a maximal clique of G. Because every vertex of G is dominated by some vertex of $G[W]$, we have that $Q \cap W \neq \emptyset$. Let Q_0 be a maximal clique of $G[W]$ that contains $Q \cap W$ and consider $f(Q_0)$. Let P be another maximal clique of G so that $P \cap Q \neq \emptyset$. Then, because every vertex of the intersection is dominated by some vertex of $G[W]$ and because P and Q are maximal cliques, we have that $P \cap Q \cap W \neq \emptyset$. Therefore $P \cap Q_0 \neq \emptyset$ and then $P \cap f(Q_0) \neq \emptyset$. Hence Q is dominated by $f(Q_0)$. Because Q was arbitrary, the result is proved. $\square$

We can now prove that dismantlable graphs are K-null, a result originally proved in Corollary 4.2 in Ref. [34].

Theorem 8.12.2. (Also see Theorem 5 in Ref. [17], which actually shows that G and H have the same K-behavior. This requires a bit more effort than what we need for this chapter, so the reader should consider Ref. [17] for further details.) *Let G be a finite graph that can be dismantled to the induced subgraph H. Then G is K-null if and only if H is K-null. In particular, dismantlable graphs are K-null.*

Proof. Because any dismantling can be viewed as a sequence of #-retractions, we are done if we can show the result for G having a #-retraction onto H. So let G have a #-retraction onto H. By iterating Theorem 8.12.1, we infer that, for every $n \in \mathbb{N}$, we have that $k^n(G)$ has a #-retraction onto an isomorphic copy of $k^n(H)$.

For the direction "$\Rightarrow$," let G be K-null. Choose n so that $k^n(G)$ is a singleton. Because an isomorphic copy of $k^n(H)$ must be a #-retract of $k^n(G)$, we conclude that $k^n(H)$ must be a singleton too. Hence H is K-null.

For the direction "$\Leftarrow$," let H be K-null. Choose n so that $k^n(H)$ is a singleton. Then $k^n(G)$ can be #-retracted to a singleton. Therefore there is a vertex in $k^n(G)$ that is adjacent to all other vertices. We conclude that all cliques of $k^n(G)$ intersect. This means that $k^{n+1}(G)$ is complete and that $k^{n+2}(G)$ is a singleton. Hence G is K-null. $\square$

For irreducible points/vertices, Theorems 8.5.1, 8.9.2, 8.10.2 and 8.12.2 show that similar theorems hold in a variety of different settings. For retractable points/vertices, we can say that, with the same procession from ordered sets to graphs to simplicial complexes to the behavior of iterated clique graphs, we have spanned an arc from a natural theorem with a quick proof (see Theorem 8.7.1) through a generalization with a more involved proof (see Theorem 8.9.2) and through a generalization for which one direction of the natural generalization is false and the other unresolved (see Example 8.10.1 and Question 8.10.1) to a situation in which the natural version of the theorem is false, as we will see in the following.

Definition 8.12.4. Let $H = (V, E)$ be a graph, let a, b be two additional vertices with $a, b \notin V$, let $V_{S^{a,b}(H)} := V \cup \{a, b\}$, let $E_{S^{a,b}(H)} := E \cup \{\{a, v\}, \{b, v\} : v \in V\}$ and let $S^{a,b}(H) := (V_{S^{a,b}(H)}, E_{S^{a,b}(H)})$. Then $S^{a,b}(H)$ is called the *suspension* or *double cone* of H with peaks a and b.

Clearly, for any graph H, in the suspension $S^{a,b}(H)$, we have that a is retractable to b (and vice versa), $S^{a,b}(H)[N(a)] = H$ and, in $S^{a,b}(H)[V_{S^{a,b}(H)} \setminus \{a\}]$, every vertex is dominated by b, so $S^{a,b}(H)[V_{S^{a,b}(H)} \setminus \{a\}]$ is dismantlable and in particular K-null. However, Figure 2 in Ref. [28] shows a graph H that is K-null and so that $S^{a,b}(H)$ is K-divergent. This also shows that there is no analogue of Theorem 5 in Ref. [17] for retractable vertices.

With a small modification of the example in Figure 2 in Ref. [28], we can show that the situation is even worse.

Example 8.12.6 (M. Pizaña). *There is a connectedly collapsible graph that is not K-null.* Consider the graph H in Figure 8.12. The vertex 7 is retractable to 1 and

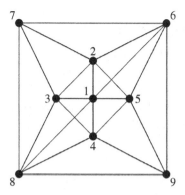

Figure 8.12 A connectedly collapsible graph with a suspension that is not K-null.

$H[N(7)]$ is a path which is dismantlable. Similarly, the vertex 9 is retractable to 1 and $H[N(9)]$ is a path which is dismantlable. Moreover, the graph $H - 7, 9$ is dismantlable, because every vertex is dominated by 1. Thus H is connectedly collapsible and so is its suspension $S^{a,b}(H)$. However, $S^{a,b}(H)$ is not K-null, because its clique graph $k(S^{a,b}(H))$ retracts onto a copy of $\mathcal{O}_4$, which can be seen as follows.

Each of $\{a,3,7,8\}$, $\{b,5,6,9\}$; $\{a,2,6,7\}$, $\{b,4,8,9\}$; $\{a,4,8,9\}$, $\{b,2,6,7\}$; $\{a,5,6,9\}$, $\{b,3,7,8\}$ is a maximal clique in $S^{a,b}(H)$. Call the set of these cliques $\mathcal{C}$. By inspection, the subgraph of $k(S^{a,b}(H))$ that is induced on $\mathcal{C}$ is isomorphic to $\mathcal{O}_4$: Pairs of nonintersecting cliques are separated by semicolons in the list.

Until the last paragraph, we now consider $\mathcal{C}$ to be a set of maximal cliques of $S^{a,b}(H)$, not as a set of vertices in $k(S^{a,b}(H))$. Suppose that there is another maximal clique B of $S^{a,b}(H)$ that intersects every one of the maximal cliques in $\mathcal{C}$. Then B must contain a or b and $B' := B \setminus \{a, b\}$ is a maximal clique in H. For every $C \in \mathcal{C}$, we have that $\{b\} \cup C \setminus \{a\} \in \mathcal{C}$ and $\{a\} \cup C \setminus \{b\} \in \mathcal{C}$. Hence B' must intersect each of $\{3,7,8\}$, $\{5,6,9\}$, $\{2,6,7\}$, and $\{4,8,9\}$. However, there is no such clique in H: Suppose, for a contradiction, that K was a maximal clique in H that intersects $\{3,7,8\}$, $\{5,6,9\}$, $\{2,6,7\}$, and $\{4,8,9\}$. If $3 \in K$, then K does not intersect $\{5,6,9\}$, because 3 is not adjacent to any of 5, 6, 9. Hence $3 \notin K$. Similarly, $2, 4, 5 \notin K$. If $7 \in K$, then $6, 7 \in K$ or $7, 8 \in K$, which, by maximality of K, implies $2 \in K$ or $3 \in K$ respectively, either of which is not possible. Thus $7 \notin K$ and we must have $8 \in K$. Now, $8, 9 \in K$ would imply $4 \in K$, which is not possible, and $1, 8 \in K$ would imply $3, 4 \in K$, which is not possible either. Hence H does not have a maximal clique that intersects $\{3,7,8\}$, $\{5,6,9\}$, $\{2,6,7\}$, and $\{4,8,9\}$. We conclude that there is no maximal clique B of $S^{a,b}(H)$ that intersects every one of the maximal cliques in $\mathcal{C}$.

Thus the set of all vertices of $k(S^{a,b}(H))$ that are adjacent to all vertices (of $k(S^{a,b}(H))$) in $\mathcal{C}$ is empty. Hence, by Lemma 8.9.2, $\mathcal{C}$ is a retract of $k(S^{a,b}(H))$.

8.13. K-Null Comparability Graphs

Now that we have a good set of examples of K-null graphs, including dismantlable graphs, we make the connection to the fixed point property. Coaffinable graphs (see Definition 8.13.1 below) include, in particular, the comparability graphs of ordered sets that have a fixed point free automorphism. As the proof of Theorem 8.13.1 will show, a coaffinable retract blocks the iterated clique graphs from shrinking to a singleton.

Definition 8.13.1. Let $G = (V, E)$ be a graph. Then a function $\phi : V \to V$ is called a *coaffination* if and only if ϕ is an automorphism so that, for all $v \in V$, we have $\phi(v) \not\sim v$ and $\phi(v) \neq v$. A graph that admits a coaffination is called *coaffinable*.

Proposition 8.13.1 [32, Proposition 5]. *Let $G = (V, E)$ be a coaffinable graph. Then $k(G)$ is coaffinable too.*

Proof. Let $\phi : V \to V$ be a coaffination. Because ϕ is an automorphism, ϕ maps maximal cliques to maximal cliques. Hence $\Phi : C \mapsto \phi[C]$ is a bijective self map of the vertices of $k(G)$ and $C_1 \cap C_2 \neq \emptyset$ if and only if $\Phi(C_1) \cap \Phi(C_2) \neq \emptyset$. This means that Φ is an automorphism of $k(G)$. Finally, suppose, for a contradiction, that there was a C with $C \cap \Phi(C) \neq \emptyset$. Let $x \in C \cap \phi[C]$. Then $x \in \phi[C]$ and, because $x \in C$, $\phi(x) \in \phi[C]$. We conclude that $x \sim_G \phi(x)$ or $x = \phi(x)$, a contradiction. □

Lemma 8.13.1 [18, Theorem 1.3]. *Let $G = (V, E)$ be a finite graph that has a simplicial endomorphism $f : V \to V$ so that, for all $v \in V$, we have $v \not\sim f(v)$ and $v \neq f(v)$. Then G has a coaffinable retract.*

Proof. Let $n := |V|$. For every vertex $x \in V$, the vertices $x, f(x), f^2(x), \ldots, f^n(x)$ are not all distinct. Hence, for every vertex $x \in V$, we have that $f^{n!}(x)$ is a vertex for which there is a $k < n$ so that $f^k\left(f^{n!}(x)\right) = f^{n!}(x)$. Moreover, if v is a vertex so that there is a $k < n$ with $f^k(v) = v$, then $f^{n!}(v) = f^{\frac{n!}{k}k}(v) = \left(f^k\right)^{\frac{n!}{k}}(v) = v$. Hence $f^{n!}$ is a retraction onto $R := f^{n!}[V]$ and $f|_R$ is a coaffination. □

Theorem 8.13.1 [18, Theorem 1.4]. *Let $G = (V, E)$ be a finite K-null graph. Then, for every simplicial endomorphism $f : V \to V$ of G, there is a vertex $v \in V$ so that $v \sim f(v)$ or $v = f(v)$.*

Proof. Suppose, for a contradiction, that G has a simplicial endomorphism $g : V \to V$ so that, for all $v \in V$, we have $v \not\sim g(v)$ and $v \neq g(v)$.

We will prove by induction that, for every $n \geq 0$, $k^n(G)$ has a coaffinable retract. For the base step, $n = 0$, by Lemma 8.13.1, $G = k^0(G)$ has a coaffinable retract. For the induction step $n \to n + 1$, let $k^n(G)$ have a coaffinable retract. Then, by Propositions 8.12.1 and 8.13.1, $k^{n+1}(G) = k(k^n(G))$ has a coaffinable retract.

However, because G is K-null, there is an n so that $k^n(G)$ has one vertex. Clearly, the graph with one vertex has no coaffinable retracts. We have arrived at a contradiction and hence, for every simplicial endomorphism $f : V \to V$ of G, there must be a vertex $v \in V$ so that $v \sim f(v)$ or $v = f(v)$. $\qquad\square$

Theorem 8.13.1 does *not* claim that being K-null implies that the graph has the fixed clique property. Example 8.9.1 shows that, in general, the conclusion of Theorem 8.13.1 does not imply the fixed clique property. M. Pizaña observed that the graph in Example 8.9.1 is not K-null with an argument that involves algebraic topology, which we can only sketch: Collapse (in the sense of Whitehead) the eight edges (and the corresponding eight triangles containing them) $1 - 3 - 5 - 7 - 1$ and $2 - 4 - 6 - 8 - 2$ to get a triangulation of the Möbius strip. This shows that the fundamental group (the first homotopy group π_1) of the clique complex is nontrivial. Because the fundamental group of graphs is preserved by the clique graph operator (see Corollary 5.2 in Ref. [29]), it follows that the graph in Example 8.9.1 is not K-null. It is not known whether the graph in Example 8.9.1 is K-bounded or K-divergent, but the orders of the first few iterated clique graphs are 12, 24, 28, 48, 116, 696, and 61020, which seems to indicate K-divergence.

Question 8.13.1. Is there a graph that does not have the fixed clique property and which is K-null? Is there a comparability graph that does not have the fixed clique property for which every simplicial endomorphism f has a vertex with $v \sim f(v)$ or $v = f(v)$?

For ordered sets, the Abian-Brown Theorem allows us to conclude that a K-null comparability graph means that the ordered set has the fixed point property.

Corollary 8.13.1. *Let P be a finite ordered set so that the comparability graph $C(P)$ of P is K-null. Then P has the fixed point property.*

Proof. Let $f : P \to P$ be order-preserving. Then f is a simplicial endomorphism of $C(P)$. By Theorem 8.13.1, there is a $p \in P$ so that $p = f(p)$ or $p \sim f(p)$ in $C(P)$. Hence $p \leq f(p)$ or $p \geq f(p)$, which means that f has a fixed point. $\qquad\square$

Rather than talking about the K-behavior of comparability graphs, for this final discussion, let "the ordered set is K-..." mean that "the ordered set's comparability graph is K-...." The paper [27] presents examples of ordered sets with the fixed point property that are K-divergent. A 4-crown-tower (see remarks after Example 8.12.5) is a K-divergent ordered set that does *not* have the fixed point property. This leaves the class of K-bounded ordered sets. A 4-crown is a K-periodic ordered set that does not have the fixed point property. We arrive at a final open question for this chapter.

Question 8.13.2. Are there ordered sets with the fixed point property whose comparability graphs are K-bounded, but not K-null?

ACKNOWLEDGMENTS

I thank Miguel Pizaña for help with Sections 8.12 and 8.13. In particular, Example 8.12.6 is due to him and the discussion before Question 8.13.1 uses his observations.

REFERENCES

1. Abian A, Fixed point theorems of the mappings of partially ordered sets. Rendiconti del Circolo Mathematico di Palermo 1971;20:139–142.
2. Abian S, Brown AB, A theorem on partially ordered sets with applications to fixed point theorems. Can. J. Math. 1961;13:78–82.
3. Baclawski K, A combinatorial proof of a fixed point property. J. Combinatorial Theory, Ser. A 2012;119:994–1013.
4. Baclawski K, Björner A, Fixed points in partially ordered sets. Adv. Math. 1979;31:263–287.
5. Bélanger MF, Constantin J, Fournier G, Graphes et ordonnés démontables. Discrete Math. 1994;130:9–17.
6. Bessière C, Debruyne R, Some practicable filtering techniques for the constraint satisfaction problem. In: International Joint Conference on Artificial Intelligence, Proceedings of the 15th IJCAI, IJCAI; 1997;412–417.
7. Bonomo F, Self-clique Helly circular-arc graphs. Discrete Math. 2006;306:595–597.
8. Chia GL, Ong PH, On self-clique graphs with given clique sizes, II. Discrete Math. 2009;309:1538–1547.
9. Carl S, Heikkilä S, Fixed Point Theory in Ordered Sets and Applications: From Differential and Integral Equations to Game Theory. New York: Springer; 2011.
10. Donalies M, Schröder B, Performance guarantees and applications for Xia's algorithm. Discrete Math. 2000;213:67–86. (Proceedings of the Banach Center Minisemester on Discrete Mathematics, week on Ordered Sets)
11. Duffus D, Poguntke W, Rival I, Retracts and the fixed point problem for finite partially ordered sets. Can. Math. Bull. 1980;23:231–236
12. Duffus D, Goddard T, The complexity of the fixed point property. Order 1996;13:209–218.
13. Escalante F, Über iterierte Cliquen-Graphen. Abh. Math. Sem. Univ. Hamburg 1973;39:58–68.
14. Farley JD, The uniqueness of the core. Order 1993;10:129–131.
15. Fofanova T, Rutkowski A, The fixed point property in ordered sets of width two. Order 1987;4:101–106.
16. Fofanova T, Rival I, Rutkowski A, Dimension 2, fixed points and dismantlable ordered sets. Order 1994;13:245–253.
17. Frías-Armenta ME, Neumann-Lara V, Pizaña MA, Dismantlings and iterated clique graphs. Discrete Math. 2004;282:263–265.
18. Hazan S, Neumann-Lara V, Fixed points of posets and clique graphs. Order 1995;13:219–225.
19. Hedetniemi ST, Slater PJ, Line graphs of triangleless graphs and iterated clique graphs. In: Lecture Notes in Mathematics, Vol. 303, Berlin: Springer; 1972; 139–147.

20. Heikkilä S, On operator and integral equations with discontinuous right-hand side. J. Math. Anal. Appl. 1990;140:200–217.
21. Heikkilä S, On fixed points through a generalized iteration method with applications to differential and integral equations involving discontinuities. Nonlinear Anal. 1990;14:413–426.
22. Heikkilä S, Lakhshmikantham V, Monotone iterative techniques for discontinuous nonlinear differential equations. New York: Marcel Dekker; 1994.
23. Hell P, Nešetril J, Graphs and Homomorphisms. Oxford: Oxford University Press (Oxford Lecture Series in Mathematics and its Applications 28); 2004.
24. Jachymski J, The contraction principle for mappings on a metric space with a graph. Proc. Am. Math. Soc. 2007;136:1359–1373.
25. Langley LJ, A recognition algorithm for orders of interval dimension 2, ARIDAM VI and VII (New Brunswick, NJ, 1991/1992). Discrete Appl. Math. 1995;60:257–266.
26. Larrión F, Neumann-Lara V, On clique divergent graphs with linear growth. Discrete Math. 2002;245:139–153.
27. Larrión F, Neumann-Lara V, Pizaña M, Clique divergent clockwork graphs and partial orders. Discrete Appl. Math. 2004;141:195–207.
28. Larrión F, Pizaña M, Villaroel-Flores R, Contractibility and the clique graph operator. Discrete Math. 2008;308:3461–3469.
29. Larrión F, Pizaña M, Villaroel-Flores R, The fundamental group of the clique graph. Eur. J. Combinatorics 2009;30:288–294.
30. Li B, Milner EC, A chain complete poset with no infinite antichain has a finite core. Order 1993;10:55–63.
31. Nadel B, Constraint satisfaction algorithms. Computational Intelligence 1989;5:188–224.
32. Neumann-Lara V, On clique-divergent graphs. Collogues Internationaux C.N.R.S. Problèmes Combinatoires et Théorie des Graphes 1976;260:313–315.
33. Neumann-Lara V, Clique divergence in graphs. Colloquia Mathematica Societatis János Bolyai; 1978; 25: Algebraic Methods in Graph Theory, 563–569.
34. Prisner E, Convergence of iterated clique graphs. Discrete Math. 1992;103:199–207.
35. Rival I, A fixed point theorem for finite partially ordered sets. J. Combinat. Theory (A) 1976;21:309–318.
36. Rossi F, van Beek P, Walsh T, Handbook of Constraint Programming. Elsevier; 2006.
37. Rutkowski A, The fixed point property for small sets. Order 1989;6:1–14.
38. Rutkowski A, Schröder B, A fixed point theorem with applications to truncated lattices. Algebra Universalis 2005;53:175–187.
39. Schröder B, Fixed point property for 11-element sets. Order 1993;10:329–347.
40. Schröder B, The uniqueness of cores for chain-complete ordered sets. Order 2000;17:207–214.
41. Schröder B, Ordered Sets – An Introduction. Boston: Birkhäuser; 2003.
42. Schröder B, The Fixed Vertex Property for Graphs. Order 2015;32:363-377
43. Schröder B, The fixed point property for ordered sets of interval dimension 2. Submitted to Order, 2015
44. Schröder B, Homomorphic Constraint Satisfaction Problem Solver. http://www.math.usm.edu/schroeder/software.htm, 2015.
45. Spinrad J, Subdivision and lattices. Order 1988;5:143–147.
46. Tsang E, Foundations of Constraint Satisfaction. New York: Academic Press; 1993.
47. vanBeek P, Dechter R, On the minimality and global consistency of row-convex constraint networks. J. Assoc. Computing Machinery 1995;42:543–561.
48. Xia W, Fixed point property and formal concept analysis. Order 1992;9:255–264.

Index

Printed in the United States
By Bookmasters